STATISTICS WITH R

Kruger Brentt Publishers

STATISTICS WITH R

Henderite Loisa Ohee
Ndakalimwe Naftal Gabriel

Kruger Brentt
Publishers

2024

Kruger Brentt Publishers UK. LTD.
Company Number 9728962

Regd. Office: 68 St Margarets Road, Edgware, Middlesex HA8 9UU

ISBN: 978-1-78715-265-6

For information on all our publications visit our website at http://krugerbrentt.com/

PREFACE

"Statistics with R," a comprehensive guide that invites you to explore the fascinating intersection of statistics and data analysis using the powerful R programming language. In this journey, we delve into the intricacies of statistical methodologies beyond the basics, providing you with the tools and knowledge needed to unlock the full potential of your data. Whether you are a student, researcher, or professional seeking to advance your statistical skills, this book offers a structured and hands-on approach to mastering intermediate-level statistical concepts in the context of real-world data analysis. From One-Way ANOVA to Multiple Linear Regression, our goal is to empower you with the skills and confidence to navigate the complexities of statistical analysis, using R as your guide. Let's embark on this enriching exploration of statistics, where theory meets practice, and data reveals its profound stories.

The book covers a range of intermediate statistical topics, starting with a brief review of fundamental statistical concepts and progressing to advanced techniques such as Two-Way ANOVA, Chi-Square Tests, and Multiple Linear Regression. Each chapter is designed to build upon the previous, fostering a progressive and cohesive learning experience. Practical application is at the core of this book. Throughout each chapter, readers will find step-by-step examples and exercises, encouraging active engagement with R programming for statistical analysis. Real-world datasets and case studies are included to provide a contextual understanding of statistical methods. The book leverages the power of R as a statistical tool. Code snippets and R scripts are integrated seamlessly into the text, allowing readers to implement statistical techniques directly and gain practical experience in using R for statistical analysis. The inclusion of case studies offers readers the opportunity to apply learned concepts in realistic scenarios. These case studies are designed to simulate real-world challenges, enabling readers to develop a problem-solving mindset and apply statistical methods to diverse situations. Complex statistical concepts are explained in a clear and accessible manner. The book is written with the assumption that readers have a foundational understanding of basic statistics, making it suitable for intermediate-level learners. We hope this book serves as a valuable resource in your journey towards mastering intermediate statistics with R. By combining theoretical knowledge with practical application, we aim to empower you to navigate the complexities of statistical analysis confidently. Whether you are analyzing experimental data, conducting research, or making data-driven decisions, the skills acquired from this book will undoubtedly enhance your ability to extract meaningful insights from the vast world of data.

This book provides a thorough exploration of intermediate statistical concepts, ensuring a well-rounded understanding of topics such as One-Way ANOVA, Two-Way ANOVA, Chi-Square Tests, and Multiple Linear Regression. Leveraging the capabilities of the R programming language, the book seamlessly integrates code snippets and R scripts throughout the text. This integration allows readers to directly implement statistical methods, fostering a practical and immersive learning experience. Complex statistical principles are explained in a clear and accessible manner. The book assumes a foundational understanding of basic statistics. This book is designed for individuals with a foundational understanding of basic statistics who are eager to progress to intermediate statistical concepts. The focus on practical application and R integration makes it a valuable resource for those engaged in statistical analysis in various professional settings. The book provides a practical and applied approach to learning intermediate statistical techniques with the R programming language. This book is ideal for students studying statistics, data science, or related fields, researchers and professionals looking to enhance their data analysis skills.

We are grateful to all those persons as well as various books, manuals, periodicals, newsletters, magazines, journals etc. that helped in the preparation of this book. In spite of the best efforts, it is possible that some errors may have occurred into the compilation and editing of the book. Further queries, constructive suggestions and criticisms for the further improvement of the book are always welcome and shall be thankfully acknowledged.

Henderite Loisa Ohee

Ndakalimwe Naftal Gabriel

CONTENTS

CHAPTER-1

GENERAL: AN OVERVIEW

1.1 STATISTICS: AN OVERVIEW

Statistics is a powerful tool that helps us make sense of the world by analyzing and interpreting data. As we move from basic statistics to the intermediate level, we delve deeper into the methods and techniques that allow us to draw more nuanced conclusions from our data. This journey is particularly exciting when paired with the versatile programming language R, which empowers us to implement statistical analyses efficiently and effectively. In this intermediate statistics course with R, we will build upon the foundational concepts you've already mastered and explore more advanced statistical methods. Our goal is to equip you with the skills needed to analyze complex datasets, make informed decisions, and communicate your findings effectively.

After you are introduced to basic statistical ideas, a wide array of statistical methods become available. The methods explored here focus on assessing (estimating and testing for) relationships between variables, sometimes when controlling for or modifying relationships based on levels of another variable – which is where statistics gets interesting and really useful. Early statistical analyses (approximately 100 years ago) were focused on describing a single variable. Your introductory statistics course should have heavily explored methods for summarizing and doing inference in situations with one group or where you were comparing results for two groups of observations. Now, we get to consider more complicated situations – culminating in a set of tools for working with multiple explanatory variables, some of which might be categorical and related to having different groups of subjects that are being compared. Throughout the methods we will cover, it will be important to retain a focus on how the appropriate statistical analysis depends on the research question and data collection process as well as the types of variables measured.

Figure 1.1 frames the topics we will discuss. Taking a broad view of the methods we will consider, there are basically two scenarios - one when the response is quantitative and one when the response is categorical. Examples of quantitative responses we will see later involve passing distance of cars for a bicycle rider (in centimeters (cm)) and body fat (percentage) . Examples of categorical variables include improvement (none, some, or marked) in a clinical trial related to arthritis symptoms or whether a student has turned in copied work (never, done this on an exam or paper, or both) . There are going to be some more nuanced aspects to all these analyses as the complexity of both sides of Figure 1.1 suggest, but note that near the bottom, each tree converges on a single procedure, using a linear model for a quantitative response variable or using a Chi-square test for a categorical response. After selecting the appropriate procedure and completing the necessary technical steps to get results for a given data set, the final step involves assessing the scope of inference and types of conclusions that are appropriate based on the design of the study.

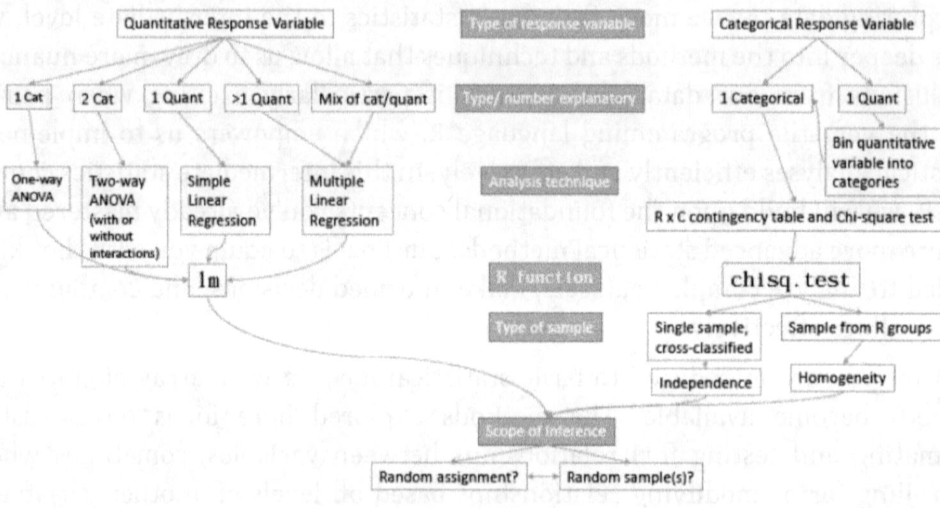

Figure 1.1: Flow chart of methods.

We will be spending most of the semester working on methods for quantitative response variables (the left side of Figure 1.1 is covered in Chapters 2, 3, 4, 6, 7, and 8), stepping over to handle the situation with a categorical response variable in Chapter 5 (right side of Figure 1.1). Chapter 9 contains case studies illustrating all the methods discussed previously, providing a final opportunity to explore additional examples that illustrate how finding a path through Figure 1.1 can lead to the appropriate analysis.

The first topics (Chapters 1, and 2) will be more familiar as we start with single and two group situations with a quantitative response. In your previous statistics course, you should have seen methods for estimating and quantifying uncertainty for the mean of a single group and for differences in the means of two groups. Once we have briefly reviewed these methods and introduced the statistical software that we will use throughout the course, we will consider the first new statistical material in Chapter 3. It involves the situation with a quantitative response variable where there are more than 2 groups to compare - this is what we call the One- Way ANOVA situation. It generalizes the 2-independent sample hypothesis test to handle situations where more than 2 groups are being studied. When we learn this method, we will begin discussing model assumptions and methods for assessing those assumptions that will be present in every analysis involving a quantitative response. The Two- Way ANOVA (Chapter 3) considers situations with two categorical explanatory variables and a quantitative response. To make this somewhat concrete, suppose we are interested in assessing differences in, say, the yield of wheat from a field based on the amount of fertilizer applied (none, low, or high) and variety of wheat (two types) . Here, yield is a quantitative response variable that might be measured in bushels per acre and there are two categorical explanatory variables, fertilizer, with three levels, and variety, with two levels. In this material, we introduce the idea of an interaction between the two explanatory variables: the relationship between one categorical variable and the mean of the response changes depending on the levels of the other categorical variable. For example, extra fertilizer might enhance the growth of one variety and hinder the growth of another so we would say that fertilizer has different impacts based on the level of variety. Given this interaction may or may not actually be present, we will consider two versions of the model in Two-Way ANOVAs, what are called the additive (no interaction) and the interaction models.

Following the methods for two categorical variables and a quantitative response, we explore a method for analyzing data where the response is categorical, called the Chi-square test in Chapter 5. This most closely matches the One-Way ANOVA situation with a single categorical explanatory variable, except now the response variable is categorical. For example, we will assess whether taking a drug (vs taking a placebo1) has an effect2 on the type of improvement the subjects demonstrate. There are two different scenarios for study design that impact the analysis technique and hypotheses tested in Chapter 5. If the explanatory variable reflects the group that subjects were obtained from, either through randomization of the treatment level to the subjects or by taking samples from separate populations, this is called a Chi-square Homogeneity Test. It is also possible to obtain a single sample from a population and then obtain information on the levels of the explanatory variable for each subject. We will analyze these results using what is called a Chi-square Independence Test. They both use the same test statistic but we use slightly different graphics and are testing different hypotheses in these two related situations. Figure 1.1 also shows that if we had a quantitative explanatory variable and a categorical response that we would need to “bin” or create categories of responses from the quantitative variable to use the Chi-square testing methods.

If the predictor and response variables are both quantitative, we start with scatter plots, correlation, and simple linear regression models (Chapters 6 and 7) – things you should have seen, at least to some degree, previously. The biggest differences here will be the depth of exploration of diagnostics and inferences for this model and discussions of transformations of variables. If there is more than one explanatory variable, then we say that we are doing multiple linear regression (Chapter 8) – the "multiple" part of the name reflects that there will be more than one explanatory variable. We use the same name if we have a mix of categorical and quantitative predictor variables but there are some new issues in setting up the models and interpreting the coefficients that we need to consider. In the situation with one categorical predictor and one quantitative predictor, we revisit the idea of an interaction. It allows us to consider situations where the estimated relationship between a quantitative predictor and the mean response varies among different levels of the categorical variable. In Chapter 9, connections among all the methods used for quantitative responses are discussed, showing that they are all just linear models . We also show how the methods discussed can be applied to a suite of new problems with a set of case studies and how that relates to further extensions of the methods.

By the end of Chapter 9 you should be able to identify, perform using the statistical software R [R Core Team, 2022], and interpret the results from each of these methods. There is a lot to learn, but many of the tools for using R and interpreting results of the analyses accumulate and repeat throughout the textbook. If you work hard to understand the initial methods, it will help you when the methods get more complicated. You will likely feel like you are just starting to learn how to use R at the end of the semester and for learning a new language that is actually an accomplishment. We will just be taking you on the first steps of a potentially long journey and it is up to you to decide how much further you want to go with learning the software.

All the methods you will learn require you to carefully consider how the data were collected, how that pertains to the population of interest, and how that impacts the inferences that can be made. The scope of inference from the bottom of Figure 1.1 is our shorthand term for remembering to think about two aspects of the study – random assignment and random sampling. In a given situation, you need to use the description of the study to decide if the explanatory variable was randomly assigned to study units (this allows for causal inferences if differences are detected) or not (so no causal statements are possible). As an example, think about two studies, one where students are randomly assigned to either get tutoring with their statistics course or not and another where the students are asked at the end of the semester whether they sought out tutoring or not. Suppose we compare the final grades in the course for the two groups (tutoring/not) and find a big difference. In the first study with random assignment, we can say the tutoring caused the differences we observed. In the second, we could only say that the tutoring was associated with differences but because students self-selected the group they ended up in, we can't say that the tutoring caused the differences. The other aspect of scope of inference concerns random sampling: If the data were obtained using a random sampling mechanism, then our inferences can be safely extended to the population that the sample was taken from. However, if we have a non-random sample, our inference can only apply to the sample collected. In the previous example, the difference would be studying a random sample of students from the population of, say, Introductory Statistics students at a university versus studying a sample of students that volunteered for the research project, maybe for extra credit in the class. We could still randomly assign them to tutoring/not but the non-random sample would only lead to conclusions about those students that volunteered. The most powerful scope of inference is when there are randomly assigned levels of explanatory variables with a random sample from a population – conclusions would be about causal impacts that would happen in the population.

By the end of this material, you should have some basic R skills and abilities to create basic ANOVA and regression models, as well as to handle Chi-square testing situations. Together, this should prepare you for future statistics courses or for other situations where you are expected to be able to identify an appropriate analysis, do the calculations and required graphics using the data set, and then effectively communicate interpretations for the methods discussed here.

1.2 GETTING STARTED IN R

You will need to download the statistical software package called R and an enhanced interface to R called RStudio [RStudio Team, 2022]. They are open source and free to download and use (and will always be that way). This means that the skills you learn now can follow you the rest of your life. R is becoming the primary language of statistics and is being adopted across academia, government, and businesses to help manage and learn from the growing volume of data being obtained. Hopefully you will get a sense of some of the power of R in this book.

The next pages will walk you through the process of getting the software downloaded and provide you with an initial experience using RStudio to do things that should look familiar even though the interface will be a new experience. Do not expect to master R quickly – it takes years (sorry!) even if you know the statistical methods being used. We will try to keep all your interactions with R code in a similar code format and that should help you in learning how to use R as we move through various methods. We will also often provide you with example code.

Everyone that learns R starts with copying other people's code and then making changes for specific applications – so expect to go back to examples from the text and focus on learning how to modify that code to work for your particular data set. Only really experienced R users "know" functions without having to check other resources. After we complete this basic introduction, Chapter 2 begins doing more sophisticated things with R, allowing us to compare quantitative responses from two groups, make some graphical displays, do hypothesis testing and create confidence intervals in a couple of different ways.

You will have two[3] downloading activities to complete before you can do anything more than read this book[4]. First, you need to download R. It is the engine that will do all the computing for us, but you will only interact with it once. Go to http://cran.rstudio.com and click on the "**Download R for. . .** " button that corresponds to your operating system. On the next page, click on "**base**" and then it will take you to a screen to download the most current version of R that is compiled for your operating system, something like "**Download R 4.2.1 for Windows**". Click on that link and then open the file you downloaded. You will need to select your preferred language (choose English so your instructor can help you), then hit "**Next**" until it starts to unpack and install the program (all the base settings will be fine). After you hit "**Finish**" you will not do anything further with R directly.

Second, you need to download RStudio. It is an enhanced interface that will make interacting with R less frustrating and allow you to directly create reports that include the code and output. To download RStudio, go near the bottom of https://www.rstudio.com/products/rstudio/download/ and select the correct version under "Installers for Supported Platforms" for your operating system. Download and then install RStudio using the installer. From this point forward, you should only open RStudio; it provides your interface with R. Note that both R and RStudio are updated frequently (up to four times a year) and if you downloaded either more than a few months previously, you should download the up-to-date versions, especially if something you are trying to do is not working. Sometimes code will not work in older versions of R and sometimes old code won't work in new versions of R.[5]

n you open RStudio, you will see a screen like Figure 1.2. The added annotation in this and the following screen-grabs is there to help you get initially oriented to the software interface. R is command-line software – meaning that in some way or another you have to create code and get it evaluated, either by entering and execute it at a command prompt or by using the RStudio interface to run the code that is stored in a file. RStudio makes the management and execution of that code more efficient than the basic version of R. In RStudio, the lower left panel is called the "console" window and is where you can type R code directly into R or where you will see the code you run and (most importantly!) where the results of your executed commands will show up. The most basic interaction with R is available once you get the cursor active at the command prompt ">" by clicking in that panel (look for a blinking vertical line). The upper left panel is for writing, saving, and running your R code either in .R script files or.Rmd (markdown) files, discussed below. Once you have code available in this window, the "Run" button will execute the code for the line that your cursor is on or for any text that you have highlighted with your mouse. The "data management" or environment panel is in the upper right, providing information on what data sets have been loaded. It also contains the "Import Dataset" button that provides the easiest way for you to read a data set into R so you can analyze it. The lower right panel contains information on the "Packages" (additional code we will download and install to add functionality to R) that are available and is where you will see plots that you make and requests for "Help" on specific functions.

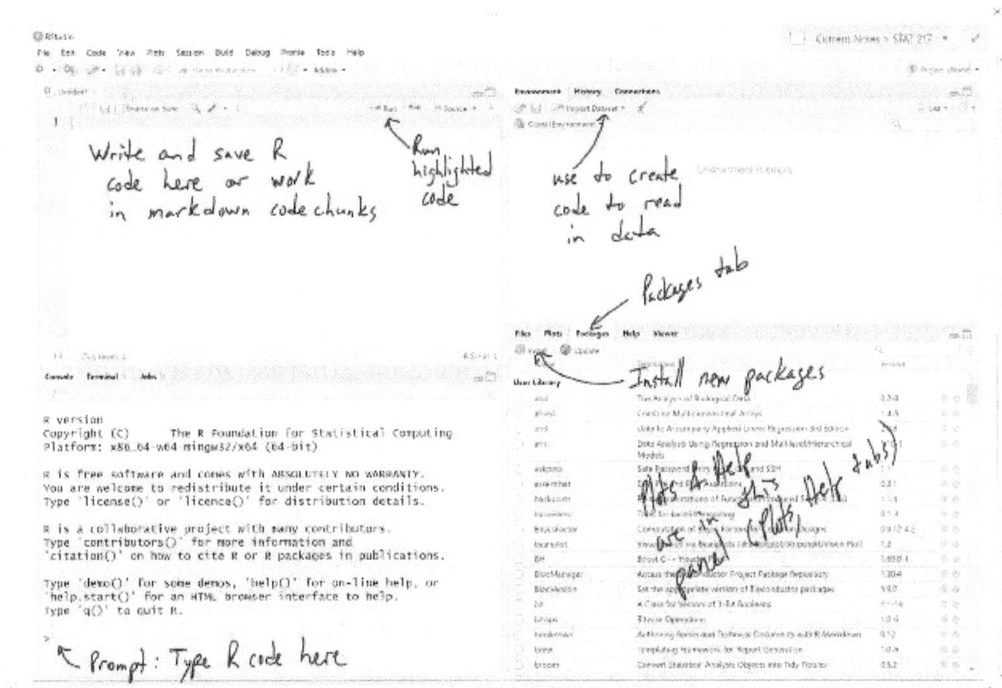

Figure 1.2: Initial RStudio layout.

To get started, we can complete some basic tasks in R using the RStudio interface. As a first interaction with R we can use it as a calculator. To do this, click near the command prompt (>) in the lower left "console" panel, type 3+4, and then hit enter. It should look like this:

```
> 3 + 4
[1] 7
```

You can do more interesting calculations, like finding the mean of the numbers -3, 5, 7, and 8 by adding them up and dividing by 4:

```
> (-3 + 5 + 7 + 8)/4
[1] 4.25
```

Note that the parentheses help R to figure out your desired order of operations. If you drop that grouping, you get a very different (and wrong!) result:

```
> -3 + 5 + 7 + 8/4
[1] 11
```

We could estimate the standard deviation similarly using the formula you might remember from introductory statistics, but that will only work in very limited situations. To use the real power of R this semester, we need to work with data sets that store the observations for our subjects in variables. Basically, we need to store observations in named vectors (one dimensional arrays) that contain a list of the observations. To create a vector containing the four numbers and assign it to a variable named variable1, we need to create a vector using the concatenate function c which means "combine the items" that follow, if they are inside parentheses and have commas separating the values, as follows:

```
> c(-3, 5, 7, 8)
[1] -3 5 7 8
```

To get this vector stored in a variable called variable1 we need to use the assignment operator, <- (read as "is defined to contain") that assigns the information on the right into the variable that you are creating on the left.

```
> variable1 <- c(-3, 5, 7, 8)
```

In R, the assignment operator, <-, is created by typing a "less than" symbol < followed by a "minus" sign (-) without a space between them. If you ever want to see what numbers are residing in an object in R, just type its name and hit enter. You can see how that variable contains the same information that was initially generated by c(-3, 5, 7, 8) but is easier to access since we just need the text for the variable name representing that vector.

```
> variable1
[1] -3 5 7 8
```

With the data stored in a variable, we can use functions such as mean and sd to find the mean and standard deviation of the observations contained in variable1:

```
> mean(variable1)
[1] 4.25
> sd(variable1)
[1] 4.99166
```

When dealing with real data, we will often have information about more than one variable. We could enter all observations by hand for each variable but this is prone to error and onerous for all but the smallest data sets. If you are to ever utilize the power of statistics in the evolving data-centered world, data management has to be accomplished in a more sophisticated way. While you can manage data sets quite effectively in R, it is often easiest to start with your data set in something like Microsoft Excel or Open Office's Calc. You want to make sure that observations are in the rows and the names of variables are in first row of the columns and that there is no "extra stuff" in the spreadsheet. If you have missing observations, they should be represented with blank cells. The file should be saved as a ".csv" file (stands for comma-separated values although Excel calls it "CSV (Comma Delimited)"), which basically strips off some of the junk that Excel adds to the necessary information in the file. Excel will tell you that this is a bad idea, but it actually creates a more stable archival format and one that R can use directly.6

The following code to read in the data set relies on an R package called readr [Wickham et al., 2022b]. Packages in R provide additional functions and data sets that are not available in the initial download of R or RStudio. To get access to the packages, first "install" (basically download) and then "load" the package. To install an R package, go to the Packages tab in the lower right panel of RStudio. Click on the Install button and then type in the name of the package in the box (here type in readr). RStudio will try to auto-complete the package name you are typing which should help you make sure you got it typed correctly. If you are working in a .Rmd file, a highlighted message may show up on the top of the file to suggest packages to install that are not present – look for this to help make sure you have the needed packages installed. This will be the first of many times that we will mention that R is case sensitive – in other words, Readr is different from readr in R syntax and this sort of thing applies to everything you do in R. You should only need to install each R package once on a given computer. If you ever see a message that R can't find a package, make sure it appears in the list in the Packages tab. If it doesn't, repeat the previous steps to install it.

Important: R is case sensitive! Readr is not the same as readr!

After installing the package, we need to load it to make it active in a given work session. Go to the command prompt and type (or copy and paste) library(readr) or require(readr):

> library(readr)

With a data set converted to a CSV file and readr installed and loaded, we need to read the data set into the active workspace. There are two ways to do this, either using the point-and-click GUI in RStudio (click the "Import Dataset" button in the upper right "Environment" panel as indicated in Figure 1.2) or modifying the read_csv function to find the file of interest. To practice this, you candownload an Excel (.xls) file from http://www.math.montana.edu/courses/s217/documents/treadmill.xls that contains observations on 31 males that volunteered for a study on methods for measuring fitness [Westfall and Young, 1993]. In the spreadsheet, you will find a data set that starts and ends with the following information (only results for Subjects 1, 2, 30, and 31 shown here):

Sub-ject	Tread-MillOx	Tread Mill- Max Pulse	RunTime	RunPulse	Rest Pulse	BodyWeight	Age
1	60.05	186	8.63	170	48	81.87	38
2	59.57	172	8.17	166	40	68.15	42
...	...	...	...	...	...	...	...
30	39.2	172	12.88	168	44	91.63	54
31	37.39	192	14.03	186	56	87.66	45

The variables contain information on the subject number (Subject), subjects' maximum treadmill oxygen consumption (TreadMillOx, in ml per kg per minute, also called maximum VO2) and maximum pulse rate (TreadMillMaxPulse, in beats per minute), time to run 1.5 miles (Run Time, in minutes), maximum pulse during 1.5 mile run (RunPulse, in beats per minute), resting pulse rate (RestPulse, beats per minute), Body Weight (BodyWeight, in kg), and Age (in years). Open the file in Excel or equivalent software and then save it as a .csv file in a location you can find on your computer. Then go to RStudio and click on File, then Import Dataset, then From Text (readr)... 7 Click "Import" and find your file. R will store the data set as an object with the same name as the .csv file. You could use another name as well, but it is often easiest just to keep the data set name in R related to the original file name. You should see some text appear in the console (lower left panel) like in Figure 1.3. The text that is created will look something like the following - if you had stored the file in a drive labeled D:, it would be:

```
treadmill <- read_csv("D:/treadmill.csv")
```

What is put inside the " will depend on the location and name of your saved. A version of the data set in what looks like a spreadsheet will appear in the upper left window due to the second line of code (View(treadmill)).

Just directly typing (or using) a line of code like this is actually the other way that we can read in files. If you choose to use the text-only interface, then you need to tell R where to look in your computer to find the data file. read_csv is a function that takes a path as an argument. To use it, specify the path to your data file, put quotes around it, and put it as the input to read_csv(...). For some examples later in the book,you will be able to copy a command like this from the text and read data sets and other code directly from the website, assuming you are connected to the internet.

To verify that you read the data set in correctly, it is always good to check its contents. We can view the first and last rows in the data set using the head and tail functions on the data set, which show the following results for the treadmill data. Note that you will sometimes need to resize the console window in RStudio to get all the columns to display in a single row which can be performed by dragging the gray bars that separate the panels.

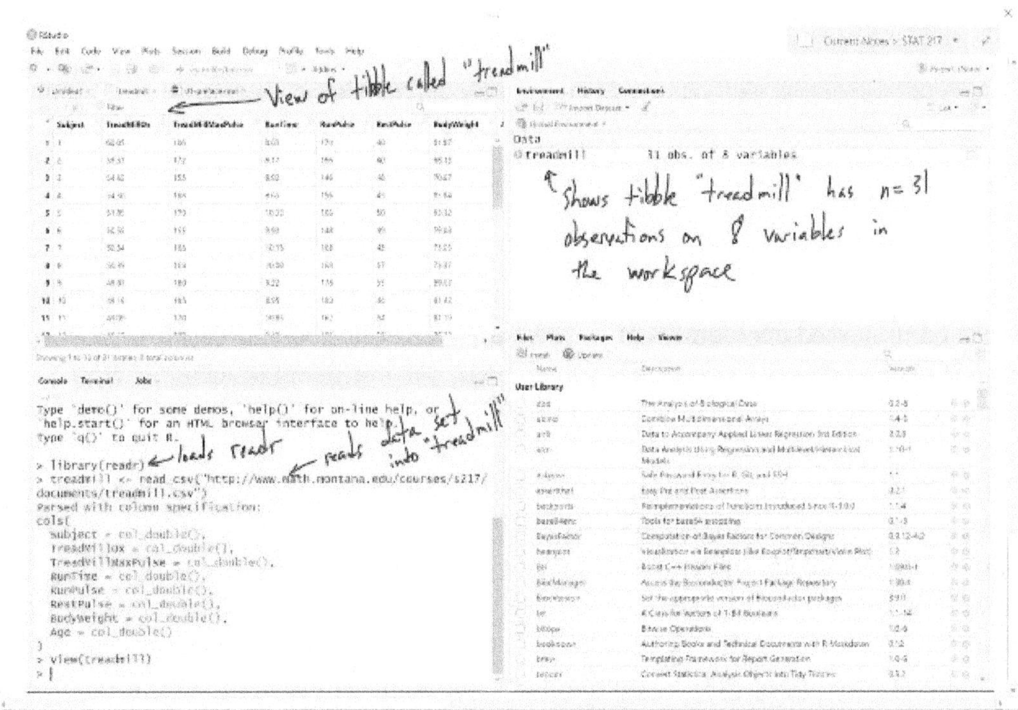

Figure 1.3: RStudio with initial data set loaded.

When you load an installed package with library, you may see a warning message about versions of thepackage and versions of R – this is usually something you can ignore. Other warning messages could be more ominous for proceeding but before getting too concerned, there are couple of basic things to check. First, double check that the package is installed (see previous steps). Second, check for typographical errors in your code – especially for mis-spellings or unintended capitalization. If you are still having issues, try repeating the installation process. Then click on the "Update" button to check for potentially newer versions of packages. If all that fails, try the cloud version of RStudio discussed before and repeat the steps there.

```
ead(treadmill)
 tibble: 6 x 8
ubject TreadMillOx TreadMillMaxPulse RunTime RunPulse RestPulse BodyWeight   Age
 <int>       <dbl>             <int>   <dbl>    <int>     <int>      <dbl> <int>
     1       60.05               186    8.63      170        48      81.87    38
     2       59.57               172    8.17      166        40      68.15    42
     3       54.62               155    8.92      146        48      70.87    50
     4       54.30               168    8.65      156        45      85.84    44
     5       51.85               170   10.33      166        50      83.12    54
     6       50.55               155    9.93      148        49      59.08    57

ail(treadmill)
 tibble: 6 x 8
ubject TreadMillOx TreadMillMaxPulse RunTime RunPulse RestPulse BodyWeight   Age
 <int>       <dbl>             <int>   <dbl>    <int>     <int>      <dbl> <int>
    26       44.61               182   11.37      178        62      89.47    44
    27       40.84               172   10.95      168        57      69.63    51
    28       39.44               176   13.08      174        63      81.42    44
    29       39.41               176   12.63      174        58      73.37    57
    30       39.20               172   12.88      168        44      91.63    54
    31       37.39               192   14.03      186        56      87.66    45
```

To help you go from basic to intermediate R usage and especially to help with more complicated problems, you will want to learn how to manage and save your R code. The best way to do this is using the upper left panel in RStudio. If you just want to manage code, then you can use what are called R Scripts, which are files that have a file extension of ".R". To start a new ".R" file to store your code, click on File, then New File, then R Script. This will create a blank page to enter and edit code – then save the file as something like "MyFileName.R" in your preferred location. Saving your code will mean that you can return to where you were working last by simply re-running the saved script file. With code in the script window, you can place the cursor on a line of code or highlight a chunk of code and hit the "Run" button8 on the upper part of the panel. It will appear in the console with results just like what you would obtain if you typed it after the command prompt and hit enter for each line. Figure 1.4 shows the screen with the code used in this section in the upper left panel, saved in a file called "Ch1.R", with the results of highlighting and executing the first section of code using the "Run" button.

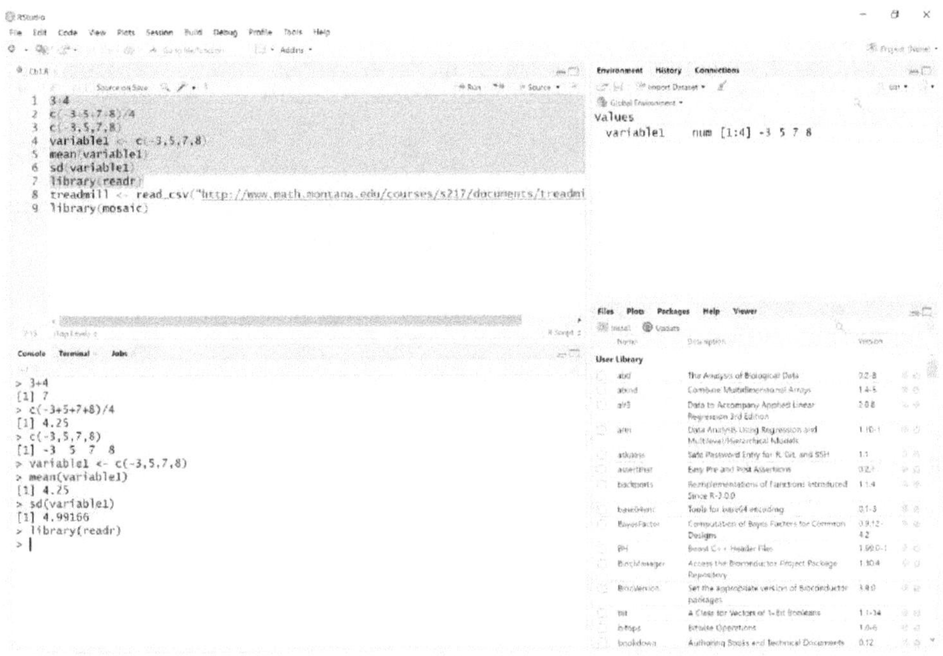

Figure 1.4: RStudio with highlighted code run.

1.3 FUNDAMENTAL SUMMARY STATISTICS, HISTOGRAMS, AND BOXPLOTS USING R

For the following material, you will need to install and load the mosaic package [Pruim et al., 2021b].

```
> library(mosaic)
```

It provides a suite of enhanced functions to aid our initial explorations. With RStudio running, the mosaic package loaded, a place to write and save code, and the treadmill data set loaded, we can (finally!) start to summarize the results of the study. The treadmill object is what R calls a tibble9 and contains columns corresponding to each variable in the spreadsheet. Every function in R will involve specifying the variable(s) of interest and how you want to use them. To access a particular variable (column) in a tibble, you can use a $ between the name of the tibble and the name of the variable of interest, generically astibblename$variablename. You can think of this as tibblename's variable name where the 's is replaced by the dollar sign. To identify the RunTime variable here it would be treadmill $RunTime. In the command line it would look like:

```
> treadmill$RunTime
[1]  8.63  8.17  8.92  8.65 10.33  9.93 10.13 10.08  9.22  8.95 10.85  9.40 11.50 10.50
[15] 10.60 10.25 10.00 11.17 10.47 11.95  9.63 10.07 11.08 11.63 11.12 11.37 10.95 13.08
[29] 12.63 12.88 14.03
```

Just as in the previous section, we can generate summary statistics using functions like mean and sd byrunning them on a specific variable:

```
> mean(treadmill$RunTime)
[1] 10.58613
> sd(treadmill$RunTime)
[1] 1.387414
```

And now we know that the average running time for 1.5 miles for the subjects in the study was 10.6 minutes with a standard deviation (SD) of 1.39 minutes. But you should remember that the mean and SD are only appropriate summaries if the distribution is roughly symmetric (both sides of the distribution are approximately the same shape and length). The mosaic package provides a useful function called favstats that provides the mean and SD as well as the 5 number summary: the minimum (min), the first quartile (Q1, the 25th percentile), the median (50th percentile), the third quartile (Q3, the 75th percentile), and the maximum (max). It also provides the number of observations (n) which was 31, as noted above, and a count of whether any missing values were encountered (missing), which was 0 here since all subjects had measurements available on this variable.

```
> favstats(treadmill$RunTime)
  min   Q1 median    Q3   max     mean       sd  n missing
 8.17 9.78  10.47 11.27 14.03 10.58613 1.387414 31       0
```

We are starting to get somewhere with understanding that the runners were somewhat fit with the worst runner covering 1.5 miles in 14 minutes (the equivalent of a 9.3 minute mile) and the best running at a 5.4 minute mile pace. The limited variation in the results suggests that the sample was obtained from a restricted group with somewhat common characteristics. When you explore the ages and weights of the subjects in the Practice Problems in Section 1.6, you will get even more information about how similar all the subjects in this study were. Researchers often publish numerical summaries of this sort of demographic information to help readers understand the subjects that they studied and that their results might apply to.

A graphical display of these results will help us to assess the shape of the distribution of run times – including considering the potential for the presence of a skew (whether the right or left tail of the distribution is noticeably more spread out, with left skew meaning that the left tail is more spread out than the right tail) and outliers (unusual observations). A histogram is a good place to start. Histograms display connected bars with counts of observations defining the height of bars based on a set of bins of values of the quantitative variable. We will apply the hist function to the RunTime variable, which produces Figure 1.5.

```
> hist(treadmill$RunTime)
```

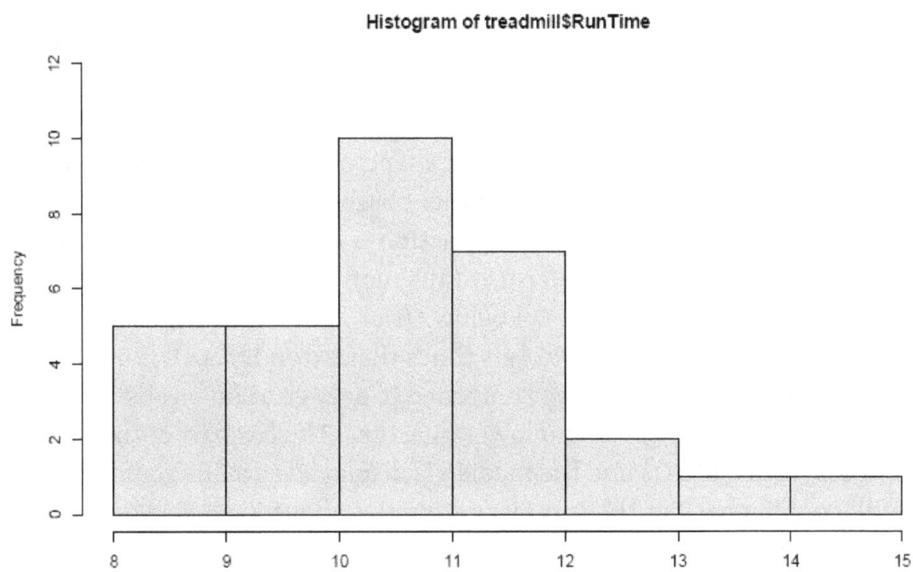

Figure 1.5: Histogram of Run Times (minutes) of n = 31 subjects in Treadmill study, bar heights are counts.

You can save this plot by clicking on the Export button found above the plot, followed by Copy to Clipboard and clicking on the Copy Plot button. Then if you open your favourite word-processing program, you should be able to paste it into a document for writing reports that include the figures. You can see the first parts of this process in the screen grab in Figure 1.6. You can also directly save the figures as separate files using Save as Image or Save as PDF and then insert them into your word processing documents.

The function hist defaults into providing a histogram on the frequency (count) scale. In most R functions, there are the default options that will occur if we don't make any specific choices but we can override the default options if we desire. One option we can modify here is to add labels to the bars to beable to see exactly how many observations fell into each bar. Specifically, we can turn the labels option "on" by making it true ("T") by adding labels = T to the previous call to the hist function, separated by a comma. Note that we will use the = sign only for changing options within functions.

```
> hist(treadmill$RunTime, labels = T)
```

Based on this histogram (Figure 1.8), it does not appear that there any outliers in the responses since there are no bars that are separated from the other observations. However, the distribution does not look symmetric and there might be a skew to the distribution. Specifically, it appears to be ***skewed right*** (the right tail is longer than the left). But histograms can sometimes mask features of the data set by binning observations and it is hard to find the percentiles accurately from the plot.

When assessing outliers and skew, the ***boxplot*** (or *Box and Whiskers* plot) can also be helpful (Figure 1.8) to describe the shape of the distribution as it displays the 5-number summary and will also indicate observations that are "far" above the middle of the observations. R's boxplot function uses the standard rule to indicate an observation as a ***potential outlier*** if it falls more than 1.5 times the ***IQR*** (Inter-Quartile Range, calculated as $Q_3 - Q_1$) below Q1 or above Q_3. The potential outliers are plotted with circles and the *Whiskers* (lines that extend from Q_1 and Q_3 typically to the minimum and maximum) are shortened to only go as far as observations that are within 1.5∗IQR of the upper and lower quartiles. The *box* part of the boxplot is a box that goes from Q1 to Q3 and the median is displayed as a line somewhere inside the box.[10] Looking back at the summary statistics above, Q1 = 9.78 and Q3 = 11.27, providing an IQR of:

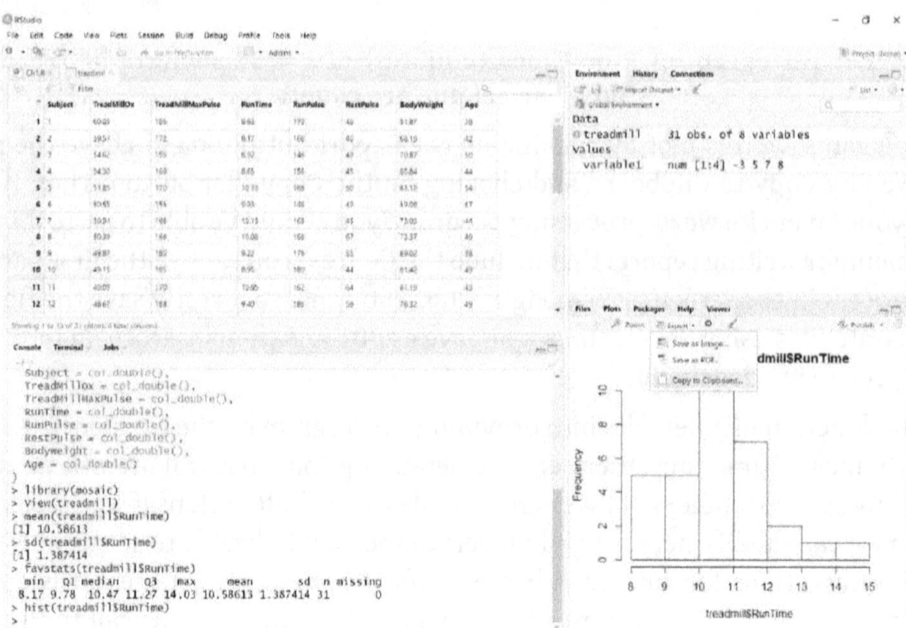

Figure 1.6: R Studio while in the process of copying the histogram.

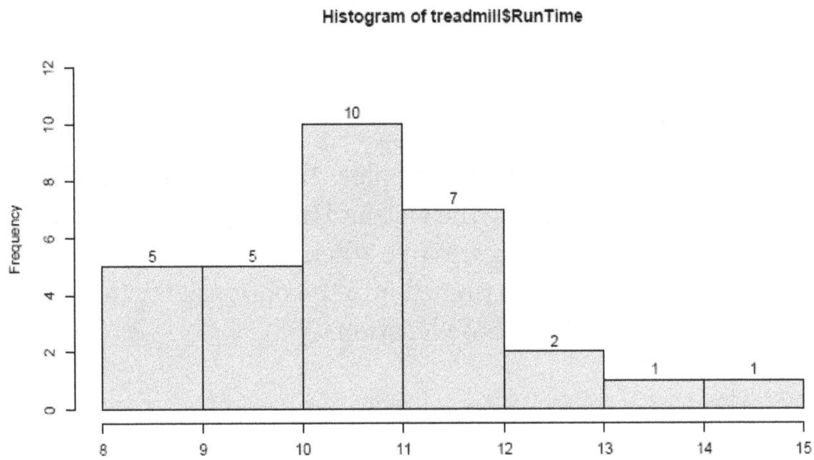

Figure 1.7: Histogram of Run Times with counts in bars labelled.

```
> IQR <- 11.27 - 9.78
> IQR
[1] 1.49
```

One observation (the maximum value of 14.03) is indicated as a potential outlier based on this result by being larger than Q3 +1.5*IQR, which was 13.505:

```
> 11.27 + 1.5*IQR
[1] 13.505
```

The boxplot also shows a slight indication of a right skew (skew towards larger values) with the distance from the minimum to the median being smaller than the distance from the median to the maximum. Additionally, the distance from Q1 to the median is smaller than the distance from the median to Q3. It is modest skew, but worth nothing.

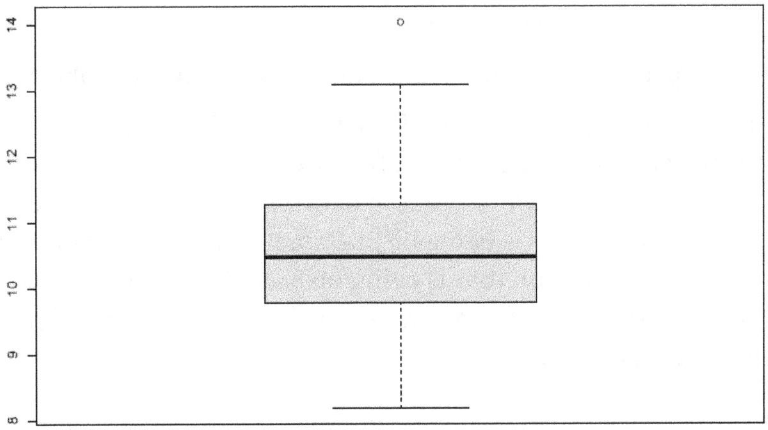

Figure 1.8: Boxplot of 1.5 mile Run Times.

```
> boxplot(treadmill$RunTime)
```

While the default boxplot is fine, it fails to provide good graphical labels, especially on the y-axis.

Additionally, there is no title on the plot. The following code provides some enhancements to the plot by using the ylab and main options in the call to boxplot, with the results displayed in Figure 1.9. When we add text to plots, it will be contained within quotes and be assigned into the options ylab (for y-axis) or main (for the title) here to put it into those locations.

```
> boxplot(treadmill$RunTime, ylab = "1.5 Mile Run Time (minutes)",
        main = "Boxplot of the Run Times of n = 31 participants")
```

Throughout the book, we will often use extra options to make figures that are easier for you to understand.

1Boxplot of the Run Times of n = 31 participants

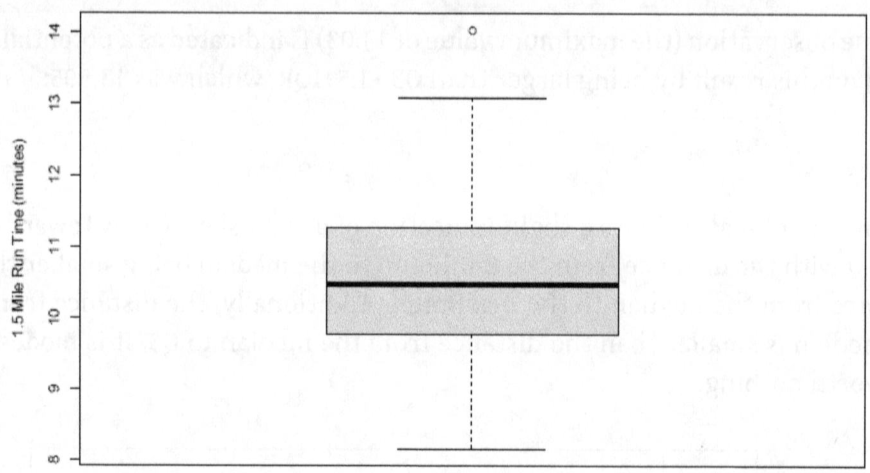

Figure 1.9: Boxplot of Run Times with improved labels.

There are often simpler versions of the functions that will suffice but the extra work to get better labelled figures is often worth it. I guess the point is that "a picture is worth a thousand words" but in data visualization, that is only true if the reader can understand what is being displayed. It is also important to think about the quality of the information that is being displayed, regardless of how pretty the graphic might be. So maybe it is better to say "a picture can be worth a thousand words" if it is well-labelled?

1.4 R MARKDOWN

The previous results were created by running the R code and then copying the results from either the console or by copying the figure and then pasting the results into the typesetting program. There is another way to use RStudio where you can have it compile the results (both output and figures) directly into a document together with other writing and the code that generated it, using what is called R Markdown (http://shiny.rstudio.com/articles/rmarkdown.html). It is basically what we used to prepare this book and what you should learn to use to do your work. From here forward, you will see a change in formatting of the R code and output as you will no longer see the command prompt (">") with the code. The output will be flagged by having two "##"'s before it. For example, the summary statistics for the *RunTime* variable from favstats function would look like when run using R Markdown:

```
favstats(treadmill$RunTime)
```

```
##   min   Q1 median    Q3   max     mean       sd  n missing
##  8.17 9.78  10.47 11.27 14.03 10.58613 1.387414 31       0
```

Statisticians (and other scientists) are starting to use R Markdown and similar methods because they provide what is called "Reproducible research" [Gandrud, 2015] where all the code and output it produced are available in a single place. This allows different researchers to run and verify results (so "reproducible results") or the original researchers to revisit their earlier work at a later date and recreate all their results exactly11. Scientific publications are currently encouraging researchers to work in this way and may someday require it. The term reproducible can also be related to whether repeated studies (with new, independent data collection stages and analyses) get the same result (also called replication) – further discussion of these terms and the implications for scientific research are discussed in Chapter 2.

In order to get some practice using R Markdown, create a sample document in this format using File -> New File -> R Markdown. . . Choose a title for your file and select the "Word" option. This will create a new file in the upper left window where we stored our .R script. Save that file to your computer. Then you can use the "Knit" button to have RStudio run the code and create a word document with the results. R Markdown documents contain basically two components, "code chunks" that contain your code and the rest of the document where you can write descriptions and interpretations of the results that code generates. The code chunks can be inserted using the "Insert" button by selecting the "R" option. Then write your code in between the ---{r} and --- lines (it should have grey highlights for those lines and white for the rest of the portions of the .Rmd document). Once you write some code inside a code chunk, you can test your code using the triangle on the upper right side of it to run all the code that resides in that chunk. Keep your write up outside of these code chunks to avoid code errors and failures to compile. Once you think your code and writing is done, you can use the "Knit" button to try to compile the file. As you are learning, you may find this challenging, so start with trying to review the sample document and knit each time you get a line of code written so you know which line was responsible for preventing the knitting from being successful. Also look around for posted examples of .Rmd files to learn how others have incorporated code with write-ups. You might even be given a template of homework or projects as .Rmd files from your instructor. After you do this a couple of times, you will find that the challenge of working with markdown files is more than matched by the simplicity of the final product and, at least to researchers, the reproducibility and documentation of work that this way of working provides.

1.5 GRAMMAR OF GRAPHICS

The previous ♥BLE MAPPINGS>)) +<GEOM_FUNCTION>(). Breaking this down, the data = ... tells the ggplot function where to look, the information inside the aes (or aesthetic) defines which variables in the data set to use and how to use them (often with x = variable1, y = variable2, etc., with x = ... for the variable on the x (horizontal) axis and y = ... for the variable on the y (vertical) axis), and the + <GEOM_FUNCTION>() defines which type of graph to make (there are geom_histogram and geom_boxplot to make the graphs discussed previously andmany, many more). Because we often have many "+"'s to include, the common practice is to hit return after the "+" and start the next layer or option on the following line for better readability. Figure 1.10 shows ahistogram of the RunTime variable made using the + geom_histogram().

```
library(ggplot2)
ggplot(data = treadmill, mapping = aes(x = RunTime)) + geom_histogram()
```

`stat_bin()` using `bins = 30`. Pick better value with `binwidth`.

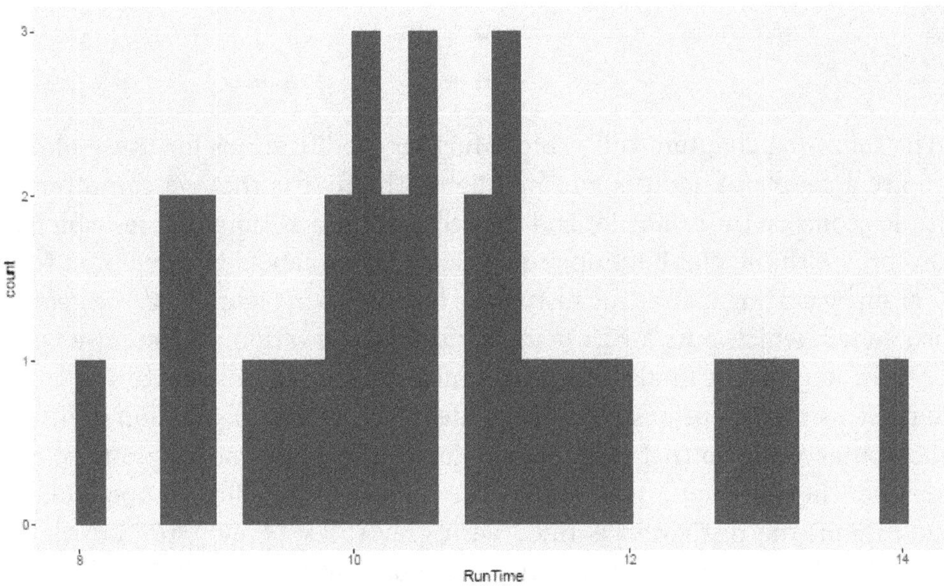

Figure 1.10: Default histogram of Run Times using ggplot.

The warning message reflects a challenge in making histograms that involves how many bins to use. Ingeom histogram, it always uses 30 bins and expects you to make your own choice, compared to hist that used a different method to try to make a better automatic choice, but there is no single right answer. So maybe we should try out other values to get a "smoother" result here, which we can do by adding the bins= ... to the + geom_histogram (), such as + geom_histogram (bins = 8) to get an 8 bin histogram in Figure 1.11.

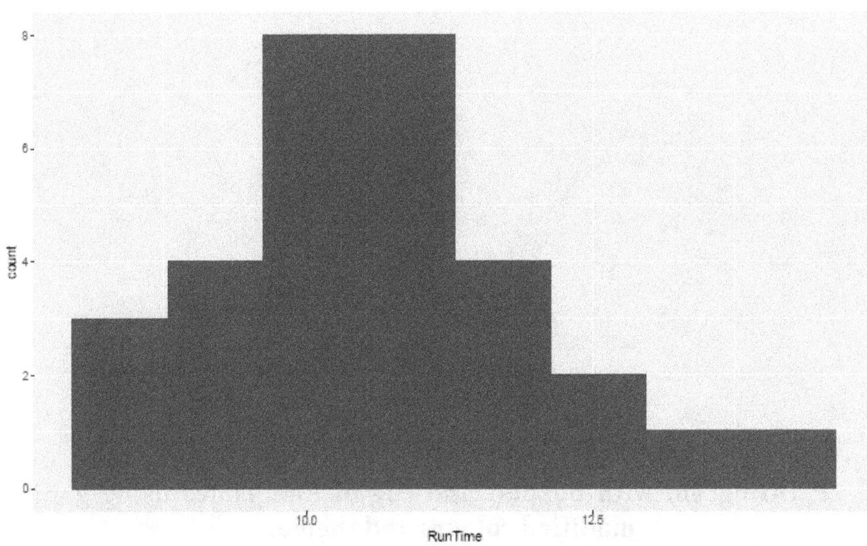

Figure 1.11: Histogram of Run Times using ggplot with 8 bins.

```
ggplot(data = treadmill, mapping = aes(x = RunTime)) +
  geom_histogram(bins = 8)
```

The following chapters will explore further modifications for these plots, but there are a couple of additions to highlight. The first is that we can often layer multiple geoms on the same plot and the order of the additions defines which layer is "on top", with the plot built up sequentially. So we can add a boxploton top of a histogram by putting it after the histogram layer. Also in Figure 1.12, the geom_rug is also added, which puts a tick mark for each observation on the lower part of the x-axis. Rug plots can also use a graphical technique called jittering to add a little noise using the options geom_rug(sides = "b", aes(y= 0), position = "jitter")13 to each observation so that multiple similar or tied observations do notplot as a single line. There are options to control the colour of individual components when we add them (the histogram is filled with grey (fill = "grey"), the boxplot is in "tomato" (colour = "tomato"), and the rug plot is in "skyblue"). Finally, the last change here is to the "theme" for the plot14 which we can include one of a suite of different layouts with themes such as + theme_bw() or + theme_light(). If you add the ggthemes package[Arnold, 2021], you can access a long list of alternative looks for your plot (seehttps://jrnold.github.io/ggthemes/reference/index.html for options there).

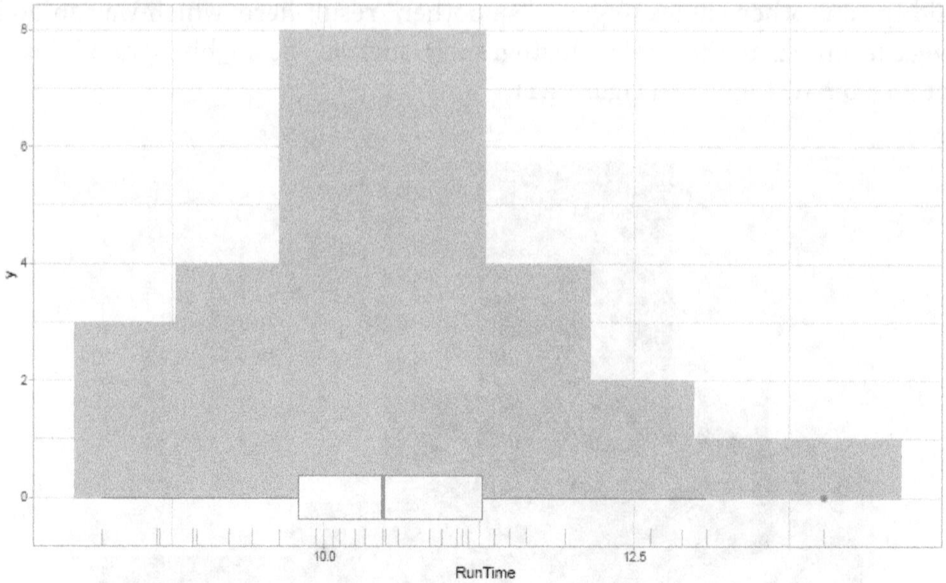

Figure 1.12: Histogram with boxplot and rug of Run Times using ggplot with modified colours and theme.

```
ggplot(data = treadmill, mapping = aes(x = RunTime)) +
  geom_histogram(fill = "grey", bins = 8) +
  geom_boxplot(color = "tomato") +
    geom_rug(color = "skyblue", sides = "b", aes(y = 0), position = "jitter") +
  theme_light()
```

1.6 EXITING RSTUDIO

Finally, when you are done with your work and attempt to exit out of RStudio, it will ask you to save your workspace. ***DO NOT DO THIS!*** It will just create a cluttered workspace and could even cause you to get incorrect results.

In fact, you should go into the Tools -> Global Options and then make sure that "Save workspace to

.RData on exit" option on the first screen you will see is set to ***Never*** . If you save your R code either as a.R or (better) an R Markdown (.Rmd) file, you can re-create any results by simply re-running that code orre-knitting the file. If you find that you have lots of "stuff" in your workspace because you accidentally saved your workspace, just run rm(list = ls()). It will delete all the data sets from your workspace.

1.7 CONCISE SUMMARY

This chapter covered getting R and RStudio downloaded and some basics of working with R via RStudio. You should be able to read a data set into R and run some basic functions, all done using the RStudio interface. If you are struggling with this, you should seek additional help with these technical issues so that you are ready for more complicated statistical methods that are going to be encountered in the following chapters. The way everyone learns R is by starting with some example code that does most of what you want to do and then you modify it. If you can complete the Practice Problems that follow, you are well on your way to learning to use R.

The statistical methods in this chapter were minimal and all should have been review. They involved a quick reminder of summarizing the center, spread, and shape of distributions using numerical summaries of the mean and SD and/or the min, Q1, median, Q3, and max and the histogram and boxplot as graphical summaries. We revisited the ideas of symmetry and skew. But the main point was really to get a start on using R via RStudio to provide results you should be familiar with from your previous statistics experience(s) and to introduce some of the code we will be building on in the next chapters.

1.8 SUMMARY OF IMPORTANT R CODE

To help you learn and use R, there is a section highlighting the most important R code used near the end of each chapter. The bold text will never change but the lighter and/or ALL CAPS text (red in the online or digital version) will need to be customized to your particular application. The sub-bullet for each function will discuss the use of the function and pertinent options or packages required. You can use this as a guide to finding the function names and some hints about options that will help you to get the code to work. You can also revisit the worked examples using each of the functions.

- **FILENAME <- read_csv("path to csv file/FILENAME.csv")**
  - Can be generated using "Import Dataset" button or by modifying this text.
  - Requires the readr package to be loaded (library(readr)) when using the code directly.
  - Imports a text file saved in the CSV format.
- **DATASETNAME$VARIABLENAME**
  - To access a particular variable in a tibble called DATASETNAME, use a $ and then the VARIABLENAME.
- **Head (DATASETNAME)**
  - Provides a list of the first few rows of the data set for all the variables in it.
- **tail(DATASETNAME)**
  - Provides a list of the last few rows of the data set for all the variables in it.
- **mean(DATASETNAME$VARIABLENAME)**
  - Calculates the mean of the observations in a variable.
- **sd(DATASETNAME$VARIABLENAME)**
  - Calculates the standard deviation of the observations in a variable.
- **favstats(DATASETNAME$VARIABLENAME)**
  - Requires the mosaic package to be loaded (library(mosaic)) after installing the package).
  - Provides a suite of numerical summaries of the observations in a variable.
- **hist(DATASETNAME$VARIABLENAME)**
  - Makes a histogram.

- boxplot(DATASETNAME$VARIABLENAME)
  - Makes a boxplot.
- ggplot(data = DATASETNAME, mapping = aes(VARIABLENAME)) + geom_histogram(bins = 10)
  - Makes a histogram with 10 bins using ggplot, requires the ggplot2 library is installed and loaded.

1.9 PRACTICE PROBLEMS

In each chapter, the last section contains some questions for you to complete to make sure you understood the material. You can download the code to answer questions 1.1 to 1.5 below at http://www.math.mon tana.edu/courses/s217/documents/Ch1.Rmd. But to practice learning R, it would be most useful for you to try to accomplish the requested tasks yourself and then only refer to the provided R code if/when you struggle. These questions provide a great venue to check your learning, often to see the methods applied to another data set, and for something to discuss in study groups, with your instructor, and at the Math Learning Center.

- Open RStudio and go to File -> New File -> R Markdown. . . to create a .Rmd. Click on the "Knit" button and see what happens. Try to complete the following questions in that document, clicking on the Knit button after you add a code chunk with code to complete each question. Part of the assignment on this question is to not get frustrated the first time you are trying this and seek out help to answer questions you have when practicing.
- Read in the treadmill data set discussed previously and find the mean and SD of the Ages (Age variable) and Body Weights (Body Weight variable). In studies involving human subjects, it is common to report a summary of characteristics of the subjects. Why does this matter? Think about how your interpretation of any study of the fitness of subjects would change if the mean age (same spread) had been 20 years older or 35 years younger.
- How does knowing about the distribution of results for *Age* and *Body Weight* help you understand the results for the Run Times discussed previously?
- The mean and SD are most useful as summary statistics only if the distribution is relatively symmetric. Make a histogram of *Age* responses and discuss the shape of the distribution (is it skewed right, skewed left, approximately symmetric?; are there outliers?). Approximately what range of ages does this study pertain to?
- The weight responses are in kilograms and you might prefer to see them in pounds. The conversion islbs = 2.205*kgs. Create a new variable in the treadmill tibble called *BWlb* using this code:

```
treadmill$BWlb <- 2.205*treadmill$BodyWeight
```

and find the mean and SD of the new variable (*BWlb*).

- Make histograms and boxplots of th90e original *Body Weight* and new *BWlb* variables, both using base R plots and using ggplot2. Discuss aspects of the distributions that changed and those that remained the same with the transformation from kilograms to pounds. What does this tell you about changing the units of a variable in terms of its distribution?

CHAPTER-2

(R)E-INTRODUCTION TO STATISTICS

INTRODUCTION: AN OVERVIEW

The previous material served to get us started in R and to get a quick review of same basic graphical and descriptive statistics. Now we will begin to engage some new material and exploit the power of R to do statistical inference. Because inference is one of the hardest topics to master in statistics, we will also review some basic terminology that is required to move forward in learning more sophisticated statistical methods. To keep this "review" as short as possible, we will not consider every situation you learned in introductory statistics and instead focus exclusively on the situation where we have a quantitative response variable measured on two groups, adding a new graphic called a "pirate-plot" to help us see the differences in the observations in the groups.

2.1 DATA WRANGLING AND DENSITY CURVES

Part of learning statistics is learning to correctly use the terminology, some of which is used colloquially differently than it is used in formal statistical settings. The most commonly "misused" statistical term is data. In statistical parlance, we want to note the plurality of data. Specifically, datum is a single measurement, possibly on multiple random variables, and so it is appropriate to say that "a datum is. . .". Once we move to discussing data, we are now referring to more than one observation, again on one, or possibly more than one, random variable, and so we need to use "data are. . . " when talking about our observations. We want to distinguish our use of the term "data" from its more colloquial1 usage that often involves treating it as singular. In a statistical setting "data" refers to measurements of our cases or units. When we summarize the results of a study (say providing the mean and SD), that information is not "data". We used our data to generate that information. Sometimes we also use the term "data set" to refer to all our observations and this is a singular term to refer to the group of observations and this makes it really easy to make mistakes on the usage of "data"2.

It is also really important to note that variables have to vary – if you measure the level of education of your subjects but all are high school graduates, then you do not have a "variable". You may not know if you have real variability in a "variable" until you explore the results you obtained.

The last, but probably most important, aspect of data is the context of the measurement. The "who, what, when, and where" of the collection of the observations is critical to the sort of conclusions we can make based on the results. The information on the study design provides information required to assess the scope of inference (SOI) of the study (see Table 2.1 for more on SOI). Generally, remember to think about the research questions the researchers were trying to answer and whether their study actually would answer those questions. There are no formulas to help us sort some of these things out, just critical thinking about the context of the measurements.

To make this concrete, consider the data collected from a study [Walker et al., 2014] to investigate whether clothing worn by a bicyclist might impact the passing distance of cars. One of the authors wore seven different outfits (outfit for the day was chosen randomly by shuffling seven playing cards) on his regular 26 km commute near London in the United Kingdom. Using a specially instrumented bicycle, they measured how close the vehicles passed to the widest point on the handlebars. The seven outfits ("conditions") that you can view at https://www.sciencedirect.com/science/article/pii/S0001457513004636 were:

- **COMMUTE:** Plain cycling jersey and pants, reflective cycle clips, commuting helmet, and bike gloves.
- **CASUAL:** Rugby shirt with pants tucked into socks, wool hat or baseball cap, plain gloves, and small backpack.
- **HIVIZ:** Bright yellow reflective cycle commuting jacket, plain pants, reflective cycle clips, commuting helmet, and bike gloves.
- **RACER:** Colourful, skin-tight, Tour de France cycle jersey with sponsor logos, Lycra bike shorts or tights, race helmet, and bike gloves.
- **NOVICE:** Yellow reflective vest with "Novice Cyclist, Pass Slowly" and plain pants, reflective cycle clips, commuting helmet, and bike gloves.
- **POLICE:** Yellow reflective vest with "POLICEwitness.com – Move Over – Camera Cyclist" and plain pants, reflective cycle clips, commuting helmet, and bike gloves.

- **POLITE:** Yellow reflective vest with blue and white checked banding and the words "POLITE notice, Pass Slowly" looking similar to a police jacket and plain pants, reflective cycle clips, commuting helmet, and bike gloves.

They collected data (distance to the vehicle in cm for each car "overtake") on between 8 and 11 rides in each outfit and between 737 and 868 "over takings" across these rides. The outfit is a categorical *predictor* or *explanatory* variable) that has seven different levels here. The distance is the *response* variable and is a quantitative variable here[3]. Note that we do not have the information on which overtake came from which ride in the data provided or the conditions related to individual overtake observations other than the distance to the vehicle (they only included over takings that had consistent conditions for the road and riding). The data are posted on my website4 at http://www.math.montana.edu/courses/s217/documents/Walker 2014_mod.csv if you want to download the file to a local directory and then import the data into R using "Import Dataset". Or you can use the code in the following code chunk to directly read the data set into R using the URL.

```
suppressMessages(library(readr))
dd <- read_csv("http://www.math.montana.edu/courses/s217/documents/Walker2014_mod.csv")
```

It is always good to review the data you have read by running the code and printing the tibble by typing the tibble name (here > dd) at the command prompt in the console, using the View function, (here View (dd)), to open a spreadsheet-like view, or using the head and tail functions to show the first and last six observations:

```
head(dd)
```

```
## # A tibble: 6 x 8
##   Condition Distance Shirt Helmet Pants Gloves ReflectClips Backpack
##   <chr>        <dbl> <chr> <chr>  <chr> <chr>  <chr>        <chr>
## 1 casual         132 Rugby hat    plain plain  no           yes
## 2 casual         137 Rugby hat    plain plain  no           yes
## 3 casual         174 Rugby hat    plain plain  no           yes
## 4 casual          82 Rugby hat    plain plain  no           yes
## 5 casual         106 Rugby hat    plain plain  no           yes
## 6 casual          48 Rugby hat    plain plain  no           yes
```

```
tail(dd)
```

```
## # A tibble: 6 x 8
##   Condition Distance Shirt       Helmet Pants Gloves ReflectClips Backpack
##   <chr>        <dbl> <chr>       <chr>  <chr> <chr>  <chr>        <chr>
## 1 racer          122 TourJersey race   lycra bike   yes          no
## 2 racer          204 TourJersey race   lycra bike   yes          no
## 3 racer          116 TourJersey race   lycra bike   yes          no
## 4 racer          132 TourJersey race   lycra bike   yes          no
## 5 racer          224 TourJersey race   lycra bike   yes          no
## 6 racer           72 TourJersey race   lycra bike   yes          no
```

Another option is to directly access specific rows and/or columns of the tibble, especially for larger datasets. In objects containing data, we can select certain rows and columns using the brackets, [..., ...], to specify the row (first element) and column (second element). For example, we can extract the datum in the fourth row and second column using dd[4,2]:

```
dd[4,2]
```

```
## # A tibble: 1 x 1
##   Distance
##      <dbl>
## 1       82
```

This provides the distance (in cm) of a pass at 82 cm. To get all of either the rows or columns, a space is used instead of specifying a particular number. For example, the information in all the columns on the fourth observation can be obtained using dd[4,]:

```
dd[4,]
```

```
## # A tibble: 1 x 8
##   Condition Distance Shirt Helmet Pants Gloves ReflectClips Backpack
##   <chr>        <dbl> <chr> <chr>  <chr> <chr>  <chr>        <chr>
## 1 casual          82 Rugby hat    plain plain  no           yes
```

So this was an observation from the casual condition that had a passing distance of 82 cm. The other columns describe some other specific aspects of the condition. To get a more complete sense of the dataset, we can extract a suite of observations from each condition using their row numbers concatenated, c(),together, extracting all columns for two observations from each of the conditions based on their rows.

```
dd[c(1, 2, 780, 781, 1637, 1638, 2374, 2375, 3181, 3182, 3971, 3972, 4839, 4840),]
```

```
## # A tibble: 14 x 8
##    Condition Distance Shirt       Helmet   Pants Gloves ReflectClips Backpack
##    <chr>        <dbl> <chr>       <chr>    <chr> <chr>  <chr>        <chr>
##  1 casual         132 Rugby       hat      plain plain  no           yes
##  2 casual         137 Rugby       hat      plain plain  no           yes
##  3 commute         70 PlainJersey commuter plain bike   yes          no
##  4 commute        151 PlainJersey commuter plain bike   yes          no
##  5 hiviz           94 Jacket      commuter plain bike   yes          no
##  6 hiviz          145 Jacket      commuter plain bike   yes          no
##  7 novice          12 Vest_Novice commuter plain bike   yes          no
##  8 novice         122 Vest_Novice commuter plain bike   yes          no
##  9 police         113 Vest_Police commuter plain bike   yes          no
## 10 police         174 Vest_Police commuter plain bike   yes          no
## 11 polite         156 Vest_Polite commuter plain bike   yes          no
## 12 polite          14 Vest_Polite commuter plain bike   yes          no
## 13 racer          104 TourJersey  race     lycra bike   yes          no
## 14 racer          141 TourJersey  race     lycra bike   yes          no
```

Now we can see the Condition variable seems to have seven different levels, the Distance variable contains the overtake distance, and then a suite of columns that describe aspects of each outfit, such as the type of shirt or whether reflective cycling clips were used or not. We will only use the "Distance" and "Condition" variables to start with.

When working with data, we should always start with summarizing the sample size. We will use n for the number of subjects in the sample and denote the population size (if available) with N . Here, the sample size is n = 5690 . In this situation, we do not have a random sample from a population (these were all of the overtakes that met the criteria during the rides) so we cannot make inferences from our sample to a larger group (other rides or for other situations like different places, times, or riders). But we can assess whether there is a causal effect5: if sufficient evidence is found to conclude that there is some difference in the responses across the conditions, we can attribute those differences to the treatments applied, since the overtake events should be same otherwise due to the outfit being randomly assigned to the rides. The story of the data set – that it was collected on a particular route for a particular rider in the UK – becomes pretty important in thinking about the ramifications of any results. Are drivers and roads in Montana or South Dakota different from drivers and roads near London? Are the road and traffic conditions likely to be different? If so, then we should not assume that the detected differences, if detected, would also exist in some other location for a different rider. The lack of a random sample here from all the overtakes in the area (or more generally all that happen around the world) makes it impossible to assume that this set of overtakes might be like others. So there are definite limitations to the inferences in the following results. But it is still interesting to see if the outfits worn caused a difference in the mean overtake distances, even though the inferences are limited to the conditions in this individual's commute. If this had been an observational study (suppose that the researcher could select their outfit), then we would have to avoid any of the "causal" language that we can consider here because the outfits were not randomly assigned to the rides. Without random assignment, the explanatory variable of outfit choice could be confounded with another characteristic of rides that might be related to the passing distances, such as wearing a particular outfit because of an expectation of heavy traffic or poor light conditions. Confounding is not the only reason to avoid causal statements with non-random assignment but the inability to separate the effect of other variables (measured or unmeasured) from the differences we are observing means that our inferences in these situations need to be carefully stated to avoid implying causal effects.

In order to get some summary statistics, we will rely on the R package called mosaic [Pruim et al., 2021b]as introduced previously. First (but only once), you need to install the package, which can be done either using the Packages tab in the lower right panel of RStudio or using the install packages function with quotes around the package name:

```
> install.packages("mosaic")
```

If you open a .Rmd file that contains code that incorporates packages and they are not installed, the bar at the top of the R Markdown document will prompt you to install those missing packages. This is the easiest way to get packages you might need installed. After making sure that any required packages are installed, use the library function around the package name (no quotes now!) to load the package, something that you need to do any time you want to use features of a package.

```
library(mosaic)
```

When you are loading a package, R might mention a need to install other packages. If the output says that it needs a package that is unavailable, then follow the same process noted above to install that package and then repeat trying to load the package you wanted. These are called package "dependencies" and are due to one package developer relying on functions that already exist in another package.

With tibbles, you have to declare categorical variables as "factors" to have R correctly handle the variables using the factor function, either creating a new variable or replacing the "character" version of the variable that is used to read in the data initially. The following code replaces the Condition character variable with a factor version of the same variable with the same name.

We use this sort of explicit declaration for either character coded (non-numeric) variables or for numerically coded variables where the numbers represent categories to force R to correctly work with the information on those variables. For quantitative variables, we do not need to declare their type and they are stored as numeric variables as long as there is no text in that column of the spreadsheet other than the variable name.

The one-at-a-time declaration of the variables as factors when there are many (here there are six more) creates repetitive and cumbersome code. There is another way of managing this and other similar related "data wrangling"[6]. To do this, we will combine using the pipe operator (%>% from the magrittr package or|> in base R) and using the mutate function from dplyr, both %>% and mutate are part of the tidyverse and start to help us write code that flows from left to right to accomplish multiple tasks. The pipe operator (%>% or |>) allows us to pass a data set to a function (sometimes more than one if you have multiple data wrangling tasks to complete – see work below) and there is a keyboard short-cut to get the combination of characters for it by using Ctrl+Shift+M on a PC or Cmd+Shift+M on a Mac. The mutate function allows us to create new columns or replace existing ones by using information from other columns, separating each additional operation by a comma (and a "return" for proper style). You will gradually see more reasons why we want to learn these functions, but for now this allows us to convert the character variables into factor variables within mutate and when we are all done to assign our final data set back in the same dd tibble that we started with.

```
dd <- dd %>% mutate(Shirt = factor(Shirt),
                    Helmet = factor(Helmet),
                    Pants = factor(Pants),
                    Gloves = factor(Gloves),
                    ReflectClips = factor(ReflectClips),
                    Backpack = factor(Backpack)
                    )
```

The first part of the code chunk (dd <-) is to save our work that follows into the dd tibble. The dd %>% mutate is translated as "take the tibble dd and apply the mutate function." Inside the mutate function, each line has a variable name = factor(variable name) that declares each variable as a factor variable with the same name as in the original tibble.

With many variables in a data set and with some preliminary data wrangling completed, it is often useful to get some quick information about all of the variables; the summary function provides useful information whether the variables are categorical or quantitative and notes if any values were missing.

```
summary(dd)
```

```
##     Condition       Distance               Shirt            Helmet         Pants          Gloves     ReflectClips Backpack
##  casual :779   Min.   :  2.0   Jacket      :737   commuter:4059   lycra: 852   bike :4911   no : 779    no :4911
##  commute:857   1st Qu.: 99.0   PlainJersey:857   hat     : 779   plain:4838   plain: 779   yes:4911    yes: 779
##  hiviz  :737   Median :117.0   Rugby       :779   race    : 852
##  novice :807   Mean   :117.1   TourJersey :852
##  police :790   3rd Qu.:134.0   Vest_Novice:807
##  polite :868   Max.   :274.0   Vest_Police:790
##  racer  :852                   Vest_Polite:868
```

The output is organized by variable, providing summary information based on the type of variable, either counts by category for categorical variables or the 5-number summary plus the mean for the quantitative variable Distance. If present, you would also get a count of missing values that are called "NAs" in R. For the first variable, called Condition and that we might more explicitly name *Outfit*, we find counts of the number of overtakes for each outfit: 779 out of 5, 690 were when wearing the casual outfit, 857 for "commute", and the other observations from the other five outfits, with the most observations when wearing the "polite" vest. We can also see that overtake distances (variable Distance) ranged from 2 cm to 274 cm with a median of 117 cm.

To accompany the numerical summaries, histograms and boxplots can provide some initial information on the shape of the distribution of the responses for the different *Outfits*. Figure 2.1 contains the histogram with a boxplot and a rug of *Distance*, all ignoring any information on which outfit was being worn. There are some additional layers and modifications in this version of the ggplot. The code uses our new pipe operator to pass our tibble into the ggplot, skipping the data = ... within ggplot(). There are some additional options modifying the title and the x- and y-axis labels inside the labs() part of the code, which will be useful for improving the labels in your plots and work across most plots made in the framework.

```
dd %>% ggplot(mapping = aes(x = Distance)) +
  geom_histogram(bins = 20, fill = "grey") +
  geom_rug(alpha = 0.1) +
  geom_boxplot(color = "tomato", width = 30) +
      # width used to scale boxplot to make it more visible
  theme_bw() +
  labs(title = "Plot of Passing Distances",
       x = "Distance (cm)",
       y = "Count")
```

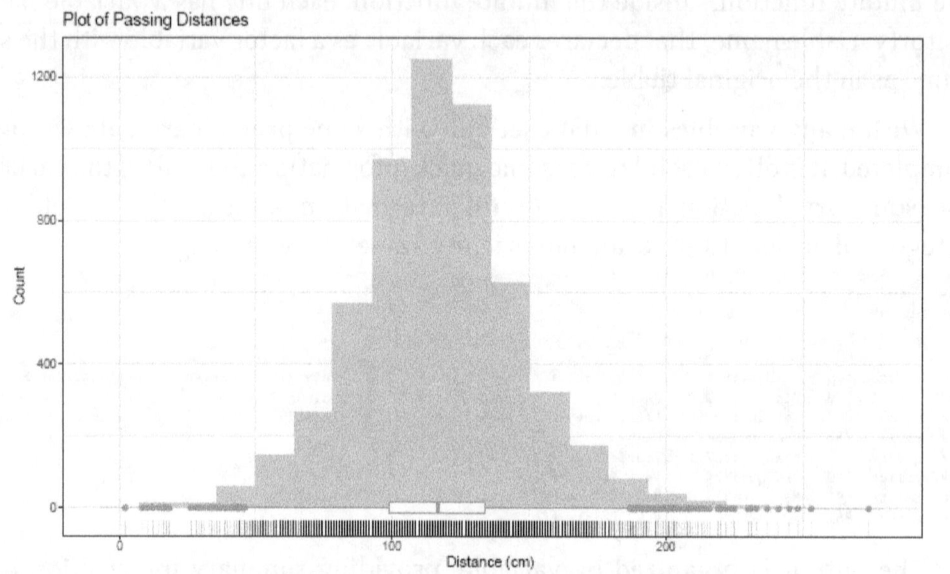

Figure 2.1: Histogram (with 20 bins), boxplot, and rug of passing distances (in cm).

Based on Figure 2.1, the distribution appears to be relatively symmetric with many observations in both tails flagged as potential outliers. Despite being flagged as potential outliers, they seem to be part of a common distribution. In real data sets, outliers are commonly encountered and the first step is to verify that they were not errors in recording (if so, fixing or removing them is easily justified). If they cannot be easily dismissed or fixed, the next step is to study their impact on the statistical analyses performed, potentially considering reporting results with and without the influential observation(s) in the results (if there are just handful). If the analysis is unaffected by the "unusual" observations, then it matters little whether they are dropped or not. If they do affect the results, then reporting both versions of results allows the reader to judge the impacts for themselves. It is important to remember that sometimes the outliers are the most interesting part of the data set. For example, those observations that were the closest would be of great interest, whether they are outliers or not.

Often when statisticians think of distributions of data, we think of the smooth underlying shape that led to the data set that is being displayed in the histogram. Instead of binning up observations and making bars in the histogram, we can estimate what is called a density curve as a smooth curve that represents the observed distribution of the responses. Density curves can sometimes help us see features of the data sets more clearly.

To understand the density curve, it is useful to initially see the histogram and density curve together. The height of the density curve is scaled so that the total area under the curve7 is 1. To make a comparable histogram, the y-axis needs to be scaled so that the histogram is also on the "density" scale which makes the bar heights adjust so that the proportion of the total data set in each bar is represented by the area in each bar (remember that area is height times width). So the height depends on the width of the bars and the totalarea across all the bars has to be 1. In the geom_histogram, its aesthetic is modified using the (cryptic8) code of (y = ..density..). The density curve is added to the histogram using the geom_density, producing the result in Figure 2.2 with added modifications for filling the density curve but using alpha = 0.1 to make the density curve fill transparent (alpha values range between 0 and 1 with lower values providing more transparency) and in purple (fill = purple). You can see how the density curve somewhat matches the histogram bars but deals with the bumps up and down and edges a little differently. We can pick out the relatively symmetric distribution using either display and will rarely make both together.

Histograms can be sensitive to the choice of the number of bars and even the cut-offs used to define the bins for a given number of bars. Small changes in the definition of cut-offs for the bins can have noticeable impacts on the shapes observed but this does not impact density curves. We have engaged the arbitrary choice of the number of bins, but we can add information on the original observations being included in each bar to better understand the choices that geom_hist is making. We can (barely) see how there are 2 observations at 2 cm (the noise added generates a wider line than for an individual observation so it is possible to see that it is more than one observation there but I had to check the data set to confirm this).

A limitation of the histogram arises at the center of the distribution where the bar that goes from approximately 110 to 120 cm suggests that the mode (peak) is in this range (but it is unclear where) but the density curve suggests that the peak is closer to 120 than 110. Both density curves and histograms can react to individual points in the tails of distributions, but sometimes in different ways.

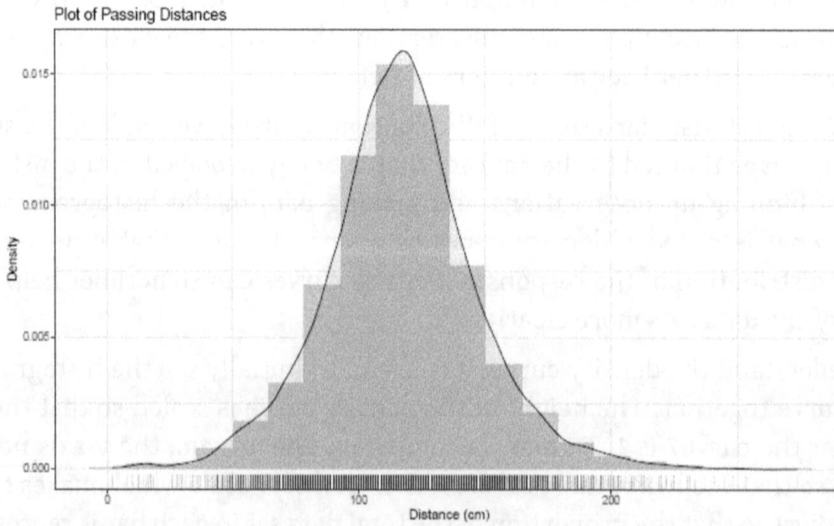

Figure 2.2: Histogram (density scaled), density curve, and rug plot of Distance responses.

```
d %>% ggplot(mapping = aes(x = Distance)) +
  geom_histogram(bins = 15, fill = "grey", aes(y = ..density..)) +
  geom_density(fill = "purple", alpha = 0.1) +
  geom_rug(alpha = 0.1) +
  theme_bw() +
  labs(title = "Plot of Passing Distances",
       x = "Distance (cm)",
       y = "Density")
```

The graphical tools we've just discussed are going to help us move to comparing the distribution of responses across more than one group. We will have two displays that will help us make these comparisons. The simplest is the side-by-side boxplot, where a boxplot is displayed for each group of interest using the same y-axis scaling. In the base R boxplot function, we can use its formula notation to see if the response(Distance) differs based on the group (Condition) by using something like Y ~ X or, here, Distance ~ Condition. We also need to tell R where to find the variables – use the last option in the command, data = DATASETNAME , to inform R of the tibble to look in to find the variables. In this example, data = dd. We will use the formula and data = ... options in almost every function we use from here forward, except in ggplot which has too many options for formulas to be useful.

Figure 2.3 contains the side-by-side boxplots showing similar distributions for all the groups, with a slightly higher median in the "police" group and some potential outliers identified in both tails of the distributions in all groups.

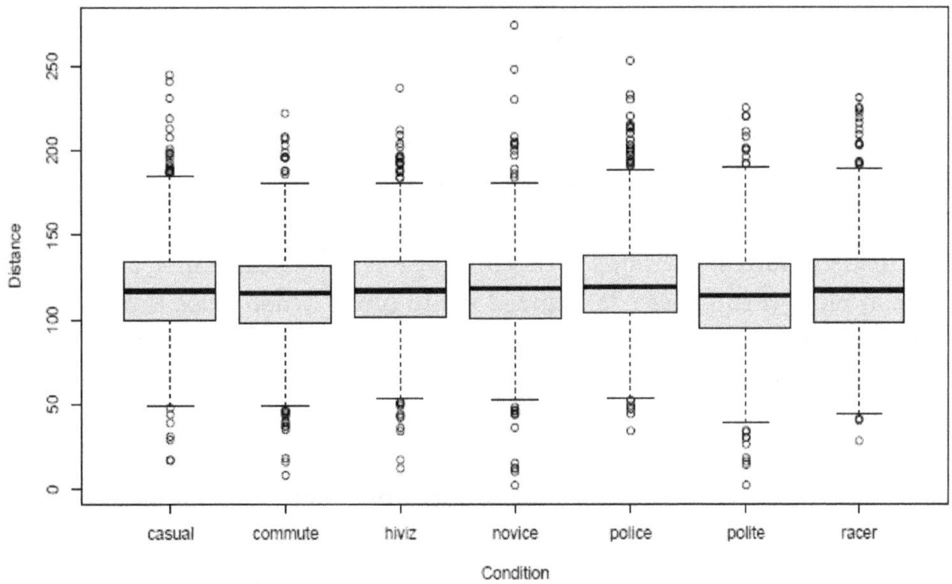

Figure 2.3: Side-by-side boxplot of distances based on outfits.

```
boxplot(Distance ~ Condition, data = dd)
```

The "~" (which is read as the tilde symbol9, which you can find in the upper left corner of your keyboard) notation will be used in two ways in this material. The formula use in R employed previously declares that the response variable here is Distance and the explanatory variable is Condition. The other use for "~" is as shorthand for "is distributed as" and is used in the context of Y ~ N (0, 1), which translates (in statistics) to defining the random variable Y as f ollowing a Normal distribution10 with mean 0 and variance of 1 (which also means that the standard deviation is 1). In the current situation, we could ask whether the Distance variable seems like it may follow a normal distribution in each group, in other words, is Distance~ N (μ, σ2)? Since the responses are relatively symmetric, it is not clear that we have a violation of the assumption of the normality assumption for the Distance variable for any of the seven groups (more later on how we can assess this and the issues that occur when we have a violation of this assumption). Remember that μ and σ are parameters where μ ("mu") is our standard symbol for the population mean and that σ ("sigma") is the symbol of the population standard deviation and σ2 is the symbol of the population variance.

2.2 PIRATE-PLOTS

An alternative graphical display for comparing multiple groups that we will use is a display called a pirate-plot[Phillips, 2017] from the yarrr package11. Figure 2.4 shows an example of a pirate-plot that provides aside-by-side display that contains the density curves, the original observations that generated the density curve as jittered points (jittered both vertically and horizontally a little), the sample mean of each group (wide bar), and vertical lines to horizontal bars that represents the confidence interval for the true mean of that group. For each group, the density curves are mirrored to aid in visual assessment of the shape of the distribution. This mirroring also creates a shape that resembles the outline of a violin with skewed distributions so versions of this display have also been called a "violin plot" or a "bean plot" (I call these "enhanced violin plots" when I use them in journal articles instead of "pirate plots"). All together this plot shows us information on the original observations, center (mean) and its confidence interval, spread, and shape of the distributions of the responses. Our inferences typically focus on the means of the groups and this plot allows us to compare those across the groups while gaining information on the shapes of the distributions of responses in each group.

To use the pirateplot function we need to install and then load the yarrr package. The function works like the boxplot used previously except that options for the type of confidence interval needs to be specified with inf.method = "ci" – otherwise you will get a different kind of interval than you learned in introductory statistics and we don't want to get caught up in trying to understand the kind of interval it makes by default. And it seems useful to add inf.disp = "line" as an additional option to add bars for theconfidence interval12. There are many other options in the function that might be useful in certain situations, but these are the only ones that are really needed to get started with pirateplots. While we could build thisplot using ggplot, the simplicity of this function keeps it a favourite way to display a quantitative variable across groups even though we lose the grammar of graphics way of modifying the plot.

Figure 2.4 suggests that the distributions are relatively symmetric which would suggest that the means and medians are similar even though only the means are displayed in these plots. In this display, none of the observations are flagged as outliers (it is not a part of this display). It is up to the consumer of the graphic to decide if observations look to be outside of the overall pattern of the rest of the observations. By plotting the observations by groups, we can also explore the narrowest (and likely most scary) overtakes in the data set. The police and racer conditions seem to have all observations over 25 cm and the most close passes were in the novice and polite outfits, including the two 2 cm passes. By displaying the original observations, we are able to explore and identify features that aggregation and summarization in plots can sometimes obfuscate. But the pirate-plots also allow you to compare the shape of the distributions (relatively symmetric and somewhat bell-shaped), variability (they look to have relatively similar variability), and the means of the groups. Our inferences are going to focus on the means but those inferences are only valid if the distributions are either approximately normal or at least have similar shapes and spreads (more on this soon).

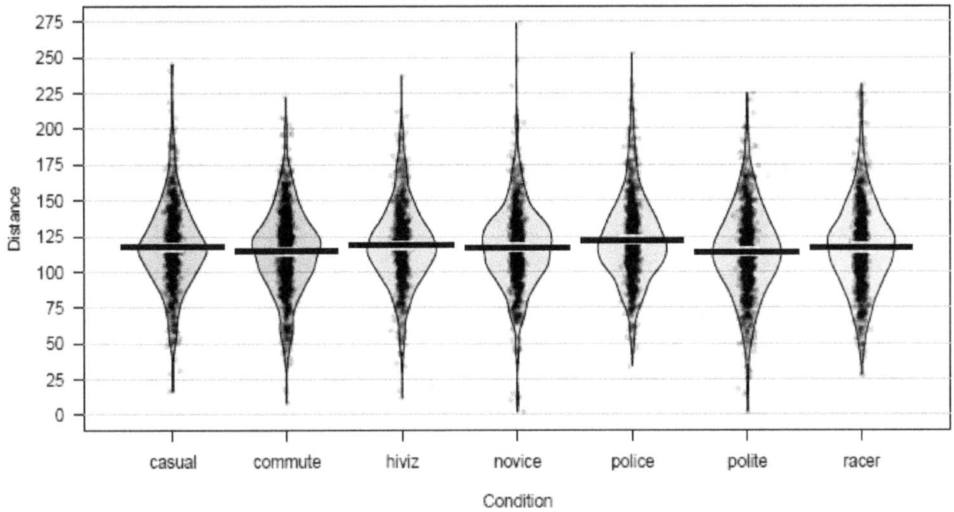

Figure 2.4: Pirate-plot of distances by outfit group. Bold horizontal lines correspond to sample mean of each group, boxes around lines (here they are very tight to the lines for the means) are the 95% confidence intervals.

```
library(yarrr)
pirateplot(Distance ~ Condition, data = dd, inf.method = "ci", inf.disp = "line")
```

It appears that the mean for *police* is higher than the other groups but that the others are not too different. But is this difference real? We will never know the answer to that question, but we can assess how likely we are to have seen a result as extreme or more extreme than our result, assuming that there is no difference in the means of the groups. And if the observed result is (extremely) unlikely to occur, then we have (extremely) strong evidence against the hypothesis that the groups have the same mean and can then conclude that there is likely a real difference. If we discover that our result was not very unlikely, given the assumption of no difference in the mean of the groups, then we can't conclude that there is a difference but also can't conclude that they are equal, just that we failed to find enough evidence against the equal means assumption to discard it as a possibility. Whether the result is unusual or not, we will want to carefully explore how big the estimated differences in the means are – is the difference in means large enough to matter to you? We would be more interested in the implications of the difference in the means when there is strong evidence against the null hypothesis that the means are equal but the size of the estimated differences should always be of some interest. To accompany the pirate-plot that displays estimated means, we need to have numerical values to compare. We can get means and standard deviations by groups easily using the same formula notation as for the plots with the mean and sd functions, if the mosaic package is loaded.

```
library(mosaic)
mean(Distance ~ Condition, data = dd)
```

```
##   casual  commute    hiviz   novice   police   polite    racer
## 117.6110 114.6079 118.4383 116.9405 122.1215 114.0518 116.7559
```

```
sd(Distance ~ Condition, data = dd)
```

```
##   casual  commute    hiviz   novice   police   polite    racer
## 29.86954 29.63166 29.03384 29.03812 29.73662 31.23684 30.60059
```

We can also use the favstats function to get those summaries and others by groups.

```
favstats(Distance ~ Condition, data = dd)
```

```
##   Condition min    Q1 median  Q3 max     mean       sd   n missing
## 1    casual  17 100.0    117 134 245 117.6110 29.86954 779       0
## 2   commute   8  98.0    116 132 222 114.6079 29.63166 857       0
## 3     hiviz  12 101.0    117 134 237 118.4383 29.03384 737       0
## 4    novice   2 100.5    118 133 274 116.9405 29.03812 807       0
## 5    police  34 104.0    119 138 253 122.1215 29.73662 790       0
## 6    polite   2  95.0    114 133 225 114.0518 31.23684 868       0
## 7     racer  28  98.0    117 135 231 116.7559 30.60059 852       0
```

Based on these results, we can see that there is an estimated difference of over 8 cm between the smallest mean (polite at 114.05 cm) and the largest mean (police at 122.12 cm). The differences among some of the other groups are much smaller, such as between casual and commute with sample means of 117.611 and 114.608 cm, respectively. Because there are seven groups being compared in this study, we will have to wait until Chapter 3 and the One-Way ANOVA test to fully assess evidence related to some difference among the seven groups. For now, we are going to focus on comparing the mean Distance between casual and commute groups – which is a two independent sample mean situation and something you should have seen before. Remember that the "independent" sample part of this refers to observations that are independently observed for the two groups as opposed to the paired sample situation that you may have explored where one observation from the first group is related to an observation in the second group (the same person with one measurement in each group (we generically call this "repeated measures") or the famous "twin" studies with one twin assigned to each group). This study has some potential violations of the "independent" sample situation (for example, repeated measurements made during a single ride), but those do not clearly fit into the matched pairs situation, so we will note this potential issue and proceed with exploring the method that assumes that we have independent samples, even though this is not true here. In Chapter 9, methods for more complex study designs like this one will be discussed briefly, but mostly this is beyond the scope of this material.

Here we are going to use the "simple" two independent group scenario to review some basic statistical concepts and connect two different frameworks for conducting statistical inference: randomization and parametric inference techniques. Parametric statistical methods involve making assumptions about the distribution of the responses and obtaining confidence intervals and/or p-values using a named distribution (like the z or t-distributions). Typically these results are generated using formulas and looking up areas under curves or cut-offs using a table or a computer. Randomization-based statistical methods use a computer to shuffle, sample, or simulate observations in ways that allow you to obtain distributions of possible results to find areas and cut-offs without resorting to using tables and named distributions. Randomization methods are what are called nonparametric methods that often make fewer assumptions (they are not free of assumptions!) and so can handle a larger set of problems more easily than parametric methods. When the assumptions involved in the parametric procedures are met by a data set, the randomization methods often provide very similar results to those provided by the parametric techniques. To be a more sophisticated statistical consumer, it is useful to have some knowledge of both of these techniques for performing statistical inference and the fact that they can provide similar results might deepen your understanding of both approaches.

To be able to work just with the observations from two of the conditions (casual and commute) we could remove all the other observations in a spreadsheet program and read that new data set back into R, but it is actually pretty easy to use R to do data management once the data set is loaded. It is also a better scientific process to do as much of your data management within R as possible so that your steps in managing the data are fully documented and reproducible. Highlighting and clicking in spreadsheet programs is a dangerous way to work and can be impossible to recreate steps that were taken from initial data set to the version that was analyzed. In R, we could identify the rows that contain the observations we want to retain and just extract those rows, but this is hard with over five thousand observations. The filter function from the dplyr package (part of the tidyverse suite of packages) is the best way to be able to focus on observations that meet a particular condition; we can "filter" the data set to retain just those rows. The filter function takes the data set via the pipe operate and then we need to define the condition we want to meet to retain those rows. Here we need to define the variable we want to work with, Condition, and then request rows that meet a condition (are %in%) and the aspects that meet that condition (here by concatenating the two levels of "casual" and "commute"), leading to code of:

dd %>% filter(Condition %in% c("casual", "commute"))

We want to save that new filtered data set into a new tibble for future work, so we can use the assignment operator (<-) to save the reduced data set into ddsub:

```
ddsub <- dd %>% filter(Condition %in% c("casual", "commute"))
```

There is also the select function that we could also use with an additional pipe operator to just focus on certain columns in the data set, here to just retain the Condition and Distance variables using:

```
ddsub <- dd %>%
  filter(Condition %in% c("casual","commute")) %>%
  select(Distance, Condition)
```

The select function shows up in multiple packages so you might need to use dplyr :: select () which tells R to use the version of select that is in dplyr. When you are working to filter or subset your data set you should always check that the correct observations were dropped either using View (ddsub) or by doing a quick summary of the Condition variable in the new tibble.

```
summary(ddsub$Condition)
```

```
##  casual commute   hiviz  novice  police  polite   racer
##     779     857       0       0       0       0       0
```

It ends up that R remembers the categories for observations that we removed even though there are 0 observations in them now and that can cause us some problems. When we remove a group of observations, we sometimes need to clean up categorical variables to just reflect the categories that are present. The factor function creates categorical variables based on the levels of the variables that are observed and is useful to run here to clean up Condition to just reflect the categories that are now present.

```
ddsub <- ddsub %>% mutate(Condition = factor(Condition))
summary(ddsub$Condition)
```

```
##  casual commute
##     779     857
```

The two categories of interest now were selected because neither looks particularly "racey" or has high visibility but could present a common choice between getting fully "geared up" for the commute or just jumping on a bike to go to work. Now if we remake the boxplots and pirate-plots, they only contain results for the two groups of interest here as seen in Figure 2.5. Note that these are available in the previous version of the plots, but now we will just focus on these two groups.

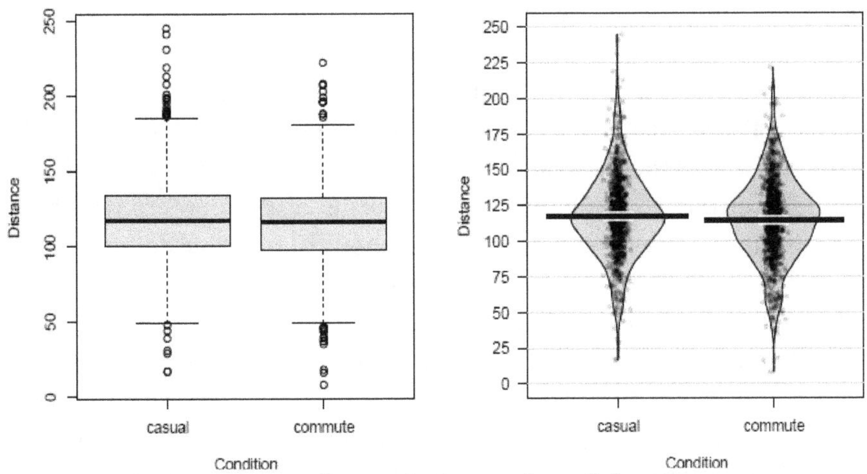

Figure 2.5: Boxplot and pirate-plot of the *Distance* responses on the reduced ddsub data set.

```
boxplot(Distance ~ Condition, data = ddsub)
pirateplot(Distance ~ Condition, data = ddsub, inf.method = "ci", inf.disp = "line")
```

The two-sample mean techniques you learned in your previous course all start with comparing the means the two groups. We can obtain the two means using the mean function or directly obtain the difference in the means using the diffmean function (both require the mosaic package). The diffmean function provides $\bar{x}$commute - $\bar{x}$casual where $\bar{x}$ (read as "x-bar") is the sample mean of observations in the subscripted group.

Note that there are two directions that you could compare the means and this function chooses to take the mean from the second group name alphabetically and subtract the mean from the first alphabetical group name. It is always good to check the direction of this calculation as having a difference of -3.003 cm versus3.003 cm could be important.

```
mean(Distance ~ Condition, data = ddsub)
```

```
##    casual  commute
## 117.6110 114.6079
```

```
diffmean(Distance ~ Condition, data = ddsub)
```

```
##  diffmean
## -3.003105
```

2.3 MODELS, HYPOTHESES, AND PERMUTATIONS FOR THE TWO SAMPLE MEAN SITUATION

There appears to be some evidence that the casual clothing group is getting higher average overtake distances than the commute group of observations, but we want to try to make sure that the difference is real – to assess evidence against the assumption that the means are the same "in the population" and possibly decide that this is not a reasonable assumption. First, a null hypothesis13 which defines a null model14 needs to be determined in terms of parameters (the true values in the population). The research question should help you determine the form of the hypotheses for the assumed population. In the two independent sample mean problem, the interest is in testing a null hypothesis of $H_0 : \mu_1 = \mu_2$ versus the alternative hypothesis of $H_A : \mu_1 \neq \mu_2$, where μ_1 is the parameter for the true mean of the first group and μ_2 is the parameter for the true mean of the second group. The alternative hypothesis involves assuming a statistical model for the ith $(i = 1, \dots, n_j)$ response from the jth $(j = 1, 2)$ group, y_{ij}, that involves modelling it as $y_{ij} = \mu_j + \varepsilon_{ij}$, where we assume that $\varepsilon_{ij} \sim N(0, \sigma^2)$. For the moment, focus on the models that either assume the means are the same (null) or different (alternative), which imply:

- **Null Model:** $y_{ij} = \mu + \varepsilon_{ij}$ There is **no** difference in **true** means for the two groups.
- **Alternative Model:** $y_{ij} = \mu_j + \varepsilon_{ij}$ There is **a** difference in **true** means for the two groups.

Suppose we are considering the alternative model for the 4th observation $(i = 4)$ from the second group $(j = 2)$, then the model for this observation is $y_{42} = \mu_2 + \varepsilon_{42}$, that defines the response as coming from the true mean for the second group plus a random error term for that observation, ε_{42}. For, say, the 5th observation from the first group $(j = 1)$, the model is $y_{51} = \mu_1 + \varepsilon_{51}$. If we were working with the null model, the mean is always the same (μ) – the group specified does not change the mean we use for that observation, so the model for y42 would be μ + ε42.

It can be helpful to think about the null and alternative models graphically. By assuming the null hypothesis is true (means are equal) and that the random errors around the mean follow a normal distribution, we assume that the truth is as displayed in the left panel of Figure 2.6 – two normal distributions with the same mean and variability. The alternative model allows the two groups to potentially have different means, such as those displayed in the right panel of Figure 2.6 where the second group has a larger mean. Note that in this scenario, we assume that the observations all came from the same distribution except that they had different means. Depending on the statistical procedure we are using, we basically are going to assume that the observations (*yij*) either were generated as samples from the null or alternative model. You can imagine drawing observations at random from the pictured distributions. For hypothesis testing, the null model is assumed to be true and then the unusualness of the actual result is assessed relative to that assumption. In hypothesis testing, we have to decide if we have enough evidence to reject the assumption that the null model (or hypothesis) is true. If we think that we have sufficient evidence to conclude that the null hypothesis is wrong, then we would conclude that the other model considered (the alternative model) is more reasonable. The researchers obviously would have hoped to encounter some sort of noticeable difference in the distances for the different outfits and have been able to find enough evidence to against the null model where the groups "look the same" to be able to conclude that they differ.

In statistical inference, null hypotheses (and their implied models) are set up as "straw men" with every interest in rejecting them even though we assume they are true to be able to assess the evidence against them. Consider the original study design here, the outfits were randomly assigned to the rides. If the null hypothesis were true, then we would have no difference in the population means of the groups. And this would apply if we had done a different random assignment of the outfits. So let's try this: assume that the null hypothesis is true and randomly re-assign the treatments (outfits) to the observations that were obtained. In other words, keep the Distance results the same and shuffle the group labels randomly. The technical term for this is doing a permutation (a random shuffling of a grouping15 variable relative to the observed responses). If the nullis true and the means in the two groups are the same, then we should be able to re-shuffle the groups to the observed Distance values and get results similar to those we actually observed. If the null is false and the means are really different in the two groups, then what we observed should differ from what we get under other random permutations and the differences between the two groups should be more noticeable in the observed data set than in (most) of the shuffled data sets. It helps to see an example of a permutation of the labels to understand what this means here.

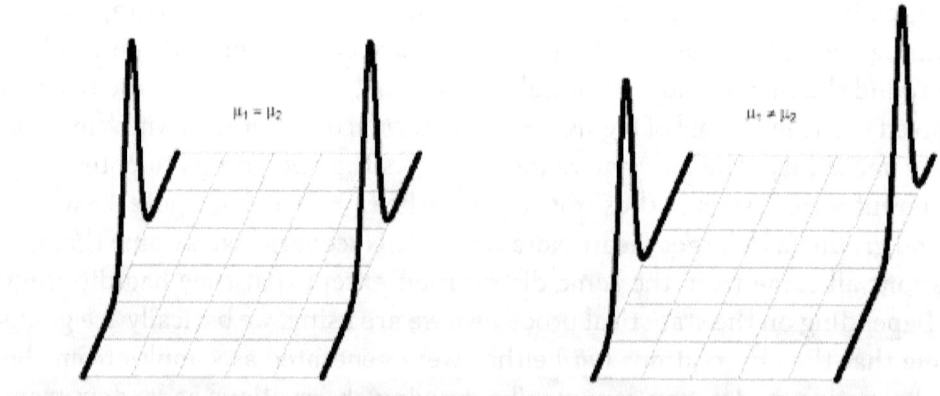

Figure 2.6: Illustration of the assumed situations under the null (left) and a single possibility that could occur if the alternative were true (right) and the true means were different. There are an infinite number of ways to make a plot like the right panel that satisfies the alternative hypothesis.

The data set we are working with is a little on the large size, especially to explore individual observations. So for the moment we are going to work with a random sample of 30 of the n = 1, 636 observations in ddsub, fifteen from each group, that are generated using the sample function. To do this16, we will use the sample function twice – once to sample from the subsetted commute observations (creating the s1 data set) and once to sample from the casual ones (creating s2). A new function for us, called rbind, is used to bindthe rows together — much like pasting a chunk of rows below another chunk in a spreadsheet program. This operation only works if the columns all have the same names and meanings both for rbind and in a spreadsheet. Together this code creates the dsample data set that we will analyze below and compare to results from the full data set. The sample means are now 135.8 and 109.87 cm for casual and commute groups, respectively, and so the difference in the sample means has increased in magnitude to -25.93 cm (commute - casual). This difference would vary based on the different random samples from the larger data set, but for the moment, pretend this was the entire data set that the researchers had collected and that we want to try to assess how unusual our sample difference was from what we might expect, if the null hypothesis that the true means are the same in these two groups was true.

```
et.seed(9432)
1 <- sample(ddsub %>% filter(Condition %in% "commute"), size = 15)
2 <- sample(ddsub %>% filter(Condition %in% "casual"), size = 15)
sample <- rbind(s1, s2)
ean(Distance ~ Condition, data = dsample)
```

```
##    casual  commute
## 135.8000 109.8667
```

In order to assess evidence against the null hypothesis of no difference, we want to permute the group labels versus the observations. In the mosaic package, the shuffle function allows us to easily perform a permutation[17]. One permutation of the treatment labels is provided in the Permuted Condition variable below. Note that the Distances are held in the same place while the group labels are shuffled.

```
Perm1 <- dsample %>%
  select(Distance, Condition) %>%
  mutate(PermutedCondition = shuffle(Condition))
# To force the tibble to print out all rows in data set -- not used often
data.frame(Perm1)
```

```
##    Distance Condition PermutedCondition
## 1       168   commute           commute
## 2       137   commute           commute
## 3        80   commute            casual
## 4       107   commute           commute
## 5       104   commute            casual
## 6        60   commute            casual
## 7        88   commute           commute
## 8       126   commute           commute
## 9       115   commute            casual
## 10      120   commute            casual
## 11      146   commute           commute
## 12      113   commute            casual
## 13       89   commute           commute
## 14       77   commute           commute
## 15      118   commute            casual
## 16      148    casual            casual
## 17      114    casual            casual
## 18      124    casual           commute
## 19      115    casual            casual
## 20      102    casual            casual
```

```
## 21     77   casual    casual
## 22     72   casual   commute
## 23    193   casual   commute
## 24    111   casual   commute
## 25    161   casual    casual
## 26    208   casual   commute
## 27    179   casual    casual
## 28    143   casual   commute
## 29    144   casual   commute
## 30    146   casual    casual
```

If you count up the number of subjects in each group by counting the number of times each label (commute, casual) occurs, it is the same in both the Condition and Permuted Condition columns (15 each). Permutations involve randomly re-ordering the values of a variable – here the Condition group labels – without changing the content of the variable. This result can also be generated using what is called sampling without replacement: sequentially select n labels from the original variable (Condition), removing each observed label and making sure that each of the original Condition labels is selected once and only once. The new, randomly selected order of selected labels provides the permuted labels. Stepping through the process helps to understand how it works: after the initial random sample of one label, there would n - 1 choices possible; on the nth selection, there would only be one label remaining to select. This makes sure that all original labels are re-used but that the order is random. Sampling without replacement is like picking names out of a hat, one-at-a-time, and not putting the names back in after they are selected. It is an exhaustive process for all the original observations. Sampling with replacement, in contrast, involves sampling from the specified list with each observation having an equal chance of selection for each sampled observation – in other words, observations can be selected more than once. This is like picking n names out of a hat that contains n names, except that every time a name is selected, it goes back into the hat – we'll use this technique in Section 2.9 to do what is called bootstrapping. Both sampling mechanisms can be used to generate inferences but each has particular situations where they are most useful. For hypothesis testing, we will use permutations (sampling without replacement) as its mechanism most closely matches the null hypotheses we will be testing.

The comparison of the pirate-plots between the real n = 30 data set and permuted version is what is really interesting (Figure 2.7). The original difference in the sample means of the two groups was -25.93 cm (commute - casual). The sample means are the statistics that estimate the parameters for the true means of the two groups and the difference in the sample means is a way to create a single number that tracks a quantity directly related to the difference between the null and alternative models. In the permuted data set, the difference in the means is 12.07 cm in the opposite direction (the commute group had a higher mean than casual in the permuted data).

```
ean(Distance ~ PermutedCondition, data = Perm1)
```

```
##    casual   commute
## 116.8000 128.8667
```

```
diffmean(Distance ~ PermutedCondition, data = Perm1)
```

```
## diffmean
## 12.06667
```

The diffmean function is a simple way to get the differences in the means, but we can also start to learn about using the lm function – that will be used for every chapter except for Chapter 5. The lm stands for ***linear model*** and, as we will see moving forward, encompasses a wide array of different models and scenarios. The ability to estimate the difference in the mean of two groups is among its simplest uses.[18] Notationally, it is very similar to other functions we have considered, lm(y ~ x, data = ...) where y is the response

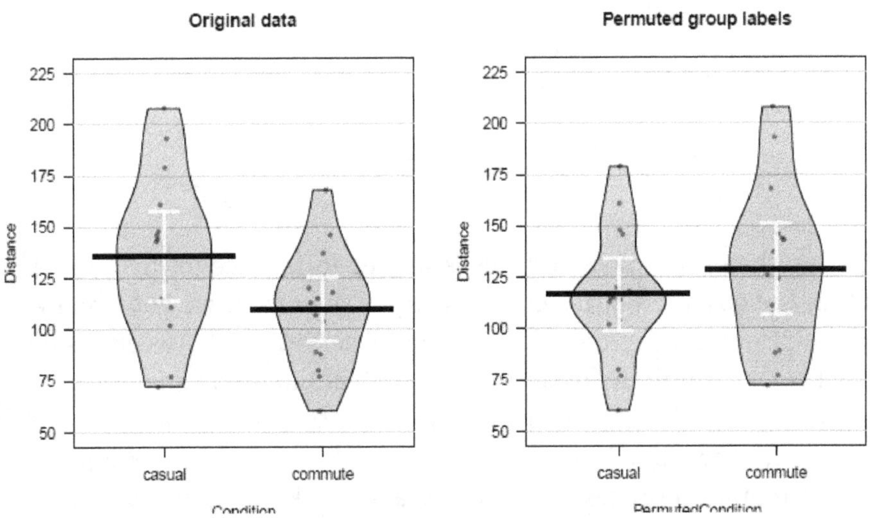

Figure 2.7: Pirate-plots of Distance responses versus actual treatment groups and permuted groups. Note how the responses are the same but that they are shuffled between the two groups differently in the permuted data set. With the smaller sample size, the 95% confidence intervals for each of the means are more clearly visible than with the original large data set.

This output is explored more in Chapter 3, but for the moment, focus on the row labelled as Condition commutein the middle of the output. In the first (Estimate) column, there is -25.933. This is a number we saw before– it is the difference in the sample means between *commute* and *casual* (*commute* - *casual*). When lm denotes a category in the row of the output (here commute), it is trying to indicate that the information to follow relates to the difference between this category and a baseline or reference category (here casual). The first ((Intercept)) row also contains a number we have seen before: -135.8 is the sample mean for the *casual* group. So the lm is generating a coefficient for the mean of one of the groups and another as the differencein the two groups19. In developing a test to assess evidence against the null hypothesis, we will focus onthe difference in the sample means. So we want to be able to extract that number from this large suite of information. It ends up that we can apply the coef function to lm models and then access that second coefficient using the bracket notation. Specifically:

```
lm1)[2]
```

```
## Conditioncommute
##         -25.93333
```

This is the same result as using the diffmean function, so either could be used here. The estimated difference in the sample means in the permuted data set of 12.07 cm is available with:

```
lmP <- lm(Distance ~ PermutedCondition, data = Perm1)
coef(lmP)[2]
```

```
## PermutedConditioncommute
##                 12.06667
```

Comparing the pirate-plots and the estimated difference in the sample means suggests that the observed difference was larger than what we got when we did a single permutation. Conceptually, permuting observations between group labels is consistent with the null hypothesis – this is a technique to generate results that we might have gotten if the null hypothesis were true since the true models for the responses are the same in the two groups if the null is true. We just need to repeat the permutation process many times and track how unusual our observed result is relative to this distribution of potential responses if the null were true. If the observed differences are unusual relative to the results under permutations, then there is evidence against the null hypothesis, and we can conclude, in the direction of the alternative hypothesis, that the true means differ. If the observed differences are similar to (or at least not unusual relative to) what we get under random shuffling under the null model, we would have a tough time concluding that there is any real difference between the groups based on our observed data set. This is formalized using the ***p-value*** as a measure of the strength of evidence against the null hypothesis and how we use it.

2.4 PERMUTATION TESTING FOR THE TWO SAMPLE MEAN SITUATION

In any testing situation, you must define some function of the observations that gives us a single number that addresses our question of interest. This quantity is called a test statistic. These often take on complicated forms and have names like t or z statistics that relate to their parametric (named) distributions so we know where to look up p-values20. In randomization settings, they can have simpler forms because we use the data set to find the distribution of the statistic under the null hypothesis and don't need to rely on a named distribution. We will label our test statistic T (for Test statistic) unless the test statistic has a commonly used name. Since we are interested in comparing the means of the two groups, we can define

$$T = \bar{x}_{\text{commute}} - \bar{x}_{\text{casual}},$$

which coincidentally is what the diffmean function and the second coefficient from the lm provided us previously. We label our ***observed test statistic*** (the one from the original data set) as

$$T_{obs} = \bar{x}_{\text{commute}} - \bar{x}_{\text{casual}},$$

which happened to be -25.933 cm here. We will compare this result to the results for the test statistic that we obtain from permuting the group labels. To denote permuted results, we will add an * to the labels:

$$T^* = \bar{x}_{\text{commute}^*} - \bar{x}_{\text{casual}^*}.$$

We then compare the *Tobs* = *x*¯commute - *x*¯casual = -25.933 to the distribution of results that are possible for the permuted results (*T* *) which corresponds to assuming the null hypothesis is true.

We need to consider lots of permutations to do a permutation test. In contrast to your introductory statistics course where, if you did this, it was just a click away, we are going to learn what was going on "under the hood" of the software you were using. Specifically, we need a for loop in R to be able to repeatedly generate the permuted data sets and record *T** for each one. Loops are a basic programming task that make randomization methods possible as well as potentially simplifying any repetitive computing task. To write a "for loop", we need to choose how many times we want to do the loop (call that B) and decide on a counter to keep track of where we are at in the loops (call that b, which goes from 1 up to B). The simplest loop just involves printing out the index, print(b) at each step. This is our first use of curly braces, { and }, that are used to group the code we want to repeatedly run as we proceed through the loop. By typing the following code in a code chunk and then highlighting it all and hitting the run button, R will go through the loop B = 5 times, printing out the counter:

```
B <- 5
for (b in (1:B)){
  print(b)
}
```

Note that when you highlight and run the code, it will look about the same with "+" printed after the first line to indicate that all the code is connected when it appears in the console, looking like this:

```
> for(b in (1:B)){
+   print(b)
+ }
```

When you run these three lines of code (or compile a .Rmd file that contains this), the console will show you the following output:

```
[1] 1
[1] 2
[1] 3
[1] 4
[1] 5
```

Instead of printing the counter, we want to use the loop to repeatedly compute our test statistic across *B* random permutations of the observations. The shuffle function performs permutations of the group labels relative to responses and the coef (lmP) [2] extracts the estimated difference in the two group means in the permuted data set. For a single permutation, the combination of shuffling Condition and finding the difference in the means, storing it in a variable called Ts is:

```
lmP <- lm(Distance ~ shuffle(Condition), data = dsample)
Ts <- coef(lmP)[2]
Ts
```

```
## shuffle(Condition)commute
##                -0.06666667
```

And putting this inside the print function allows us to find the test statistic under 5 different permutations easily:

```
B <- 5
for (b in (1:B)){
  lmP <- lm(Distance ~ shuffle(Condition), data = dsample)
  Ts <- coef(lmP)[2]
  print(Ts)
}
```

```
## shuffle(Condition)commute
##                      -1.4
## shuffle(Condition)commute
##                  1.133333
## shuffle(Condition)commute
##                  20.86667
## shuffle(Condition)commute
##                  3.133333
## shuffle(Condition)commute
##                 -2.333333
```

Finally, we would like to store the values of the test statistic instead of just printing them out on each pass through the loop. To do this, we need to create a variable to store the results, let's call it Tstar. We know that we need to store B results so will create a vector[21] of length *B*, which contains *B* elements, full of missing values (NA) using the matrix function with the nrow option specifying the number of elements:

```
Tstar <- matrix(NA, nrow = B)
Tstar
```

```
##      [,1]
## [1,]   NA
## [2,]   NA
## [3,]   NA
## [4,]   NA
## [5,]   NA
```

Now we can run our loop *B* times and store the results in Tstar.

```
for (b in (1:B)){
  lmP <- lm(Distance ~ shuffle(Condition), data = dsample)
  Tstar[b] <- coef(lmP)[2]
}
# Print out the results stored in Tstar with the next line of code
```

```
Tstar
```

```
##               [,1]
## [1,] -5.400000
## [2,] -3.266667
## [3,] -7.933333
## [4,] 13.133333
## [5,] -6.466667
```

Five permutations are still not enough to assess whether our *Tobs* of -25.933 is unusual and we need to do many permutations to get an accurate assessment of the possibilities under the null hypothesis. It is common practice to consider something like 1,000 permutations. The Tstar vector when we set *B* to be large, sayB = 1000, contains the permutation distribution for the selected test statistic under[22] the null hypothesis –what is called the ***null distribution*** of the statistic. The null distribution is the distribution of possible values of a statistic under the null hypothesis. We want to visualize this distribution and use it to assess how unusual our *Tobs* result of -25.933 cm was relative to all the possibilities under permutations (under the null hypothesis). So we repeat the loop, now with *B* = 1000 and generate a histogram (modified to add counts to the bars using stat_bin[23]), density curve, and summary statistics of the results:

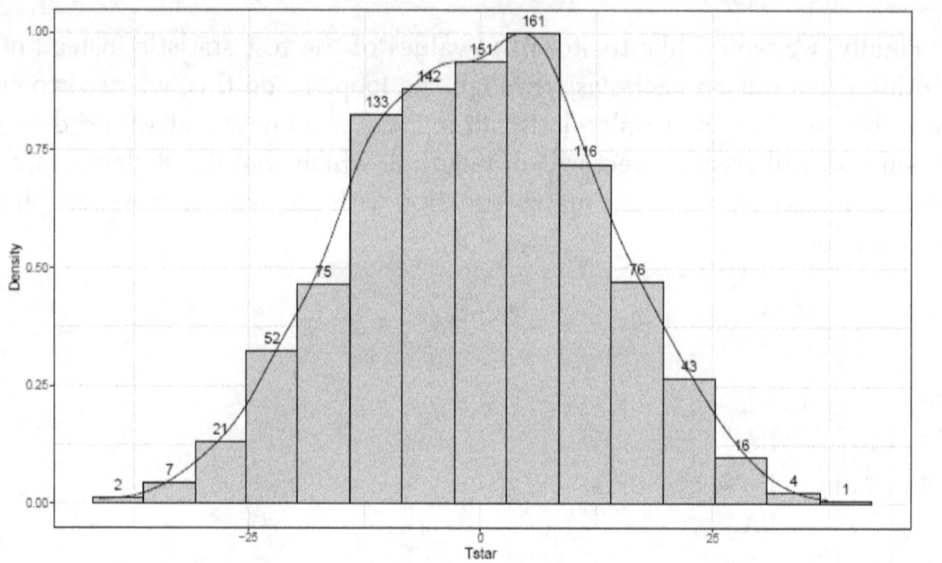

Figure 2.8: Histogram (left, with counts in bars) and density curve (right) of values of test statistic for *B* = 1,000 permutations.

```
B <- 1000
Tstar <- matrix(NA, nrow = B)
for (b in (1:B)){
  lmP <- lm(Distance ~ shuffle(Condition), data = dsample)
  Tstar[b] <- coef(lmP)[2]
}
```

```
tibble(Tstar) %>% ggplot(aes(x = Tstar)) +
  geom_histogram(aes(y = ..ncount..), bins = 15, col = 1, fill = "skyblue", center = 0) +
  geom_density(aes(y = ..scaled..)) +
  theme_bw() +
  labs(y = "Density") +
  stat_bin(aes(y = ..ncount.., label = ..count..), bins = 15, geom = "text", vjust = -0.75)

favstats(Tstar)
```

```
##         min        Q1     median  Q3      max       mean       sd    n missing
##   -41.26667 -10.06667 -0.3333333 8.6 37.26667 -0.5054667 13.17156 1000       0
```

Figure 2.8 contains visualizations of T* and the favstats summary provides the related numerical summaries. Our observed Tobs of -25.933 seems somewhat unusual relative to these results with only 30 T* values smaller than -25 based on the histogram. We need to make more specific comparisons of the permuted results versus our observed result to be able to clearly decide whether our observed result is really unusual.

To make the comparisons more concrete, first we can enhance the previous graphs by adding the value of the test statistic from the real data set, as shown in Figure 2.9, using the geom_vline function to draw a vertical line at our Tobs value specified in the xintercept option. Notice the order of the parameters. The code for the vertical line is before the code for the bin counts. This order is preferred so that the counts are still readable if the vertical line and a bin count are in the same horizontal position.

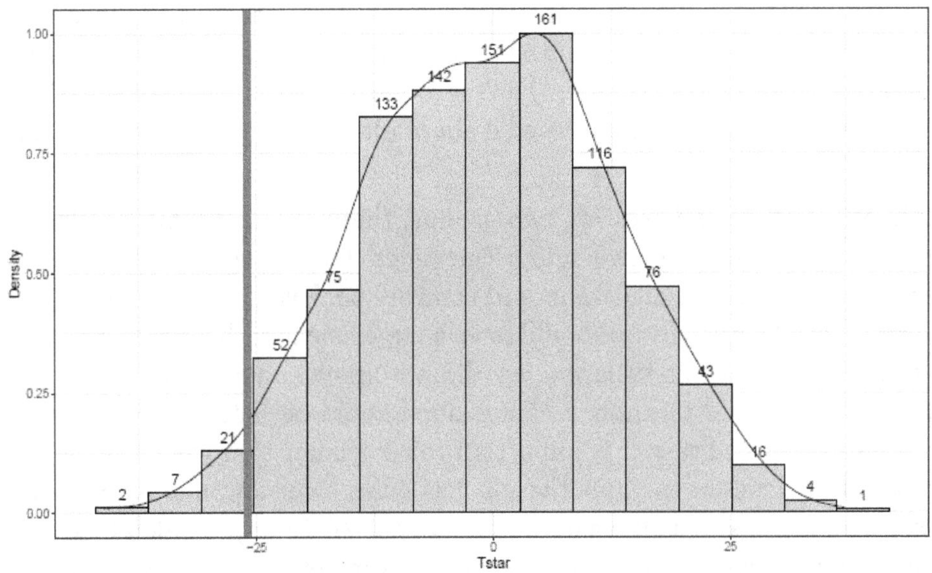

Figure 2.9: Histogram and density curve of values of test statistic for 1,000 permutations with bold vertical line for the value of observed test statistic.

```
Tobs <- -25.933
tibble(Tstar) %>% ggplot(aes(x = Tstar)) +
  geom_histogram(aes(y = ..ncount..), bins = 15, col = 1, fill = "skyblue", center = 0) +
  geom_density(aes(y = ..scaled..)) +
  theme_bw() +
  labs(y = "Density") +
  geom_vline(xintercept = Tobs, col = "red", lwd = 2) +
  stat_bin(aes(y = ..ncount.., label = ..count..), bins = 15, geom = "text", vjust = -0.75)
```

Second, we can calculate the exact number of permuted results that were as small or smaller than what we observed. To calculate the proportion of the 1,000 values that were as small or smaller than what we observed, we will use the pdata function. To use this function, we need to provide the distribution of values to compare to the cut-off (Tstar), the cut-off point (Tobs), and whether we want calculate the proportion that are below (left of) or above (right of) the cut-off (lower.tail = T option provides the proportion of values to the left of (below) the cut-off of interest).

```
pdata(Tstar, Tobs, lower.tail = T)[[1]]
```

```
## [1] 0.027
```

The proportion of 0.027 tells us that 27 of the 1,000 permuted results (2.7%) were as small or smaller than what we observed. This type of work is how we can generate p-values using permutation distributions. P-values, as you should remember, are the probability of getting a result as extreme as or more extreme than what we observed, given that the null is true. Finding only 27 permutations of 1,000 that were as small or smaller than our observed result suggests that it is hard to find a result like what we observed if there really were no difference, although it is not impossible.

When testing hypotheses for two groups, there are two types of alternative hypotheses, one-sided or two-sided. One-sided tests involve only considering differences in one-direction (like μ1 > μ2) and are performed when researchers can decide a priori24 which group should have a larger mean if there is going to be any sort of difference. In this situation, we did not know enough about the potential impacts of the outfits to know which group should be larger than the other so should do a two-sided test. It is important to remember that you can't look at the responses to decide on the hypotheses. It is often safer and moreconservative25 to start with a two-sided alternative (HA : μ1 /= μ2). To do a 2-sided test, find the areasmaller than what we observed as above (or larger if the test statistic had been positive). We also needto add the area in the other tail (here the right tail) similar to what we observed in the right tail. Some statisticians suggest doubling the area in one tail but we will collect information on the number that were as or more extreme than the same value in the other tail26. In other words, we count the proportion below-25.933 and over 25.933. So we need to find how many of the permuted results were larger than or equal to 25.933 cm to add to our previous proportion. Using pdata with -Tobs as the cut-off and lower.tail = F

provides this result:

```
pdata(Tstar, -Tobs, lower.tail = F)[[1]]
```

```
## [1] 0.017
```

So the p-value to test our null hypothesis of no difference in the true means between the groups is 0.027 + 0.017, providing a p-value of 0.044. Figure 2.10 shows both cut-offs on the histogram and density curve.

```
tibble(Tstar) %>% ggplot(aes(x = Tstar)) +
  geom_histogram(aes(y = ..ncount..), bins = 15, col = 1, fill = "skyblue", center = 0) +
  geom_density(aes(y = ..scaled..)) +
  theme_bw() +
  labs(y = "Density") +
  geom_vline(xintercept = c(-1,1)*Tobs, col = "red", lwd = 2) +
  stat_bin(aes(y = ..ncount.., label = ..count..), bins = 15,
           geom = "text", vjust = -0.75)
```

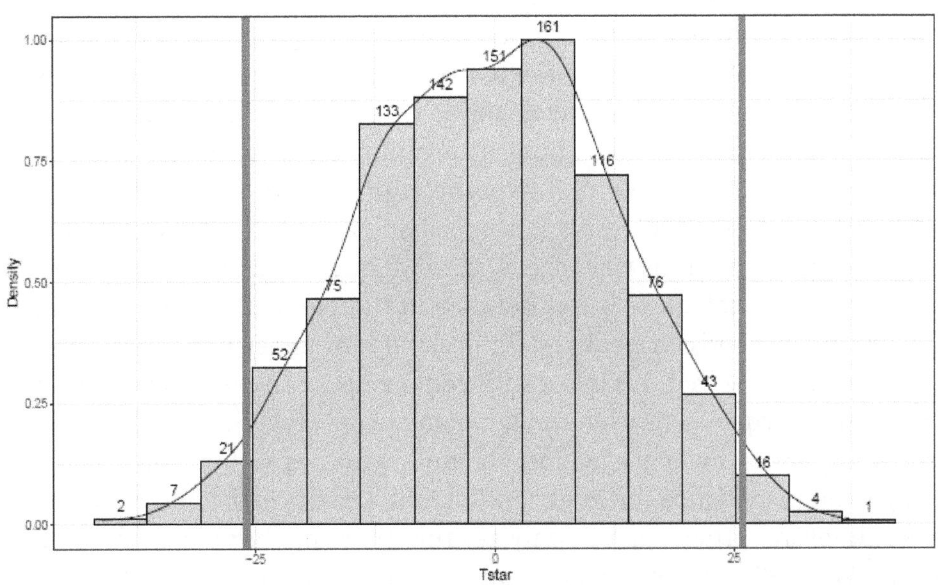

Figure 2.10: Histogram and density curve of values of test statistic for 1,000 permutations with bold lines for value of observed test statistic (-25.933) and its opposite value (25.933) required for performing the two-sided test.

In general, the one-sided test p-value is the proportion of the permuted results that are as extreme or more extreme than observed in the direction of the alternative hypothesis (lower or upper tail, remembering that this also depends on the direction of the difference taken). For the two-sided test, the p-value is the proportion of the permuted results that are less than or equal to the negative version of the observed statistic and greater than or equal to the positive version of the observed statistic. Using absolute values (| |), we can simplify this: the two-sided p-value is the proportion of the |permuted statistics| that are as large or larger than |observed statistic|. This will always work and finds areas in both tails regardless of whether the observed statistic is positive or negative. In R, the abs function provides the absolute value and we can again use pdata to find our p-value in one line of code:

```
pdata(abs(Tstar), abs(Tobs), lower.tail = F)[[1]]
```

```
## [1] 0.044
```

We will encourage you to think through what might constitute strong evidence against your null hypotheses and then discuss how strong you feel the evidence is against the null hypothesis in the p-value that you obtained. Basically, p-values present a measure of evidence against the null hypothesis, with smaller values presenting more evidence against the null. They range from 0 to 1 and you should interpret them on a graded scale from strong evidence (close to 0) to little evidence to no evidence (1). We will discuss the use of a fixed significance level below as it is still commonly used in many fields and is necessary to discuss to think about the theory of hypothesis testing, but, for the moment, we can say that there is moderate evidence against the null hypothesis presented by having a p-value of 0.044 because our observed result is somewhat rare relative to what we would expect if the null hypothesis was true. And so we might conclude (in the direction of the alternative) that there is a difference in the population means in the two groups, but that depends on what you think about how unusual that result was. It is also reasonable to feel that this is not sufficient evidence to conclude that there is a difference in the true means even though many people feel that p-values less than 0.05 are fairly strong evidence against the null hypothesis. If you do not rate this as strong enough evidence (or in general obtain weak evidence) to conclude that there is a difference, then you can only say that there might not be a difference in the means. We can't conclude that the null hypothesis is true – we just failed to find enough evidence to be sure that it is wrong. It might still be wrong but we couldn't detect it, either as a mistake because of an unusual sample from our population, or because our sample size was not large enough to detect the size of difference in the populations, or results with larger p-values could happen because there really isn't a difference. We don't know which of these might be the truth and certainly don't know that the null hypothesis is true even if the p-value obtained is 127.

Before we move on, let's note some interesting features of the permutation distribution of the difference in the sample means shown in Figure 2.10.

- It is basically centered at 0. Since we are performing permutations assuming the null model is true, we are assuming that $\mu 1 = \mu 2$ which implies that $\mu 1 - \mu 2 = 0$. This also suggests that 0 should be the center of the permutation distribution and it was.
- It is approximately normally distributed. This is due to the Central Limit Theorem28, where the sampling distribution (distribution of all possible results for samples of this size) of the difference in sample means ($\bar{x}1 - \bar{x}2$) becomes more normally distributed as the sample sizes increase. With 15observations in each group, we have no guarantee to have a relatively normal looking distribution of the difference in the sample means but with the distributions of the original observations looking somewhat normally distributed, the sampling distribution of the sample means likely will look fairly normal. This result will allow us to use a parametric method to approximate this sampling distribution under the null model if some assumptions are met, as we'll discuss below.
- Our observed difference in the sample means (-25.933) is a fairly unusual result relative to the rest of these results but there are some permuted data sets that produce more extreme differences in the sample means. When the observed differences are really large, we may not see any permuted results that are as extreme as what we observed. When pdata gives you 0, the p-value should be reported to be smaller than 0.001 (not 0!) if B is 1,000 since it happened in less than 1 in 1,000 tries but does occur once – in the actual data set. This applies to any p-values when they are very small – just report them as less than 0.001, or 0.0001 if you prefer that next smaller upper limit, when they are under these values.
- Since our null model is not specific about the direction of the difference, considering a result like ours but in the other direction (25.933 cm) needs to be included. The observed result seems to put about the same area in both tails of the distribution but it is not exactly the same. The small difference in the tails is a useful aspect of this approach compared to the parametric method discussed below as it accounts for potential asymmetry in the sampling distribution.

Earlier, we decided that the p-value provided moderate evidence against the null hypothesis. You should use your own judgment about whether the p-value obtain is sufficiently small to conclude that you think the null hypothesis is wrong. Remembering that the p-value is the probability you would observe a result like you did (or more extreme), assuming the null hypothesis is true; this tells you that the smaller the p-value is, the more evidence you have against the null. Figure 2.11 provides a diagram of some suggestions for the graded p-value interpretation that you can use. The next section provides a more formal review of the hypothesis testing infrastructure, terminology, and some of things that can happen when testing hypotheses. P-values have been (validly) criticized for the inability of studies to be reproduced, for the bias in publications to only include studies that have small p-values, and for the lack of thought that often accompanies using a fixed significance level to make decisions (and only focusing on that decision). To alleviate some of these criticisms, we recommend reporting the strength of evidence of the result based on the p-value and also reporting and discussing the size of the estimated results (with a measure of precision of the estimated difference). We will explore the implications of how p-values are used in scientific research in Section 2.8.

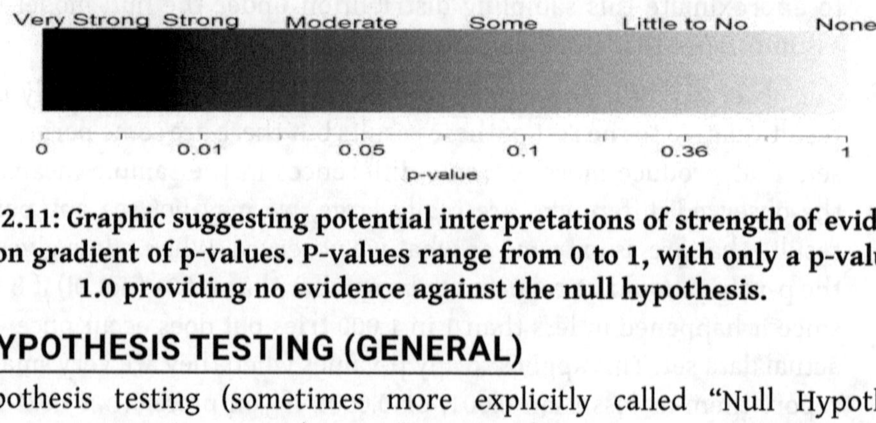

Figure 2.11: Graphic suggesting potential interpretations of strength of evidence based on gradient of p-values. P-values range from 0 to 1, with only a p-value of 1.0 providing no evidence against the null hypothesis.

2.5 HYPOTHESIS TESTING (GENERAL)

In hypothesis testing (sometimes more explicitly called "Null Hypothesis Significance Testing" or NHST), it is formulated to answer a specific question about a population or true parameter(s) using a statistic based on a data set. In your previous statistics course, you (hopefully) considered one-sample hypotheses about population means and proportions and the two-sample mean situation we are focused on here. Hypotheses relate to trying to answer the question about whether the population mean overtake distances between the two groups are different, with an initial assumption of no difference.

NHST is much like a criminal trial with a jury where you are in the role of a jury member. Initially, the defendant is assumed innocent. In our situation, the true means are assumed to be equal between the groups. Then evidence is presented and, as a juror, you analyze it. In statistical hypothesis testing, data are collected and analyzed. Then you have to decide if we had "enough" evidence to reject the initial assumption ("innocence" that is initially assumed). To make this decision, you want to have thought about and decided on the standard of evidence required to reject the initial assumption. In criminal cases, "beyond a reasonable doubt" is used. Wikipedia's definition (https://en.wikipedia.org/wiki/Reasonable_doubt) suggests that this standard is that "there can still be a doubt, but only to the extent that it would not affect a reasonable person's belief regarding whether or not the defendant is guilty". In civil trials, a lower standard called a "preponderance of evidence" is used. Based on that defined and pre-decided (a priori) measure, you decide that the defendant is guilty or not guilty. In statistics, the standard is set by choosing a significance level, α, and then you compare the p-value to it. In this approach, if the p-value is less than α, we reject the null hypothesis. The choice of the significance level is like the variation in standards of evidence between criminal and civil trials – and in all situations everyone should know the standards required for rejecting the initial assumption before any information is "analyzed". Once someone is found guilty, then there is the matter of sentencing which is related to the impacts ("size") of the crime. In statistics, this is similar to the estimated size of differences and the related judgments about whether the differences are practically important or not. If the crime is proven beyond a reasonable doubt but it is a minor crime, then the sentence will be small. With the same level of evidence and a more serious crime, the sentence will be more dramatic. This latter step is more critical than the p-value as it directly relates to actions to be taken based on the research but unfortunately p-values and the related decisions get most of the attention.

There are some important aspects of the testing process to note that inform how we interpret statistical hypothesis test results. When someone is found "not guilty", it does not mean "innocent", it just means that there was not enough evidence to find the person guilty "beyond a reasonable doubt". Not finding enough evidence to reject the null hypothesis does not imply that the true means are equal, just that there was not enough evidence to conclude that they were different. There are many potential reasons why we might fail to reject the null, but the most common one is that our sample size was too small (which is related to having too little evidence). Other reasons include simply the variation in taking a random sample from the population(s). This randomness in samples and the differences in the sample means also implies that p-values are random and can easily vary if the data set had been slightly different. This also relates to the suggestion of using a graded interpretation of p-values instead of the fixed α usage – if the p-value is an estimated quantity, is there really any difference between p-values of 0.049 and 0.051? We probably shouldn't think there is a big difference in results for these two p-values even though the standard NHST reject/fail to reject the null approach considers these as completely different results. So where does that leave us? Interpret the p-values using strength of evidence against the null hypothesis, remembering that smaller (but not really small) p-values can still be interesting. And if you think the p-value is small enough, then you can reject the null hypothesis and conclude that the alternative hypothesis is a better characterization of the truth – and then make sure to estimate and think about the size of the differences.

Throughout this material, we will continue to re-iterate the distinctions between parameters and statistics and want you to be clear about the distinctions between estimates based on the sample and inferences for the population or true values of the parameters of interest. Remember that statistics are summaries of the sample information and parameters are characteristics of populations (which we rarely know). In the two-sample mean situation, the sample means are always at least a little different – that is not an interesting conclusion. What is interesting is whether we have enough evidence to feel like we have proven that the population or true means differ "beyond a reasonable doubt".

The scope of any inferences is constrained based on whether there is a random sample (RS) and/or random assignment (RA). Table 2.1 contains the four possible combinations of these two characteristics of a given study. Random assignment of treatment levels to subjects allows for causal inferences for differences that are observed – the difference in treatment levels is said to cause differences in the mean responses. Random sampling (or at least some sort of representative sample) allows inferences to be made to the population of interest. If we do not have RA, then causal inferences cannot be made. If we do not have a representative sample, then our inferences are limited to the sampled subjects.

A simple example helps to clarify how the scope of inference can change based on the study design. Suppose we are interested in studying the GPA of students. If we had taken a random sample from, say, Intermediate Statistics students in a given semester at a university, our scope of inference would be the population of students in that semester taking that course. If we had taken a random sample from the entire population of students at that school, then the inferences would be to the entire population of students in that semester. These are similar types of problems but the two populations are very different and the group you are trying to make conclusions about should be noted carefully in your results – it does matter! If we did not have a representative sample, say the students could choose to provide this information or not and some chose not to, then we can only make inferences to volunteers. These volunteers might differ in systematic ways from the entire population of Intermediate Statistics students (for example, they are proud of their GPA) so we cannot safely extend our inferences beyond the group that volunteered.

Table 2.1: Scope of inference summary.

Random Sampling/ Random Assignment	Random Assignment (RA) - Yes (controlled experiment)	Random Assignment (RA) - No (observational study
Random Sampling (RS) - Yes (or some method that results in a representative sample of population of interest)	Because we have RS, we can generalize inferences to the population the RS was taken from. Because we have RA we can assume the groups were equivalent on all aspects except for the treatment and can establish causal inference.	Can generalize inference to population the RS was taken from but cannot establish causal inference (no RA - cannot isolate treatment variable as only difference among groups, could be confounding variables).
Random Sampling (RS) - No (usually a convenience sample)	Cannot generalize inference to the population of interest because the sample was not random and could be biased - may not be "representative" of the population of interest. Can establish causal inference due to RA ! the inference from this type of study applies only to the sample.	Cannot generalize inference to the population of interest because the sample was not random and could be biased - may not be "representative" of the population of interest. Cannot establish causal inference due to lack of RA of the treatment.

To consider the impacts of RA versus results from purely observational studies, we need to be comparing groups. Suppose that we are interested in differences in the mean GPAs for different sections of Intermediate Statistics and that we take a random sample of students from each section and compare the results and find evidence of some difference. In this scenario, we can conclude that there is some difference in the population of these statistics students but we can't say that being in different sections caused the differences in the mean GPAs. Now suppose that we randomly assigned every student to get extra training in one of three different study techniques and found evidence of differences among the training methods. We could conclude that the training methods caused the differences in these students. These conclusions would only apply to Intermediate Statistics students at this university in this semester and could not be generalized to a larger population of students. If we took a random sample of Intermediate Statistics students (say only 10 from each section) and then randomly assigned them to one of three training programs and found evidence of differences, then we can say that the training programs caused the differences. But we can also say that we have evidence that those differences pertain to the population of Intermediate Statistics students in that semester at this university. This seems similar to the scenario where all the students participated in the training programs except that by using random sampling, only a fraction of the population needs to actually be studied to make inferences to the entire population of interest – saving time and money.

A quick summary of the terminology of hypothesis testing is useful at this point. The null hypothesis(H0) states that there is no difference or no relationship in the population. This is the statement of no effect or no difference and the claim that we are trying to find evidence against in NHST. In this chapter, H0: μ1 = μ2.When doing two-group problems, you always need to specify which group is 1 and which one is 2 because the order does matter. The alternative hypothesis (H1 or HA) states a specific difference between parameters. This is the research hypothesis and the claim about the population that we often hope to demonstrate is more reasonable to conclude than the null hypothesis. In the two-group situation, we can have one-sided alternatives HA : μ1 > μ2 (greater than) or HA : μ1 < μ2 (less than) or, the more common, two-sided alternative HA : μ1 /= μ2 (not equal to). We usually default to using two-sided tests because we often do not know enough to know the direction of a difference a priori, especially in more complicated situations. The sampling distribution under the null is the distribution of all possible values of a statistic under the assumption that H0 is true. It is used to calculate the p-value, the probability of obtaining a result as extreme or more extreme (defined by the alternative) than what we observed given that the null hypothesis is true. We will find sampling distributions using nonparametric approaches (like the permutation approach used previously) and parametric methods (using "named" distributions like the t, F, and $\chi 2$).

Small p-values are evidence against the null hypothesis because the observed result is unlikely due to chance if H0 is true. Large p-values provide little to no evidence against H0 but do not allow us to conclude that the null hypothesis is correct – just that we didn't find enough evidence to think it was wrong. The level of significance is an a priori definition of how small the p-value needs to be to provide "enough" (sufficient)evidence against H0. This is most useful to prevent sliding the standards after the results are found but you can interpret p-values as strength of evidence against the null hypothesis without employing the fixed significance level. If using a fixed significance level, we can compare the p-value to the level of significance to decide if the p-value is small enough to constitute sufficient evidence to reject the null hypothesis. We use α to denote the level of significance and most typically use 0.05 which we refer to as the 5% significance level. We can compare the p-value to this level and make a decision, focusing our interpretation on the strength of evidence we found based on the p-value from very strong too little to none. If we are using the strict version of NHST, the two options for decisions are to either reject the null hypothesis if the p-value $\leq \alpha$ or fail to reject the null hypothesis if the p-value $> \alpha$. When interpreting hypothesis testing results, remember that the p-value is a measure of how unlikely the observed outcome was, assuming that the null hypothesis is true. It is NOT the probability of the data or the probability of either hypothesis being true. The p-value, simply, is a measure of evidence against the null hypothesis.

Although we want to use graded evidence to interpret p-values, there is one situation where thinking about comparisons to fixed α levels is useful for understanding and studying statistical hypothesis testing. The specific definition of α is that it is the probability of rejecting H0 when H0 is true, the probability of what is called a Type I error . Type I errors are also called false rejections or false detections. In the two-group mean situation, a Type I error would be concluding that there is a difference in the true means between the groups when none really exists in the population. In the courtroom setting, this is like falsely finding someone guilty. We don't want to do this very often, so we use small values of the significance level, allowing us to control the rate of Type I errors at α. We also have to worry about Type II errors, which are failing to reject the null hypothesis when it's false. In a courtroom, this is the same as failing to convict a truly guilty person. This most often occurs due to a lack of evidence that could be due to a small sample size or merely just an unusual sample from the population. You can use the Table 2.2 to help you remember all the possibilities.

Table 2.2: Table of decisions and truth scenarios in a hypothesis testing situation. But we never know the truth in a real situation.

	H0 True	H0 False
FTR H0 Reject H0	Correct decision Type I error	Type II error Correct decision

In comparing different procedures or in planning studies, there is an interest in studying the rate or probability of Type I and II errors. The probability of a Type I error was defined previously as α, the significance level. The power of a procedure is the probability of rejecting the null hypothesis when it is false. Power is defined as

Power = 1 – Probability(Type II error) = Probability(Reject $H_0|H_0$ is false),

or, in words, the probability of detecting a difference when it actually exists. We want to use a statistical procedure that controls the Type I error rate at the pre-specified level and has high power to detect false null hypotheses. Increasing the sample size is one of the most commonly used methods for increasing the power in a given situation. Sometimes we can choose among different procedures and use the power of the procedures to help us make that selection. Note that there are many ways *H0* can be false and the power changes based on how false the null hypothesis actually is. To make this concrete, suppose that the true mean overtake distances differed by either 1 or 30 cm in previous example. The chances of rejecting the null hypothesis are much larger when the group means actually differ by 30 cm than if they differ by just 1 cm, given the same sample size. The null hypothesis is false in both cases. Similarly, for a given difference in the true means, the larger the sample, the higher the power of the study to actually find evidence of a difference in the groups. We will see this difference when we return to using the entire overtake data set instead of the sample of $n = 30$ used to illustrate the permutation procedures.

After making a decision (was there enough evidence to reject the null or not), we want to make the conclusions specific to the problem of interest. If we reject *H*0, then we can conclude that there was sufficient evidence at the α-level that the null hypothesis is wrong (and the results point in the direction of the alternative). If we fail to reject *H*0 (FTR *H*0), then we can conclude that there was insufficient evidence at the α-level to say that the null hypothesis is wrong. We are **NOT** saying that the null is correct and we **NEVER** accept the null hypothesis. We just failed to find enough evidence to say it's wrong. If we find sufficient evidence to reject the null, then we need to revisit the method of data collection and design of the study to discuss the scope of inference. Can we discuss causality (due to RA) and/or make inferences to a larger group than those in the sample (due to RS)?

To perform a hypothesis test, there are some steps to remember to complete to make sure you have thought through and reported all aspects of the results.

Outline of 6+ steps to perform a Hypothesis Test

Preliminary steps:

- * Define research question (RQ) and consider study design – what question can the data collected address?
- * What graphs are appropriate to visualize the data?
- * What model/statistic (T) is needed to address RQ?
  - Write the null and alternative hypotheses.
  - Plot the data and assess the "Validity Conditions" for the procedure being used (discussed below).
  - Find the value of the appropriate test statistic and p-value for your hypotheses.
  - Write a conclusion specific to the problem based on the p-value, reporting the strength of evidence against the null hypothesis (include test statistic, its distribution under the null hypothesis, and p-value).

- Report and discuss an estimate of the size of the differences, with confidence interval(s) if appropriate.
- Scope of inference discussion for results.

2.6 CONNECTING RANDOMIZATION (NONPARAMETRIC) AND PARAMETRIC TESTS

In developing statistical inference techniques, we need to define the test statistic, T , that measures the quantity of interest. To compare the means of two groups, a statistic is needed that measures their differences. In general, for comparing two groups, the choice is simple – a difference in the means often works well and is a natural choice. There are other options such as tracking the ratio of means or possibly the difference in medians. Instead of just using the difference in the means, we also could "standardize" the difference in the means by dividing by an appropriate quantity that reflects the variation in the difference in the means. All of these are valid and can sometimes provide similar results – it ends up that there are many possibilities for testing using the randomization (nonparametric) techniques introduced previously. Parametric statistical methods focus on means because the statistical theory surrounding means is quite a bit easier (not easy, just easier) than other options. There are just a couple of test statistics that you can use and end up with named distributions to use for generating inferences. Randomization techniques allow inference for other quantities (such as ratios of means or differences in medians) but our focus here will be on using randomization for inferences on means to see the similarities with the more traditional parametric procedures used in these situations.

In two-sample mean situations, instead of working just with the difference in the means, we often calculate a test statistic that is called the equal variance two-independent samples t-statistic. The test statistic is

$$t = \frac{\bar{x}_1 - \bar{x}_2}{s_p\sqrt{\frac{1}{n_1} + \frac{1}{n_2}}},$$

where S12 and s22 are the sample variances for the two groups, n1 and n2 are the sample sizes for the two groups, and the pooled sample standard deviation,

$$s_p = \sqrt{\frac{(n_1 - 1)s_1^2 + (n_2 - 1)s_2^2}{n_1 + n_2 - 2}}.$$

The t-statistic keeps the important comparison between the means in the numerator that we used before and standardizes (re-scales) that difference so that t will follow a t-distribution (a parametric "named" distribution) if certain assumptions are met. But first we should see if standardizing the difference in the means had an impact on our permutation test results. It ends up that, while not too obvious, the summary of the lm we fitearlier contains this test statistic 29. Instead of using the second model coefficient that estimates the differencein the means of the groups, we will extract the test statistic from the table of summary output that is in the coef object in the summary – using $ to reference the coef information only. In the coef object in the summary, results related to the Condition Commute are again useful for the comparison of two groups.

```
:y(lm1)$coef
```

```
##                    Estimate Std. Error   t value     Pr(>|t|)
## (Intercept)       135.80000   8.862996 15.322133 3.832161e-15
## Conditioncommute -25.93333  12.534169 -2.069011 4.788928e-02
```

The first column of numbers contains the estimated difference in the sample means (-25.933 here) that was used before. The next column is the Std. Error column that contains the standard error (SE) of the $s_p\sqrt{\frac{1}{n_1}+\frac{1}{n_2}}$ and also the denominator used to form the t-test statistic (12.53 here). It will be a common theme in this material to take the ratio of the estimate (-25.933) to its SE (12.53) to generate test statistics, which provides -2.07 – this is the "standardized" estimate of the difference in the means. It is also a test statistic (T) that we can use in a permutation test. This value is in the second row and third column of summary(lm1)$coef and to extract it the bracket notation is again employed. Specifically we want to extract summary(lm1)$coef[2,3] and using it and its permuted data equivalents to calculate a p-value. Since we are doing a two-sided test, the code resembles the permutation test code in Section 2.4 with the new *t*-statistic replacing the difference in the sample means that we used before.

```
Tobs <- summary(lm1)$coef[2,3]
Tobs
```

[1] -2.069011

```
B <- 1000
set.seed(406)
Tstar <- matrix(NA, nrow = B)
for (b in (1:B)){
  lmP <- lm(Distance ~ shuffle(Condition), data = dsample)
  Tstar[b] <- summary(lmP)$coef[2,3]
}
pdata(abs(Tstar), abs(Tobs), lower.tail = F)
```

[1] 0.041

The permutation distribution in Figure 2.12 looks similar to the previous results with slightly different x-axis scaling. The observed t-statistic was -2.07 and the proportion of permuted results that were as or more extreme than the observed result was 0.041. This difference is due to a different set of random permutations being selected. If you run permutation code, you will often get slightly different results each time you run it. If you are uncomfortable with the variation in the results, you can run more than B = 1,000 permutations (say 10,000) and the variability in the resulting p-values will be reduced further. Usually this uncertainty will not cause any substantive problems – but do not be surprised if your results vary if you use different random number seeds.

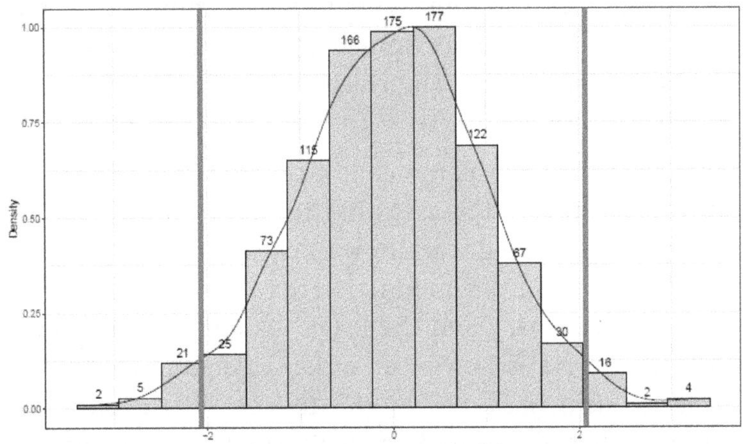

Figure 2.12: Permutation distribution of the *t*-statistic.

```
tibble(Tstar) %>% ggplot(aes(x = Tstar)) +
  geom_histogram(aes(y = ..ncount..), bins = 15, col = 1, fill = "skyblue", center = 0) +
  geom_density(aes(y = ..scaled..)) +
  theme_bw() +
  labs(y = "Density") +
  geom_vline(xintercept = c(-1,1)*Tobs, col = "red", lwd = 2) +
  stat_bin(aes(y = ..ncount.., label = ..count..), bins = 15,
           geom = "text", vjust = -0.75)
```

The parametric version of these results is based on using what is called the two-independent sample t-test. There are actually two versions of this test, one that assumes that variances are equal in the groups and one that does not. There is a rule of thumb that if the ratio of the larger standard deviation over the smaller standard deviation is less than 2, the equal variance procedure is OK. It ends up that this assumption is less important if the sample sizes in the groups are approximately equal and more important if the groups contain different numbers of observations. In comparing the two potential test statistics, the procedure that assumes equal variances has a complicated denominator (see the formula above for t involving sp) but a simple formula for degrees of freedom (df) for the t-distribution (df = n1 + n2 - 2)that approximates the distribution of the test statistic, t, under the null hypothesis. The procedure that assumes unequal variances has a simpler test statistic and a very complicated degrees of freedom formula. The equal variance procedure is equivalent to the methods we will consider in Chapters 3 and 4 so that will be our focus for the two group problem and is what we get when using the lm model to estimate the differences in the group means. The unequal variance version of the two-sample t-test is available in the t.test function if needed.

If the assumptions for the equal variance t-test and the null hypothesis are true, then the sampling

distribution of the test statistic should follow a t-distribution with n1 + n2 - 2 degrees of freedom (so the total sample size, n, minus 2). The t-distribution is a bell-shaped curve that is more spread out for smaller values of degrees of freedom as shown in Figure 2.13. The t-distribution looks more and more like a standardnormal distribution (N (0, 1)) as the degrees of freedom increase.

To get the p-value for the parametric t-test, we need to calculate the test statistic and df , then look up the areas in the tails of the t-distribution relative to the observed t-statistic. We'll learn how to use R to do this below, but for now we will allow the summary of the lm function to take care of this. In the Condition Commute row of the summary and the Pr(>|t|) column, we can find the p-value associated with the test statistic. We can either calculate the degrees of freedom for the t-distribution using n1 + n2 - 2 = 15 + 15 - 2 = 28 or explore the full suite of the model summary that is repeated below. In the first row below the Condition Commute row, it reports "...28 degrees of freedom" and these are the same df that are needed to report and look up for any of the t-statistics in the model summary.

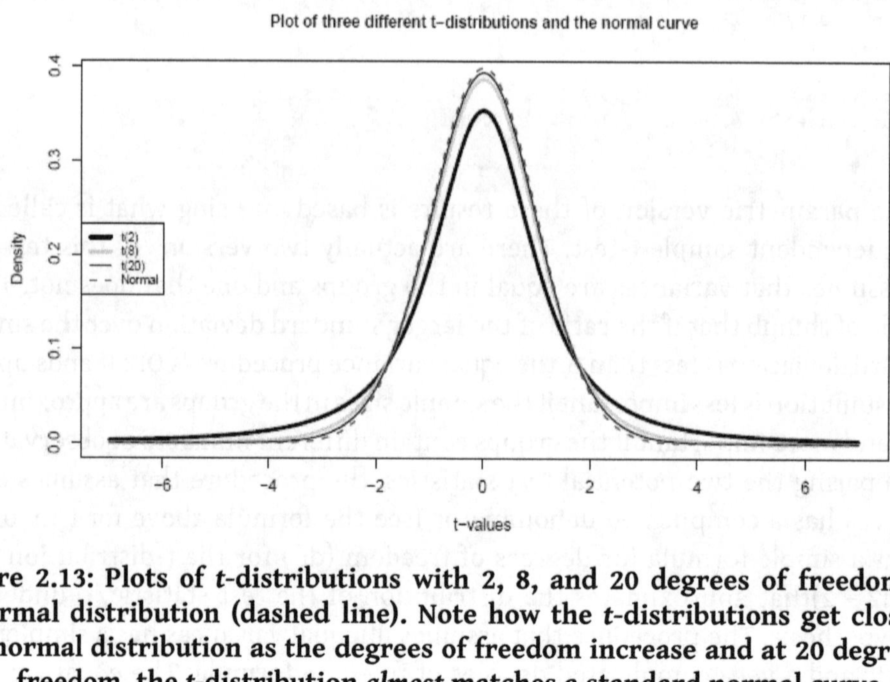

Figure 2.13: Plots of *t*-distributions with 2, 8, and 20 degrees of freedom and a normal distribution (dashed line). Note how the *t*-distributions get closer to the normal distribution as the degrees of freedom increase and at 20 degrees of freedom, the *t*-distribution *almost* matches a standard normal curve.

```
summary(lm1)
```

```
##
## Call:
## lm(formula = Distance ~ Condition, data = dsample)
##
## Residuals:
##     Min      1Q  Median      3Q     Max
## -63.800 -21.850   4.133  15.150  72.200
##
## Coefficients:
##                    Estimate Std. Error t value Pr(>|t|)
## (Intercept)         135.800      8.863  15.322 3.83e-15
## Conditioncommute    -25.933     12.534  -2.069   0.0479
##
## Residual standard error: 34.33 on 28 degrees of freedom
## Multiple R-squared:  0.1326, Adjusted R-squared:  0.1016
## F-statistic: 4.281 on 1 and 28 DF,  p-value: 0.04789
```

So the parametric *t*-test gives a p-value of 0.0479 from a test statistic of -2.07. The p-value is very similar to the two permutation results found before. The reason for this similarity is that the permutation distribution looks like a *t*-distribution with 28 degrees of freedom. Figure 2.14 shows how similar the two distributions happened to be here, where the only difference in shape is near the peak of the distributions with a slight difference of the permutation distribution to shift to the right.

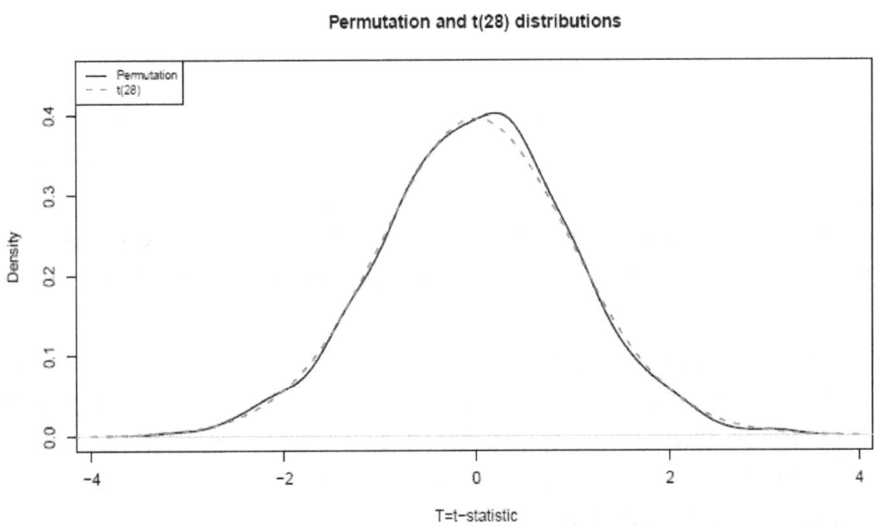

Figure 2.14: Plot of permutation and *t*-distribution with *df* = 28. Note the close match in the two distributions, especially in the tails of the distributions where we are obtaining the p-values.

In your previous statistics course, you might have used an applet or a table to find p-values such as what was provided in the previous R output. When not directly provided in the output of a function, R can be used to look up p-values[30] from named distributions such as the *t*-distribution. In this case, the distribution of the test statistic under the null hypothesis is a *t*(28) or a *t* with 28 degrees of freedom. The pt function is used to get p-values from the *t*-distribution in the same manner that pdata could help us to find p-values from the permutation distribution. We need to provide the df = ... and specify the tail of the distribution of interest using the lower.tail option along with the cutoff of interest. If we want the area to the left of-2.07:

```
-2.069, df = 28, lower.tail = T)
```

[1] 0.02394519

And we can double it to get the p-value that was in the output, because the *t*-distribution is symmetric:

```
2*pt(-2.069, df = 28, lower.tail = T)
```

[1] 0.04789038

More generally, we could always make the test statistic positive using the absolute value (abs), find the area to the right of it (lower.tail = F), and then double that for a two-sided test p-value:

```
2*pt(abs(-2.069), df = 28, lower.tail = F)
```

[1] 0.04789038

Permutation distributions do not need to match the named parametric distribution to work correctly, although this happened in the previous example. The parametric approach, the *t*-test, requires certain conditions to be true (or at least not be clearly violated) for the sampling distribution of the statistic to follow the named distribution and provide accurate p-values. The conditions for the *t*-test are:

- **Independent observations:** Each observation obtained is unrelated to all other observations. To assess this, consider whether anything in the data collection might lead to clustered or related observations that are unrelated to the differences in the groups. For example, was the same person measured more than once[31]?
- **Equal variances** in the groups (because we used a procedure that assumes equal variances! – there is another procedure that allows you to relax this assumption if needed. . .). To assess this, compare the standard deviations and variability in the pirate-plots and see if they look noticeably different.

Be particularly critical of this assessment if the sample sizes differ greatly between groups.

- **Normal distributions** of the observations in each group. We'll learn more diagnostics later, but the pirate-plots are a good place to start to help you look for potential skew or outliers. If you find skew and/or outliers, that would suggest a problem with the assumption of normality as normal distributions are symmetric and extreme observations occur very rarely.

For the permutation test, we relax the third condition and replace it with:

- **Similar distributions for the groups:** The permutation approach allows valid inferences as long as the two groups have similar shapes and only possibly differ in their centers. In other words, the distributions need not look normal for the procedure to work well, but they do need to look similar.

In the bicycle overtake study, the independent observation condition is violated because of multiple measurements taken on the same ride. The fact that the same rider was used for all observations is not really a violation of independence here because there was only one subject used. If multiple subjects had been used, then that also could present a violation of the independence assumption. This violation is important to note as the inferences may not be correct due to the violation of this assumption and more sophisticated statistical methods would be needed to complete this analysis correctly. The equal variance condition does not appear to be violated. The standard deviations are 28.4 vs 39.4, so this difference is not "large" according to the rule of thumb noted above (ratio of SDs is about 1.4). There is also little evidence in the pirate-plots to suggest a violation of the normality condition for each of the groups (Figure 2.5). Additionally, the shapes look similar for the two groups so we also could feel comfortable using the permutation approach based on its version of condition (3) above. Note that when assessing assumptions, it is important to never state that assumptions are met – we never know the truth and can only look at the information in the sample to look for evidence of problems with particular conditions. Violations of those conditions suggest a need for either more sophisticated statistical tools32 or possibly transformations of the response variable (discussed in Chapter 7).

The permutation approach is resistant to impacts of violations of the normality assumption. It is not resistant to impacts of violations of any of the other assumptions. In fact, it can be quite sensitive to unequal variances as it will detect differences in the variances of the groups instead of differences in the means. Its scope of inference is the same as the parametric approach. It also provides similarly inaccurate conclusions in the presence of non-independent observations as for the parametric approach. In this example, we discover that parametric and permutation approaches provide very similar inferences, but both are subject to concerns related to violations of the independent observations condition. And we haven't directly addressed the size and direction of the differences, which is addressed in the coming discussion of confidence intervals.

For comparison, we can also explore the original data set of all n = 1, 636 observations for the two outfits. The estimated difference in the means is -3.003 cm (commute minus casual), the standard error is 1.472, the t-statistic is -2.039 and using a t-distribution with 1634 df , the p-value is 0.0416. The estimated difference in the means is much smaller but the p-value is similar to the results for the sub-sample we analyzed. The SE is much smaller with the large sample size which corresponds to having higher power to detect smaller differences.

```
lm_all <- lm(Distance ~ Condition, data = ddsub)
summary(lm_all)
```

```
##
## Call:
## lm(formula = Distance ~ Condition, data = ddsub)
##
## Residuals:
##      Min       1Q   Median       3Q      Max
## -106.608  -17.608    0.389   16.392  127.389
##
## Coefficients:
##                  Estimate Std. Error t value Pr(>|t|)
## (Intercept)       117.611      1.066 110.357   <2e-16
## Conditioncommute   -3.003      1.472  -2.039   0.0416
##
## Residual standard error: 29.75 on 1634 degrees of freedom
## Multiple R-squared:  0.002539,   Adjusted R-squared:  0.001929
## F-statistic:  4.16 on 1 and 1634 DF,  p-value: 0.04156
```

The permutations take a little more computing power with almost two thousand observations to shuffle, but this is manageable on a modern laptop as it only has to be completed once to fill in the distribution of the test statistic under 1,000 shuffles. And the p-value obtained is a close match to the parametric result at 0.045 for the permutation version and 0.042 for the parametric approach. So we would get similar inferences for strength of evidence against the null with either the smaller data set or the full data set but the estimated size of the differences is quite a bit different. It is important to note that other random samples from the larger data set would give different p-values and this one happened to match the larger set more closely than one might expect in general.

```
Tobs <- summary(lm_all)$coef[2,3]
Tobs
```

```
## [1] -2.039491
```

```
B <- 1000
set.seed(406)
Tstar <- matrix(NA, nrow = B)
for (b in (1:B)){
  lmP <- lm(Distance ~ shuffle(Condition), data = ddsub)
  Tstar[b] <- summary(lmP)$coef[2,3]
}
pdata(abs(Tstar), abs(Tobs), lower.tail = F)
```

[1] 0.045

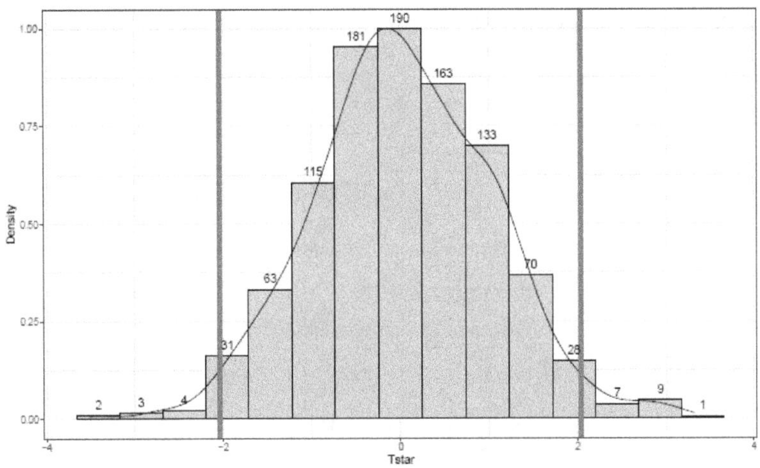

Figure 2.15: Permutation distribution of the *t*-statistic for *n* = 1, 636 overtake data set.

2.7 SECOND EXAMPLE OF PERMUTATION TESTS

In every chapter, the first example, used to motivate and explain the methods, is followed with a "worked" example where we focus just on the results. In a previous semester, some of the Intermediate Statistics (STAT 217) students at Montana State University (***n*** = 79) provided information on their *Sex*[33], *Age*, and current cumulative *GPA*. We might be interested in whether Males and Females had different average GPAs. First, we can take a look at the difference in the responses by groups based on the output and as displayed in Figure 2.16.

```
s217 <- read_csv("http://www.math.montana.edu/courses/s217/documents/s217.csv")
library(mosaic)
library(yarrr)
```

```
mean(GPA ~ Sex, data = s217)
```

```
##        F        M
## 3.338378 3.088571
```

```
favstats(GPA ~ Sex, data = s217)
```

```
##    Sex  min  Q1 median   Q3 max     mean        sd  n missing
## 1    F 2.50 3.1  3.400 3.70   4 3.338378 0.4074549 37       0
## 2    M 1.96 2.8  3.175 3.46   4 3.088571 0.4151789 42       0
```

```
boxplot(GPA ~ Sex, data = s217)
pirateplot(GPA ~ Sex, data = s217, inf.method = "ci", inf.disp = "line")
```

In these data, the distributions of the GPAs look to be left skewed. The Female GPAs look to be slightly higher than for Males (0.25 GPA difference in the means) but is that a "real" difference? We need our inference tools to more fully assess these differences.

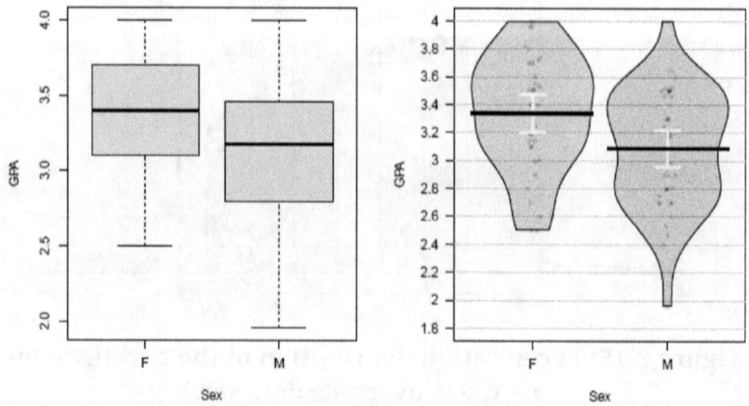

Figure 2.16: Side-by-side boxplot and pirate-plot of GPAs of Intermediate Statistics students by gender.

First, we can try the parametric approach:

```
lm_GPA <- lm(GPA ~ Sex, data = s217)
summary(lm_GPA)
```

```
##
## Call:
## lm(formula = GPA ~ Sex, data = s217)
##
## Residuals:
##      Min       1Q   Median       3Q      Max
## -1.12857 -0.28857  0.06162  0.36162  0.91143
##
## Coefficients:
##             Estimate Std. Error t value Pr(>|t|)
## (Intercept)  3.33838    0.06766  49.337  < 2e-16
## SexM        -0.24981    0.09280  -2.692  0.00871
##
## Residual standard error: 0.4116 on 77 degrees of freedom
## Multiple R-squared:  0.08601,    Adjusted R-squared:  0.07414
## F-statistic: 7.246 on 1 and 77 DF,  p-value: 0.008713
```

So the test statistic was observed to be $t = 2.69$ and it hopefully follows a $t(77)$ distribution under the null hypothesis. This provides a p-value of 0.008713 that we can trust if the conditions to use this procedure are at least not clearly violated. Compare these results to the permutation approach, which relaxes that normality assumption, with the results that follow. In the permutation test, $T = -2.692$ and the p-value is 0.011 which is a little larger than the result provided by the parametric approach. The general agreement of the two approaches, again, provides some re-assurance about the use of either approach when there are not dramatic violations of validity conditions.

```
B <- 1000
Tobs <- summary(lm_GPA)$coef[2,3]
Tstar <- matrix(NA, nrow = B)
for (b in (1:B)){
  lmP <- lm(GPA ~ shuffle(Sex), data = s217)
  Tstar[b] <- summary(lmP)$coef[2,3]
}
pdata(abs(Tstar), abs(Tobs), lower.tail = F)[[1]]
```

[1] 0.011

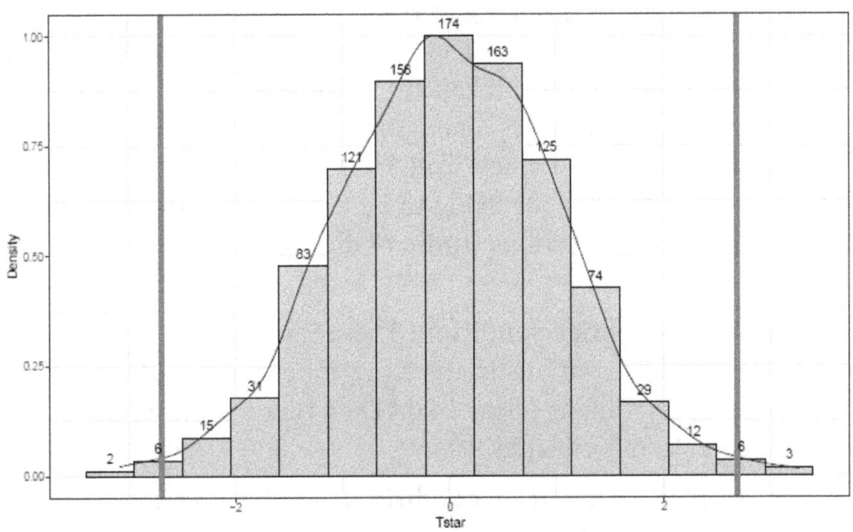

Figure 2.17: Histogram and density curve of permutation distribution of test statistic for Intermediate Statistics student GPAs.

```
tibble(Tstar) %>% ggplot(aes(x = Tstar)) +
  geom_histogram(aes(y = ..ncount..), bins = 15, col = 1, fill = "skyblue", center = 0) +
  geom_density(aes(y = ..scaled..)) +
  theme_bw() +
  labs(y = "Density") +
  geom_vline(xintercept = c(-1,1)*Tobs, col = "red", lwd = 2) +
  stat_bin(aes(y = ..ncount.., label = ..count..), bins = 15,
           geom = "text", vjust = -0.75)
```

Here is a full write-up of the results using all 6+ hypothesis testing steps, using the permutation results for the grade data:

The research question involves exploring differences in GPAs between males and females. With data collected from both groups, we should be able to assess this RQ. The pirate-plot with GPAs by gender is a useful visualization. We could use either differences in the sample means or the *t*-statistic for the test statistic here.

- **Write the null and alternative hypotheses:**
  - $H_0 : \mu male = \mu female$
    - where μmale is the true mean GPA for males and μfemale is true mean GPA for females.
    - $H_A : \mu male /= \mu female$
- **Plot the data and assess the "Validity Conditions" for the procedure being used:**
  - **Independent observations condition:** It does not appear that this assumption is violated because there is no reason to assume any clustering or grouping of responses that might create dependence in the observations. The only possible consideration is that the observations were taken from different sections and there could be some differences among the sections. However, for overall GPA there is not too much likelihood that the overall GPAs would vary greatly so this not likely to be a big issue. However, it is possible that certain sections (times of day) attract students with different GPA levels.
  - **Equal variance condition:** There is a small difference in the range of the observations in the two groups but the standard deviations are very similar (close to 0.41) so there is little evidence that this condition is violated.
  - **Similar distribution condition:** Based on the side-by-side boxplots and pirate-plots, it appears that both groups have slightly left-skewed distributions, which could be problematic for the parametric approach. The two distributions are not exactly alike but they are similar enough that the permutation approach condition is not clearly violated.
- **Find the value of the appropriate test statistic and p-value for your hypotheses:**
  - $T = -2.69$ from the previous R output.

- p-value = 0.011 from the permutation distribution results.
- This means that there is about a 1.1% chance we would observe a difference in mean GPA (female-male or male-female) of 0.25 points or more if there in fact is no difference in true mean GPA between females and males in Intermediate Statistics in a particular semester.

⦿ **Write a conclusion specific to the problem based on the p-value:**

- There is strong evidence against the null hypothesis of no difference in the true mean GPA between males and females for the Intermediate Statistics students in this semester and so we conclude that there is a difference in the mean GPAs between males and females in these students.

⦿ **Report and discuss an estimate of the size of the differences, with confidence interval(s) if appropriate.**

- Females were estimated to have a higher mean GPA by 0.25 points. The next section discusses confidence intervals that we could add to this result to quantify the uncertainty in this estimate since an estimate without any idea of its precision is only a partial result. This difference of 0.25 on a GPA scale does not seem like a very large difference in the means even though we were able to detect a difference in the groups.

⦿ **Scope of inference:**

- Because this was not a randomized experiment in our explanatory variable, we can't say that the difference in gender causes the difference in mean GPA. Because it was not a random sample from a larger population (they were asked to participate but not required to and not all the students did participate), our inferences only pertain the Intermediate Statistics students that responded to the survey in that semester.

2.8 REPRODUCIBILITY CRISIS: MOVING BEYOND P < 0.05, PUBLICATION BIAS, AND MULTIPLE TESTING ISSUES

In the previous examples, some variation in p-values was observed as different methods (parametric, non- parametric) were applied to the same data set and in the permutation approach, the p-values can vary as well from one set of permutations to another. P-values also vary based on randomness in the data that were collected - take a different (random) sample and you will get different data and a different p-value. We want the best estimate of a p-value we can obtain, so should use the best sampling method and inference technique that we can. But it is just an estimate of the evidence against the null hypothesis. These sources of variability make fixed α NHST especially worry-some as sampling variability could take a p-value from just below to just above α and this would lead to completely different inferences if the only focus is on rejecting the null hypothesis at a fixed significance level. But viewing p-values on a gradient from extremely strong (close to 0) to no (1) evidence against the null hypothesis, p-values of, say, 0.046 and 0.054 provide basically the same evidence against the null hypothesis. The fixed α decision-making is tied into the use of the terminology of "significant results" or, slightly better, "statistically significant results" that are intended to convey that there was sufficient evidence to reject the null hypothesis at some pre-decided α level. You will notice that this is the only time that the "s-word" (significant) is considered here.

The focus on p-values has been criticized for a suite of reasons [Wasserstein and Lazar, 2016]. There are situations when p-values do not address the question of interest or the fact that a small p-value was obtained is so un-surprising that one wonders why it was even reported. For example, in Smith [Smith, 2014] the researcher considered bee sting pain ratings across 27 different body locations34. I don't think anyone would be surprised to learn that there was strong evidence against the null hypothesis of no difference in the true mean pain ratings across different body locations. What is really of interest are the differences in the means – especially which locations are most painful and how much more painful those locations were than others, on average.

As a field, Statistics is trying to encourage a move away from the focus on p-values and the use of the term "significant", even when modified by "statistically". There are a variety of reasons for this change. Science (especially in research going into academic journals and in some introductory statistics books) has taken to using p-value < 0.05 and rejected null hypotheses as the only way to "certify" that a result is interesting. It has (and unfortunately still is) hard to publish a paper with a primary result with a p-value that is higher than 0.05, even if the p-value is close to that "magical" threshold. One thing that is lost when using that strict cut-off for decisions is that any p-value that is not exactly 1 suggests that there is at least some evidence against the null hypothesis in the data and that evidence is then on a continuum from none to very strong. And that p-values are both a function of the size of the difference and the sample size. It is easy to get small p-values for small size differences with large data sets. A small p-value can be associated with an unimportant (not practically meaningful) size difference. And large p-values, especially in smaller sample situations, could be associated with very meaningful differences in size even though evidence is not strong against the null hypothesis. It is critical to always try to estimate and discuss the size of the differences, whether a large or small p-value is encountered.

There are some other related issues to consider in working with p-values that help to illustrate some of the issues with how p-values and "statistical significance" are used in practice. In many studies, researchers have a suite of outcome variables that they measure on their subjects. For example, in an agricultural experiment they might measure the yield of the crops, the protein concentration, the digestibility, and other characteristics of the crops. In various "omics" fields such as genomics, proteomics, and metabolomics, responses for each subject on hundreds, thousands, or even millions of variables are considered and a p-value may be generated for each of those variables. In education, researchers might be interested in impacts on grades (as in the previous discussion) but we could also be interested in reading comprehension, student interest in the subject, and the amount of time spent studying, each as response variables in their own right. In each of these situations it means that we are considering not just one null hypothesis and assessing evidence against it, but are doing it many times, from just a few to millions of repetitions. There are two aspects of this process and implications for research to explore further: the impacts on scientific research of focusing solely on "statistically significant" results and the impacts of considering more than one hypothesis test in the same study.

There is the systematic bias in scientific research that has emerged from scientists having a difficult time publishing research if p-values for their data are not smaller than 0.05. This has two implications. Many researchers have assumed that results with "large" p-values are not interesting – so they either exclude these results from papers (they put them in *their* file drawer instead of into their papers – the so-called "file-drawer" bias) or reviewers reject papers because they did not have small p-values to support their discussions (only results with small p-values are judged as being of interest for publication – the so-called "publication bias"). Some also include bias from researchers only choosing to move forward with attempting to publish results if they are in the same direction that the researchers expect/theorized as part of this problem – ignoring results that contradict their theories is an example of "confirmation bias" but also would hinder the evolution of scientific theories to ignore contradictory results. But since most researchers focus on p-values and not on estimates of size (and direction) of differences, that will be our focus here.

We will use some of our new abilities in R to begin to study some of the impacts of systematically favouring only results with small p-values using a "simulation study" inspired by the explorations in Schneck [2017]. Specifically, let's focus on the bicycle passing data. We start with assuming that there really is no difference in the two groups, so the true mean is the same in both groups, the variability is the same around the means in the two groups, and all responses follow normal distributions. This is basically like the permutation idea where we assumed the group labels could be equivalently swapped among responses if the null hypothesis were true except that observations will be generated by a normal distribution instead of shuffling the original observations among groups. This is a little stronger assumption than in the permutation approach but makes it possible to study Type I error rates, power, and to explore a process that is similar to how statistical results are generated and used in academic research settings.

Now let's suppose that we are interested in what happens when we do ten independent studies of the same research question. You could think of this as ten different researchers conducting their own studies of the same topic (say passing distance) or ten times the same researchers did the same study or (less obviously) a researcher focusing on ten different response variables in the same study[35]. Now suppose that one of two things happens with these ten unique response variables – we just report one of them (any could be used, but suppose the first one is selected) OR we only report the one of the ten with the smallest p-value. This would correspond to reporting the results of *a* study or to reporting the "most significant" of ten tries at (or in) the same study – either because nine researchers decided not to publish/ got their papers rejected by journals or because one researcher put the other nine results into their drawer of "failed studies" and never even tried to report the results.

The following code generates one realization of this process to explore both the p-values that are created and the estimated differences. To simulate new observations with the null hypothesis true, there are two new ideas to consider. First, we need to fit a model that makes the means the same in both groups. This is called the "mean-only" model and is implemented with lm(y ~ 1, data = ...), with the ~ 1 indicating that no predictor variable is used and that a common mean is considered for all observations. Note that this notation also works in the favstats function to get summary statistics for the response variable without splitting it apart based on a grouping variable. In the $n = 30$ passing distance data set, the mean of all the observations is 116.04 cm and this estimate is present in the (Intercept) row in the lm_commonmean model summary.

```
lm_commonmean <- lm(Distance ~ 1, data = ddsub)
summary(lm_commonmean)
```

```
##
## Call:
## lm(formula = Distance ~ 1, data = ddsub)
##
## Residuals:
##      Min       1Q   Median       3Q      Max
## -108.038  -17.038   -0.038   16.962  128.962
##
## Coefficients:
##             Estimate Std. Error t value Pr(>|t|)
## (Intercept) 116.0379     0.7361   157.6   <2e-16
##
## Residual standard error: 29.77 on 1635 degrees of freedom
```

```
favstats(Distance ~ 1, data = ddsub)
```

```
##   1 min Q1 median  Q3 max     mean       sd    n missing
## 1 1   8 99    116 133 245 116.0379 29.77388 1636       0
```

The second new R code needed is the simulate function that can be applied to lm-objects; it generates a new data set that contains the same number of observations as the original one but assumes that all the aspects of the estimated model (mean(s), variance, and normal distributions) are true to generate the new observations. In this situation that implies generating new observations with the same mean (116.04) and standard deviation (29.77, also found as the "residual standard error" in the model summary). The new responses are stored in SimDistance in ddsub and then plotted in Figure 2.18.

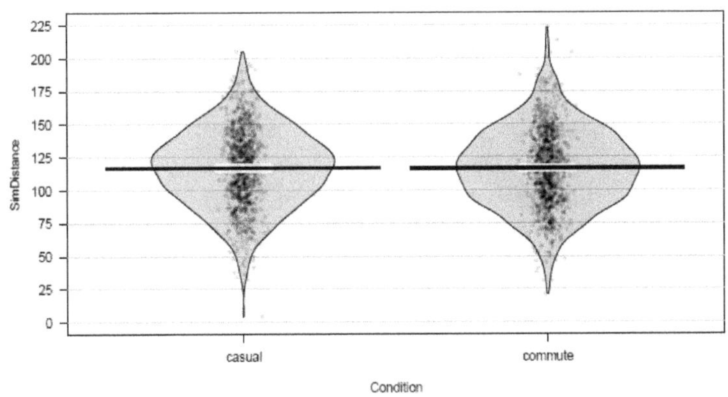

Figure 2.18: Pirate-plot of a simulated data set that assumes the same mean for both groups. The means in the two groups are very similar.

The following code chunk generates one run through generating ten data sets as the loop works through the index c, simulates a new set of responses (ddsub$SimDistance), fits a model that explores the difference in the means of the two groups (lm_sim), and extracts the ten p-values (stored in pval10) and estimated difference in the means (stored in diff10). The smallest p-value of the ten p-values (min(pval10)) is 0.00576. By finding the value of diff10 where pval10 is equal to (==) the min(pval10), the estimated difference in the means from the simulated responses that produced the smallest p-value can be extracted. The difference was -4.17 here. As in the previous initial explorations of permutations, this is just one realization of this process and it needs to be repeated many times to study the impacts of using (1) the first realization of the responses to estimate the difference and p-value and (2) the result with the smallest p-value from ten different realizations of the responses to estimate the difference and p-value. In the following code, we addedoctothorpes (#)36 and then some text to explain what is being calculated. In computer code, octothorpes provide a way of adding comments that tell the software (here R) to ignore any text after a "#" on a given line. In the colour version of the text, comments are even more clearly distinguished.

```
# For one iteration through generating 10 data sets:
diff10 <- pval10 <- matrix(NA, nrow = 10) #Create empty vectors to store 10 results
set.seed(222)
# Create 10 data sets, keep estimated differences and p-values in diff10 and pval10
for (c in (1:10)){
  ddsub$SimDistance <- simulate(lm_commonmean)[[1]]
  # Estimate two group model using simulated responses
  lm_sim <- lm(SimDistance ~ Condition, data = ddsub)
  diff10[c] <- coef(lm_sim)[2]
  pval10[c] <- summary(lm_sim)$coef[2,4]
}

tibble(pval10, diff10)
```

```
## # A tibble: 10 x 2
##     pval10[,1] diff10[,1]
##          <dbl>      <dbl>
##  1     0.735      -0.492
##  2     0.326       1.44
##  3     0.158      -2.06
##  4     0.265      -1.66
##  5     0.153       2.09
##  6     0.00576    -4.17
##  7     0.915       0.160
##  8     0.313      -1.50
##  9     0.983       0.0307
## 10     0.268      -1.69
```

```
min(pval10) #Smallest of 10 p-values
```

```
## [1] 0.005764602
```

```
diff10[pval10 == min(pval10)] #Estimated difference for data set with smallest p-value
```

```
## [1] -4.170526
```

In these results, the first data set shows little evidence against the null hypothesis with a p-value of 0.735 and an estimated difference of -0.49. But if you repeat this process and focus just on the "top" p-value result, you think that there is moderate evidence against the null hypothesis with a p-value from the sixth data set due to its p-value of 0.0057. Remember that these are all data sets simulated with the null hypothesis being true, so we should not reject the null hypothesis. But we would expect an occasional false detection (Type I error – rejecting the null hypothesis when it is true) due to sampling variability in the data sets. But by exploring many results and selecting a single result from that suite of results (and not accounting for that selection process in the results), there is a clear issue with exaggerating the strength of evidence. While not obvious yet, we also create an issue with the estimated mean difference in the groups that is demonstrated below.

To fully explore the impacts of either the office drawer or publication bias (they basically have the same impacts on published results even though they are different mechanisms), this process must be repeated many times. The code is a bit more complex here, as the previous code that created ten data sets needs to be replicated B = 1,000 times and four sets of results stored (estimated mean differences and p-values for the first data set and the smallest p-value one). This involves a loop that is very similar to our permutation loop but with more activity inside that loop, with the code for generating and extracting the realization of ten results repeated B times. Figure 2.19 contains the results for the simulation study. In the left plot that contains the p-values we can immediately see some important differences in the distribution of p-values. In the "first" result, the p-values are evenly spread from 0 to 1 – this is what happens when the null hypothesis is true and you simulate from that scenario one time and track the p-values. A good testing method should make a mistake at the α-level at a rate around α (a 5% significance level test should make a mistake 5% of the time). If the p-values are evenly spread from 0 to 1, then about 0.05 will be between 0 and 0.05 (think of areas in rectangles with a height of 1 where the total area from 0 to 1 has to add up to 1). But when a researcher focuses only on the top result of ten, then the p-value distribution is smashed toward 0. Using favstats on each distribution of p-values shows that the median for the p-values from taking the first result is around 0.5 but for taking the minimum of ten results, the median p-value is 0.065. So half the results are at the "moderate" evidence level or better when selection of results is included. This gets even worse as more results are explored but seems quite problematic here.

The estimated difference in the means also presents an interesting story. When just reporting the first result, the distribution of the estimated means in panel b of Figure 2.19 shows a symmetric distribution that is centered around 0 with results extending just past ± 4 in each tail. When selection of results is included, only more extreme estimated differences are considered and no results close to 0 are even reported. There are two modes here around ± 2.5 and multiple results close to ± 5 are observed. Interestingly, the mean of both distributions is close to 0 so both are "unbiased" 37 estimators but the distribution for the estimated difference from the selected "top" result is clearly flawed and would not give correct inferences for differences when the null hypothesis is correct. If a one-sided test had been employed, the selection of the top result would result is a clearly biased estimator as only one of the two modes would be selected. The presentation of these results is a great example of why pirate-plots are better than boxplots as a boxplot of these results would not allow the viewer to notice the two distinct groups of results.

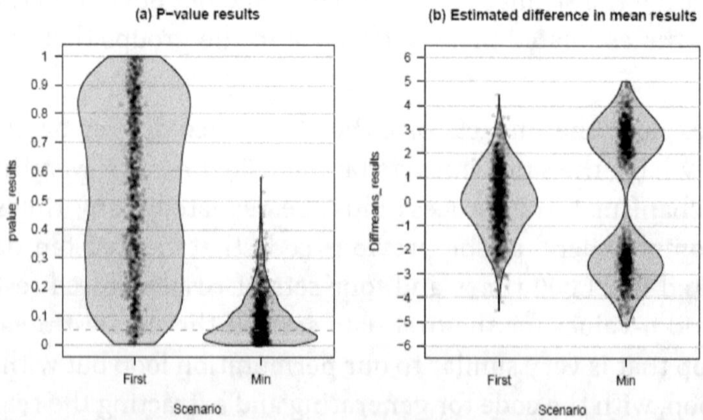

Figure 2.19: Pirate-plot of a simulation study results. Panel (a) contains the B = 1,000 p-values and (b)contains the B = 1,000 estimated differences in the means. Note that the estimated means and confidence intervals normally present in pirate-plots are suppressed here with inf.f.o = 0, inf.b.o = 0, avg.line.o= 0 because these plots are being used to summarize simulation results instead of an original data set.

```
# Simulation study of generating 10 data sets and either using the first
# or "best p-value" result:
set.seed(1234)

B <- 1000 # # of simulations
# To store results
Diffmeans <- pvalues <- Diffmeans_Min <- pvalues_Min <- matrix(NA, nrow = B)
for (b in (1:B)){ #Simulation study loop to repeat process B times
  # Create empty vectors to store 10 results for each b
  diff10 <- pval10 <- matrix(NA, nrow = 10)
  for (c in (1:10)){ #Loop to create 10 data sets and extract results
    ddsub$SimDistance <- simulate(lm_commonmean)[[1]]
    # Estimate two group model using simulated responses
    lm_sim <- lm(SimDistance ~ Condition, data = ddsub)
    diff10[c] <- coef(lm_sim)[2]
    pval10[c] <- summary(lm_sim)$coef[2,4]
  }

  pvalues[b] <- pval10[1] #Store first result p-value
  Diffmeans[b] <- diff10[1] #Store first result estimated difference

  pvalues_Min[b] <- min(pval10) #Store smallest p-value
  Diffmeans_Min[b] <- diff10[pval10 == min(pval10)] #Store est. diff of smallest p-value

}

# Put results together
results <- tibble(pvalue_results = c(pvalues,pvalues_Min),
                  Diffmeans_results = c(Diffmeans, Diffmeans_Min),
                  Scenario = rep(c("First", "Min"), each = B))

par(mfrow = c(1,2)) #Plot results
pirateplot(pvalue_results ~ Scenario, data = results, inf.f.o = 0, inf.b.o = 0,
           avg.line.o = 0, main = "(a) P-value results")
abline(h = 0.05, lwd = 2, col = "red", lty = 2)
pirateplot(Diffmeans_results ~ Scenario, data = results, inf.f.o = 0, inf.b.o = 0,
           avg.line.o = 0, main = "(b) Estimated difference in mean results")
```

```
# Numerical summaries of results
favstats(pvalue_results ~ Scenario, data = results)
```

```
##    Scenario          min         Q1    median        Q3       max       mean         sd    n missing
## 1     First 0.0017051496 0.27075755 0.5234412 0.7784957 0.9995293 0.51899179 0.28823469 1000       0
## 2       Min 0.0005727895 0.02718018 0.0646370 0.1273880 0.5830232 0.09156364 0.08611836 1000       0
```

```
favstats(Diffmeans_results ~ Scenario, data = results)
```

```
##    Scenario       min         Q1     median       Q3      max       mean       sd    n missing
## 1     First -4.531864 -0.8424604 0.07360378 1.002228 4.458951 0.05411473 1.392940 1000       0
## 2       Min -5.136510 -2.6857436 1.24042295 2.736930 5.011190 0.03539750 2.874454 1000       0
```

Generally, the challenge in this situation is that if you perform many tests (ten were the focus before) at the same time (instead of just one test), you inflate the Type I error rate across the tests. We can define the family-wise error rate as the probability that at least one error is made on a set of tests or, more compactly, Pr(At least 1 error is made) where Pr() is the probability of an event occurring. The family-wise error is meant to capture the overall situation in terms of measuring the likelihood of making a mistake if we consider many tests, each with some chance of making their own mistake, and focus on how often we make at least one error when we do many tests. A quick probability calculation shows the magnitude of the problem. If we start with a 5% significance level test, then Pr(Type I error on one test) = 0.05 and the Pr(no errors made on one test) = 0.95, by definition. This is our standard hypothesis testing situation. Now, suppose we have m independent tests, then

Pr(make at least 1 Type I error given all null hypotheses are true)

= 1 - Pr(no errors made)

= 1 - 0.95^m.

Figure 2.20 shows how the probability of having at least one false detection grows rapidly with the number of tests, m. The plot stops at 100 tests since it is effectively a 100% chance of at least one false detection. It might seem like doing 100 tests is a lot, but, as mentioned before, some researchers consider situations where millions of tests are considered. Researchers want to make sure that when they report a "significant" result that it is really likely to be a real result and will show up as a difference in the next data set they collect. Some researchers are now collecting multiple data sets to use in a single study and using one data set to identify interesting results and then using a validation or test data set that they withheld from initial analysis to try to verify that the first results are also present in that second data set. This also has problems but the only way to develop an understanding of a process is to look across a suite of studies and learn from that accumulation of evidence. This is a good start but needs to be coupled with complete reporting of all results, even those that have p-values larger than 0.05 to avoid the bias identified in the previous simulation study.

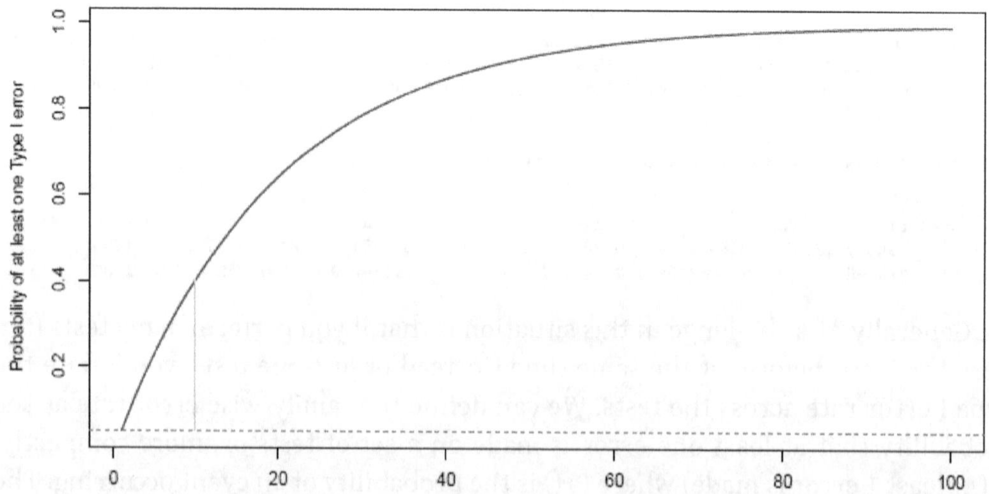

Figure 2.20: Plot of family-wise error rate (bold solid line) as the number of tests performed increases. Dashed line indicates 0.05 and grey solid line highlights the probability of at least on error on m = 10 tests.

All hope is not lost when multiple tests are being considered in the same study or by a researcher and exploring more than one result need not lead to clearly biased and flawed results being reported. To account for multiple testing in the same study/analysis, there are many approaches that adjust results to acknowledge that multiple tests are being considered. A simple approach called the "Bonferroni Correction" [Bland and Altman, 1995] is a good starting point for learning about these methods. It works to control the family-wise error rate of a suite of tests by either dividing α by the number of tests (α/m) or, equivalently and more usefully, multiplying the p-value by the number of tests being considered ($p - valueadjusted = p - value \cdot m$or 1 if $p - value \cdot m > 1$). The "Bonferroni adjusted p-values" are then used as regular p-values to assess evidence against each null hypothesis but now accounting for exploring many of them together. There are some assumptions that this adjustment method makes that make it to generally be a conservative adjustment method. In particular, it assumes that all m tests are independent of each other and that the null hypothesis was true for all m tests conducted. While all p-values should be reported in this situation when considering ten results, the impacts of using a Bonferroni correction are that the resulting p-values are not driving inflated Type I error rates even if the smallest p-value is the main focus of the results. The correction also provides a suggestion of decreasing evidence in the first test result because it is now incorporated in considering ten results instead of one.

The following code repeats the simulation study but with the p-values adjusted for multiple testing within each simulation but does not repeat tracking the estimated differences in the means as this is not impacted by the p-value adjustment process. The p.adjust function provides Bonferroni corrections to a vector of p-values (here ten are collected together) using the bonferroni method option (p.adjust(pval10, method = "bonferroni")) and then stores those results. Figure 2.21 shows the results for the first result and minimum result again, but now with these corrections incorporated. The plots may look a bit odd, but in the first data set, so many of the first data sets had p-values that were "large" that they were adjusted to have p-values of 1 (so no evidence against the null once we account for multiple testing). The distribution for the minimum p-value results with adjustment more closely resembles the distribution of the first result p-values from Figure 2.19, except for some minor clumping up at adjusted p-values of 1.

```
# Simulation study of generating 10 data sets and either using the first
# or "best p-value" result:
set.seed(1234)

B <- 1000 # # of simulations
pvalues <- pvalues_Min <- matrix(NA, nrow = B) #To store results
for (b in (1:B)){ #Simulation study loop to repeat process B times
  # Create empty vectors to store 10 results for each b
  pval10 <- matrix(NA, nrow = 10)
  for (c in (1:10)){ #Loop to create 10 data sets and extract results
    ddsub$SimDistance <- simulate(lm_commonmean)[[1]]
    # Estimate two group model using simulated responses
    lm_sim <- lm(SimDistance ~ Condition, data = ddsub)
    pval10[c] <- summary(lm_sim)$coef[2,4]
  }

  pval10 <- p.adjust(pval10, method = "bonferroni")

  pvalues[b] <- pval10[1] #Store first result adjusted p-value

  pvalues_Min[b] <- min(pval10) #Store smallest adjusted p-value
}

# Put results together
results <- tibble(pvalue_results = c(pvalues, pvalues_Min),
                  Scenario = rep(c("First", "Min"), each = B))

pirateplot(pvalue_results ~ Scenario, data = results, inf.f.o = 0, inf.b.o = 0,
           avg.line.o = 0, main = "P-value results")
abline(h = 0.05, lwd = 2, col = "red", lty = 2)
```

By applying the pdata function to the two groups of results, we can directly assess how many of each type ("First" or "Min") resulted in p-values less than 0.05. It ends up that if we adjust for ten tests and just focus on the first result, it is really hard to find moderate or strong evidence against the null hypothesis as only 3 in 1,000 results had adjusted p-values less than 0.05. When the focus is on the "best" (or minimum) p-value result when ten are considered and adjustments are made, 52 out of 1,000 results (0.052) show at least moderate evidence against the null hypothesis. This is the rate we would expect from a well-behaved hypothesis test when the null hypothesis is true – that we would only make a mistake 5% of the time when α is 0.05.

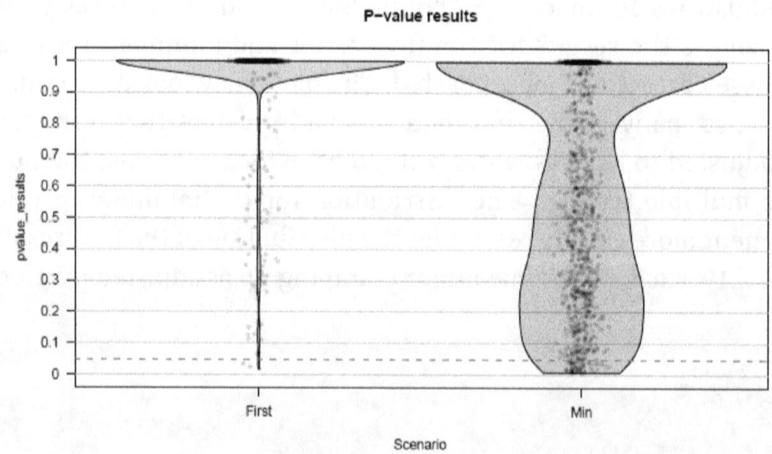

Figure 2.21: Pirate-plot of a simulation study results of p-values with Bonferroni correction.

```
# Numerical summaries of results
favstats(pvalue_results ~ Scenario, data = results)
```

```
##   Scenario         min        Q1  median Q3 max      mean        sd    n missing
## 1    First 0.017051496 1.0000000 1.00000  1   1 0.9628911 0.1502805 1000       0
## 2      Min 0.005727895 0.2718018 0.64637  1   1 0.6212932 0.3597701 1000       0
```

```
# Proportion of simulations with adjusted p-values less than 0.05
pdata(pvalue_results ~ Scenario, data = results, .05, lower.tail = T)
```

```
##   Scenario pdata_v
## 1    First   0.003
## 2      Min   0.052
```

So adjusting for multiple testing is suggested when multiple tests are being considered "simultaneously". The Bonferroni adjustment is easy but also crude and can be conservative in applications, especially when the number of tests grows very large (think of multiplying all your p-values by m = 1,000,000). So other approaches are considered in situations with many tests (there are six other options in the p.adjust function and other functions for doing similar things in R) and there are other approaches that are customized for particular situations with one example discussed in Chapter 3. The biggest lesson as a statistics student to take from this is that all results are of interest and should be reported and that adjustment of p-values should be considered in studies where many results are being considered. If you are reading results that seem to have walked discretely around these issues you should be suspicious of the real strength of their evidence.

While it wasn't used here, the same general code used to explore this multiple testing issue could be used to explore the power of a particular procedure. If simulations were created from a model with a difference in the means in the groups, then the null hypothesis would have been false and the rate of correctly rejecting the null hypothesis could be studied. The rate of correct rejections is the *power* of a procedure for a chosen version of a true alternative hypothesis (there are many ways to have it be true and you have to choose one to study power) and simply switching the model being simulated from would allow that to be explored. We could also use similar code to compare the power and Type I error rates of parametric versus permutation procedures or to explore situations where an assumption is not true. The steps would be similar – decide on what you need to simulate from and track a quantity of interest across repeated simulated data sets.

2.9 CONFIDENCE INTERVALS AND BOOTSTRAPPING

Up to this point the focus has been on hypotheses, p-values, and estimates of the size of differences. But so far this has not explored inference techniques for the size of the difference. Confidence intervals provide an interval where we are% confident that the true parameter lies. The idea of "confidence" is that if we repeated randomly sampling from the same population and made a similar confidence interval, the collection of all these confidence intervals would contain the true parameter at the specified confidence level (usually 95%). We only get to make one interval and so it either has the true parameter in it or not, and we don't know the truth in real situations.

Confidence intervals can be constructed with parametric and a nonparametric approaches. The nonpara- metric approach will be using what is called bootstrapping and draws its name from "pull yourself up by your bootstraps" where you improve your situation based on your own efforts. In statistics, we make our situation or inferences better by re-using the observations we have by assuming that the sample represents the population. Since each observation represents other similar observations in the population that we didn't get to measure, if we sample with replacement to generate a new data set of size n from our data set (also of size n) it mimics the process of taking repeated random samples of size n from our population of interest. This process also ends up giving us useful sampling distributions of statistics even when our standard normality assumption is violated, similar to what we encountered in the permutation tests. Bootstrapping is especially useful in situations where we are interested in statistics other than the mean (say we want a confidence interval for a median or a standard deviation) or when we consider functions of more than one parameter and don't want to derive the distribution of the statistic (say the difference in two medians). Here, bootstrapping is used to provide more trustworthy inferences when some of our assumptions (especially normality) might be violated for our parametric confidence interval procedure.

To perform bootstrapping, the resample function from the mosaic package will be used. We can apply this function to a data set and get a new version of the data set by sampling new observations *with replacement* from the original one[38]. The new, bootstrapped version of the data set (called dsample_BTS below) contains a new variable called orig.id which is the number of the subject from the original data set. By summarizing how often each of these id's occurred in a bootstrapped data set, we can see how the re-sampling works. The table function will count up how many times each observation was used in the bootstrap sample, providing a row with the id followed by a row with the count[39]. In the first bootstrap sample shown, the 1st, 14th, and 26th observations were sampled twice, the 9th and 28th observations were sampled four times, and the 4th, 5th, 6th, and many others were not sampled at all. Bootstrap sampling thus picks some observations multiple times and to do that it has to ignore some[40] observations.

```
set.seed(406)
dsample_BTS <- resample(dsample)
```

```
table(as.numeric(dsample_BTS$orig.id))
```

```
##
##  1  2  3  7  8  9 10 11 12 13 14 16 18 19 23 24 25 26 27 28 30
##  2  1  1  1  1  4  1  1  1  1  2  1  1  1  1  1  1  2  1  4  1
```

Like in permutations, one randomization isn't enough. A second bootstrap sample is also provided to help you get a sense of what bootstrap data sets contain. It did not select observations two through five but did select eight others more than once. You can see other variations in the resulting re-sampling of subjects with the most sampled observation used four times. With $n = 30$, the chance of selecting any observation for any slot in the new data set is 1/30 and the expected or mean number of appearances we expect to see for an observation is the number of random draws times the probably of selection on each so $30 * 1/30 = 1$. So we expect to see each observation in the bootstrap sample on average once but random variability in the samples then creates the possibility of seeing it more than once or not all.

```
dsample_BTS2 <- resample(dsample)
table(as.numeric(dsample_BTS2$orig.id))
```

```
##
## 1 6 7 8 9 10 11 12 13 16 17 20 22 23 24 25 26 28 30
## 2 2 1 1 2 1 4 1 3 1 1 1 2 2 1 1 2 1 1
```

We can use the two results to get an idea of distribution of results in terms of number of times observations might be re-sampled when sampling with replacement and the variation in those results, as shown in Figure 2.22. We could also derive the expected counts for each number of times of re-sampling when we start with all observations having an equal chance and sampling with replacement but this isn't important for using bootstrapping methods.

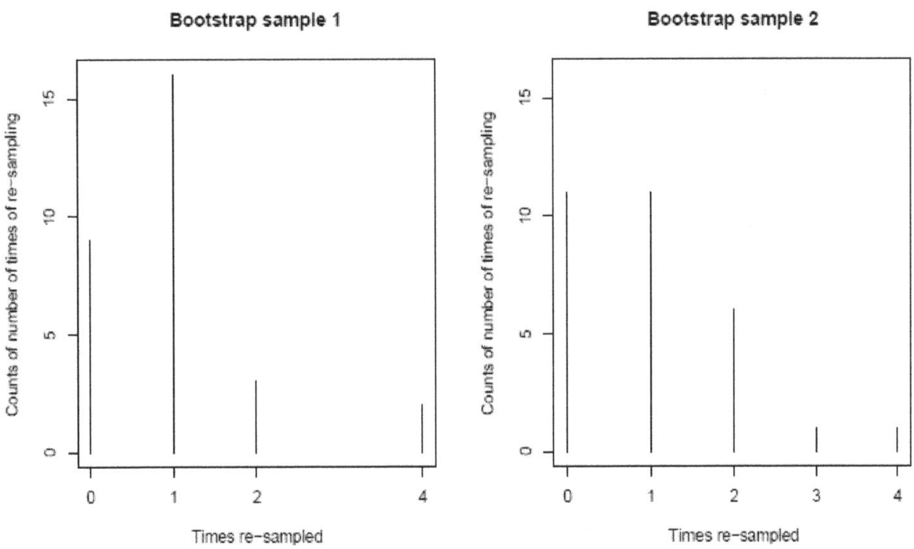

Figure 2.22: Counts of number of times of observation (or not observed for times re-sampled of 0) for two bootstrap samples.

The main point of this exploration was to see that each run of the resample function provides a new version of the data set. Repeating this B times using another for loop, we will track our quantity of interest, say T , in all these new "data sets" and call those results T*. The distribution of the bootstrapped T* statistics tells us about the range of results to expect for the statistic. The middle % of the T*'s provides a %bootstrap confidence interval41 for the true parameter – here the difference in the two population means.

To make this concrete, we can revisit our previous examples, starting with the dsample data created before and our interest in comparing the mean passing distances for the *commuter* and *casual* outfit groups in the n = 30 stratified random sample that was extracted. The bootstrapping code is very similar to the permutation code except that we apply the resample function to the entire data set used in lm as opposed to the shuffle function that was applied only to the explanatory variable.

```
lm1 <- lm(Distance ~ Condition, data = dsample)
Tobs <- coef(lm1)[2]; Tobs
```

Conditioncommute

-25.93333

```
B <- 1000
set.seed(1234)
Tstar <- matrix(NA, nrow = B)
for (b in (1:B)){
  lmP <- lm(Distance ~ Condition, data = resample(dsample))
  Tstar[b] <- coef(lmP)[2]
}
```

```
favstats(Tstar)
```

min Q1 median Q3 max mean sd n missing

-66.96429 -34.57159 -25.65881 -17.12391 17.17857 -25.73641 12.30987 1000 0

```
tibble(Tstar) %>% ggplot(aes(x = Tstar)) +
  geom_histogram(aes(y = ..ncount..), bins = 15, col = 1, fill = "skyblue", center = 0) +
  geom_density(aes(y = ..scaled..)) +
  theme_bw() +
  labs(y = "Density") +
  geom_vline(xintercept = Tobs, col = "red", lwd = 2) +
  stat_bin(aes(y = ..ncount.., label = ..count..), bins = 15,
           geom = "text", vjust = -0.75)
```

In this situation, the observed difference in the mean passing distances is -25.933 cm (*commute* - *casual*), which is the bold vertical line in Figure 2.23. The bootstrap distribution shows the results for the difference in the sample means when fake data sets are re-constructed by sampling from the original data set with replacement. The bootstrap distribution is approximately centered at the observed value (difference in the sample means) and is relatively symmetric.

The permutation distribution in the same situation (Figure 2.10) had a similar shape but was centered at 0. Permutations create sampling distributions based on assuming the null hypothesis is true, which is useful for hypothesis testing. Bootstrapping creates distributions centered at the observed result, which is the sampling distribution "under the alternative" or when no null hypothesis is assumed; bootstrap distributions are useful for generating confidence intervals for the true parameter values.

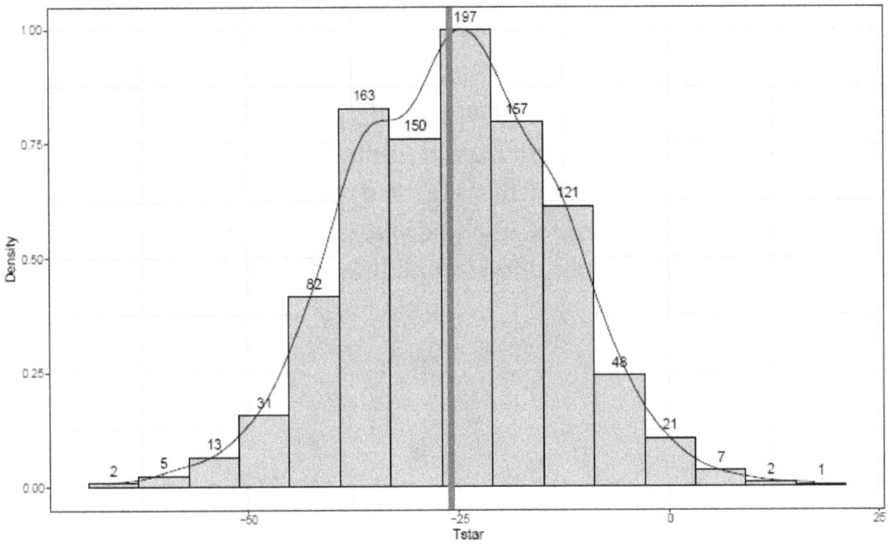

Figure 2.23: Histogram and density curve of bootstrap distributions of difference in sample mean Distances with vertical line for the observed difference in the means of -25.933.

To create a 95% bootstrap confidence interval for the difference in the true mean distances (μcommute -μcasual), select the middle 95% of results from the bootstrap distribution. Specifically, find the 2.5th percentile and the 97.5th percentile (values that put 2.5 and %97.5 of the results to the left) in the bootstrap distribution, which leaves %95 in the middle for the confidence interval. To find percentiles in a distribution in R, functions are of the form q[Name of distribution], with the function qt extracting percentiles from a *t*-distribution (examples below). From the bootstrap results, use the qdata function on the Tstar results that contain the bootstrap distribution of the statistic of interest.

```
qdata(Tstar, 0.025)
```

2.5%

-50.0055

```
qdata(Tstar, 0.975)
```

97.5%

-2.248774

These results tell us that the 2.5th percentile of the bootstrap distribution is at 50.006- cm and the 97.5th percentile is at 2.249- cm. We can combine these results to provide a %95 confidence for μcommute - μcasaualthat is between -50.01

and -2.25 cm. This interval is interpreted as with any confidence interval, that we are 95% confident that the difference in the true mean distances (*commute* minus *casual* groups) is between-50.01 and -2.25 cm. Or we can switch the direction of the comparison and say that we are 95% confident that the difference in the true means is between 2.25 and 50.01 cm (*casual* minus *commute*). This result would be incorporated into step 5 of the hypothesis testing protocol to accompany discussing the size of the estimated difference in the groups or used as a result of interest in itself. Both percentiles can be obtained in one line of code using:

```
quantiles <- qdata(Tstar, c(0.025,0.975))
```

```
quantiles
```

```
##        2.5%      97.5%
## -50.005502  -2.248774
```

Figure 2.24 displays those same percentiles on the bootstrap distribution residing in Tstar.

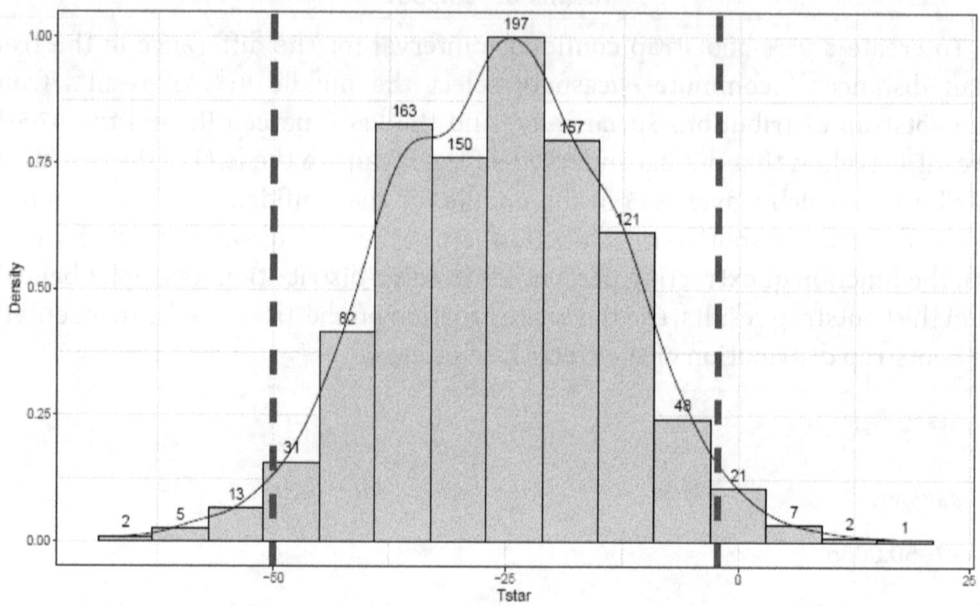

Figure 2.24: Histogram and density curve of bootstrap distribution with 95% bootstrap confidence intervals displayed (bold, dashed vertical lines).

```
tibble(Tstar) %>% ggplot(aes(x = Tstar)) +
  geom_histogram(aes(y = ..ncount..), bins = 15, col = 1, fill = "skyblue", center = 0) +
  geom_density(aes(y = ..scaled..)) +
  theme_bw() +
  labs(y = "Density") +
  geom_vline(xintercept = quantiles, col = "blue", lwd = 2, lty = 2) +
  stat_bin(aes(y = ..ncount.., label = ..count..), bins = 15,
        geom = "text", vjust = -0.75)
```

Although confidence intervals can exist without referencing hypotheses, we can revisit our previous hypotheses and see what this confidence interval tells us about the test of H0 : μcommute = μcasual. This null hypothesis is equivalent to testing H0 : μcommute - μcasual = 0, that the difference in the true means is equal to 0 cm. And the difference in the means was the scale for our confidence interval, which did not contain 0 cm. The 0 cm values is an interesting reference value for the confidence interval, because here it is the value where the true means are equal to each other (have a difference of 0 cm). In general, if our confidence interval does not contain 0, then it is saying that 0 is not one of the likely values for the difference in the true means at the selected confidence level. This implies that we should reject a claim that they are equal. This provides the same inferences for the hypotheses that we considered previously using both parametric and permutation approaches using a fixed α approach where α = 1 - confidence level.

The general summary is that we can use confidence intervals to test hypotheses by assessing whether the reference value under the null hypothesis is in the confidence interval (suggests insufficient evidence against *H*0 to reject it, at least at the α level and equivalent to having a p-value larger than α) or outside the confidence interval (sufficient evidence against *H*0 to reject it and equivalent to having a p-value that is less than α). P-values are more informative about hypotheses (measure of evidence against the null hypothesis)but confidence intervals are more informative about the size of differences, so both offer useful information and, as shown here, can provide consistent conclusions about hypotheses. But it is best practice to use p-values to assess evidence against null hypotheses and confidence intervals to do inferences for the size of differences.

As in the previous situation, we also want to consider the parametric approach for comparison purposes and to have that method available, especially to help us understand some methods where we will only consider parametric inferences in later chapters. The parametric confidence interval is called the equal variance, two-sample t confidence interval and additionally assumes that the populations being sampled from are normally distributed instead of just that they have similar shapes in the bootstrap approach. The parametric method leads to using a t-distribution to form the interval with the degrees of freedom for the t-distribution of n - 2 although we can obtain it without direct reference to this distribution using the confint function applied to the lm model. This function generates two confidence intervals and the one in the second row is the one we are interested as it pertains to the difference in the true means of the two groups. The parametric 95% confidence interval here is from -51.6 to -0.26 cm which is a bit different in width from the nonparametric bootstrap interval that was from -50.01 and -2.25 cm.

```
confint(lm1)
```

```
##                       2.5 %      97.5 %
## (Intercept)       117.64498 153.9550243
## Conditioncommute -51.60841  -0.2582517
```

The bootstrap interval was narrower by almost 4 cm and its upper limit was much further from 0. The bootstrap CI can vary depending on the random number seed used and additional runs of the code produced intervals of (-49.6, -2.8), (-48.3, -2.5), and (-50.9, -1.1) so the differences between the parametric and nonparametric approaches was not just due to an unusual bootstrap distribution. It is not entirely clear why the two intervals differ but there are slightly more results in the left tail of Figure 2.24 than in the right tail and this shifts the 95% confidence slightly away from 0 as compared to the parametric approach. All intervals have the same interpretation, only the methods for calculating the intervals and the assumptions differ. Specifically, the bootstrap interval can tolerate different distribution shapes other than normal and still provide intervals that work well42. The other assumptions are all the same as for the hypothesis test, where we continue to assume that we have independent observations with equal variances for the two groups and maintain concerns about inferences here due to the violation of independence in these responses.

The formula that lm is using to calculate the parametric equal variance, two-sample t-based confidence interval is:

$$\bar{x}_1 - \bar{x}_2 \mp t^*_{df} s_p \sqrt{\frac{1}{n_1} + \frac{1}{n_2}}$$

In this situation, the df is again n1 + n2 - 2 (the total sample size - 2) and sp = $\sqrt{\frac{(n_1-1)s_1^2+(n_2-1)s_2^2}{n_1+n_2-2}}$. The t*df

is a multiplier that comes from finding the percentile from the *t*-distribution that puts C% in the middle of the distribution with C being the confidence level. It is important to note that this *t** has nothing to do with the previous test statistic *t*. It is confusing and students first engaging these two options often happily take the result from a test statistic calculation and use it for a multiplier in a *t*-based confidence interval – try to focus on which *t* you are interested in before you use either. Figure 2.25 shows the t-distribution with 28 degrees of freedom and the cut-offs that put 95% of the area in the middle.

For 95% confidence intervals, the multiplier is going to be close to 2 and anything else is a likely indication of a mistake. We can use R to get the multipliers for confidence intervals using the qt function in a similar fashion to how qdata was used in the bootstrap results, except that this new value must be used in the previous confidence interval formula. This function produces values for requested percentiles, so if we want to put 95% in the middle, we place 2.5% in each tail of the distribution and need to request the 97.5th percentile. Because the t-distribution is always symmetric around 0, we merely need to look up the value for the 97.5th percentile and know that the multiplier for the 2.5th percentile is just $-t^*$. The t^* multiplier to form the confidence interval is 2.0484 for a 95% confidence interval when the $df = 28$ based on the results from qt:

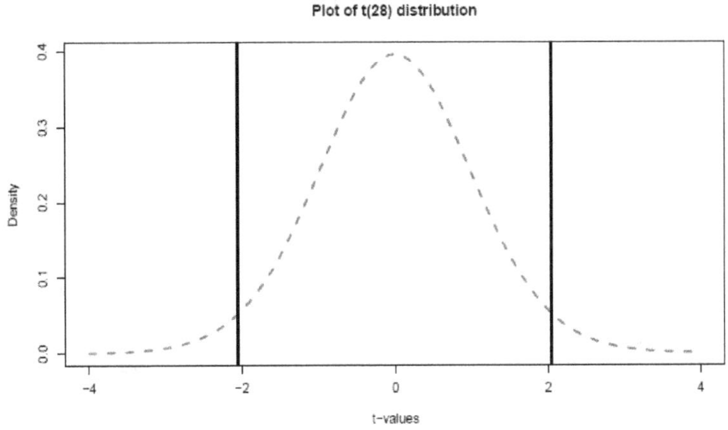

Figure 2.25: Plot of $t(28)$ with cut-offs for putting 95% of distribution in the middle that delineate the t^*multiplier to make a 95% confidence interval.

```
qt(0.975, df = 28)
```

[1] 2.048407

Note that the 2.5th percentile is just the negative of this value due to symmetry and the real source of the minus in the minus/plus in the formula for the confidence interval.

```
qt(0.025, df = 28)
```

[1] -2.048407

We can also re-write the confidence interval formula into a slightly more general forms as

$$\bar{x}_1 - \bar{x}_2 \mp t^*_{df} SE_{\bar{x}_1 - \bar{x}_2} \text{ OR } \bar{x}_1 - \bar{x}_2 \mp ME$$

where $SE_{\bar{x}_1 - \bar{x}_2} = s_p\sqrt{\frac{1}{n_1}+\frac{1}{n_2}}$ and ME = . The SE is available in the lm model summary for the line related to the difference in groups in the "Std. Error" column. In some situations, researchers will report the standard error (SE) or margin of error (ME) as a method of quantifying the uncertainty in a statistic. The SE is an estimate of the standard deviation of the statistic (here x¯1 - x¯2) and the ME is

an estimate of the precision of a statistic that can be used to directly form a confidence interval. The ME depends on the choice of confidence level although 95% is almost always selected.

To finish this example, R can be used to help you do calculations much like a calculator except with much more power "under the hood". You have to make sure you are careful with using ( ) to group items and remember that the asterisk (*) is used for multiplication. We need the pertinent information which is available from the favstats output repeated below to calculate the confidence interval "by hand"[43] using R.

```
favstats(Distance ~ Condition, data = dsample)
```

```
##    Condition min    Q1 median    Q3 max     mean       sd  n missing
## 1     casual  72 112.5    143 154.5 208 135.8000 39.36133 15       0
## 2    commute  60  88.5    113 123.0 168 109.8667 28.41244 15       0
```

Start with typing the following command to calculate *sp* and store it in a variable named sp:

```
sp <- sqrt(((15 - 1)*(39.36133^2) + (15 - 1)*(28.4124^2))/(15 + 15 - 2))
sp
```

[1] 34.32622

Then calculate the confidence interval that confint provided using:

```
109.8667 - 135.8 + c(-1,1)*qt(0.975, df = 28)*sp*sqrt(1/15 + 1/15)
```

[1] -51.6083698 -0.2582302

Or using the information from the model summary:

```
-25.933 + c(-1,1)*qt(0.975, df = 28)*12.534
```

[1] -51.6077351 -0.2582649

The previous results all use c(-1, 1) times the margin of error to subtract and add the ME to the difference in the sample means (109.8667 - 135.8), which generates the lower and then upper bounds of the confidence interval. If desired, we can also use just the last portion of the calculation to find the margin of error, which is 25.675 here.

```
qt(0.975, df = 28)*sp*sqrt(1/15 + 1/15)
```

[1] 25.67507

For the entire $n = 1,636$ data set for these two groups, the results are obtained using the following code. The estimated difference in the means is -3 cm (*commute* minus *casual*). The *t*-based 95% confidence interval is from -5.89 to -0.11.

```
lm_all <- lm(Distance ~ Condition, data = ddsub)
confint(lm_all) #Parametric 95% CI
```

```
##                        2.5 %      97.5 %
## (Intercept)       115.520697 119.7013823
## Conditioncommute   -5.891248  -0.1149621

## Conditioncommute
##        -3.003105

##         2.5%         97.5%
## -5.81626474 -0.07606663
```

The bootstrap 95% confidence interval is from -5.816 to -0.076. With this large data set, the differences between parametric and permutation approaches decrease and they essentially equivalent here. The bootstrap distribution (not displayed) for the differences in the sample means is relatively symmetric and centered around the estimated difference of 6 cm. So using all the observations we would be 95% confident that the true mean difference in overtake distances (*commute* - *casual*) is between -5.82 and -0.08 cm, providing additional information about the estimated difference in the sample means of 6 cm.

```
Tobs <- coef(lm_all)[2]; Tobs

B <- 1000
set.seed(1234)
Tstar <- matrix(NA, nrow = B)
for (b in (1:B)){
  lmP <- lm(Distance ~ Condition, data = resample(ddsub))
  Tstar[b] <- coef(lmP)[2]
}
```

```
qdata(Tstar, c(0.025, 0.975))
```

```
##          2.5%       97.5%
## -5.81626474 -0.07606663
```

2.10 BOOTSTRAP CONFIDENCE INTERVALS FOR DIFFERENCE IN GPAS

We can now apply the new confidence interval methods on the STAT 217 grade data. This time we start with the parametric 95% confidence interval "by hand" in R and then use lm to verify our result. The favstats output provides us with the required information to calculate the confidence interval, with the estimated difference in the sample mean GPAs of 3.338 - 3.0886 = 0.2494:

```
favstats(GPA ~ Sex, data = s217)
```

```
##   Sex  min  Q1 median   Q3 max     mean        sd  n missing
## 1   F 2.50 3.1  3.400 3.70   4 3.338378 0.4074549 37       0
## 2   M 1.96 2.8  3.175 3.46   4 3.088571 0.4151789 42       0
```

The *df* are 37 + 42 - 2 = 77. Using the SDs from the two groups and their sample sizes, we can calculate *sp*:

```
sp <- sqrt(((37 - 1)*(0.4075^2) + (42 - 1)*(0.41518^2))/(37 + 42 - 2))
sp
```

[1] 0.4116072

The margin of error is:

```
qt(0.975, df = 77)*sp*sqrt(1/37 + 1/42)
```

[1] 0.1847982

All together, the 95% confidence interval is:

```
3.338 - 3.0886 + c(-1,1)*qt(0.975, df = 77)*sp*sqrt(1/37 + 1/42)
```

[1] 0.0646018 0.4341982

So we are 95% confident that the difference in the true mean GPAs between females and males (females minus males) is between 0.065 and 0.434 GPA points. We get a similar result from confint on lm, except that lm switched the direction of the comparison from what was done "by hand" above, with the estimated mean difference of -0.25 GPA points (male - female) and similarly switched CI:

```
lm_GPA <- lm(GPA ~ Sex, data = s217)
summary(lm_GPA)
```

```
##
## Call:
## lm(formula = GPA ~ Sex, data = s217)
##
## Residuals:
##      Min       1Q   Median       3Q      Max
## -1.12857 -0.28857  0.06162  0.36162  0.91143
##
## Coefficients:
##             Estimate Std. Error t value Pr(>|t|)
## (Intercept)  3.33838    0.06766  49.337  < 2e-16
## SexM        -0.24981    0.09280  -2.692  0.00871
##
## Residual standard error: 0.4116 on 77 degrees of freedom
## Multiple R-squared:  0.08601,    Adjusted R-squared:  0.07414
## F-statistic: 7.246 on 1 and 77 DF,  p-value: 0.008713
```

```
confint(lm_GPA)
```

```
##                  2.5 %      97.5 %
## (Intercept)  3.2036416  3.47311517
## SexM        -0.4345955 -0.06501838
```

Note that we can easily switch to 90% or 99% confidence intervals by simply changing the percentile in qt or changing the level option in the confint function.

```
qt(0.95, df = 77) #For 90% confidence and 77 df
```

[1] 1.664885

```
qt(0.995, df = 77) #For 99% confidence and 77 df
```

[1] 2.641198

```
confint(lm_GPA, level = 0.9) #90% confidence interval
```

```
##                    5 %        95 %
## (Intercept)  3.2257252  3.45103159
## SexM        -0.4043084 -0.09530553
```

```
confint(lm_GPA, level = 0.99) #99% confidence interval
```

```
##                  0.5 %       99.5 %
## (Intercept)  3.1596636  3.517093108
## SexM        -0.4949103 -0.004703598
```

As a review of some basic ideas with confidence intervals make sure you can answer the following questions:

- What is the impact of increasing the confidence level in this situation?
- What happens to the width of the confidence interval if the size of the SE increases or decreases?
- What about increasing the sample size – should that increase or decrease the width of the interval?

All the general results you learned before about impacts to widths of CIs hold in this situation whether we are considering the parametric or bootstrap methods...

To finish this example, we will generate the comparable bootstrap 90% confidence interval using the bootstrap distribution in Figure 2.26.

```
'obs <- coef(lm_GPA)[2]; Tobs
```

```
##        SexM
## -0.2498069
```

```
B <- 1000
set.seed(1234)
Tstar <- matrix(NA, nrow = B)

for (b in (1:B)){
  lmP <- lm(GPA ~ Sex, data = resample(s217))
  Tstar[b] <- coef(lmP)[2]
}

quantiles <- qdata(Tstar, c(0.05, 0.95))
quantiles
```

```
##          5%         95%
## -0.39290566 -0.09622185
```

The output tells us that the 90% confidence interval is from -0.393 to -0.096 GPA points. The bootstrap distribution with the observed difference in the sample means and these cut-offs is displayed in Figure 2.26 using this code:

```
›ble(Tstar) %>% ggplot(aes(x = Tstar)) +
;eom_histogram(aes(y = ..ncount..), bins = 15, col = 1, fill = "grey",
               center = 0) +
;eom_density(aes(y = ..scaled..)) +
:heme_bw() +
.abs(y = "Density") +
;eom_vline(xintercept = quantiles, col = "blue", lwd = 2, lty = 2) +
;eom_vline(xintercept = Tobs, col = "red", lwd = 2) +
;tat_bin(aes(y = ..ncount.., label = ..count..), bins = 15,
        geom = "text", vjust = -0.75)
```

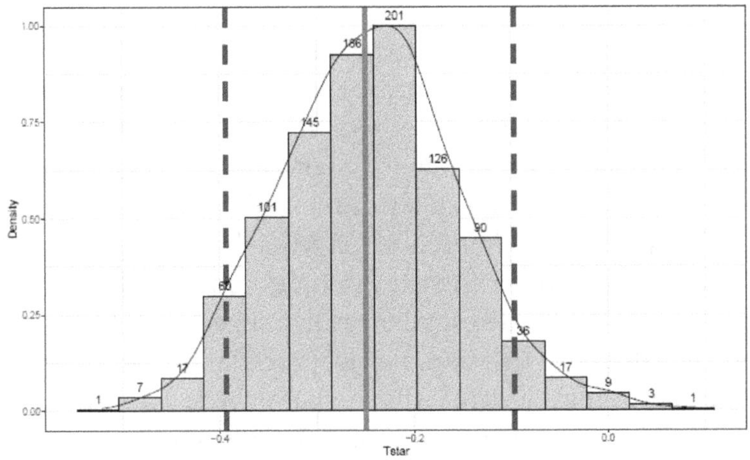

Figure 2.26: Histogram and density curve of bootstrap distribution of difference in sample mean GPAs (male minus female) with observed difference (solid vertical line) and quintiles that delineate the 90% confidence intervals (dashed vertical lines).

In the previous output, the parametric 90% confidence interval is from -0.404 to -0.095, suggesting similar results again from the two approaches. Based on the bootstrap CI, we can say that we are 90% confident that the difference in the true mean GPAs for STAT 217 students is between -0.393 to -0.096 GPA points (male minus females). This result would be usefully added to step 5 in the 6+ steps of the hypothesis testing protocol with an updated result of:

Report and discuss an estimate of the size of the differences, with confidence interval(s) if appropriate.

- Females were estimated to have a higher mean GPA by 0.3 points (90% bootstrap confidence interval: 0.096 to 0.393). This difference of 0.3 on a GPA scale does not seem like a very large difference in the means even though we were able to detect a difference in the groups.

Throughout the text, pay attention to the distinctions between parameters and statistics, focusing on the differences between estimates based on the sample and inferences for the population of interest in the form of the parameters of interest. Remember that statistics are summaries of the sample information and parameters are characteristics of populations (which we rarely know). And that our inferences are limited to the population that we randomly sampled from, if we randomly sampled.

2.11 CONCISE SUMMARY

In this chapter, we reviewed basic statistical inference methods in the context of a two-sample mean problem using linear models and the lm function. You were introduced to using R to do enhanced visualizations(pirate-plots), permutation testing, and generate bootstrap confidence intervals as well as obtaining parametric *t*-test and confidence intervals. You should have learned how to use a for loop for doing the nonparametric inferences and the lm and confint functions for generating parametric inferences. In the examples considered, the parametric and nonparametric methods provided similar results, suggesting that the assumptions were not too violated for the parametric procedures. When parametric and nonparametric approaches disagree, the nonparametric methods are likely to be more trustworthy since they have less restrictive assumptions but can still make assumptions and can have problems.

When the noted conditions are violated in a hypothesis testing situation, the Type I error rates can be inflated, meaning that we reject the null hypothesis more often than we have allowed to occur by chance. Specifically, we could have a situation where our assumed 5% significance level test might actually reject the null when it is true 20% of the time. If this is occurring, we call a procedure ***liberal*** (it rejects too easily) and if the procedure is liberal, how could we trust a small p-value to be a "real" result and not just an artifact of violating the assumptions of the procedure? Likewise, for confidence intervals we hope that our 95% confidence level procedure, when repeated, will contain the true parameter 95% of the time. If our assumptions are violated, we might actually have an 80% confidence level procedure and it makes it hard to trust the reported results for our observed data set. Statistical inference relies on a belief in the methods underlying our inferences. If we don't trust our assumptions, we shouldn't trust the conclusions to perform the way we want them to. As sample sizes increase and/or violations of conditions lessen, then the procedures will perform better. In Chapter 3, some new tools for doing diagnostics are introduced to help us assess how and how much those validity conditions are violated.

It is good to review how to report hypothesis test conclusions and compare those for when we have strong, moderate, or weak evidence. Suppose that we are doing parametric inferences with lm for differences between groups A and B, are extracting the *t*-statistics, have 15 degrees of freedom, and obtain the following test statistics and p-values:

- **t_{15} = 3.5, p-value = 0.0016:**There is strong evidence against the null hypothesis of no difference in the true means of the response between A and B (t_{15} = 3.5, p-value = 0.0016), so we would conclude that there is a difference in the true means.

- **t_{15} = 1.75, p-value = 0.0503:**There is moderate evidence against the null hypothesis of no difference in the true means of the response between A and B (t_{15} = 1.75, p-value = 0.0503), so we would conclude that there is likely[44] a difference in the true means.
- **t_{15} = 0.75, p-value = 0.232:**There is weak evidence against the null hypothesis of no difference in the true means of the response between A and B (t_{15} = 0.75, p-value = 0.232), so we would onclude that there is likely not a difference in the true means.

The last conclusion also suggests an action to take when we encounter weak evidence against null hypotheses– we could potentially model the responses using the null model since we couldn't prove it was wrong. We would take this action knowing that we could be wrong, but the "simpler" model that the null hypothesis suggests is often an attractive option in very complex models, such as what we are going to encounter in the coming chapters, especially in Chapters 5 and 8.

2.12 SUMMARY OF IMPORTANT R CODE

The main components of R code used in this chapter follow with components to modify in lighter and/or ALL CAPS text, remembering that any R packages mentioned need to be installed and loaded for this code to have a chance of working:

- summary(DATASETNAME)
  - Provides numerical summaries of all variables in the data set.
- summary(lm(Y ~ X, data = DATASETNAME))
  - Provides estimate, SE, test statistic, and p-value for difference in second row of coefficient table.
- confint(lm(**Y ~ X, data = DATASETNAME), level = 0.95)**
  - Provides 95% confidence interval for difference in second row of output.
- 2*pt(abs(Tobs), df = DF, lower.tail = F)
  - Finds the two-sided test p-value for an observed 2-sample t-test statistic of Tobs.
- hist(DATASETNAME$Y)
  - Makes a histogram of a variable named Y from the data set of interest.
- boxplot(Y ~ X, data = DATASETNAME)
  - Makes a boxplot of a variable named Y for groups in X from the data set.
- pirateplot(Y ~ X, data = DATASETNAME, inf.method = "ci", inf.disp = "line")
  - Requires the yarrr package is loaded.

- Makes a pirate-plot of a variable named Y for groups in X from the data set with estimated means and 95% confidence intervals for each group.
- Add theme = 2 if the confidence intervals extend outside the density curves and you can't see how far they extend.

◉ mean(Y ~ X, data = DATASETNAME); sd(Y ~ X, data = DATASETNAME)

- This usage of mean and sd requires the mosaic package.
- Provides the mean and sd of responses of Y for each group described in X.

◉ favstats(Y ~ X, data = DATASETNAME)

- Provides numerical summaries of Y by groups described in X.

◉ Tobs <- coef(lm(Y ~ X, data = DATASETNAME))[2]; Tobs

```
B <- 1000
Tstar <- matrix(NA, nrow = B)
for (b in (1:B)){
lmP <- lm(Y ~ shuffle(X), data = DATASETNAME)
Tstar[b] <- coef(lmP)[2]
}
```

◉ Code to run a for loop to generate 1000 permuted versions of the test statistic using the shuffle function and keep track of the results in Tstar

◉ pdata(Tstar, abs(Tobs), lower.tail = F)[[1]]

- Finds the proportion of the permuted test statistics in Tstar that are less than -|Tobs| or greater than |Tobs|, useful for finding the two-sided test p-value.

◉ Tobs <- coef(lm(Y ~ X, data = DATASETNAME))[2]; Tobs

```
B <- 1000
Tstar <- matrix(NA, nrow = B) for (b in (1:B)){
lmP <- lm(Y ~ X, data = resample(DATASETNAME))
Tstar[b] <- coef(lmP)[2]
}
```

- Code to run a for loop to generate 1000 bootstrapped versions of the data set using the resample function and keep track of the results of the statistic in Tstar.

◉ **qdata(Tstar, c(0.025, 0.975))**

- Provides the values that delineate the middle 95% of the results in the bootstrap distribution (Tstar).

2.13 PRACTICE PROBLEMS

2.1 Overtake Distance Analysis: The tests for the overtake distance data were performed with two-sided alternatives and so two-sided areas used to find the p-values. Suppose that the researchers expected that the average passing distance would be less (closer) for the commute clothing than for the casual clothing group. Repeat obtaining the permutation-based p-value for the one-sided test for either the full or smaller sample data set. Hint: Your p-value should be just about half of what it was before and in the direction of the alternative.

2.2 HELP Study Data Analysis: Load the HELPrct data set from the mosaic Data package [Pruim et al., 2021a] (you need to install the mosaic Data package once to be able to load it). The HELP study was a clinical trial for adult inpatients recruited from a detoxification unit. Patients with no primary care physician were randomly assigned to receive a multidisciplinary assessment and a brief motivational intervention orusual care and various outcomes were observed. Two of the variables in the data set are sex, a factor with levels male and female and daysanysub which is the time (in days) to first use of any substance post-detox. We are interested in the difference in mean number of days to first use of any substance post-detox between males and females. There are some missing responses and the following code will produce favstats with the missing values and then provide a data set that by applying the drop_na() function to the piped data setremoves any observations with missing values.

```
library(mosaicData)
data(HELPrct)
# Just focus on two variables
HELPrct2 <- HELPrct %>% select(daysanysub, sex)
# Removes subjects (complete rows) with any missing values
HELPrct3 <- HELPrct2 %>% drop_na()
favstats(daysanysub ~ sex, data = HELPrct2)
favstats(daysanysub ~ sex, data = HELPrct3)
```

2.2.1 Based on the results provided, how many observations were missing for males and females? Missing values here likely mean that the subjects didn't use any substances post-detox in the time of the study but might have at a later date – the study just didn't run long enough. This is called ***censoring***. What is the problem with the numerical summaries here if the missing responses were all something larger than the largest observation?

2.2.2 Make a pirate-plot and a boxplot of daysanysub ~ sex using the HELPrct3 data set created above. Compare the distributions, recommending parametric or nonparametric inferences.

Generate the permutation results and write out the 6+ steps of the hypothesis test.

2.2.4 Interpret the p-value for these results.

2.2.5 Generate the parametric test results using lm, reporting the test-statistic, its distribution under the null hypothesis, and compare the p-value to those observed using the permutation approach.

2.2.6 Make and interpret a 95% bootstrap confidence interval for the difference in the means.

CHAPTER-3

ONE-WAY ANOVA

3.1 SITUATION

In Chapter 2, tools for comparing the means of two groups were considered. More generally, these methods are used for a quantitative response and a categorical explanatory variable (group) which had two and only two levels. The complete overtake distance data set actually contained seven groups (Figure 3.1) with the outfit for each commute randomly assigned. In a situation with more than two groups, we have two choices. First, we could rely on our two group comparisons, performing tests for every possible pair (*commute* vs *casual*, *casual* vs *highviz*, *commute* vs *highviz*, . . . , *polite* vs *racer*), which would entail 21 different comparisons. But this would engage multiple testing issues and inflation of Type I error rates if not accounted for in some fashion. We would also end up with 21 p-values that answer detailed questions but none that addresses a simple but initially useful question – is there a difference somewhere among the pairs of groups or, under the null hypothesis, are all the true group means the same? In this chapter, we will learn a new method, called Analysis of Variance, ANOVA, or sometimes AOV that directly assesses evidence against the null hypothesis of no difference and then possibly leading to the ability to conclude that there is some overall difference in the means among the groups. This version of an ANOVA is called a One-Way ANOVA since there is just one1 grouping variable. After we perform our One-Way ANOVA test for overall evidence of some difference, we will revisit the comparisons similar to those considered in Chapter 2 to get more details on specific differences among all the pairs of groups – what we call pair-wise comparisons. We will augment our previous methods for comparing two groups with an adjusted method for pairwise comparisons to make our results valid called Tukey's Honest Significant Difference.

To make this more concrete, we return to the original overtake data, making a pirate-plot (Figure 3.1) as well as summarizing the overtake distances by the seven groups using favstats.

```
library(mosaic)
library(readr)
library(yarrr)
dd <- read_csv("http://www.math.montana.edu/courses/s217/documents/Walker2014_mod.csv")
dd <- dd %>% mutate(Condition = factor(Condition))
```

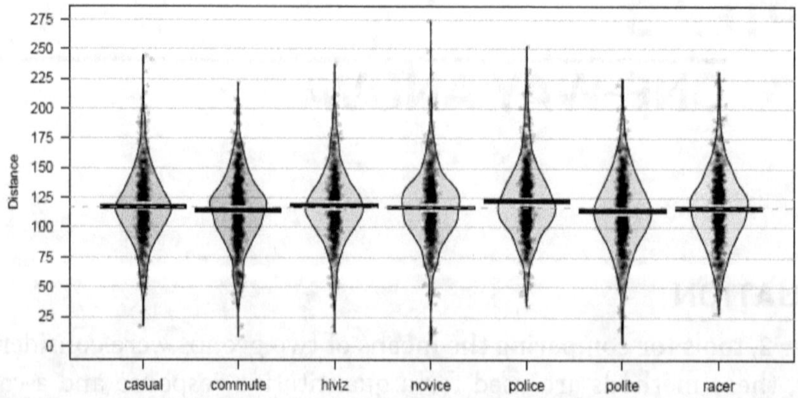

Figure 3.1: Pirate-plot of the overtake distances for the seven groups with group mean (bold lines with boxes indicating 95% confidence intervals) and the overall sample mean (dashed line) of 117.1 cm added.

```
pirateplot(Distance ~ Condition, data = dd, inf.method = "ci", inf.disp = "line")
abline(h = mean(dd$Distance), lwd = 2, col = "green", lty = 2) # Adds overall mean to plot

favstats(Distance ~ Condition, data = dd)
```

```
##   Condition min    Q1 median  Q3 max     mean       sd   n missing
## 1    casual  17 100.0    117 134 245 117.6110 29.86954 779       0
## 2   commute   8  98.0    116 132 222 114.6079 29.63166 857       0
## 3     hiviz  12 101.0    117 134 237 118.4383 29.03384 737       0
## 4    novice   2 100.5    118 133 274 116.9405 29.03812 807       0
## 5    police  34 104.0    119 138 253 122.1215 29.73662 790       0
## 6    polite   2  95.0    114 133 225 114.0518 31.23684 868       0
## 7     racer  28  98.0    117 135 231 116.7559 30.60059 852       0
```

There are slight differences in the sample sizes in the seven groups with between 737 and 868 observations, providing a data set has a total sample size of N = 5, 690. The sample means vary from 114.05 to 122.12cm. In Chapter 2, we found moderate evidence regarding the difference in commute and casual. It is less clear whether we might find evidence of a difference between, say, commute and novice groups since we are comparing means of 114.05 and 116.94 cm. All the distributions appear to have similar shapes that are generally symmetric and bell-shaped and have relatively similar variability. The police vest group of observations seems to have highest sample mean, but there are many open questions about what differences might really exist here and there are many comparisons that could be considered.

3.2 LINEAR MODEL FOR ONE-WAY ANOVA (CELL MEANS AND REFERENCE- CODING)

We introduced the statistical model $y_{ij} = \mu_j + \varepsilon_{ij}$ in Chapter 2 for the situation with $j = 1$ or 2 to denote a situation where there were two groups and, for the model that is consistent with the alternative hypothesis, the means differed. Now there are seven groups and the previous model can be extended to this new situation by allowing j to be 1, 2, 3, . . . , 7. As before, the linear model assumes that the responses follow a normal distribution with the model defining the mean of the normal distributions and all observations have the same variance. Linear models assume that the parameters for the mean in the model enter linearly. This last condition is hard to explain at this level of material – it is sufficient to know that there are models where the parameters enter the model nonlinearly and that they are beyond the scope of this function and this material and you won't run into them in most statistical models. By employing this general "linear" modelling methodology, we will be able to use the same general modelling framework for the methods in Chapters 3, 4, 6, 7, and 8.

As in Chapter 2, the null hypothesis defines a situation (and model) where all the groups have the same mean. Specifically, the null hypothesis in the general situation with J groups ($J \geq 2$) is to have all the true group means equal,

$$H0 : \mu 1 = \ldots = \mu J .$$

This defines a model where all the groups have the same mean so it can be defined in terms of a single mean,μ, for the ith observation from the jth group as $y_{ij} = \mu + \varepsilon_{ij}$. This is not the model that most researchers want to be the final description of their study as it implies no difference in the groups. There is more caution required to specify the alternative hypothesis with more than two groups. The alternative hypothesis needs to be the logical negation of this null hypothesis of all groups having equal means; to make the null hypothesis false, we only need one group to differ but more than one group could differ from the others. Essentially, there are many ways to "violate" the null hypothesis so we choose some delicate wording for the alternative hypothesis when there are more than 2 groups. Specifically, we state the alternative as

HA: Not all μ_j are equal or, in words, at least one of the true means differs among the J groups. You might be attracted to trying to say that all means are different in the alternative but we do not put this strict a requirement in place to reject the null hypothesis. The alternative model allows all the true group means to differ but does require that they are actually all different with the model written as:

$$y_{ij} = \mu_j + \varepsilon_{ij} .$$

This linear model states that the response for the ith observation in the jth group, y_{ij}, is modelled with a group j ($j = 1, \ldots, J$) population mean, μ_j, and a random error for each subject in each group, ε_{ij}, that we assume follows a normal distribution and that all the random errors have the same variance, σ^2. We can write the assumption about the random errors, often called the normality assumption, as $\varepsilon_{ij} \sim N(0, \sigma^2)$.There is a second way to write out this model that allows extension to more complex models discussed below, so we need a name for this version of the model. The model written in terms of the μ_j 's is called the cell means model and is the easier version of this model to understand.

One of the reasons we learned about pirate-plots is that it helps us visually consider all the aspects of this model. In Figure 3.1, we can see the bold horizontal lines that provide the estimated (sample) group means. The bigger the differences in the sample means (especially relative to the variability around the means), the more evidence we will find against the null hypothesis. You can also see the null model on the plot that assumes all the groups have the same mean as displayed in the dashed horizontal line at 117.1 cm (the R code below shows the overall mean of Distance is 117.1). While the hypotheses focus on the means, the model also contains assumptions about the distribution of the responses - specifically that the distributions are normal and that all the groups have the same variability, which do not appear to be clearly violated in this situation.

```
mean(dd$Distance)
```

[1] 117.126

There is a second way to write out the One-Way ANOVA model that provides a framework for extensions to more complex models described in Chapter 4 and beyond. The other parameterization (way of writing out or defining) of the model is called the reference-coded model since it writes out the model in terms of a baseline group and deviations from that baseline or reference level. The reference-coded model for the ith subject in the jth group is $y_{ij} = \alpha + \tau_j + \varepsilon_{ij}$ where α ("alpha") is the true mean for the baseline group(usually first alphabetically) and the τ_j (tau j) are the deviations from the baseline group for group j. The deviation for the baseline group, τ_1, is always set to 0 so there are really just deviations for groups 2 through J. The equivalence between the reference-coded and cell means models can be seen by considering the mean for the first, second, and J th groups in both models:

	Cell means:	Reference-coded:
Group 1 :	μ^1	α
Group 2 :	μ^2	$\alpha + \tau^2$
...	...	...
Group J :	μ^J	$\alpha + \tau J$

The hypotheses for the reference-coded model are similar to those in the cell means coding except that they are defined in terms of the deviations, τj . The null hypothesis is that there is no deviation from the baseline for any group – that all the τj 's = 0,

H0: $\tau 2 = \ldots = \tau J = 0$.

The alternative hypothesis is that at least one of the deviations ($j = 2, \ldots, J$) is not 0,

HA: Not all τj equal 0, for $j = 2, \ldots, J$.

In this chapter, you are welcome to use either version (unless we instruct you otherwise) but we have to use the reference-coding in subsequent chapters. The next task is to learn how to use R's linear model, lm, function to get estimates of the parameters2 in each model, but first a quick review of these new ideas:

Cell Means Version

- ***H0*** : $\mu 1 = \ldots = \mu J$ *HA* : Not all μj equal
- **Null hypothesis in words:** No difference in the true means among the groups.
- **Null model:** $yij = \mu + \varepsilon ij$
- **Alternative hypothesis in words:** At least one of the true means differs among the groups.
- **Alternative model:** $yij = \mu j + \varepsilon ij$.

Reference-coded Version

- **H_0 :** $\tau 2 = \ldots = \tau J = 0$ HA : Not all τj equal 0, for $j = 2, \ldots, J$
- **Null hypothesis in words:** No deviation of the true mean for any groups from the baseline group.
- **Null model:** $yij = \alpha + \varepsilon ij$
- **Alternative hypothesis in words:** At least one of the true deviations is different from 0 or that at least one group has a different true mean than the baseline group.
- **Alternative model:** $yij = \alpha + \tau j + \varepsilon ij$

In order to estimate the models discussed above, the lm function is used[3]. The lm function continues to use the same format as previous functions and in Chapter 2 , lm(Y ~ X, data = dataset name). It ends up that lm generates the reference-coded version of the model by default (The developers of R thought it was that important!). But we want to start with the cell means version of the model, so we have to override the standard technique and add a "-1" to the formula interface to tell R that we want to the cell means coding. Generally, this looks like lm(Y ~ X - 1, data = dataset name). Once we fit a model in R, the summary function run on the model provides a useful "summary" of the model coefficients and a suite of other potentially interesting information. For the moment, we will focus on the estimated model coefficients, so only those lines are provided. When fitting the cell means version of the One-Way ANOVA model, you will find a row of output for each group relating estimating the μj 's. The output contains columns for an estimate (Estimate), standard error (Std. Error), *t*-value (t value), and p-value (Pr(>|t|)). We'll explore which of these are of interest in these models below, but focus on the estimates of the parameters that the function provides in the first column ("Estimate") of the coefficient table and compare these results to what was found using favstats.

```
lm1 <- lm(Distance ~ Condition - 1, data = dd)
summary(lm1)$coefficients
```

In general, we denote estimated parameters with a hat over the parameter of interest to show that it is an estimate. For the true mean of group j, μj , we estimate it with μj, which is just the sample mean for group j, x¯j . The model suggests an estimate for each observation that we denote as yij that we will also call a fitted value based on the model being considered. The same estimate is used for all observations in the each group in this model. R tries to help you to sort out which row of output corresponds to which group by appending the group name with the variable name. Here, the variable name was Condition and the first group alphabetically was casual, so R provides a row labelled Condition casual with an estimate of 117.61. The sample means from the seven groups can be seen to directly match the favstats results presented previously.

The reference-coded version of the same model is more complicated but ends up giving the same results once we understand what it is doing. It uses a different parameterization to accomplish this, so has different model output. Here is the model summary:

```
lm2 <- lm(Distance ~ Condition, data = dd)
summary(lm2)$coefficients
```

```
##                       Estimate Std. Error     t value     Pr(>|t|)
## (Intercept)        117.6110398   1.071873 109.7247845 0.000000000
## Conditioncommute    -3.0031051   1.480964  -2.0278039 0.042626835
## Conditionhiviz       0.8272234   1.537302   0.5381008 0.590528548
## Conditionnovice     -0.6705193   1.502651  -0.4462242 0.655452292
## Conditionpolice      4.5104792   1.510571   2.9859423 0.002839115
## Conditionpolite     -3.5591965   1.476489  -2.4105807 0.015958695
## Conditionracer      -0.8551713   1.483032  -0.5766371 0.564207492
```

The estimated model coefficients are $\hat{a} + \hat{\tau}_2 = 117.61$ cm, $\hat{\tau}_2 = -3.00$ cm, $+ \hat{\tau}_3 = 0.83$ cm, and so on up to $+ \hat{\tau}_7 = -0.86$cm, where R selected group 1 for casual, 2 for commute, 3 for hiviz, all the way up to group 7 for racer. The way you can figure out the baseline group (group 1 is casual here) is to see which category label is not present in the reference-coded output. The baseline level is typically the first group label alphabetically,but you should always check this4. Based on these definitions, there are interpretations available for each coefficient. For $\hat{a} = 117.61$ cm, this is an estimate of the mean overtake distance for the casual outfit group. $\hat{\tau}_2 = -3.00$ cm is the deviation of the commute group's mean from the causal group's mean (specifically, it is3.00 cm lower and was a quantity we explored in detail in Chapter 2 when we just focused on comparing casual and commute groups). $\hat{\tau}_3 = 0.83$ cm tells us that the hiviz group mean distance is 0.83 cm higher than the casual group mean and $\hat{\tau}_7 = -0.86$ says that the racer sample mean was 0.86 cm lower than for the casual group. These interpretations are interesting as they directly relate to comparisons of groups with the baseline and lead directly to reconstructing the estimated means for each group by combining the baseline and a pertinent deviation as shown in Table 3.1.

Table 3.1: Constructing group mean estimates from the reference-coded linear model estimates.

Group	Formula	Estimates
casual	$\hat{a}$	117.61 cm
commute	$\hat{a} + \hat{\tau}_2$	117.61 - 3.00 = 114.61 cm
hiviz	$\hat{a} + \hat{\tau}_3$	117.61 + 0.83 = 118.44 cm
novice	$\hat{a} + \hat{\tau}_4$	117.61 - 0.67 = 116.94 cm
police	$\hat{a} + \hat{\tau}_5$	117.61 + 4.51 = 122.12 cm
polite	$\hat{a} + \hat{\tau}_6$	117.61 - 3.56 = 114.05 cm
racer	$\hat{a} + \hat{\tau}_7$	117.61 - 0.86 = 116.75 cm

We can also visualize the results of our linear models using what are called ***term-plots*** or ***effect-plots***

(from the effects package; [Fox et al., 2022c]) as displayed in Figure 3.2. We don't want to use the word "effect" for these model components unless we have random assignment in the study design so we generically call these ***term-plots*** as they display terms or components from the model in hopefully useful ways to aid in model interpretation even in the presence of complicated model parameterizations. The word "effect" has a causal connotation that we want to avoid as much as possible in non-causal (so non-randomly assigned) situations. Term-plots take an estimated model and show you its estimates along with 95% confidence intervals generated by the linear model. These confidence intervals may differ from the confidence intervals in the pirate-plots since the pirate-plots make them for each group separately and term-plots are combining information across groups via the estimated model and then doing inferences for individual group means. To make term-plots, you need to install and load the effects package and then use plot(all Effects(...))functions together on the lm object called lm2 that was estimated above. You can find the correspondence between the displayed means and the estimates that were constructed in Table 3.1.

```
library(effects)
plot(allEffects(lm2))
```

In order to assess overall evidence against having the same means for the all groups (vs having at least one mean different from the others), we compare either of the previous models (cell means or reference-coded) to a null model based on the null hypothesis of $H0 : \mu 1 = \ldots = \mu J$, which implies a model of $yij = \mu + \varepsilon ij$ in the cell means version where μ is a common mean for all the observations. We will call this the mean-only model since it only has a single mean in it. In the reference-coded version of the model, we have a null hypothesis of ***H0*** $: \tau 2 = \ldots = \tau J = 0$, so the "mean-only" model is $yij = \alpha + \varepsilon ij$ with α having the same definition as μ for the cell means model – it forces a common value for the mean for all the groups. Moving from the reference-coded model to the mean only model is also an example of a situation where we move from a "full" model to a "reduced" model by setting some coefficients in the "full" model to 0 and, by doing this, get a simpler or "reduced" model. Simple models can be good as they are easier to interpret, but having a model for *J* groups that suggests no difference in the groups is not a very exciting result in most, but not all, situations[5]. In order for R to provide results for the mean-only model, we remove the grouping variable, Condition, from the model formula and just include a "1". The (Intercept) row of the output provides theestimate for the mean-only model as a reduced model from either the cell means or reference-coded models when we assume that the mean is the same for all groups:

This model provides an estimate of the common mean for all observations of $117.13 = \hat{a} =$ cm. This valueμ α also is the dshed horizontal line in the pirate-plot in Figure 3.1. Some people call this mean-only model estimate the "grand" or "overall" mean and notationally is represented as $\bar{\bar{y}}$.

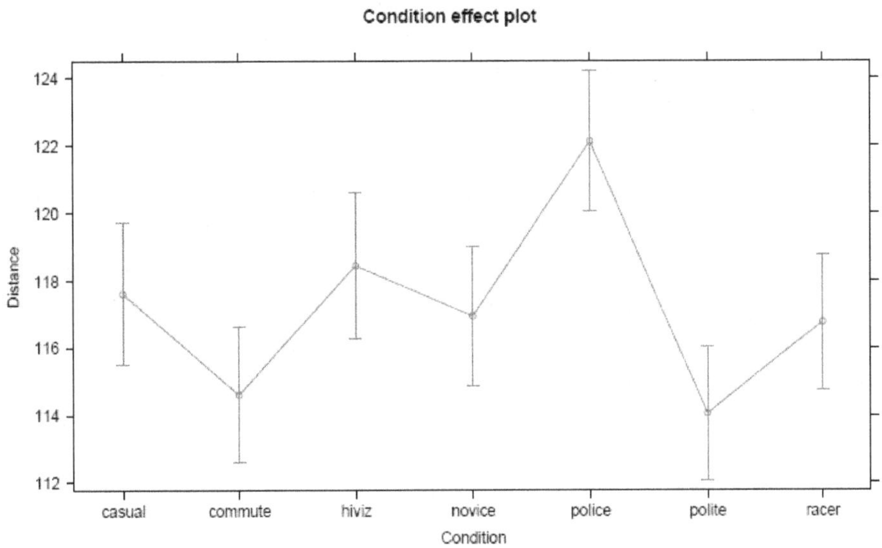

Figure 3.2: Plot of the estimated group mean distances from the reference-coded model for the overtake data from the effects package.

```
lm3 <- lm(Distance ~ 1, data = dd)
summary(lm3)$coefficients
```

```
##             Estimate Std. Error t value Pr(>|t|)
## (Intercept)  117.126  0.3977533 294.469        0
```

3.3 ONE-WAY ANOVA SUMS OF SQUARES, MEAN SQUARES, AND F-TEST

The previous discussion showed two ways of parameterizing models for the One-Way ANOVA model and getting estimates from output but still hasn't addressed how to assess evidence related to whether the observed differences in the means among the groups is "real". In this section, we develop what is called the ANOVA F-test that provides a method of aggregating the differences among the means of 2 or more groups and testing (assessing evidence against) our null hypothesis of no difference in the means vs the alternative. In order to develop the test, some additional notation is needed. The sample size in each group is denoted $n_j$and the total sample size is $N = \Sigma n_j = n_1 + n_2 + \ldots + n_J$ where Σ (capital sigma) means "add up over whatever follows". An estimated residual (e_{ij}) is the difference between an observation, y_{ij} , and the model estimate, $\mu_{ij} = \mu_j$, for that observation, $y_{ij} - y_{ij} = e_{ij}$. It is basically what is left over that the mean part of the model (μ_j) does not explain. It is also a window into how "good" the model might be because it reflects what the model was unable to explain.

Consider the four different fake results for a situation with four groups (J = 4) displayed in Figure 3.3. Which of the different results shows the most and least evidence of differences in the means? In trying to answer this, think about both how different the means are (obviously important) and how variable the results are around the mean. These situations were created to have the same means in Scenarios 1 and 2 as well as matching means in Scenarios 3 and 4. In Scenarios 1 and 2, the differences in the means is smaller than in the other two results. But Scenario 2 should provide more evidence of what little difference is present than Scenario 1 because it has less variability around the means. The best situation for finding group differences here is Scenario 4 since it has the largest difference in the means and the least variability around those means. Our test statistic somehow needs to allow a comparison of the variability in the means to the overall variability to help us get results that reflect that Scenario 4 has the strongest evidence of a difference (most variability in the means and least variability around those means) and Scenario 1 would have the least evidence (least variability in the means and most variability around those means).

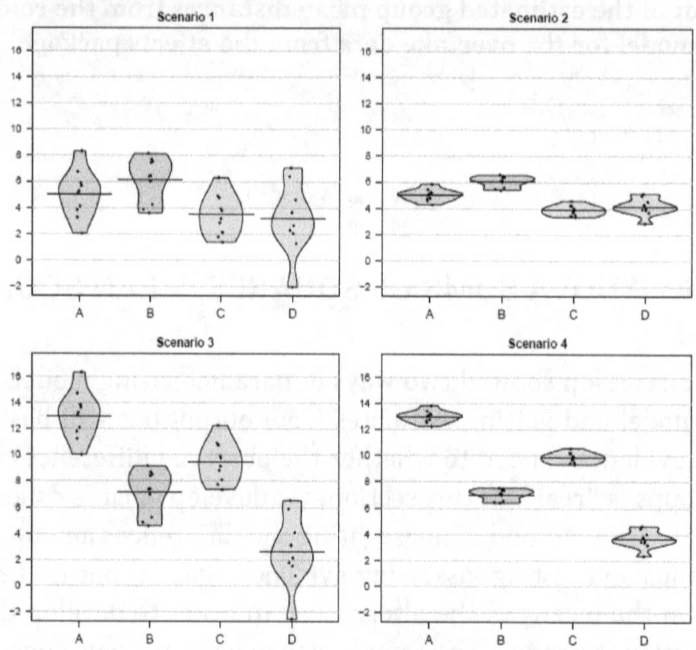

Figure 3.3: Demonstration of different amounts of difference in means relative to variability. Scenarios have the same means in rows and same variance around means in columns of plot. Confidence intervals not reported in the pirate-plots.

The statistic that allows the comparison of relative amounts of variation is called the ANOVA F-statistic. It is developed using sums of squares which are measures of total variation like those that are used in the numerator of the standard deviation (ΣN (yi - y¯)2) that took all the observations, subtracted the mean, squared the differences, and then added up the results over all the observations to generate a measure of total variability. With multiple groups, we will focus on decomposing that total variability (Total Sums of Squares) into variability among the means (we'll call this Explanatory Variable A's Sums of Squares) and variability in the residuals or errors (Error Sums of Squares). We define each of these quantities in the One-Way ANOVA situation as follows:

- $\mathbf{SS_{Total}}$ = Total Sums of Squares = $\Sigma_{j=1}^{J}\Sigma_{i=1}^{n_j}(y_{ij}-\bar{\bar{y}})^2$
  - This is the total variation in the responses around the grand mean (y¯¯, the estimated mean for all the observations and available from the mean-only model).
  - By summing over all nj observations in each group, $\Sigma_{i=1}^{n_j}(\ )$, and then adding those results up across the groups, $\Sigma_{j=1}^{J}(\ )$, we accumulate the variation across all N observations.
  - Note: this is the residual variation if the null model is used, so there is no further decomposition possible for that model.
  - This is also equivalent to the numerator of the sample variance, $\Sigma_1^N(y_i-\bar{y})^2$ which is what you get when you ignore the information on the potential differences in the groups.
- $\mathbf{SS_A}$ = Explanatory Variable A's Sums of Squares = $\Sigma_{j=1}^{J}\Sigma_{i=1}^{n_j}(y_{ij}-\bar{\bar{y}})^2$
  - This is the variation in the group means around the grand mean based on the explanatory variable A.
  - This is also called sums of squares for the treatment, regression, or model.
- $\mathbf{SS_E}$ = Error (Residual) Sums of Squares = ΣJ
  - This is the variation in the responses around the group means.
  - Also called the sums of squares for the residuals, especially when using the second version of the formula, which shows that it is just the squared residuals added up across all the observations.

The possibly surprising result given the mass of notation just presented is that the total sums of squares is ALWAYS equal to the sum of explanatory variable A's sum of squares and the error sums of squares,

$$\mathrm{SS_{Total}}=\mathrm{SS_A}+\mathrm{SS_E}.$$

This result is called the sums of squares decomposition formula. The equality implies that if the SS_A goes up, then the SS_E must go down if SS_{Total} remains the same. We use these results to build our test statistic and organize this information in what is called an ANOVA table. The ANOVA table is generated using the anova function applied to the reference-coded model, lm_2:

```
lm2 <- lm(Distance ~ Condition, data = dd)
anova(lm2)
```

```
## Analysis of Variance Table
##
## Response: Distance
##              Df  Sum Sq Mean Sq F value    Pr(>F)
## Condition     6   34948  5824.7  6.5081 7.392e-07
## Residuals 5683 5086298   895.0
```

Note that the ANOVA table has a row labelled Condition, which contains information for the grouping variable (we'll generally refer to this as explanatory variable A but here it is the outfit group that was randomly assigned), and a row labelled Residuals, which is synonymous with "Error". The Sums of Squares (SS) are available in the Sum Sq column. It doesn't show a row for "Total" but the $SS_{Total} = SS_A + SS_E = 5,121,246$.

```
34948 + 5086298
```

```
## [1] 5121246
```

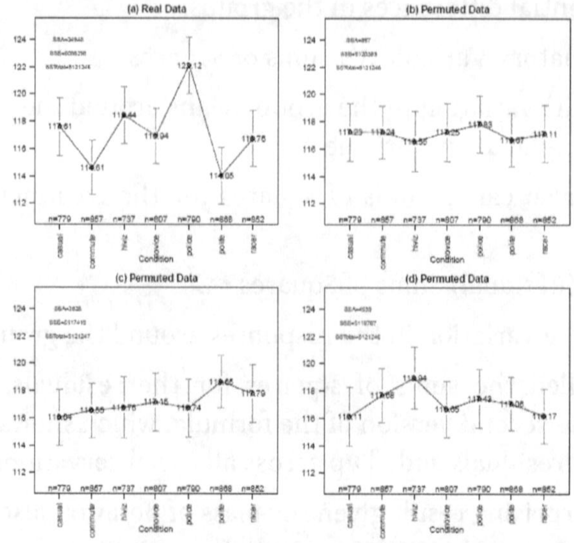

Figure 3.4: Plot of means and 95% confidence intervals for the three groups for the real overtake data (a)and three different permutations of the outfit group labels to the same responses in (b), (c), and (d). Note that SS_{Total} is always the same but the different amounts of variation associated with the means (SS_A) or the errors (SS_E) changes in permutation.

It may be easiest to understand the *sums of squares decomposition* by connecting it to our permutation ideas. In a permutation situation, the total variation (*SS*Total) cannot change – it is the same responses varying around the same grand mean. However, the amount of variation attributed to variation among the means and in the residuals can change if we change which observations go with which group. In Figure 3.4 (panel a), the means, sums of squares, and 95% confidence intervals for each mean are displayed for the seven groups from the original overtake data. Three permuted versions of the data set are summarized in panels (b), (c), and (d). The SSA is 34948 in the real data set and between 857 and 4539 in the permuted data sets. If you had to pick among the plots for the one with the most evidence of a difference in the means, you hopefully would pick panel (a). This visual "unusualness" suggests that this observed result is unusual relative to the possibilities under permutations, which are, again, the possibilities tied to having the null hypothesis being true. But note that the differences here are not that great between these three permuted data sets and the real one. It is likely that at least some might have selected panel (d) as also looking like it shows some evidence of differences, although the variation in the means in the real data set is clearly more pronounced than in this or the other permutations.

One way to think about SS_A is that it is a function that converts the variation in the group means into a single value. This makes it a reasonable test statistic in a permutation testing context. By comparing the observed $SS_A = 34948$ to the permutation results of 857, 3828, and 4539 we see that the observed result is much more extreme than the three alternate versions. In contrast to our previous test statistics where positive and negative differences were possible, SSA is always positive with a value of 0 corresponding to no variation in the means. The larger the SS_A, the more variation there is in the means. The permutation p-value for the alternative hypothesis of some (not of greater or less than!) difference in the true means of the groups will involve counting the number of permuted SS_A results that are as large or larger than what we observed.

To do a permutation test, we need to be able to calculate and extract the SSA value. In the ANOVA table, it is the second number in the first row; we can use the bracket, [,], referencing to extract that number from the ANOVA table that anova produces with anova(lm(Distance ~ Condition, data = dd))[1, 2]. We'll store the observed value of SS_A in Tobs, reusing some ideas from Chapter 2.

```
Tobs <- anova(lm(Distance ~ Condition, data = dd))[1,2]; Tobs
```

```
## [1] 34948.43
```

The following code performs the permutations B = 1,000 times using the shuffle function, builds up a vector of results in Tobs, and then makes a plot of the resulting permutation distribution:

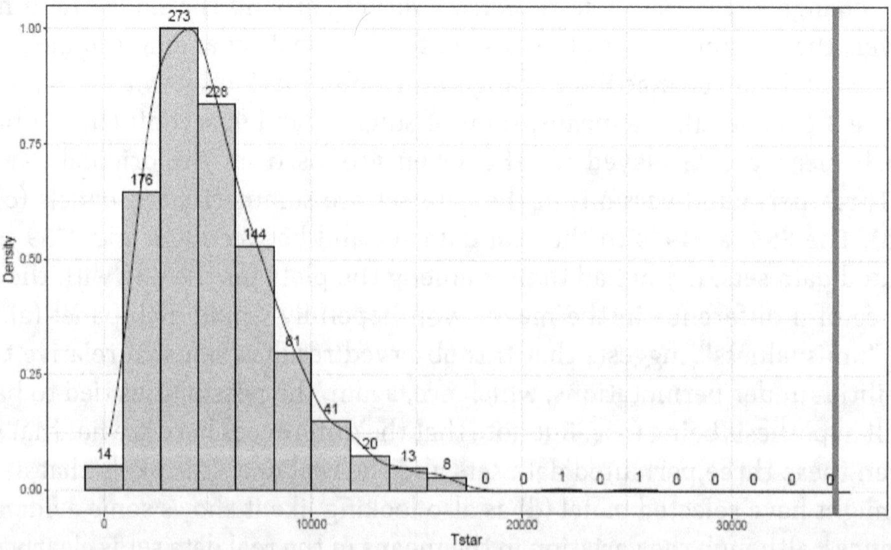

Figure 3.5: Histogram and density curve of permutation distribution of SSA with the observed value of SSA displayed as a bold, vertical line. The proportion of results that are as large or larger than the observed value of SSA provides an estimate of the p-value.

```
B <- 1000
Tstar <- matrix(NA, nrow = B)
for (b in (1:B)){
  Tstar[b] <- anova(lm(Distance ~ shuffle(Condition), data = dd))[1,2]
  }
tibble(Tstar) %>% ggplot(aes(x = Tstar)) +
  geom_histogram(aes(y = ..ncount..), bins = 20, col = 1, fill = "skyblue") +
  geom_density(aes(y = ..scaled..)) +
  theme_bw() +
  labs(y = "Density") +
  geom_vline(xintercept = Tobs, col = "red", lwd = 2) +
  stat_bin(aes(y = ..ncount.., label = ..count..), bins = 20,
        geom = "text", vjust = -0.75)
```

The right-skewed distribution (Figure 3.5) contains the distribution of SS^*_A's under permutations (where all the groups are assumed to be equivalent under the null hypothesis). The observed result is larger than all of the SS^*_A's. The proportion of permuted results that exceed the observed value is found using pdata as before, except only for the area to the right of the observed result. We know that Tobs will always be positive so no absolute values are required here.

```
pdata(Tstar, Tobs, lower.tail = F)[[1]]
```

```
## [1] 0
```

Because there were no permutations that exceeded the observed value, the p-value should be reported as p-value < 0.001 (less than 1 in 1,000) and not 0. This suggests very strong evidence against the null hypothesis of no difference in the true means. We would interpret this p-value as saying that there is less than a 0.1% chance of getting a SSA as large or larger than we observed, given that the null hypothesis is true.

It ends up that some nice parametric statistical results are available (if our assumptions are met) for the ratio of estimated variances, the estimated variances are called Mean Squares. To turn sums of squares into mean square (variance) estimates, we divide the sums of squares by the amount of free information available. For example, remember the typical variance estimator introductory statistics, $\Sigma_1^N (y_i - \bar{y})^2/(N-1)$?Your instructor probably spent some time trying various approaches to explaining why the denominator is the sample size minus 1. The most useful explanation for our purposes moving forward is that we "lose" one piece of information to estimate the mean and there are N deviations around the single mean so we divide by N - 1. The main point is that the sums of squares were divided by something and we got an estimator for the variance, in that situation for the observations overall.

Now consider $SS_E = \Sigma_{j=1}^J \Sigma_{i=1}^{n_j} (y_{ij} - \bar{y}_j)^2$which still has *N* deviations but it varies around the *J* means, so the Mean Square Error = $MS_E = SS_E/(N-J)$.

Basically, we lose J pieces of information in this calculation because we have to estimate J means. The similar calculation of the Mean Square for variable A (MSA) is harder to see in the formula$(SS_A = \Sigma_{j=1}^J n_j(\bar{y}_i - \bar{\bar{y}})^2)$, but the same reasoning can be used to understand the denominator for forming MSA: there are J means that vary around the grand mean so

$$MS_A = SS_A/(J-1).$$

In summary, the two mean squares are simply:

- MSA = SSA/(J - 1), which estimates the variance of the group means around the grand mean.
- MSError = SSError/(N - *J*), which estimates the variation of the errors around the group means.

These results are put together using a ratio to define the ANOVA F-statistic (also called the F-ratio)as:

$$F = MS_A/MS_{Error}.$$

If the variability in the means is "similar" to the variability in the residuals, the statistic would have a value around 1. If that variability is similar then there would be no evidence of a difference in the means. If the MSA is much larger than the MSE

, the F -statistic will provide evidence against the null hypothesis. The "size" of the F -statistic is formalized by finding the p-value. The F -statistic, if assumptions discussed below are not violated and we assume the null hypothesis is true, follows what is called an F -distribution. The F-distribution is a right-skewed distribution whose shape is defined by what are called the numerator degrees of freedom (J - 1) and the denominator degrees of freedom (N - J). These names correspond to the values that we used to calculate the mean squares and where in the F -ratio each mean square was used; F -distributions are denoted by their degrees of freedom using the convention of F (numerator df, denominator df). Some examples of different F -distributions are displayed for you in Figure 3.6.

The characteristics of the F-distribution can be summarized as:

- Right skewed,
- Nonzero probabilities for values greater than 0,
- Its shape changes depending on the numerator DF and denominator DF, and
- Always use the right-tailed area for p-values.

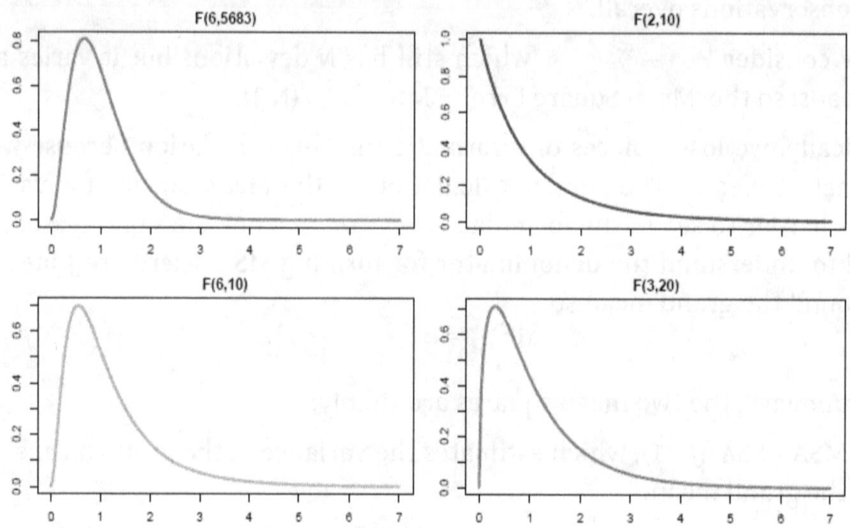

Figure 3.6: Density curves of four different *F* -distributions. Upper left is an *F* (6, 5683), upper right is *F* (2, 10), lower left is *F* (6, 10), and lower right is *F* (3, 20). P-values are found using the areas to the right of the observed *F* -statistic value in all F-distributions.

Now we are ready to discuss an ANOVA table since we know about each of its components. Note the general format of the ANOVA table is in Table 3.2[6]:

Table 3.2: General One-Way ANOVA table.

Source	DF	Sums of Squares	Mean Squares	F-ratio	P-value
Variable A	$J-1$	SS_A	$MS_A = SS_A/(J-1)$	$F = MS_A/MS_E$	Right tail of $F(J-1, N-J)$
Residuals	$N-J$	SS_E	$MS_E = SS_E/(N-J)$		
Total	$N-1$	SS_{Total}			

The table is oriented to help you reconstruct the F -ratio from each of its components. The output from R is similar although it does not provide the last row and sometimes switches the order of columns in different functions we will use. The R version of the table for the type of outfit effect (Condition) with $J = 7$ levels and $N = 5,690$ observations, repeated from above, is:

```
anova(lm2)
```

```
## Analysis of Variance Table
##
## Response: Distance
##              Df  Sum Sq Mean Sq F value        Pr(>F)
## Condition     6   34948  5824.7  6.5081 0.0000007392
## Residuals 5683 5086298   895.0
```

The p-value from the F -distribution is 0.0000007 so we can report it[7] as a p-value < 0.0001. We can verify this result using the observed F -statistic of 6.51 (which came from taking the ratio of the two mean squares, F = 5824.74/895) which follows an F (6, 5683) distribution if the null hypothesis is true and some other assumptions are met. Using the pf function provides us with areas in the specified F -distribution with the df1 provided to the function as the numerator *df* and df2 as the denominator *df* and lower.tail = F reflecting our desire for a right tailed area.

```
6.51, df1 = 6, df2 = 5683, lower.tail = F)
```

```
## [1] 0.0000007353832
```

The result from the F -distribution using this parametric procedure is similar to the p-value obtained using permutations with the test statistic of the SSA, which was < 0.0001. The F -statistic obviously is another potential test statistic to use as a test statistic in a permutation approach, now that we know about it. We should check that we get similar results from it with permutations as we did from using SSA as a permutation-test test statistic. The following code generates the permutation distribution for the F -statistic (Figure 3.7) and assesses how unusual the observed F -statistic of 6.51 was in this permutation distribution. The only change in the code involves moving from extracting *SSA* to extracting the *F*-ratio which is in the 4th column of the anova output:

```
Tobs <- anova(lm(Distance ~ Condition, data = dd))[1,4]; Tobs
```

```
## [1] 6.508071
```

```
B <- 1000
Tstar <- matrix(NA, nrow = B)
for (b in (1:B)){
  Tstar[b] <- anova(lm(Distance ~ shuffle(Condition), data = dd))[1,4]
}

pdata(Tstar, Tobs, lower.tail = F)[[1]]
```

```
## [1] 0
```

```
tibble(Tstar) %>% ggplot(aes(x = Tstar)) +
  geom_histogram(aes(y = ..ncount..), bins = 20, col = 1, fill = "skyblue") +
  geom_density(aes(y = ..scaled..)) +
  theme_bw() +
  labs(y = "Density") +
  geom_vline(xintercept = Tobs, col = "red", lwd = 2) +
```

```
  stat_bin(aes(y = ..ncount.., label = ..count..), bins = 20,
           geom = "text", vjust = -0.75)
```

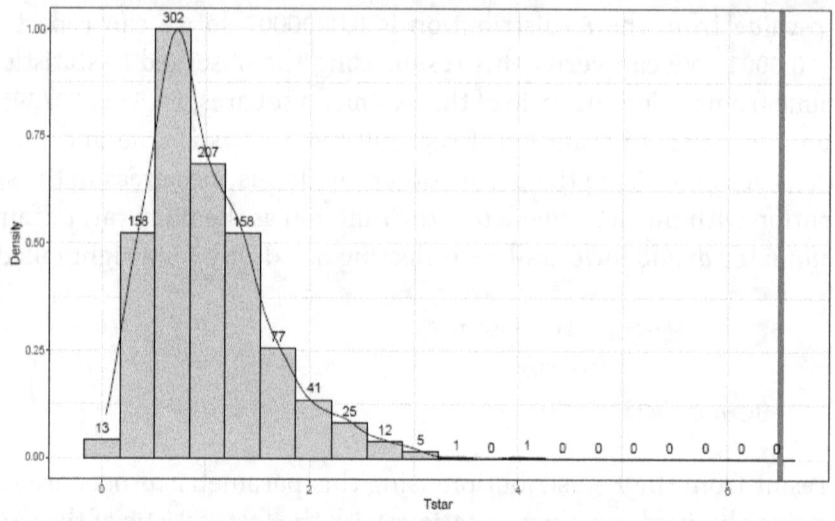

Figure 3.7: Histogram and density curve of the permutation distribution of the F-statistic with bold, vertical line for the observed value of the test statistic of 6.51.

The permutation-based p-value is again at less than 1 in 1,000, which matches the other results closely. The first conclusion is that using a test statistic of either the *F*-statistic or the *SSA* provide similar permutation results. However, we tend to favour using the *F*-statistic because it is more commonly used in reporting ANOVA results, not because it is any better in a permutation context.

It is also interesting to compare the permutation distribution for the F-statistic and the parametric $F(6, 6583)$ distribution (Figure 3.8). They do not match perfectly but are quite similar. Some the differences around 0 are due to the behaviour of the method used to create the density curve and are not really a problem for the methods. The similarity in the two curves explains why both methods would give similar p-value results for almost any test statistic value. In some situations, the correspondence will not be quite so close.

So how can we rectify this result (p-value < 0.0001) and the Chapter 2 result that reported moderate evidence against the null hypothesis of no difference between commute and casual with a p-value ≈ 0.04?I selected the two groups to compare in Chapter 2 because they were somewhat far apart but not too far apart. I could have selected police and polite as they are furthest apart and just focused on that difference. "Cherry-picking" a comparison when many are present, especially one that is most different, without accounting for this choice creates a false sense of the real situation and inflates the Type I error rate because of the selection[8]. If the entire suite of pairwise comparisons are considered, this result may lose some of its luster. In other words, if we consider the suite of 21 pair-wise differences (and the tests) implicit in comparing all of them, we may need really strong evidence against the null in at least some of the pairs to suggest overall differences. In this situation, the hiviz and casual groups are not that different from each other so their difference does not contribute much to the overall F -test. In Section 3.6, we will revisit this topic and consider a method that is statistically valid for performing all possible pair-wise comparisons that is also consistent with our overall test results.

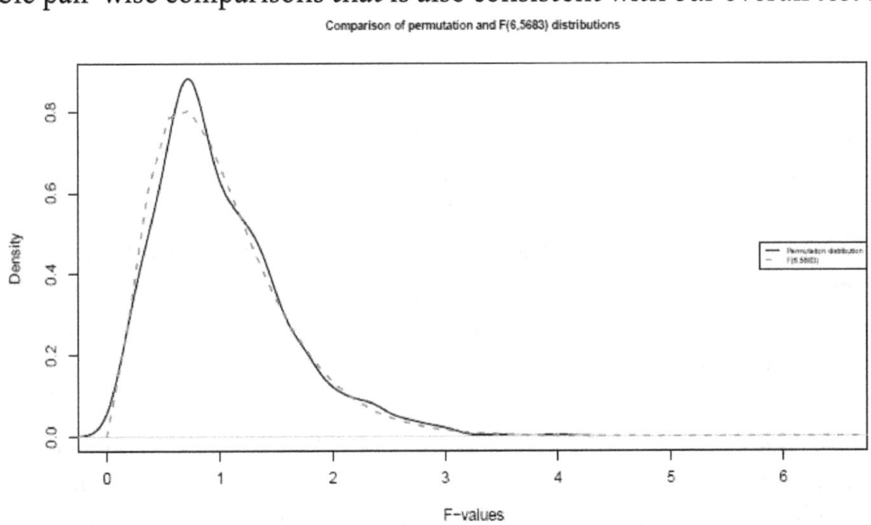

Figure 3.8: Comparison of $F(6, 6583)$ (dashed line) and permutation distribution (solid line).

3.4 ANOVA MODEL DIAGNOSTICS INCLUDING QQ-PLOTS

The requirements for a One-Way ANOVA F -test are similar to those discussed in Chapter 2, except that there are now J groups instead of only 2. Specifically, the linear model assumes:

- Independent observations,
- Equal variances, and
- Normal distributions.

For assessing equal variances across the groups, it is best to use plots to assess this. We can use pirate-plots to compare the spreads of the groups, which were provided in Figure 3.1. The spreads (both in terms of extrema and rest of the distributions) should look relatively similar across the groups for you to suggest that there is no evidence of a problem with this assumption. You should start with noting how clear or big the violation of the conditions might be but remember that there will always be some differences in the variation among groups even if the true variability is exactly equal in the populations. In addition to our direct plotting, there are some diagnostic plots available from the lm function that can help us more clearly assess potential violations of the assumptions.

We can obtain a suite of four diagnostic plots by using the plot function on any linear model object that we have fit. To get all the plots together in four panels we need to add the par(mfrow = c(2,2)) commandto tell R to make a graph with 4 panels[9].

```
par(mfrow = c(2,2))
plot(lm2, pch = 16)
```

There are two plots in Figure 3.9 with useful information for assessing the equal variance assumption. The "Residuals vs Fitted" panel in the top left panel displays the residuals ($e_{ij} = y_{ij} - \hat{y}_{ij}$) on the y-axis and thefitted values ($\hat{y}_{ij}$) on the x-axis. This allows you to see if the variability of the observations differs across the groups as a function of the mean of the groups, because all the observations in the same group get the same fitted value – the mean of the group. In this plot, the points seem to have fairly similar spreads at the fitted values for the seven groups with fitted values at 114 up to 122 cm. The "Scale-Location" plotin the lower left panel has the same x-axis of fitted values but the y-axis contains the square-root of the absolute value of the standardized residuals. The standardization scales the residuals to have a variance of 1 so help you in other displays to get a sense of how many standard deviations you are away from the mean in the residual distribution. The absolute value transforms all the residuals into a magnitude scale (removing direction) and the square-root helps you see differences in variability more accurately. The visual assessment is similar in the two plots – you want to consider whether it appears that the groups have somewhat similar or noticeably different amounts of variability. If you see a clear funnel shape (narrow (less variability) on the left or right and wide (more variability) at the right or left) in the Residuals vs Fitted and/or an increase or decrease in the height of the upper edge of points in the Scale-Location plot that may indicate a violation of the constant variance assumption. Remember that some variation across the groups is expected, does not suggest a violation of a validity conditions, and means that you can proceed with trusting your inferences, but large differences in the spread are problematic for all the procedures that involve linear models. When discussing these results, you want to discuss how clearly the differences in variation are and whether that shows a *clear* violation of the condition of equal variance for all observations. Like in hypothesis testing, you can never prove that an assumption is true based on a plot "looking OK", but you can say that there is no clear evidence that the condition is violated!

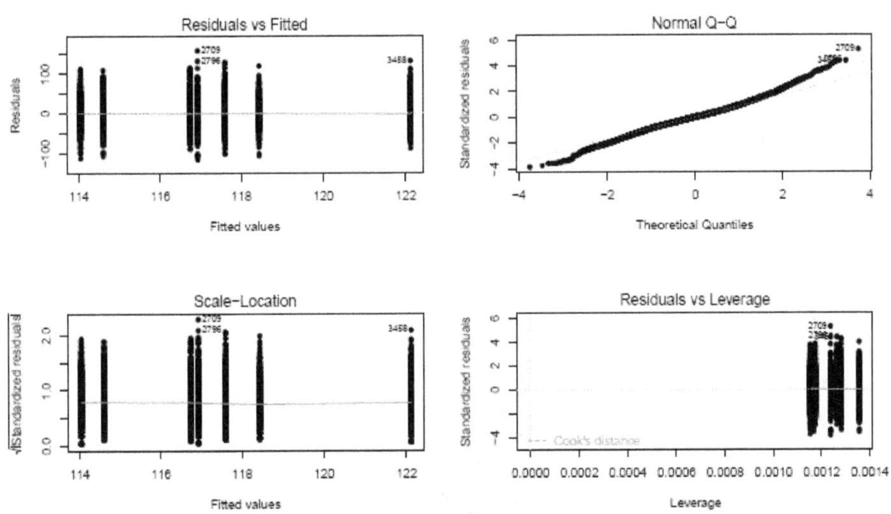

Figure 3.9: Default diagnostic plots for the full overtake data linear model.

The linear model also assumes that all the random errors (ε_{ij}) follow a normal distribution. To gain insight into the validity of this assumption, we can explore the original observations as displayed in the pirate-plots, mentally subtracting off the differences in the means and focusing on the shapes of the distributions of observations in each group. Each group should look approximately normal to avoid a concern on this assumption. These plots are especially good for assessing whether there is a skew or are outliers present in each group. If either skew or clear outliers are present, by definition, the normality assumption is violated. But our assumption is about the distribution of all the errors after removing the differences in the means and so we want an overall assessment technique to understand how reasonable our assumption might be overall for our model. The residuals from the entire model provide us with estimates of the random errors and if the normality assumption is met, then the residuals all-together should approximately follow a normal distribution. The Normal QQ-Plot in the upper right panel of Figure 3.9 also provides a direct visual assessment of how well our residuals match what we would expect from a normal distribution. Outliers, skew, heavy and light-tailed aspects of distributions (all violations of normality) show up in this plot once you learn to read it – which is our next task. To make it easier to read QQ-plots, it is nice to start with just considering histograms and/or density plots of the residuals and to see how that maps into this new display. We can obtain the residuals from the linear model using the residuals function on any linear model object. Figure 3.10 makes both a histogram and density curve of these residuals. It shows that they have a subtle right skew present (right half of the distribution is a little more spread out than the left, so the skew is to the right) once we accounted for the different means in the groups but there are no apparent outliers.

```
par(mfrow = c(1,2))
dd <- dd %>% mutate(eij = residuals(lm2)) #Adds residuals to dd

dd %>% ggplot(aes(x = eij)) +
  geom_histogram(aes(y = ..ncount..), bins = 25, col = 1, fill = "tomato") +
  geom_density(aes(y = ..scaled..)) +
  theme_bw() +
  labs(y = "Density",
       x = "Residuals",
       title = "Histogram of residuals")
```

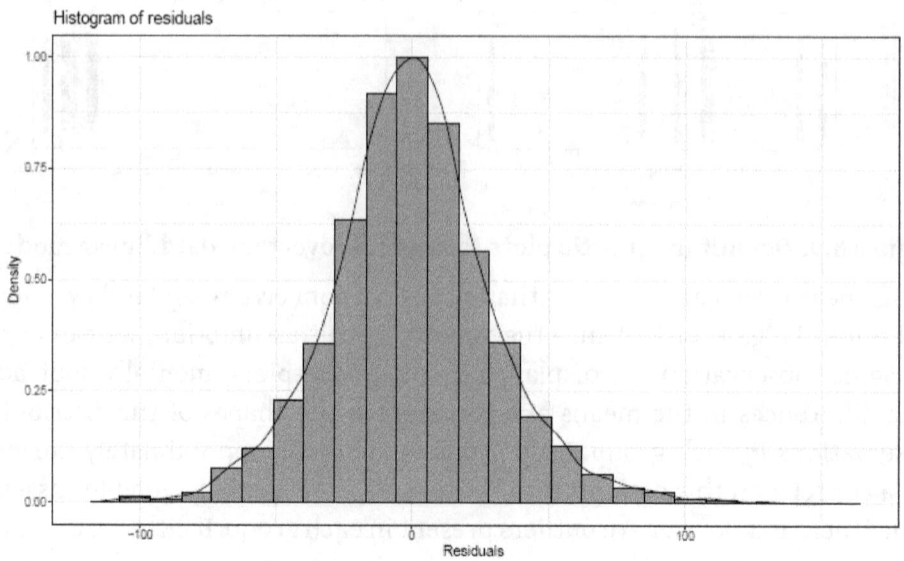

Figure 3.10: Histogram and density curve of the linear model raw residuals from the overtake data linear model.

A Quantile-Quantile plot (QQ-plot) shows the "match" of an observed distribution with a theoretical distribution, almost always the normal distribution. They are also known as Quantile Comparison, Normal Probability, or Normal Q-Q plots, with the last two names being specific to comparing results to a normal distribution. In this version10, the QQ-plots display the value of observed percentiles in the residual distribution on the y-axis versus the percentiles of a theoretical normal distribution on the x-axis. If the observed distribution of the residuals matches the shape of the normal distribution, then the plotted points should follow a 1-1 relationship. The 1-1 line is based on the Q1 (25th) and Q3 (75th) percentiles in the distributions to avoid impacts of the tails on the line you are using to compare the two distributions, with points added to the plot using geom_qq and the reference (1-1) line added with stat_qq_line. If the points follow the displayed straight line then that suggests that the residuals have a similar shape to a normal distribution. Some variation is expected around the line and some patterns of deviation are worse than others for our models, so you need to go beyond saying "it does not match a normal distribution". Be specific about the type of deviation you are detecting (right or left skew, heavy tails, multi-modal, etc.) and how clear or obvious that deviation is. And to do that, we need to practice interpreting some QQ-plots.

```
qq1 <- dd %>% ggplot(aes(sample = eij)) +
  geom_qq() +
  stat_qq_line() +
  theme_bw() +
  labs(title = "QQ-Plot of residuals")
den1 <- dd %>% ggplot(mapping = aes(x = eij)) +
  geom_density(color = "darkcyan") +
  labs(title = "Density plot of residuals",
       y = "Density",
       x = "Residuals") +
  theme_bw()
grid.arrange(qq1, den1, ncol = 2)
```

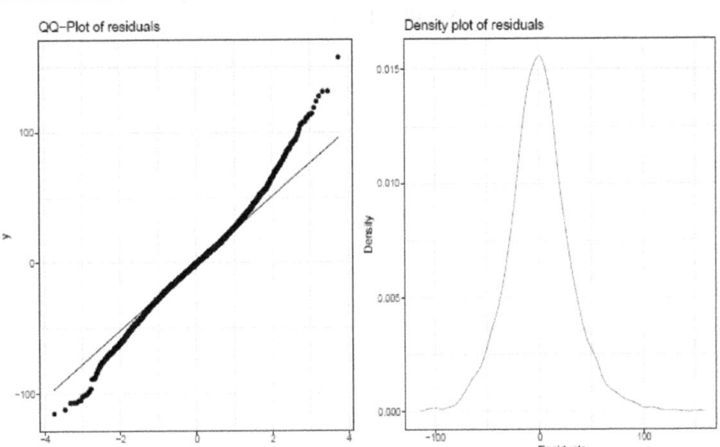

Figure 3.11: QQ-plot and density plot of residuals from the overtake data linear model.

The QQ-plot of the linear model residuals from Figure 3.9 is extracted and enhanced a little to make Figure 3.11 so we can just focus on it. We know from looking at the histogram that this is a (very) slightly right skewed distribution. Either version of the QQ-plots we will work with place the observed residuals on the y-axis and the expected results for a normal distribution on the x-axis. In some plots, the standardized11residuals are used (Figure 3.9) and in others the raw residuals are used (Figure 3.11) to compare the residual distribution to a normal (Gaussian) one. Both the upper and lower tails (upper tail in the upper right and the lower tail in the lower right of the plot) show some separation from the 1-1 line. The separation in the upper tail is more clear and these positive residuals are higher than the line "predicts" if the distribution had been normal. Being higher than the line in the right tail means being bigger than expected and so more spread out in that direction than a normal distribution should be. The left tail for the negative residuals also shows some separation from the line to have more extreme (here more negative) than expected, suggesting a little extra spread in the lower tail than suggested by a normal distribution. If the two sides had been similarly far from the 1-1 line, then we would have a symmetric and heavy-tailed distribution. Here, the slight 3.9 (the 2709th observation in the data set) as a potentially interesting point that falls in the far right-tail of positive residuals with a raw residual of almost 160 cm. This is a great opportunity to review what residuals are and how they are calculated for this observation. First, we can extract the row for this observation and find that it was a novice vest observation with a distance of 274 cm (that is almost 9 feet). The fitted value for this observation can be obtained using the fitted function on the estimated lm – which here is just the sample mean of the group of the observations (novice) of 116.94 cm. The residual is stored in the 2,709th value of eij or can be calculated by taking 274 minus the fitted value of 116.94. Given the large magnitude of this passing distance (it was the maximum distance observed in the Distance variable), it is not too surprising that it ends up as the largest positive residual.

```
9, c(1:2)]
```

```
## # A tibble: 1 x 2
##   Condition Distance
##   <fct>        <dbl>
## 1 novice         274
```

```
fitted(lm2)[2709]
```

```
##     2709
## 116.9405
```

```
dd$eij[2709]
```

```
##      2709
## 157.0595
```

```
274 - 116.9405
```

```
## [1] 157.0595
```

Generally, when both tails deviate on the same side of the line (forming a sort of quadratic curve, especially in more extreme cases), that indicates a skewed residual distribution (the one above has a very minor skew so this does not occur) and presence of a skew is evidence of a violation of the normality assumption. To see some different potential shapes in QQ-plots, six different data sets are displayed in Figures 3.12 and 3.13. In each row, a QQ-plot and associated density curve are displayed. If the points form a pattern where all are above the 1-1 line in the lower and upper tails as in Figure 3.12(a), then the pattern is a right skew, more extreme and easy to see than in the previous real data set. If the points form a pattern where they are below the 1-1 line in both tails as in Figure 3.12(c), then the pattern is identified as a left skew. Skewed residual distributions (either direction) are problematic for models that assume normally distributed responses but not necessarily for our permutation approaches if all the groups have similar skewed shapes. The other problematic pattern is to have more spread than a normal curve as in Figure 3.12(e) and (f). This shows up with the points being below the line in the left tail (more extreme negative than expected by the normal) and the points being above the line for the right tail (more extreme positive than the normal predicts). We call these distributions heavy-tailed which can manifest as distributions with outliers in both tails or just a bit more spread out than a normal distribution.

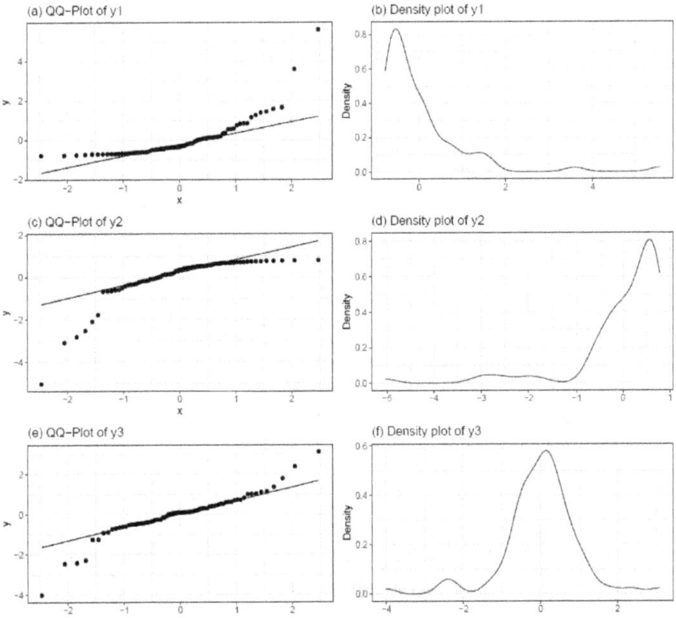

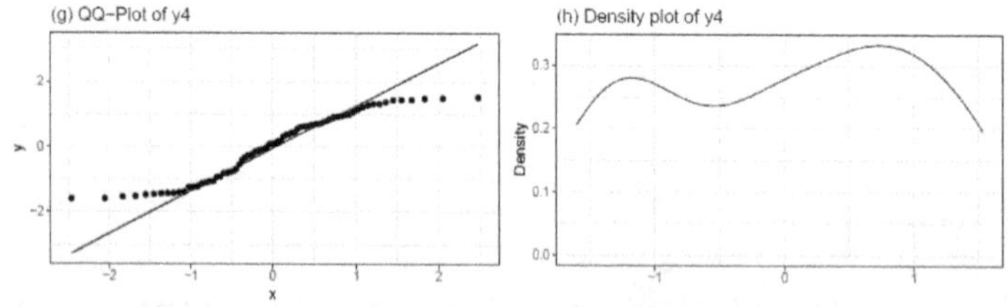

Figure 3.12: QQ-plots and density curves of four simulated distributions with different shapes.

Heavy-tailed residual distributions can be problematic for our models as the variation is greater than what the normal distribution can account for and our methods might under-estimate the variability in the results. The opposite pattern with the left tail above the line and the right tail below the line suggests less spread (light-tailed) than a normal as in Figure 3.12(g) and (h). This pattern is relatively harmless and you can proceed with methods that assume normality safely as they will just be a little conservative. For any of the patterns, you would note a potential violation of the normality assumption and then proceed to describe the type of violation and how clear or extreme it seems to be.

Finally, to help you calibrate expectations for data that are actually normally distributed, two data sets simulated from normal distributions are displayed in Figure 3.13. Note how neither follows the line exactly but that the overall pattern matches fairly well. You have to allow for some variation from the line in real data sets and focus on when there are really noticeable issues in the distribution of the residuals such as those displayed above. Again, you will never be able to prove that you have normally distributed residuals even if the residuals are all exactly on the line, but if you see QQ-plots as in Figure 3.12 you can determine that there is clear evidence of violations of the normality assumption.

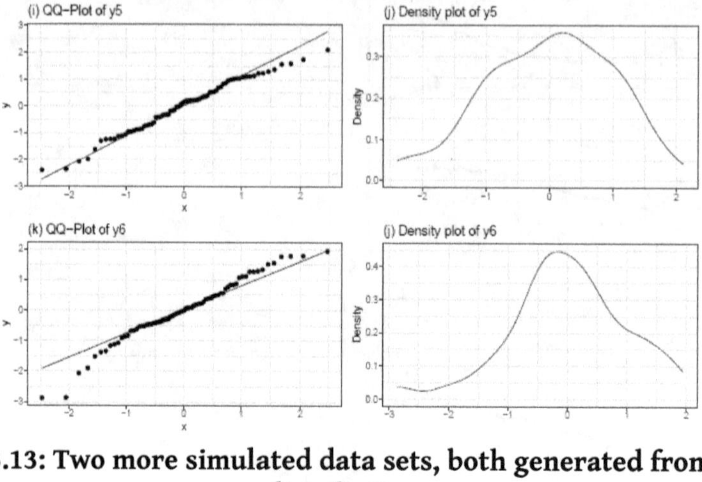

Figure 3.13: Two more simulated data sets, both generated from normal distributions.

The last issues with assessing the assumptions in an ANOVA relates to situations where the methods are more or less resistant[12] to violations of assumptions. In simulation studies of the performance of the F -test, researchers have found that the parametric ANOVA F -test is more resistant to violations of the assumptions of the normality and equal variance assumptions if the design is balanced. A balanced design occurs when each group is measured the same number of times. The resistance decreases as the data set becomes less balanced, as the sample sizes in the groups are more different, so having close to balance is preferred to a more imbalanced situation if there is a choice available. There is some intuition available here – it makes some sense that you would have better results in comparing groups if the information available is similar in all the groups and none are relatively under-represented. We can check the number of observations in each group to see if they are equal or similar using the tally function from the mosaic package. This function is useful for being able to get counts of observations, especially for cross-classifying observations on two variables that is used in Chapter 5. For just a single variable, we use tally(~ x, data = ...):

```
brary(mosaic)
lly(~ Condition, data = dd)
```

```
## Condition
##  casual commute   hiviz  novice  police  polite   racer
##     779     857     737     807     790     868     852
```

So the sample sizes do vary among the groups and the design is not balanced, but all the sample sizes are between 737 and 868 so it is (in percentage terms at least) not too far from balanced. It is better than having, say, 50 in one group and 1,200 in another. This tells us that the *F* -test should have some resistance to violations of assumptions. We also get more resistance to violation of assumptions as our sample sizes increase. With such a large data set here and only minor concerns with the normality assumption, the inferences generated for the means should be trustworthy and we will get similar results from parametric and nonparametric procedures. If we had only 15 observations per group and a slightly skewed residual distribution, then we might want to appeal to the permutation approach to have more trustworthy results, even if the design were balanced.

3.5 GUINEA PIG TOOTH GROWTH ONE-WAY ANOVA EXAMPLE

A second example of the One-way ANOVA methods involves a study of length of odonto blasts (cells that are responsible for tooth growth) in 60 Guinea Pigs (measured in microns) from Crampton [1947] and is available in base R using data(ToothGrowth). N = 60 Guinea Pigs were obtained from a local breeder and each received one of three dosages (0.5, 1, or 2 mg/day) of Vitamin C via one of two delivery methods, Orange Juice (*OJ*) or ascorbic acid (the stuff in vitamin C capsules, called VC below) as the source of Vitamin C in their diets. Each guinea pig was randomly assigned to receive one of the six different treatment combinations possible (OJ at 0.5 mg, OJ at 1 mg, OJ at 2 mg, VC at 0.5 mg, VC at 1 mg, and VC at 2 mg). The animals were treated similarly otherwise and we can assume lived in separate cages and only one observation was taken for each guinea pig, so we can assume the observations are independent[13]. We need to create a variable that combines the levels of delivery type (OJ, VC) and the dosages (0.5, 1, and 2) to use our One-Way ANOVA on the six levels. The interaction function can be used create a new variable that is based on combinations of the levels of other variables. Here a new variable is created in the ToothGrowth tibble that we called Treat using the interaction function that provides a six-level grouping variable for our One-Way ANOVA to compare the combinations of treatments. To get a sense of the pattern of observations in the data set, the counts in supp (supplement type) and dose are provided and then the counts in the new categorical explanatory variable, Treat.

```
data(ToothGrowth) #Available in Base R
library(tibble)
ToothGrowth <- as_tibble(ToothGrowth) #Convert data.frame to tibble
library(mosaic)
```

```
tally(~ supp, data = ToothGrowth) #Supplement Type (VC or OJ)
```

```
## supp
## OJ VC
## 30 30
```

```
tally(~ dose, data = ToothGrowth) #Dosage level
```

```
# Creates a new variable Treat with 6 levels using mutate and interaction:
ToothGrowth <- ToothGrowth %>% mutate(Treat = interaction(supp, dose))

# New variable that combines supplement type and dosage
tally(~ Treat, data = ToothGrowth)
```

```
## Treat
## OJ.0.5 VC.0.5   OJ.1   VC.1   OJ.2   VC.2
##     10     10     10     10     10     10
```

The tally function helps us to check for balance; this is a balanced design because the same number of guinea pigs (nj = 10 for j = 1, 2, . . . , 6) were measured in each treatment combination.

With the variable Treat prepared, the first task is to visualize the results using pirate-plots[14] (Figure 3.14) and generate some summary statistics for each group using favstats.

```
favstats(len ~ Treat, data = ToothGrowth)
```

```
##     Treat  min     Q1 median     Q3  max  mean       sd  n missing
## 1 OJ.0.5  8.2  9.700  12.25 16.175 21.5 13.23 4.459709 10       0
## 2 VC.0.5  4.2  5.950   7.15 10.900 11.5  7.98 2.746634 10       0
## 3   OJ.1 14.5 20.300  23.45 25.650 27.3 22.70 3.910953 10       0
## 4   VC.1 13.6 15.275  16.50 17.300 22.5 16.77 2.515309 10       0
## 5   OJ.2 22.4 24.575  25.95 27.075 30.9 26.06 2.655058 10       0
## 6   VC.2 18.5 23.375  25.95 28.800 33.9 26.14 4.797731 10       0
```

```
pirateplot(len ~ Treat, data = ToothGrowth, inf.method = "ci", inf.disp = "line",
           ylab = "Odontoblast Growth in microns", point.o = .7)
```

Figure 3.14 suggests that the mean tooth growth increases with the dosage level and that *OJ* might lead to higher growth rates than *VC* except at a dosage of 2 mg/day. The variability around the means looks to be small relative to the differences among the means, so we should expect a small p-value from our *F*-test. The design is balanced as noted above (n_j = 10 for all six groups) so the methods are somewhat resistant to impacts from potential non-normality and non-constant variance but we should still assess the patterns in the plots, especially with smaller sample sizes in each group. There is some suggestion of non-constant variance in the plots but this will be explored further below when we can remove the difference in the means and combine all the residuals together. There might be some skew in the responses in some of the groups (for example in *OJ.0.5* a right skew may be present and in *OJ.1* a left skew) but there are only 10 observations per group so visual evidence of skew in the pirate-plots could be generated by impacts of very few of the observations. This actually highlights an issue with residual explorations: when the sample sizes are small, our assumptions matter more than when the sample sizes are large, but when the sample sizes are small, we don't have much information to assess the assumptions and come to a clear conclusion.

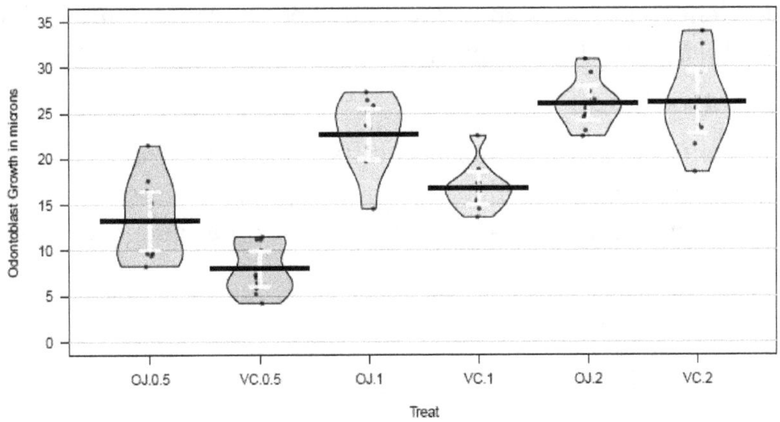

Figure 3.14: Pirate-plot of odontoblast growth responses for the six treatment level combinations.

Now we can apply our 6+ steps for performing a hypothesis test with these observations.

0. The research question is about differences in odontoblast growth across these combinations of treatments and they seem to have collected data that allow this to explored. A pirate-plot would be a good start to displaying the results and understanding all the combinations of the predictor variable.

1. Hypotheses

$$H_0: \mu_{OJ0.5} = \mu_{VC0.5} = \mu_{OJ1} = \mu_{VC1} = \mu_{OJ2} = \mu_{VC2}$$

vs

$$H_A: \text{Not all } \mu j \text{ equal}$$

- The null hypothesis could also be written in reference-coding as below since OJ.0.5 is chosen as the baseline group (discussed below).
  - $H_0: \tau_{VC0.5} = \tau_{OJ1} = \tau_{VC1} = \tau_{OJ2} = \tau_{VC2} = 0$
- The alternative hypothesis can be left a bit less specific:
  - H_A : Not all τj equal 0 for $j = 2, \ldots, 6$

2. Plot the data and assess validity conditions:

- **Independence:**
  - This is where the separate cages note above is important. Suppose that there were cages that contained multiple animals and they competed for food or could share illness or levels of activity. The animals in one cage might be systematically different from the others and this "clustering" of observations would present a potential violation of the independence assumption.

  If the experiment had the animals in separate cages, there is no clear dependency in the design of the study and we can assume15 that there is no problem with this assumption.

- **Constant variance:**
  - There is some indication of a difference in the variability among the groups in the pirate-plots but the sample size was small in each group. We need to fit the linear model to get the other diagnostic plots to make an overall assessment.

```
m2 <- lm(len ~ Treat, data = ToothGrowth)
par(mfrow = c(2,2))
plot(m2, pch = 16)
```

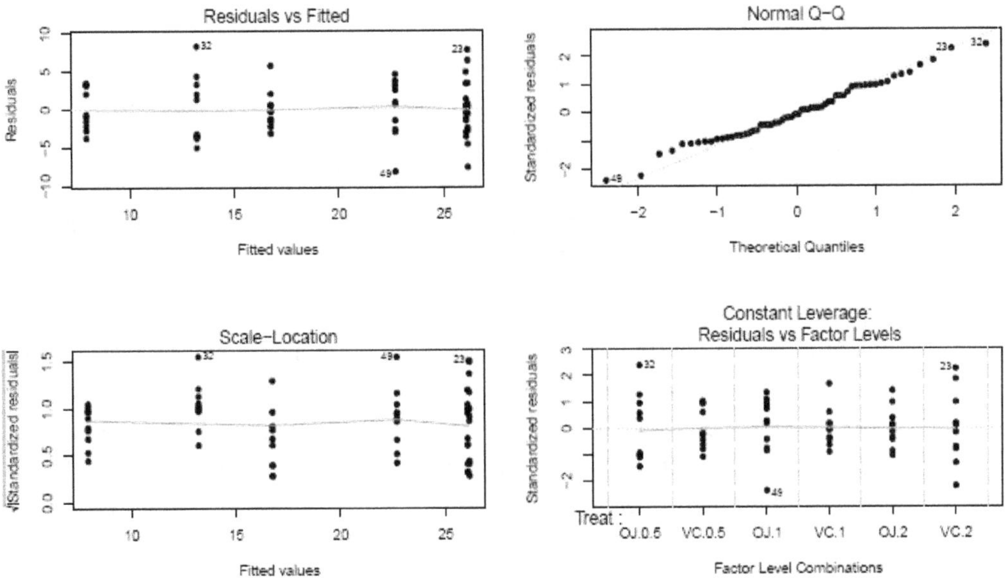

Figure 3.15: Diagnostic plots for the odontoblast growth model.

- The Residuals vs Fitted panel in Figure 3.15 shows some difference in the spreads but the spread is not that different among the groups.
- The Scale-Location plot also shows just a little less variability in the group with the smallest fitted value but the spread of the groups looks fairly similar in this alternative presentation related to assessing equal variance.
- Put together, the evidence for non-constant variance is not that strong and we can proceed comfortably that there is at least not a clear issue with this assumption. Because of the balanced design, we also get a little more resistance to violation of the equal variance assumption.

⦿ **Normality of residuals:**

- The Normal Q-Q plot shows a small deviation in the lower tail but nothing that we wouldn't expect from a normal distribution. So there is no evidence of a problem with the normality assumption based on the upper right panel of Figure 3.15. Because of the balanced design, we also get a little more resistance to violation of the normality assumption.

3. Calculate the test statistic and find the p-value:

⦿ The ANOVA table for our model follows, providing an F-statistic of 41.557:

```
## Analysis of Variance Table
##
```

```
## Response: len
##            Df  Sum Sq Mean Sq F value    Pr(>F)
## Treat       5 2740.10  548.02  41.557 < 2.2e-16
## Residuals 54  712.11   13.19
```

- There are two options here, especially since it seems that our assumptions about variance and normality are not violated (note that we do not say "met" – we just have no clear evidence against them). The parametric and nonparametric approaches should provide similar results here.
- The parametric approach is easiest – the p-value comes from the previous ANOVA table as < 2e-16. First, note that this is in scientific notation that is a compact way of saying that the p-value here is 2.2 ∗ 10-16 or 0.00000000000000022. When you see 2.2e-16 in R output, it also means that the calculation is at the numerical precision limits of the computer. What R is really trying to report is that this is a very small number. When you encounter p-values that are smaller than 0.0001, you should just report that the p-value < 0.0001. Do not report that it is 0 as this gives the false impression that there is no chance of the result occurring when it is just a really small probability. This p-value came from an F (5, 54) distribution (this is the distribution of the test statistic if the null hypothesis is true) with an F -statistic of 41.56.
- The nonparametric approach is not too hard so we can compare the two approaches here as well. The permutation p-value is reported as 0. This should be reported as p-value < 0.001 since we did 1,000 permutations and found that none of the permuted F -statistics, F∗, were larger than the observed F -statistic of 41.56. The permuted results do not exceed 6 as seen in Figure 3.16, so the observed result is really unusual relative to the null hypothesis. As suggested previously, the parametric and nonparametric approaches should be similar here and they were.

```
Tobs <- anova(lm(len ~ Treat, data = ToothGrowth))[1,4]; Tobs
```

```
## [1] 41.55718
```

```
par(mfrow = c(1,2))
B <- 1000
Tstar <- matrix(NA, nrow = B)
for (b in (1:B)){
  Tstar[b] <- anova(lm(len ~ shuffle(Treat), data = ToothGrowth))[1,4]
}
pdata(Tstar, Tobs, lower.tail = F)[[1]]
```

```
## [1] 0
```

```
tibble(Tstar) %>% ggplot(aes(x = Tstar)) +
  geom_histogram(aes(y = ..ncount..), bins = 25, col = 1, fill = "skyblue") +
  geom_density(aes(y = ..scaled..)) +
  theme_bw() +
  labs(y = "Density") +
  geom_vline(xintercept = Tobs, col = "red", lwd = 2) +
  stat_bin(aes(y = ..ncount.., label = ..count..), bins = 25,
           geom = "text", vjust = -0.75)
```

Conclusion:

There is strong evidence (F = 41.56, permutation p-value < 0.001) against the null hypothesis that the different treatments (combinations of OJ/VC and dosage levels) have the same true mean odontoblast growth for these guinea pigs, so we would conclude that the treatments cause at least one of the combinations to have a different true mean.

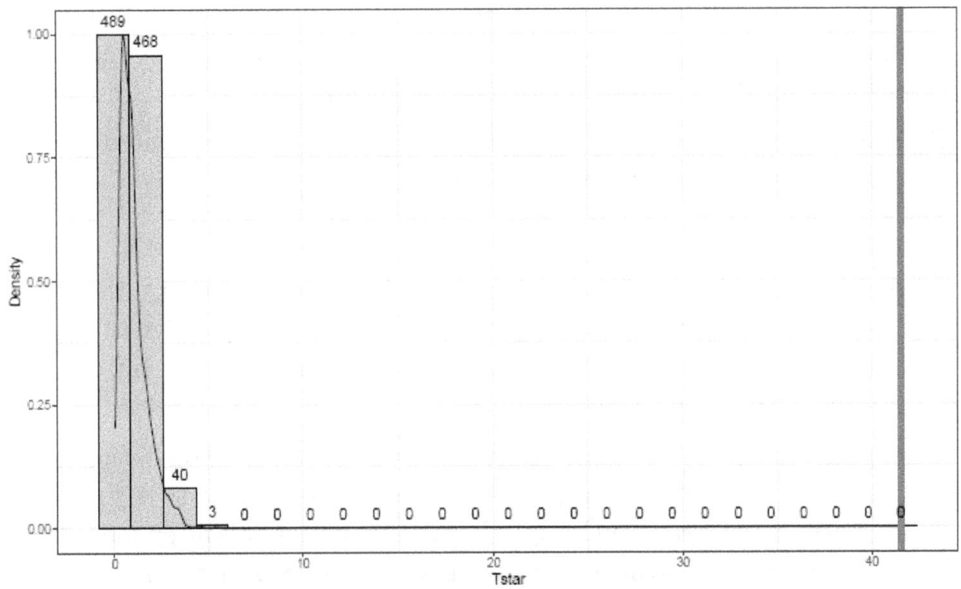

Figure 3.16: Histogram and density curve of permutation distribution for F -statistic for odontoblast growth data. Observed test statistic in bold, vertical line at 41.56.

- We can make the causal statement of the treatment causing differences because the treatments were randomly assigned but these inferences only apply to these guinea pigs since they were not randomly selected from a larger population.
- Remember that we are making inferences to the population or true means and not the sample means and want to make that clear in any conclusion. When there is not a random sample from a population it is more natural to discuss the true means since we can't extend to the population values.

- The alternative is that there is some difference in the true means – be sure to make the wording clear that you aren't saying that all the means differ. In fact, if you look back at Figure 3.14, the means for the 2 mg dosages look almost the same so we will have a tough time arguing that all groups differ. The F-test is about finding evidence of some difference *somewhere* among the true means. The next section will provide some additional tools to get more specific about the source of those detected differences and allow us to get at estimates of the differences we observed to complete our interpretation.

Discuss size of differences:

- It appears that increasing dose levels are related to increased odontoblast growth and that the differences in dose effects change based on the type of delivery method. The difference between 7 and 26 microns for the average length of the cells could be quite interesting to the researchers. This result is harder for me to judge and likely for you than the average distances of cars to bikes but the differences could be very interesting to these researchers.
- The "size" discussion can be further augmented by estimated pair-wise differences using methods discussed below.

Scope of inference

- We can make a causal statement of the treatment causing differences in the responses because the treatments were randomly assigned but these inferences only apply to these guinea pigs since they were not randomly selected from a larger population.
    - Remember that we are making inferences to the population or true means and not the sample means and want to make that clear. When there is not a random sample from a population it is often more natural to discuss the true means since we can't extend the results to the population values.

Before we leave this example, we should revisit our model estimates and interpretations. The default model parameterization uses reference-coding. Running the model summary function on m2 provides the estimated coefficients:

```
##              Estimate Std. Error   t value     Pr(>|t|)
## (Intercept)     13.23   1.148353 11.520847 3.602548e-16
## TreatVC.0.5     -5.25   1.624017 -3.232726 2.092470e-03
## TreatOJ.1        9.47   1.624017  5.831222 3.175641e-07
## TreatVC.1        3.54   1.624017  2.179781 3.365317e-02
## TreatOJ.2       12.83   1.624017  7.900166 1.429712e-10
## TreatVC.2       12.91   1.624017  7.949427 1.190410e-10
```

For some practice with the reference-coding used in these models, let's find the estimates (fitted values) for observations for a couple of the groups. To work with the parameters, you need to start with determining the baseline category that was used by considering which level is not displayed in the output. The levels function can list the groups in a categorical variable and their coding in the data set. The first level is usually the baseline category but you should check this in the model summary as well.

```
levels(ToothGrowth$Treat)
```

```
## [1] "OJ.0.5" "VC.0.5" "OJ.1"   "VC.1"   "OJ.2"   "VC.2"
```

There is a VC.0.5 in the second row of the model summary, but there is no row for OJ.0.5 and so this must be the baseline category. That means that the fitted value or model estimate for the *OJ* at 0.5 mg/dayαgroup is the same as the (Intercept) row or$\hat{\alpha}$, estimating a mean tooth growth of 13.23 microns when the pigs get OJ at a 0.5 mg/day dosage level. You should always start with working on the baseline level in a reference-coded model. To get estimates for any other group, then you can use the (Intercept) estimate and add the deviation (which could be negative) for the group of interest. For VC.0.5, the estimated mean tooth growth is $\hat{\alpha} + \hat{\tau}_2 = \hat{\alpha} + \hat{\tau}_{VC0.5} = 13.23 + (-5.25) = 7.98$ microns. It is also potentially interesting to directly interpret the estimated difference (or deviation) between OJ.0.5 (the baseline) and VC.0.5 (group 2)τVC0.5that is $\hat{\tau}_{VC0.5} = -5.25$: we estimate that the mean tooth growth in VC.0.5 is 5.25 microns shorter than it is in OJ.0.5. This and many other direct comparisons of groups are likely of interest to researchers involved in studying the impacts of these supplements on tooth growth and the next section will show us how to do that (correctly!).

The reference-coding is still going to feel a little uncomfortable so the comparison to the cell means model and exploring the effect plot can help to reinforce that both models patch together the same estimated means for each group. For example, we can find our estimate of 7.98 microns for the VC0.5 group in the output and Figure 3.17. Also note that Figure 3.17 is the same whether you plot the results from m2 or m3.

```
m3 <- lm(len ~ Treat - 1, data = ToothGrowth)
summary(m3)
```

```
##
## Call:
## lm(formula = len ~ Treat - 1, data = ToothGrowth)
##
## Residuals:
##    Min     1Q Median     3Q    Max
##  -8.20  -2.72  -0.27   2.65   8.27
##
## Coefficients:
##              Estimate Std. Error t value Pr(>|t|)
## TreatOJ.0.5    13.230      1.148  11.521 3.60e-16
## TreatVC.0.5     7.980      1.148   6.949 4.98e-09
## TreatOJ.1      22.700      1.148  19.767  < 2e-16
## TreatVC.1      16.770      1.148  14.604  < 2e-16
## TreatOJ.2      26.060      1.148  22.693  < 2e-16
## TreatVC.2      26.140      1.148  22.763  < 2e-16
##
## Residual standard error: 3.631 on 54 degrees of freedom
## Multiple R-squared:  0.9712, Adjusted R-squared:  0.968
## F-statistic:   303 on 6 and 54 DF,  p-value: < 2.2e-16
```

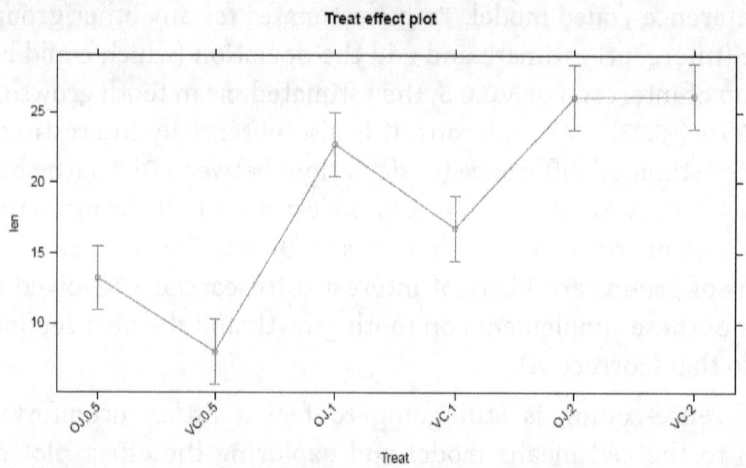

Figure 3.17: Effect plot of the One-Way ANOVA model for the odontoblast growth data, with rotated x-axis text for enhanced readability of the levels of the predictor variable.

```
plot(allEffects(m2), rotx = 45)
```

3.6 MULTIPLE (PAIR-WISE) COMPARISONS USING TUKEY'S HSD AND THE COMPACT LETTER DISPLAY

With evidence against all the true means being equal and concluding that not all are equal, many researchers want to explore which groups show evidence of differing from one another. This provides information on the source of the overall difference that was detected and detailed information on which groups differed from one another. Because this is a shot-gun/unfocused sort of approach, some people think it is an over-used procedure. Others feel that it is an important method of addressing detailed questions about group comparisons in a valid and safe way. For example, we might want to know if OJ is different from VC at the 0.5 mg/day dosage level and these methods will allow us to get an answer to this sort of question. It also will test for differences between the OJ.0.5 and VC.2 groups and every other pair of levels that you can construct (15 total!). This method actually takes us back to the methods in Chapter 2 where we compared the means of two groups except that we need to deal with potentially many pair-wise comparisons, making an adjustment to account for that inflation in Type I errors that occurs due to many tests being performed at the same time. A commonly used method to make all the pair-wise comparisons that includes a correction for doing this is called Tukey's Honest Significant Difference (Tukey's HSD) method16. The name suggests that not using it could lead to a dishonest answer and that it will give you an honest result. It is more that if you don't do some sort of correction for all the tests you are performing, you might find some spurious17 results. There are other methods that could be used to do a similar correction and also provide "honest" inferences; we are just going to learn one of them. Tukey's method employs a different correction from the Bonferroni method discussed in Chapter 2 but also controls the family-wise error rate across all the pairs being compared.

In pair-wise comparisons between all the pairs of means in a One-Way ANOVA, the number of tests is based on the number of pairs. We can calculate the number of tests using J choose 2,to get thenumber of unique pairs of size 2 that we can make out of J individual treatment levels. We don't need to explore the combinatorics formula for this, as the choose function in R can give us the answers:

So if you have three groups (the smallest number where we have to worry about more than one pair), there are three unique pairs to compare. For six groups, like in the Guinea Pig study, we have to consider 15 tests to compare all the unique pairs of groups and with seven groups, there are 21 tests. Once there are more than two groups to compare, it seems like we should be worried about inflated family-wise error rates. Fortunately, the Tukey's HSD method controls the family-wise error rate at your specified level (say 0.05) across any number of pair-wise comparisons. This means that the overall rate of at least one Type I error across all the tests is controlled at the specified significance level, often 5%. To do this, each test must use a slightly more conservative cut-off than if just one test is performed and the procedure helps us figure out how much more conservative we need to be.

```
choose(3, 2)
```

```
## [1] 3
```

```
choose(4, 2)
```

```
## [1] 6
```

```
choose(5, 2)
```

```
## [1] 10
```

```
choose(6, 2)
```

```
## [1] 15
```

```
choose(7, 2)
```

```
## [1] 21
```

Tukey's HSD starts with focusing on the difference between the groups with the largest and smallest means $(\bar{y}_{max} - \bar{y}_{min})$. If $(\bar{y}_{max} - \bar{y}_{min}) \leq$ Margin of Error for the difference in the means, then all other pair wise differences, say $|\bar{y}_j - \bar{y}_{j'}|$, for two groups *j* and *j*1, will be less than or equal to that margin of error. This also means that any confidence intervals for any difference in the means will contain 0. Tukey's HSD selects a critical value so that $(\bar{y}_{max} - \bar{y}_{min})$ will be less than the margin of error in 95% of data sets drawn from populations with a common mean. This implies that in 95% of data sets in which all the population means are the same, all confidence intervals for differences in pairs of means will contain 0. Tukey's HSD provides confidence intervals for the difference in true means between groups *j* and *j*1, *μj* - *μj1*, for all pairs where *j*≠*j*1, using

$$(\bar{y}_j - \bar{y}_{j'}) \mp \frac{q^*}{\sqrt{2}}\sqrt{\text{MS}_E\left(\frac{1}{n_j} + \frac{1}{n_{j'}}\right)}$$

where $\frac{q^*}{\sqrt{2}}\sqrt{\text{MS}_E\left(\frac{1}{n_j} + \frac{1}{n_{j'}}\right)}$ is the margin of error for the intervals. The distribution used to find the multiplier, q*, for the confidence intervals is available in the qtukey function and generally provides a slightly larger multiplier than the regular t* from our two-sample t-based confidence interval discussed in Chapter 2. The formula otherwise is very similar to the one used in Chapter 2 with the SE for the difference in the means based on a measure of residual variance (here MSE) times $\left(\frac{1}{n_j} + \frac{1}{n_{j'}}\right)$ which weights the results based on the relative sample sizes in the groups.

We will use the confint, cld, and plot functions applied to output from the glht function (all from the multcomp package; Hothorn et al. [2008], [Hothorn et al., 2022]) to get the required comparisons fromour ANOVA model. Unfortunately, its code format is a little complicated – but there are just two places to modify the code: include the model name and after mcp (stands for multiple comparison procedure) in the linfct option, you need to include the explanatory variable name as VARIABLENAME = "Tukey". The lastpart is to get the Tukey HSD multiple comparisons run on our explanatory variable18. Once we obtain the intervals using the confint function or using plot applied to the stored results, we can use them to test H0 : μj = μj1 vs HA : μj ≠ μj1 by assessing whether 0 is in the confidence interval for each pair. If 0 is in the interval, then there is weak evidence against the null hypothesis for that pair, so we do not detect a difference in that pair and do not conclude that there is a difference. If 0 is not in the interval, then we have strong evidence against H0 for that pair, detect a difference, and conclude that there is a difference in that pair at the specified family-wise significance level. You will see a switch to using the word "detection" to describe null hypotheses that we find strong evidence against as it can help to compactly write up these complicated results. The following code provides the numerical and graphical19 results of applying Tukey's HSD to the linear model for the Guinea Pig data:

```
library(multcomp)
Tm2 <- glht(m2, linfct = mcp(Treat = "Tukey"))

confint(Tm2)
```

```
##
##   Simultaneous Confidence Intervals
##
## Multiple Comparisons of Means: Tukey Contrasts
##
##
## Fit: lm(formula = len ~ Treat, data = ToothGrowth)
##
## Quantile = 2.955
## 95% family-wise confidence level
##
##
## Linear Hypotheses:
```

```
##                        Estimate lwr      upr
## VC.0.5 - OJ.0.5 == 0   -5.2500 -10.0490  -0.4510
## OJ.1 - OJ.0.5 == 0      9.4700   4.6710  14.2690
## VC.1 - OJ.0.5 == 0      3.5400  -1.2590   8.3390
## OJ.2 - OJ.0.5 == 0     12.8300   8.0310  17.6290
## VC.2 - OJ.0.5 == 0     12.9100   8.1110  17.7090
## OJ.1 - VC.0.5 == 0     14.7200   9.9210  19.5190
## VC.1 - VC.0.5 == 0      8.7900   3.9910  13.5890
## OJ.2 - VC.0.5 == 0     18.0800  13.2810  22.8790
## VC.2 - VC.0.5 == 0     18.1600  13.3610  22.9590
## VC.1 - OJ.1 == 0       -5.9300 -10.7290  -1.1310
## OJ.2 - OJ.1 == 0        3.3600  -1.4390   8.1590
## VC.2 - OJ.1 == 0        3.4400  -1.3590   8.2390
## OJ.2 - VC.1 == 0        9.2900   4.4910  14.0890
## VC.2 - VC.1 == 0        9.3700   4.5710  14.1690
## VC.2 - OJ.2 == 0        0.0800  -4.7190   4.8790
```

```
old.par <- par(mai = c(1,2,1,1)) #Makes room on the plot for the group names
plot(Tm2)
```

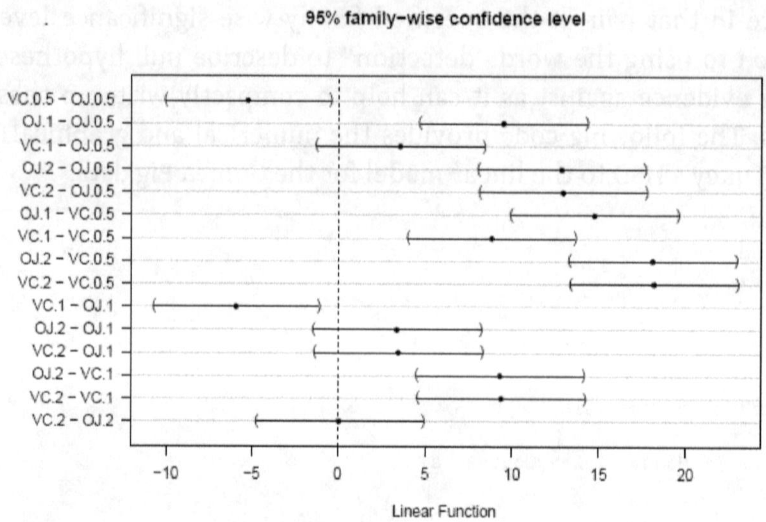

Figure 3.18: Graphical display of pair-wise comparisons from Tukey's HSD for the Guinea Pig data. Any confidence intervals that do not contain 0 provide strong evidence against the null hypothesis of no difference in the true means for that pair of groups.

Figure 3.18 contains confidence intervals for the difference in the means for all 15 pairs of groups. For example, the first row in the plot contains the confidence interval for comparing VC.0.5 and OJ.0.5 (VC.0.5 minus OJ.0.5). In the numerical output, you can find that this 95% family-wise confidence interval goes from-10.05 to -0.45 microns (lwr and upr in the numerical output provide the CI endpoints). This interval does not contain 0 since its upper end point is -0.45 microns and so we can now say that there is strong evidence against the null hypothesis of no difference in this pair and that we detect that OJ and VC have different true mean growth rates at the 0.5 mg dosage level. We can go further and say that we are 95% confident that the difference in the true mean tooth growth between VC.0.5 and OJ.0.5 (VC.0.5-OJ.0.5) is between-10.05 and -0.45 microns, after adjusting for comparing all the pairs of groups. The center of this CI is -5.25which is $\widehat{\tau}_2$and the estimate difference between VC.0.5 and the baseline category of OJ.0.5. That means we can get an un-adjusted 95% confidence interval from the confint function to compare to this adjusted CI.

The interval that does not account for all the comparisons goes from -8.51 to -1.99 microns (second row outconfint output), showing the increased width needed in Tukey's interval to control the family-wise errorrate when many pairs are being compared. With 14 other intervals, we obviously can't give them all this much attention...

```
confint(m2)
```

```
##                   2.5 %    97.5 %
## (Intercept) 10.9276907 15.532309
## TreatVC.0.5 -8.5059571 -1.994043
## TreatOJ.1    6.2140429 12.725957
## TreatVC.1    0.2840429  6.795957
## TreatOJ.2    9.5740429 16.085957
## TreatVC.2    9.6540429 16.165957
```

If you put all these pair-wise tests together, you can generate an overall interpretation of Tukey's HSD results that discusses sets of groups that are not detectably different from one another and those groups that were distinguished from other sets of groups. To do this, start with listing out the groups that are not detectably different (CIs contain 0), which, here, only occurs for four of the pairs. The CIs that contain 0 are for the pairs VC.1 and OJ.0.5, OJ.2 and OJ.1, VC.2 and OJ.1, and, finally, VC.2 and OJ.2. So VC.2, OJ.1, and OJ.2 are all not detectably different from each other and VC.1 and OJ.0.5 are also not detectably different. If you look carefully, VC.0.5 is detected as different from every other group. So there are basically three sets of groups that can be grouped together as "similar": VC.2, OJ.1, and OJ.2; VC.1 and OJ.0.5; and VC.0.5. Sometimes groups overlap with some levels not being detectably different from other levels that belong to different groups and the story is not as clear as it is in this case. An example of this sort of overlap is seen in the next section.

There is a method that many researchers use to more efficiently generate and report these sorts of results that is called a compact letter display (CLD, Piepho [2004])20. The cld function can be applied to the results from glht to generate the CLD that we can use to provide a "simple" summary of the sets of groups. In this discussion, we define a set as a union of different groups that can contain one or more members and the member of these groups are the different treatment levels.

```
cld(Tm2)
```

```
## OJ.0.5 VC.0.5    OJ.1    VC.1    OJ.2    VC.2
##    "b"     "a"     "c"     "b"     "c"     "c"
```

Groups with the same letter are not detectably different (are in the same set) and groups that are detectably different get different letters (are in different sets). Groups can have more than one letter to reflect "overlap" between the sets of groups and sometimes a set of groups contains only a single treatment level (VC.0.5 is a set of size 1). Note that if the groups have the same letter, this does not mean they are the same, just that there is insufficient evidence to declare a difference for that pair. If we consider the previous output for the CLD, the "a" set contains VC.0.5, the "b" set contains OJ.0.5 and VC.1, and the "c" set contains OJ.1, OJ.2, and VC.2. These are exactly the groups of treatment levels that we obtained by going through all fifteen pair wise results.

One benefit of this work is that the CLD letters can be added to a plot (such as the pirate-plot) to help fully report the results and understand the sorts of differences Tukey's HSD detected. The code with text involves placing text on the figure. In the text function, the x and y axis locations are specified (x-axis goes from 1 to 6 for the 6 categories) as well as the text to add (the CLD here). Some trial and error for locations may be needed to get the letters to be easily seen in a given pirate-plot. Figure 3.19 enhances the discussion by showing that the "**a**" group with VC.0.5 had the lowest average tooth growth, the "**b**" group had intermediate tooth growth for treatments OJ.0.5 and VC.1, and the highest growth rates came from OJ.1, OJ.2, and VC.2. Even though VC.2 had the highest average growth rate, we are not able to prove that its true mean is any higher than the other groups labelled with "**c**". Hopefully the ease of getting to the story of the Tukey's HSD results from a plot like this explains why it is common to report results using these methods instead of reporting 15 confidence intervals for all the pair-wise differences, either in a table or the plot.

```
# Options theme = 2,inf.f.o = 0,point.o = .5 added to focus on CLD
pirateplot(len ~ Treat, data = ToothGrowth, ylab = "Growth (microns)", inf.method = "ci",
           inf.disp = "line", theme = 2, inf.f.o = 0.3, point.o = .5)
# CLD added to second bean (x = 2) at height of y = 10
text(x = 2, y = 10,"a", col = "blue", cex = 1.5)
# Adds "b" to first and fourth bean
text(x = c(1,4), y = c(15,18), "b", col = "red", cex = 1.5)
text(x = c(3,5,6), y = c(25,28,28), "c", col = "green", cex = 1.5) #Add "c" to three beans
```

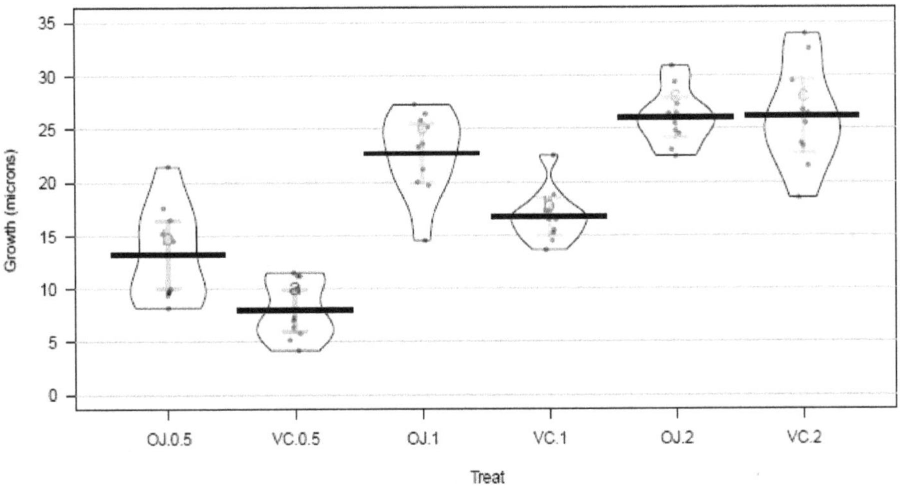

Figure 3.19: Pirate-plot of odontoblast growth by group with Tukey's HSD compact letter display. Note some extra pirate-plot options are used to enhance focus on the CLD results.

There are just a couple of other details to mention on this set of methods. First, note that we interpret the set of confidence intervals simultaneously: We are 95% confident that ALL the intervals contain the respective differences in the true means (this is a family-wise interpretation). These intervals are adjusted from our regular two-sample t intervals that came from lm from Chapter 2 to allow this stronger interpretation. Specifically, they are wider. Second, if sample sizes are unequal in the groups, Tukey's HSD is conservative and provides a family-wise error rate that is lower than the nominal (or specified) level. In other words, it fails less often than expected and the intervals provided are a little wider than needed, containing all the pair wise differences at higher than the nominal confidence level of (typically) 95%. Third, this is a parametric approach and violations of normality and constant variance will push the method in the other direction, potentially making the technique dangerously liberal. Nonparametric approaches to this problem are also possible, but will not be considered here.

Tukey's HSD results can also be displayed as p-values for each pair-wise test result. This is a little less common but can allow you to directly assess the strength of evidence for a particular pair instead of using the detected/not result that the family-wise CIs provide. But the family-wise CIs are useful for exploring the size of the differences in the pairs and we need to simplify things to detect/not in these situations because there are so many tests. But if you want to see the Tukey HSD p-values, you can use

```
summary(Tm2)
```

```
##
##   Simultaneous Tests for General Linear Hypotheses
##
## Multiple Comparisons of Means: Tukey Contrasts
##
##
## Fit: lm(formula = len ~ Treat, data = ToothGrowth)
##
## Linear Hypotheses:
##                      Estimate Std. Error t value Pr(>|t|)
## VC.0.5 - OJ.0.5 == 0   -5.250      1.624  -3.233  0.02424
## OJ.1 - OJ.0.5 == 0      9.470      1.624   5.831  < 0.001
## VC.1 - OJ.0.5 == 0      3.540      1.624   2.180  0.26411
## OJ.2 - OJ.0.5 == 0     12.830      1.624   7.900  < 0.001
## VC.2 - OJ.0.5 == 0     12.910      1.624   7.949  < 0.001
## OJ.1 - VC.0.5 == 0     14.720      1.624   9.064  < 0.001
## VC.1 - VC.0.5 == 0      8.790      1.624   5.413  < 0.001
## OJ.2 - VC.0.5 == 0     18.080      1.624  11.133  < 0.001
## VC.2 - VC.0.5 == 0     18.160      1.624  11.182  < 0.001
## VC.1 - OJ.1 == 0       -5.930      1.624  -3.651  0.00739
## OJ.2 - OJ.1 == 0        3.360      1.624   2.069  0.31868
## VC.2 - OJ.1 == 0        3.440      1.624   2.118  0.29372
## OJ.2 - VC.1 == 0        9.290      1.624   5.720  < 0.001
## VC.2 - VC.1 == 0        9.370      1.624   5.770  < 0.001
## VC.2 - OJ.2 == 0        0.080      1.624   0.049  1.00000
## (Adjusted p values reported -- single-step method)
```

These reinforce the strong evidence for many of the pairs and less strong evidence for four pairs that were not detected to be different. So these p-values provide another method to employ to report the Tukey's HSD results – you would only need to report and explore the confidence intervals or the p-values, not both.

Tukey's HSD does not require you to find a small p-value from your overall *F* -test to employ the methods but if you apply it to situations with p-values larger than your *a priori* significance level, you are unlikely to find any pairs that are detected as being different. Some statisticians suggest that you shouldn't employ follow-up tests such as Tukey's HSD when there is not much evidence against the overall null hypothesis. If you needed to use a permutation approach for your overall F-test, there are techniques for generating multiple-comparison adjusted permutation confidence intervals, but they are beyond the scope of this material. Using the tools here there are two options. First, you can subset the data set and do pair wise two-sample t-tests for all combinations of pairs of levels and apply a Bonferroni correction for the p-values that this would generate (this is more conservative than employing Tukey's adjustments). Another alternative to be able to employ Tukey's HSD as discussed here is to try to use a transformation on the response variable (things like logs or square-roots) so that the parametric approach is reasonable to use; transformations are discussed in Sections 7.5 and 7.6.

3.7 PAIR-WISE COMPARISONS FOR THE OVERTAKE DATA

In our previous work with the overtake data, the overall ANOVA test led to a conclusion that there is some difference in the true means across the seven groups with a p-value < 0.001 giving very strong evidence against the null hypothesis of them all being equal. The original authors followed up their overall F -test with comparing every pair of outfits using one of the other methods for multiple testing adjustments available in the p.adjust function and detected differences between the *police* outfit and all others except for *hiviz* and no other pairs had p-values less than 0.05 using their approach. We will employ the Tukey's HSD approach to address the same exploration and get basically the same results as they obtained, as well as estimated differences in the means in all the pairs of groups.

The code is similar21 to the previous example focusing on the Condition variable for the 21 pairs to compare. To make these results easier to read and generally to make all the results with seven groups easier to understand, we can sort the levels of the explanatory based on the values in the response, using something like the means or medians of the responses for the groups. This does not change the analyses (the F -statistic and all pair-wise comparisons are the same), it just sorts them to be easier to discuss. Note that it might change the baseline group so would impact the reference-coded model even though the fitted values are the same. Specifically, we can use the reorder function based on the mean using something like reorder(FACTORVARIABLE, RESPONSEVARIABLE, FUN = mean), with our pipe and mutate functions used to modify the Condition variable. I like to put this "reordered" factor into a new variable so I can always go back to the other version if I want it but you could also re-write the original version with this modification – this only impacts the underlying order of the factor levels, not the entries for the observations themselves. The code here creates Condition 2 and checks the levels for it and the original Condition variable, which shows the change in the order of the levels of the two factor variables:

```
dd <- dd %>% mutate(Condition2 = reorder(Condition, Distance, FUN = mean))

levels(dd$Condition)
```

```
## [1] "casual"  "commute" "hiviz"   "novice"  "police"  "polite"  "racer"
```

```
levels(dd$Condition2)
```

```
## [1] "polite"  "commute" "racer"   "novice"  "casual"  "hiviz"   "police"
```

And to verify that this worked, we can compare the means based on Condition and Condition2, and now it is even more clear which groups have the smallest and largest mean passing distances:

```
mean(Distance ~ Condition, data = dd)
```

```
##   casual  commute    hiviz   novice   police   polite    racer
## 117.6110 114.6079 118.4383 116.9405 122.1215 114.0518 116.7559
```

```
mean(Distance ~ Condition2, data = dd)
```

```
##   polite  commute    racer   novice   casual    hiviz   police
## 114.0518 114.6079 116.7559 116.9405 117.6110 118.4383 122.1215
```

In Figure 3.20, the 95% family-wise confidence intervals are displayed. There are only five pairs that have confidence intervals that do not contain 0 and all contain comparisons of the police group with others. So there is a detectable difference between police and polite, commute, racer, novice, and casual. The police versus casual comparison is hard to see whether 0 is in the interval or not in the plot, but the confidence interval goes from 0.06 to 8.97 cm (look at the results from confint), so suggests sufficient evidence to detecta difference in these groups (barely!) at the 5% family-wise significance level.

```
lm2 <- lm(Distance ~ Condition2, data = dd)
library(multcomp)
TmOV <- glht(lm2, linfct = mcp(Condition2 = "Tukey"))
```

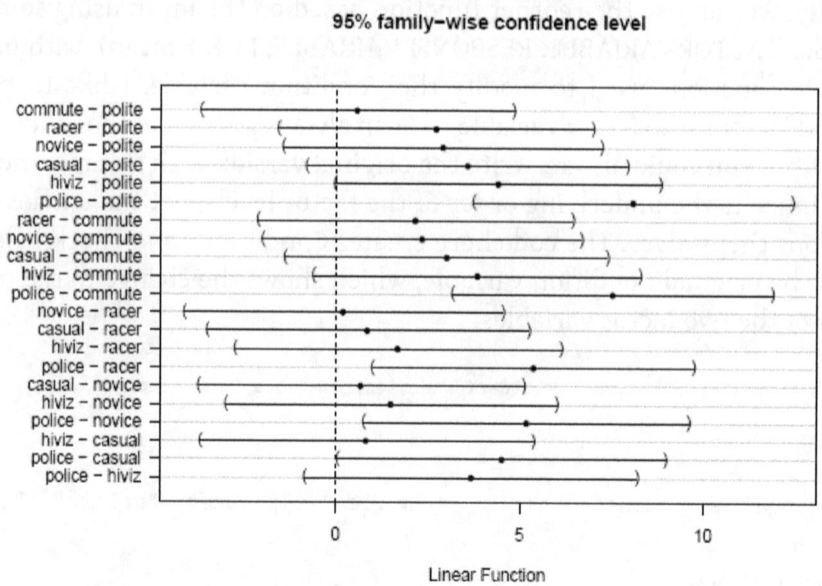

Figure 3.20: Tukey's HSD confidence interval results at the 95% family-wise confidence level for the overtake distances linear model using the new Condition2 explanatory variable.

```
##
##   Simultaneous Confidence Intervals
##
## Multiple Comparisons of Means: Tukey Contrasts
##
##
## Fit: lm(formula = Distance ~ Condition2, data = dd)
##
## Quantile = 2.9486
## 95% family-wise confidence level
##
##
## Linear Hypotheses:
##                       Estimate lwr      upr
## commute - polite == 0  0.55609 -3.69182  4.80400
## racer - polite == 0    2.70403 -1.55015  6.95820
## novice - polite == 0   2.88868 -1.42494  7.20230
## casual - polite == 0   3.55920 -0.79441  7.91281
## hiviz - polite == 0    4.38642 -0.03208  8.80492
## police - polite == 0   8.06968  3.73207 12.40728
## racer - commute == 0   2.14793 -2.11975  6.41562
## novice - commute == 0  2.33259 -1.99435  6.65952
## casual - commute == 0  3.00311 -1.36370  7.36991
## hiviz - commute == 0   3.83033 -0.60118  8.26183
## police - commute == 0  7.51358  3.16273 11.86443
## novice - racer == 0    0.18465 -4.14844  4.51774
## casual - racer == 0    0.85517 -3.51773  5.22807
## hiviz - racer == 0     1.68239 -2.75512  6.11991
## police - racer == 0    5.36565  1.00868  9.72262
## casual - novice == 0   0.67052 -3.76023  5.10127
## hiviz - novice == 0    1.49774 -2.99679  5.99227
## police - novice == 0   5.18100  0.76597  9.59603
## hiviz - casual == 0    0.82722 -3.70570  5.36015
## police - casual == 0   4.51048  0.05637  8.96458
## police - hiviz == 0    3.68326 -0.83430  8.20081
```

```
cld(TmOv, abseps = 0.1)
```

```
##  polite commute   racer  novice  casual   hiviz  police
##     "a"     "a"     "a"     "a"     "a"    "ab"     "b"
```

```
# Makes room on the plot for the group names, the second number of 2.5 is most
# often adjusted: larger values provide more room on the left of the plot.
# Order is Bottom, Left, Top, Right (clockwise starting from the bottom).
old.par <- par(mai = c(1,2.5,1,1))
plot(TmOv)
```

The CLD also reinforces the previous discussion of which levels were detected as different and elucidates the other aspects of the results. Specifically, police is in a group with hiviz only (group "b", not detectably different). But hiviz is also in a group with all the other levels so also is in group "a". Figure 3.21 adds the CLD to the pirate-plot with the sorted means to help visually present these results with the original data, reiterating the benefits of sorting factor levels to make these plots easier to read. To wrap up this example (finally), we can see that we found that there was clear evidence against the null hypothesis of no difference in the true means, so concluded that there was some difference. The follow-up explorations show that we can really only suggest that the police outfit has detectably different mean distances and that is only for five of the six other levels. So if you are bike commuter (in the UK near London?), you are left to consider the size of this difference. The biggest estimated mean difference was 8.07cm (3.2 inches) between police and polite. Do you think it is worth this potential extra average distance, especially given the wide variability in the distances, to make and then wear this vest? It is interesting that this result is found but it also is a fairly minimal size of a difference. It required an extremely large data set to detect these differences because the differences in the means are not very large relative to the variability in the responses. It seems like there might be many other reasons for why overtake distances vary that were not included our suite of predictors (they explored traffic volume in the paper as one other factor but we don't have that in our data set) or maybe it is just unexplainably variable. But it makes me wonder whether it matters what I wear when I bike and whether it has an impact that matters for average overtake distances – even in the face of these "statistically significant" results. But maybe there is an impact on the "close calls" as you can see some differences in the lower tails of the distributions across the groups. The authors looked at the rates of "closer" overtakes by classifying the distances as either less than 100cm (39.4 inches) as closer or not and also found some interesting results. Chapter 5 discusses a method called a Chi-square test of Homogeneity that would be appropriate here and allow for an analysis of the rates of closer passes and this study is revisited in the Practice Problems (Section 5.14) there. It ends up showing that rates of "closer passes" are smallest in the police group.

```
pirateplot(Distance ~ Condition2, data = dd, ylab = "Distance (cm)", inf.method = "ci",
           inf.disp = "line", theme = 2)
text(x = 1:5,y = 200,"a",col = "blue",cex = 1.5) #CLD added
text(x = 5.9,y = 210,"a",col = "blue",cex = 1.5)
text(x = 6.1,y = 210,"b",col = "red",cex = 1.5)
text(x = 7,y = 215,"b",col = "red",cex = 1.5)
```

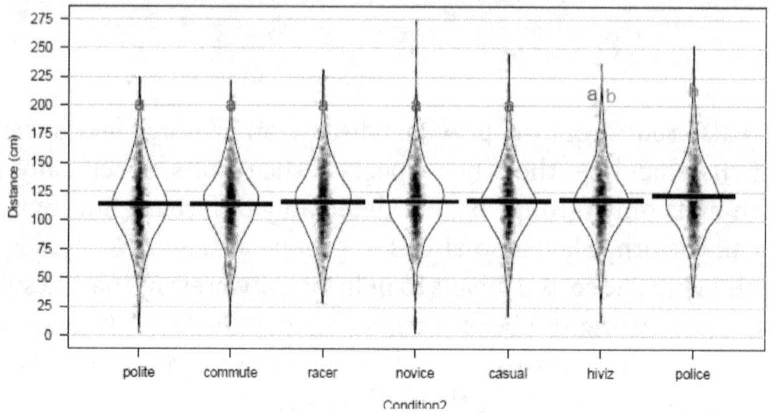

Figure 3.21: Pirate-plot of overtake distances by group, sorted by sample means with Tukey's HSD CLD displayed.

3.8 CONCISE SUMMARY

In this chapter, we explored methods for comparing a quantitative response across J groups (J ≥ 2), with what is called the One-Way ANOVA procedure. The initial test is based on assessing evidence against a null hypothesis of no difference in the true means for the J groups. There are two different methods for estimating these One-Way ANOVA models: the cell means model and the reference-coded versions of the model. There are times when either model will be preferred, but for the rest of the text, the reference coding is used (sorry!). The ANOVA F -statistic, often presented with underlying information in the ANOVA table, provides a method of assessing evidence against the null hypothesis either using permutations or via the F -distribution. Pair-wise comparisons using Tukey's HSD provide a method for comparing all the groups and are a nice complement to the overall ANOVA results. A compact letter display was shown that enhanced the interpretation of Tukey's HSD result.

In the Guinea Pig example, we are left with some lingering questions based on these results. It appears that the effect of dosage changes as a function of the delivery method (OJ, VC) because the size of the differences between OJ and VC change for different dosages. These methods can't directly assess the question of whether the effect of delivery method is the same or not across the different dosages. In Chapter 4, the two variables, Dosage and Delivery method are modelled as two separate variables so we can consider their effects both separately and together. This allows more refined hypotheses, such as Is the effect of delivery method the same for all dosages?, to be tested. This will introduce new models and methods for analyzing data where there are two factors as explanatory variables in a model for a quantitative response variable in what is called the Two-Way ANOVA.

3.9 SUMMARY OF IMPORTANT R CODE

The main components of R code used in this chapter follow with components to modify in lighter and/or ALL CAPS text, remembering that any R packages mentioned need to be installed and loaded for this code to have a chance of working:

- MODELNAME <- lm(Y ~ X, data = DATASETNAME)
  - Probably the most frequently used command in R.
  - Here it is used to fit the reference-coded One-Way ANOVA model with Y as the response variable and X as the grouping variable, storing the estimated model object in MODELNAME. Remember that X should be defined as a factor variable.
- MODELNAME <- lm(Y ~ X - 1, data = DATASETNAME)
  - Fits the cell means version of the One-Way ANOVA model.

- summary(MODELNAME)
  - Generates model summary information including the estimated model coefficients, SEs, t-tests, and p-values.
- anova(MODELNAME)
  - Generates the ANOVA table but **must only be run on the reference-coded version of the model.**
  - Results are incorrect if run on the cell means model since the reduced model under the null is that the mean of all the observations is 0!
- pf(FSTATISTIC, df1 = NUMDF, df2 = DENOMDF, lower.tail = F)
  - Finds the p-value for an observed *F*-statistic with NUMDF and DENOMDF degrees of freedom.
- Tobs <- anova(lm(Y ~ X, data = DATASETNAME))[1,4]; Tobs

```
B <- 1000
Tstar <- matrix(NA, nrow = B)
 for (b in (1:B)){
Tstar[b] <- anova(lm(Y ~ shuffle(X), data = DATASETNAME))[1,4]
}
```

- pdata(Tstar, Tobs, lower.tail = F)[[1]]
  - Code to run a for loop to generate 1000 permuted F-statistics, store and calculate the permutation- based p-value from Tstar.
- par(mfrow = c(2,2)); plot(MODELNAME)
  - Generates four diagnostic plots including the Residuals vs Fitted and Normal Q-Q plot.
- plot(allEffects(MODELNAME))
  - Requires the effects package be loaded.
  - Plots the estimated model component.

Tm2 <- glht(MODELNAME, linfct = mcp(X = "Tukey")); confint(Tm2); plot(Tm2); summary(Tm2); cld(Tm2)

- Requires the multcomp package to be installed and loaded.
- Can only be run on the reference-coded version of the model.
- Generates the text output and plot for Tukey's HSD as well as the compact letter display information.

3.10 PRACTICE PROBLEMS

3.1 **Cholesterol Analysis** For the first practice problems, you will work with the cholesterol data set from the multcomp package that was used to generate the

Tukey's HSD results. To load the data set and learnmore about the study, use the following code:

```
library(multcomp)
data(cholesterol)
library(tibble)
cholesterol <- as_tibble(cholesterol)
help(cholesterol)
```

3.1.1. Graphically explore the differences in the changes in Cholesterol levels for the five levels using pirate-plots.

3.1.2. Is the design balanced? Generate R output to support this assessment.

3.1.3. Complete all 6+ steps of the hypothesis test using the parametric F -test, reporting the ANOVA table and the distribution of the test statistic under the null. When you discuss the scope of inference, make sure you note that the treatment levels were randomly assigned to volunteers in the study.

3.1.4. Generate the permutation distribution and find the p-value. Compare the parametric p-value to the permutation test results.

3.1.5. Perform Tukey's HSD on the data set. Discuss the results – which pairs were detected as different and which were not? Bigger reductions in cholesterol are good, so are there any levels you would recommend or that might provide similar reductions?

3.1.6. Find and interpret the CLD and compare that to your interpretation of results from 3.1.5.

3.2 Sting Location Analysis: These data come from [Smith, 2014] where the author experimented on himself by daily stinging himself five times on randomly selected body locations over the course of months. You can read more about this fascinating (and cringe inducing) study at https://peerj.com/articles/338/. The following code gets the data prepared for analysis by removing the observations he took each day on how painful it was to sting himself on his forearm before and after the other three observations that were of interest each day of the study. This is done with a negation (using "!" of the %in% which identifies rows related to the two daily forearm locations (Forearm and Forearm1) to leave all the rows in the data set for any levels of Body_Location that were not in these two levels. This is easier than trying to list all 24 other levels, then Body_Location variable is re-factored to clean out its unused levels, and finally the reorderfunction is used to order the levels based on the sample mean pain rating – and the results of these steps are stored in the sd_fixedR tibble.

```
library(readr)
sd_fixed <- read_csv("http://www.math.montana.edu/courses/s217/documents/stingdata_fixed.csv")

sd_fixedR <- sd_fixed %>%
             filter(!(Body_Location %in% c("Forearm", "Forearm1"))) %>%
             mutate(Body_Location = factor(Body_Location),
                    Body_Location = reorder(Body_Location, Rating, FUN = mean)
                   )
```

3.2.1. Graphically explore the differences in the pain ratings (Rating) across the different Body_Location levels using boxplots and pirate-plots. How are boxplots misleading for representing these data? **Hint:** look for discreteness in the responses.

3.2.2. Is the design balanced?

3.2.3. How does taking 3 measurements that are of interest each day lead to a violation of the independence assumption here?

3.3.4. Complete all 6+ steps of the hypothesis test using the parametric F-test, reporting the ANOVA table and the distribution of the test statistic under the null. For the scope of inference use the information that the sting locations were randomly assigned but only one person (the researcher) participated in the study.

3.3.5. Generate the permutation distribution and find the p-value. Compare the parametric p-value to the permutation test results.

3.3.6. Generate an effects plot (use something like plot(all Effects (lm_model), rotx = 45) to rotate the x-axis text 45 degrees so you can read it!). Which of the locations did he find most painful on average?

3.3.7. Generated our standard panel of diagnostic plots. In the QQ-Plot, you should see a stair-step pattern that presents a violation of the normality assumption that we have not see before. Look at your answer to and try to explain why this pattern is present.

3.3.8 Often we might consider Tukey's pair wise comparisons given the initial result here. How many levels are there in Body_Location in the filtered data set? How many pairs would be compared if we tried Tukey's– calculate this using the choose function?

CHAPTER-4

TWO-WAY ANOVA

INTRODUCTION: AN OVERVIEW

Two-way analysis of variance (ANOVA) is a statistical method that provides a comprehensive framework for examining the influence of two independent categorical variables on a dependent variable. This analytical technique is a powerful tool widely employed in experimental and observational research to assess the interactive effects and individual contributions of two factors on the variability observed in the response variable. By extending the principles of one-way ANOVA, which evaluates the impact of a single categorical variable, two-way ANOVA allows researchers to explore the nuanced interplay between two distinct factors, shedding light on their joint and independent influences. This method is particularly valuable in uncovering patterns, differences, or interactions that may not be evident when examining each factor in isolation. As we delve into the intricacies of two-way ANOVA, we embark on a journey to disentangle the complex relationships between variables, providing a deeper understanding of the dynamics at play in diverse fields such as psychology, biology, economics, and beyond.

4.1 SITUATION

In this chapter, we extend the One-Way ANOVA to situations with two factors or categorical explanatory variables in a method that is generally called the Two-Way ANOVA. This allows researchers to simultaneously study two variables that might explain variability in the responses and explore whether the impacts of one explanatory variable change depending on the level of the other explanatory variable. In some situations, each observation is so expensive that researchers want to use a single study to explore two different sets of research questions in the same round of data collection. For example, a company might want to study factors that affect the number of defective products per day and are interested in the impacts of two different types of training programs and three different levels of production quotas. These methods would allow engineers to compare the training programs, production quotas, and see if the training programs "work differently" for different production quotas. In a clinical trials context, it is well known that certain factors can change the performance of certain drugs. For example, different dosages of a drug might have different benefits or side-effects on men, versus women or children or even for different age groups in adults. When the impact of one factor on the response changes depending on the level of another factor, we say that the two explanatory variables interact. It is also possible for both factors to be related to differences in the mean responses and not interact. For example, suppose there is a difference in the response variable means between young and old subjects and a difference in the responses among various dosages, but the effect of increasing the dosage is the same for both young and old subjects. This is an example of what is called an additive type of model. In general, the world is more complicated than the single factor models we considered in Chapter 3 can account for, especially in observational studies, so these models allow us to start to handle more realistic situations.

Consider the following "experiment" where we want to compare the strength of different brands of paper towels when they are wet. The response variable will be the time to failure in seconds (a continuous response variable) when a weight is placed on the towel held at the four corners. We are interested in studying the differences between brands and the impact of different amounts of water applied to the towels.

- **Predictors (Explanatory Variables):** A: Brand (2 brands of interest, named B1 and B2) and B: Number of Drops of water (10, 20, 30 drops).
- **Response:** Time to failure (in seconds) of a towel (y) with a weight sitting in the middle of the towel.

4.2 DESIGNING A TWO-WAY EXPERIMENT AND VISUALIZING RESULTS

Ideally, we want to randomly assign the levels of each factor so that we can attribute causality to any detected effects and to reduce the chances of *confounding*, where the differences we think are due to one explanatory variable might be due to another variable that varied with the this explanatory variable of interest. Because there are two factors, we would need to design a random assignment scheme to select the levels of both variables. For example, we could randomly select a brand and then randomly select the number of drops to apply from the levels chosen for each measurement. Or we could decide on how many observations we want at each combination of the two factors (ideally having them all equal so the design is ***balanced***) and then randomize the order of applying the different combinations of levels.

Why might it be important to randomly apply the brand and number of drops in an experiment? There are situations where the order of observations can be related to changes in the responses and we want to be able to eliminate the order of observations from being related to the levels of the factors – otherwise the order of observations and levels of the factors would be confounded. For example, suppose that the area where the experiment is being performed becomes wet over time and the later measurements have extra water that gets onto the paper towels and they tend to fail more quickly. If all the observations for the second brand were done later in the study, then the order of observations impacts could make the second brand look worse. If the order of measurements to be made is randomized, then even if there is some drift in the responses over the order of observations it should still be possible to see the differences in the randomly assigned effects. If the study incorporates repeated measurements on human or animal subjects, randomizing the order of treatments they are exposed to can alleviate impacts of them "learning" through the study or changing just due to being studied, something that we would not have to worry about with paper towels.

In observational studies, we do not have the luxury of random assignment, that is, we cannot randomly assign levels of the treatment variables to our subjects, so we cannot guarantee that the only differences between the groups are based on the differences in the explanatory variables. As discussed before, because we can't control which level of the variables are assigned to the subjects, we cannot make causal inferences and have to worry about other variables being the real drivers of the results. Although we can never establish causal inference with observational studies, we can generalize our results to a larger population if we have a representative (ideally random) sample from our population of interest.

It is also possible that we might have studies where some of the variables are randomly assigned and others that are not randomly assignable. The most common versions of this are what we sometimes call subject "demographics", such as gender, income, race, etc. We might be performing a study where we can randomly assign treatments to these subjects but might also want to account for differences based on income level, which we can't assign. In these cases, the scope of inference gets complicated – differences seen on randomized variables can be causally interpreted but you have to be careful to not say that the demographics caused differences. Suppose that a randomly assigned drug dosage is found to show positive differences in older adults and negative changes in younger adults. We could say that the dosage causes the increases in older adults and decreases in younger ones, but we can't say that age caused the differences in the responses – it just modified how the drug works and what the drug caused to happen in the responses.

Even when we do have random assignment of treatments it is important to think about who/what is included in the sample. To get back to the paper towel example, we are probably interested in more than the sheets of the rolls we have to work with. If we could randomly select the studied paper towels from all paper towels made by each brand, our conclusions could be extended to those populations. That probably would not be practical, but trying to make sure that the towels are representative of all made by each brand by checking for defects and maybe picking towels from a few different rolls would be a good start to being able to extend inferences beyond the tested towels. But if you were doing this study in the factory, it might be possible to randomly sample from the towels produced, at least over the course of a day.

Once random assignment and random sampling is settled, the final aspect of study design involves deciding on the number of observations that should be made. The short (glib) answer is to take as many as you can afford. With more observations comes higher power to detect differences if they exist, which is a desired attribute of all studies. It is also important to make sure that you obtain multiple observations at each combination of the treatment levels, which are called replicates. Having replicate measurements allows estimation of the mean for each combination of the treatment levels as well as estimation and testing for an interaction. And we always prefer1 having balanced designs because they provide resistance to violation of With two categorical explanatory variables, there are now five possible scenarios for the truth. Different situations are created depending on whether there is an interaction between the two variables, whether both variables are important but do not interact, or whether either of the variables matter at all. Basically, there are five different possible outcomes in a randomized Two-Way ANOVA study, listed in order of increasing model complexity:

- Neither A or B has an effect on the responses (nothing causes differences in responses).
- A has an effect, B does not (only A causes differences in responses).
- B has an effect, A does not (only B causes differences in responses).
- Both A and B have effects on response but no interaction (A and B both cause differences in responses but the impacts are *additive*).
- Effect of A on response differs based on the levels of B, the opposite is also true (means for levels of response across A are different for different levels of B, or, simply, A and B interact in their effect on the response).

To illustrate these five potential outcomes, we will consider a fake version of the paper towel example. It ended up being really messy and complicated to actually perform the experiment as described so these data were simulated. The hope is to use this simple example to illustrate some of the Two-Way ANOVA possibilities. The first step is to understand what has been observed (number observations at each combination of factors) and look at some summary statistics across all the "groups". The data set is available via the following link:

```
library(readr)
pt <- read_csv("http://www.math.montana.edu/courses/s217/documents/pt.csv")
pt <- pt %>% mutate(drops = factor(drops),
                    brand = factor(brand)
                    )
```

The data set contains five observations per combination of treatment levels as provided by the tally function. To get counts for combinations of the variables, use the general formula of tally(x1 ~ x2, data = ...) – noting that the order of x1 and x2 doesn't matter here:

```
library(mosaic)
tally(brand ~ drops, data = pt)
```

```
##        drops
## brand 10 20 30
##    B1  5  5  5
##    B2  5  5  5
```

The sample sizes in each of the six treatment level combinations of Brand and Drops [(B1, 10), (B1, 20), (B1, 30), (B2, 10), (B2, 20), (B2, 30)] are njk = 5 for jth level of Brand (j = 1, 2) and kth level of Drops (k = 1, 2, 3). The tally function gives us an R by C contingency table with R = 2 rows (B1, B2) and C = 3 columns (10, 20, and 30). We'll have more fun with R by C tables in Chapter 5 – here it helps us to see the sample size in each combination of factor levels. The favstats function also helps us dig into the results for all combinations of factor levels. The notation involves putting both factor variables after the "~" with a "+" between them. In the output, the first row contains summary information for the 5 observations for Brand B1 and Drops amount 10. It also contains the sample size in the n column, although here it rolled into a new set of rows with the standard deviations of each combination.

```
favstats(responses ~ brand + drops, data = pt)
```

```
##   brand.drops       min        Q1   median       Q3      max     mean        sd n missing
## 1       B1.10 0.3892621 1.3158737 1.906436 2.050363 2.333138 1.599015 0.7714970 5       0
## 2       B2.10 2.3078095 2.8556961 3.001147 3.043846 3.050417 2.851783 0.3140764 5       0
## 3       B1.20 0.3838299 0.7737965 1.516424 1.808725 2.105380 1.317631 0.7191978 5       0
## 4       B2.20 1.1415868 1.9382142 2.066681 2.838412 3.001200 2.197219 0.7509989 5       0
## 5       B1.30 0.2387500 0.9804284 1.226804 1.555707 1.829617 1.166261 0.6103657 5       0
## 6       B2.30 0.5470565 1.1205102 1.284117 1.511692 2.106356 1.313946 0.5686485 5       0
```

The next step is to visually explore the results across the combinations of the two explanatory variables. The pirate-plot can be extended to handle these sorts of two-way situations using a formula that is something like y ~ A * B. The x-axis in the pirate-plot shows two rows of labels based on the two categories and the unique combinations of those categories are directly related to a displayed distribution of responses and mean and confidence interval. For example, in Figure 4.1, the Brand with levels of *B1* and *B2* is the first row of x-axis labels and they are repeated across the three levels of Drops. In reading these plots, look for differences in the means across the levels of the first row variable (Brand) for each level of the second row variable (Drops) and then focus on whether those differences change across the levels of the second variable

that is an interaction as the differences in differences change. Specifically, start with comparing the two brands at each amount of water. Do the brands seem different? Certainly for 10 drops of water the two look different but not for 30 drops, suggesting a different impact of brands based on the amount of water present. We can also look for combinations of factors that produce the highest or lowest responses in this display. It appears that the time to failure is highest in the low water drop groups but as the water levels increase, the time to failure falls and the differences in the two brands seem to decrease. The fake data seem to have relatively similar amounts of variability and distribution shapes except for 10 drops and brand B2

remembering that there are only 5 observations available for describing the shape of responses for each combination. These data were simulated using a normal distribution with constant variance if that gives you some extra confidence in assessing these model assumptions.

```
library(yarrr)
set.seed(12)
pirateplot(responses ~ brand * drops, data = pt, xlab = "Drops", ylab = "Time",
        inf.method = "ci", inf.disp = "line", theme = 2, point.o = 1)
```

The pirate-plots can handle situations where both variables have more than two levels but it can sometimes get a bit cluttered to actually display the data when our analysis is going to focus on means of the responses. The means for each combination of levels that you can find in the favstats output are more usefully used in what is called an interaction plot. Interaction plots display the mean responses (y-axis) versus levels of one predictor variable on the x-axis, adding points and separate lines for each level of the other predictor variable. Because we don't like any of the available functions in R, we wrote our own function. It is available two ways. The easiest, if it works, is to install and load the catstats R package [Greenwood et al., 2022]. If you are working on a local RStudio installation, the first step involves installing the remotes [Csárdi et al., 2021] R package and then loading it – this will allow you to install catstats from our github2 repository (you can type "3" during the installation to avoid updating other packages when you do this step).

```
# To install the catstats R package (just the first time!):
library(remotes)
remotes::install_github("greenwood-stat/catstats")
```

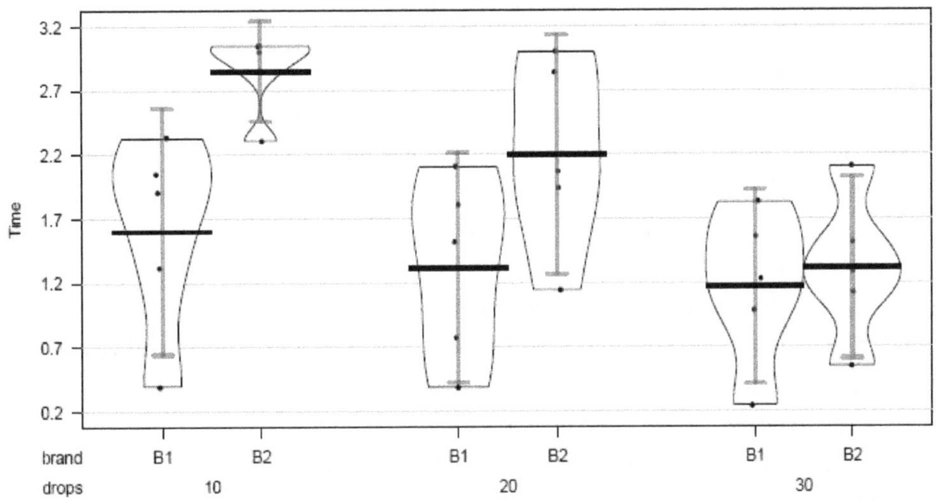

Figure 4.1: Pirate-plot of paper towel data by Brand (first row of *x*-axis) and Drops (second row of *x*-axis).

After that step, you can load catstats like any other R package, using library(catstats). Some users have experienced issues with getting this package installed, so you can also download the needed intplot and intplotarray functions3 using:

```
source("http://www.math.montana.edu/courses/s217/documents/intplotfunctions_v3.R")
```

The intplot function allows a formula interface like Y ~ X1 * X2 and provides the means ± 1 SE (vertical bars) and adds a legend to help make everything clear.

```
intplot(responses ~ brand * drops, data = pt)
```

Interaction plots can always be made two different ways by switching the order of the variables. Figure 4.2 contains Drops on the x-axis and Figure 4.3 has Brand on the x-axis. Typically putting the variable with more levels on the x-axis will make interpretation easier, but not always. Try both and decide on the one that you like best.

```
intplot(responses ~ drops * brand, data = pt)
```

The formula in this function builds on our previous notation and now we include both predictor variables with an "*" between them. Using an asterisk between explanatory variables is one way of telling R to include an interaction between the variables. While the interaction may or may not be present, the interaction plot helps us to explore those potential differences.

There are a variety of aspects of the interaction plots to pay attention to. Initially, the question to answer is whether it appears that there is an interaction between the predictor variables. When there is an interaction, you will see non-parallel lines in the interaction plot. You want to look from left to right in the plot and assess whether the lines connecting the means are close to parallel, relative to the amount of variability in the estimated means as represented by the SEs in the bars. If it seems that there is clear visual evidence of non-parallel lines, then the interaction is likely worth considering (we will use a hypothesis test to formally assess this – see the discussion below). If the lines look to be close to parallel, then there probably isn't an interaction between the variables. Without an interaction present, that means that the differences in the response across levels of one variable doesn't change based on the levels of the other variable and vice-versa. This means that we can consider the main effects of each variable on their own4. Main effects are much like the results we found in Chapter 3 where we can compare means across levels of a single variable except that there are results for two variables to extract from the model. With the presence of an interaction, it is complicated to summarize how each variable is affecting the response variable because their impacts change depending on the level of the other factor. And plots like the interaction plot provide us with useful information on the pattern of those changes.

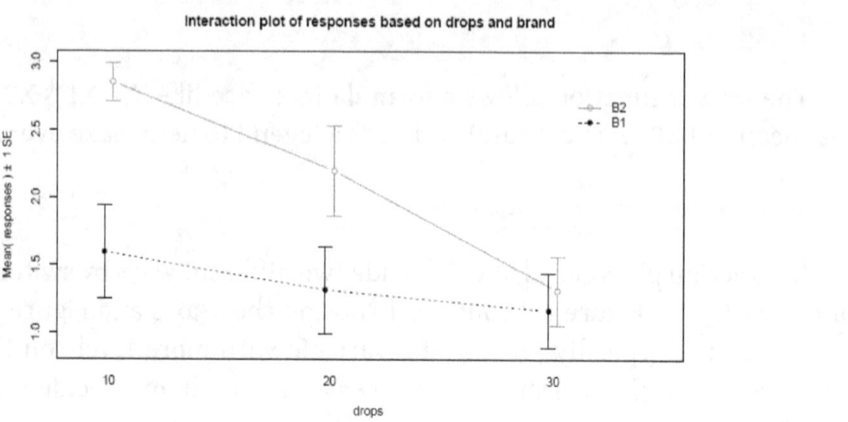

Figure 4.2: Interaction plot of the paper towel data with Drops on the x-axis and different lines based on Brand.

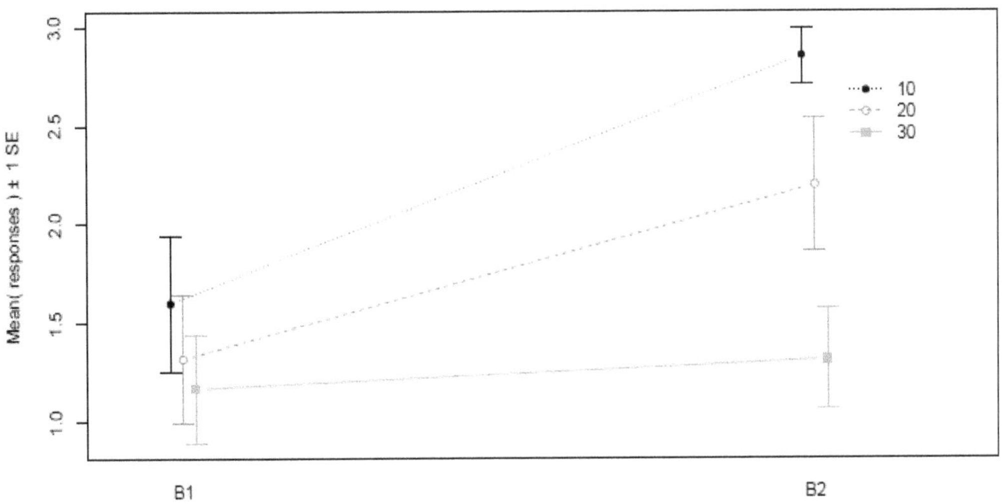

Figure 4.3: Interaction plot of paper towel data with Brand on the x-axis and lines based on Drops.

If the lines are not parallel, then focus in on comparing the levels of one variable as the other variable changes. Remember that the definition of an interaction is that the differences among levels of one variable depends on the level of the other variable being considered. "Visually" this means comparing the size of the differences in the lines from left to right. In Figures 4.2 and 4.3, the effect of amount of water changes based on the brand being considered. In Figure 4.3, the three lines represent the three water levels. The difference between the brands (left to right, *B1* to *B2*) is different depending on how much water was present. It appears that Brand *B2* lasted longer at the lower water levels but that the difference between the two brands dropped as the water levels increased. The same story appears in Figure 4.2. As the water levels increase (left to right, 10 to 20 to 30 drops), the differences between the two brands decrease. Of the two versions, Figure 4.2 is probably easier to read here. Sometimes it is nice to see the interaction plot made both ways simultaneously, so you can also use the intplotarray function, which provides Figure 4.4. This plot also adds pirate-plots to the off-diagonals so you can explore the main effects of each variable, if that is reasonable. The interaction plots can be used to identify the best and worst mean responses for combinations of the treatment levels. For example, 10 Drops and Brand *B2* lasts longest, on average, and 30 Drops with Brand *B1* fails fastest, on average. In any version of the plot here, the lines do not appear to be parallel suggesting that further exploration of the interaction appears to be warranted.

```
intplotarray(responses ~ drops * brand, data = pt)
```

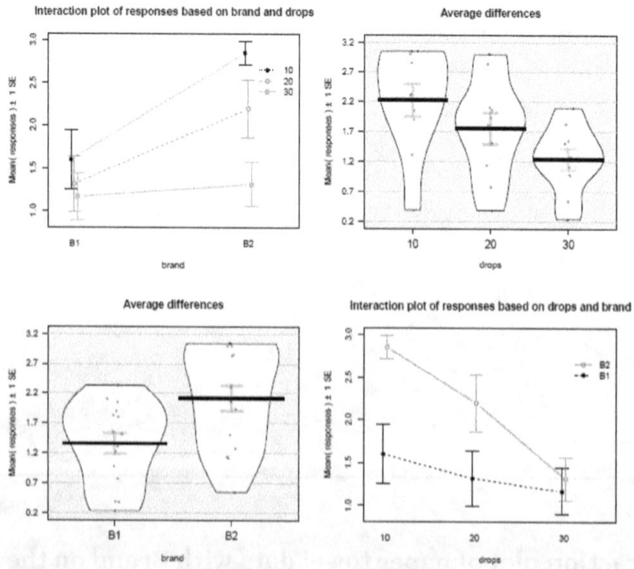

Figure 4.4: Interaction plot array of paper towel data with two different versions of interaction plots and pirate-plots of the responses versus each explanatory variable.

Before we get to the hypothesis tests to formally make this assessment (you knew some sort of p-value was coming, right?), we can visualize the 5 different scenarios that could characterize the sorts of results you could observe in a Two-Way ANOVA situation. Figure 4.5 shows 4 of the 5 scenarios. In panel (a), when there are no differences from either variable (Scenario 1), it provides relatively parallel lines and basically no differences either across Drops levels (x-axis) or Brand (lines). Data such as these would likely result in little to no evidence related to a difference in brands, water levels, or any interaction between them in this data set.

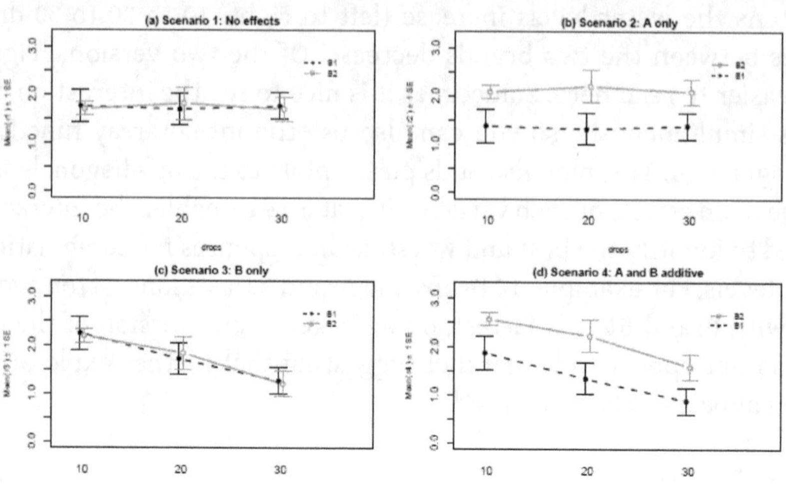

Figure 4.5: Interaction plots of four possible scenarios in the paper towel study.

Scenario 2 (Figure 4.5 panel (b)) incorporates differences based on factor A (here that is Brand) but no real difference based on the Drops or any interaction. This results in a clear shift between the lines for the means of the Brands but little to no changes in the level of those lines across water levels. These lines are relatively parallel. We can see that Brand B2 is better than Brand B1 but that is all we can show with these sorts of results.

Scenario 3 (Figure 4.5 panel (c)) flips the important variable to B (Drops) and shows decreasing average times as the water levels increase. Again, the interaction panels show near parallel-ness in the lines and really just show differences among the levels of the water. In both Scenarios 2 and 3, we could use a single variable and drop the other from the model, getting back to a One-Way ANOVA model, without losing any important information.

Scenario 4 (Figure 4.5 panel (d)) incorporates effects of A and B, but they are additive. That means that the effect of one variable is the same across the levels of the other variable. In this experiment, that would mean that Drops has the same impact on performance regardless of brand and that the brands differ but each type of difference is the same regardless of levels of the other variable. The interaction plot lines are more or less parallel but now the brands are clearly different from each other. The plot shows the decrease in performance based on increasing water levels and that Brand B2 is better than Brand B1. Additive effects show the same difference in lines from left to right in the interaction plots.

Finally, Scenario 5 (Figure 4.6) involves an interaction between the two variables (Drops and Brand). There are many ways that interactions can present but the main thing is to look for clearly non-parallel lines. As noted in the previous discussion, the Drops effect appears to change depending on which level of Brand is being considered. Note that the plot here described as Scenario 5 is the same as the initial plot of the results in Figure 4.2.

The typical modelling protocol is to start with assuming that Scenario 5 is a possible description of the results, related to fitting what is called the interaction model, and then attempt to simplify the model (to the additive model) if warranted. We need a hypothesis test to help decide if the interaction is "real". We start with assuming there is no interaction between the two factors in their impacts on the response and assess evidence against that null hypothesis. We need a hypothesis test because the lines will never be exactly parallel in real data and, just like in the One-Way ANOVA situation, the amount of variation around the lines impacts the ability of the model to detect differences, in this case of an interaction.

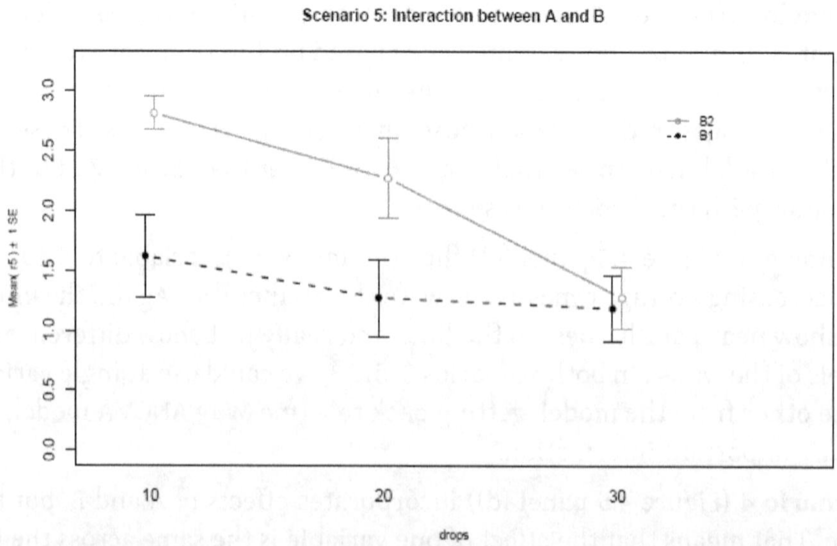

Figure 4.6: Interaction plot of Scenario 5 where it appears that an interaction is present.

4.3 TWO-WAY ANOVA MODELS AND HYPOTHESIS TESTS

To assess interactions with two variables, we need to fully describe models for the additive and interaction scenarios and then develop a method for assessing evidence of the need for different aspects of the models. First, we need to define the notation for these models:

- yijk is the ith response from the group for level j of factor A and level k of factor B
- *j* = 1,. . . , *J J* is the number of levels of A
- *k* = 1,. . . , *K K* is the number of levels of B
- *i* = 1,. . . , *njk njk* is the sample size for level *j* of factor A and level *k* of factor B
- *N* = Σ*j* Σ*k njk* is the total sample size (sum of the number of observations across all *JK* groups)

We need to extend our previous discussion of reference-coded models to develop a Two-Way ANOVA model. We start with the Two-Way ANOVA interaction model:

$$y_{ijk} = \alpha + \tau_j + \gamma_k + \omega_{jk} + \varepsilon_{ijk},$$

where α is the baseline group mean (for level 1 of A and level 1 of B), τj is the deviation for the main effect of A from the baseline for levels 2, . . . , J , γk (gamma k) is the deviation for the main effect of B from the baseline for levels 2, . . . , K, and ωjk (omega jk) is the adjustment for the interaction effect for level j of factor A and level k of factor B for j = 1, . . . , J and k = 1, . . . , K. In this model, $\tau 1$, $\gamma 1$, and $\omega 11$ are all fixed at 0 because α is the mean for the combination of the baseline levels of both variables and so no adjustments are needed. Additionally, any ωjk 's that contain the baseline category of either factor A or B are also set to 0 and the model for these levels just involves τj or γk added to the intercept. Exploring the R output will help clarify which coefficients are present or set to 0 (so not displayed) in these models. As in Chapter 3, R will typically choose the baseline categories alphabetically but now it is choosing a baseline for both variables and so our detective work will be doubled to sort this out.

If the interaction term is not important, usually based on the interaction test presented below, the ωjk 's can be dropped from the model and we get a model that corresponds to Scenario 4 above. Scenario 4 is where there are two main effects in the model but no interaction between them. The additive Two-Way model is

$$y_{ijk} = \alpha + \tau_j + \gamma_k + \varepsilon_{ijk},$$

where each component is defined as in the interaction model. The difference between the interaction and additive models is setting all the ωjk 's to 0 that are present in the interaction model. When we set parameters to 0 in models it removes them from the model. Setting parameters to 0 is also how we will develop our hypotheses to test for an interaction, by assessing evidence against a null hypothesis that all ωjk 's = 0.

The interaction test hypotheses are

- ***H0:*** No interaction between A and B on response in population ⇔ All ωjk 's = 0.
- ***HA:*** Interaction between A and B on response in population ⇔ At least one ωjk /= 0.

To perform this test, a new ANOVA F -test is required (presented below) but there are also hypotheses relating to the main effects of A (τj 's) and B (γk 's). If you decide that there is sufficient evidence against the null hypothesis that no interaction is present to conclude that one is likely present, then it is dangerous to ignore the interaction and test for the main effects because important main effects can be masked by interactions (examples later). It is important to note that, by definition, both variables matter if an interaction is found to be important so the main effect tests may not be very interesting in an interaction model. If the interaction is found to be important based on the test and so is retained in the model, you should focus on the interaction model (also called the full model) in order to understand and describe the form of the interaction among the variables.

If the interaction test does not return a small p-value and you decide that you do not have enough evidence against the null hypothesis to suggest that the interaction is needed, the interaction can be dropped from the model. In this situation, we would re-fit the model and focus on the results provided by the additive model – performing tests for the two additive main effects. For the first, but not last time, we encounter a model with more than one variable and more than one test of potential interest. In models with multiple variables at similar levels (here both are main effects), we are interested in the results for each variable given that the other variable is in the model. In many situations, including more than one variable in a model changes the results for the other variable even if those variables do not interact. The reason for this is more clear in Chapter 8 and really only matters here if we have unbalanced designs, but we need to start adding a short modifier to our discussions of main effects – they are the results *conditional on* or *adjusting for* or, simply, *given*, the other variable(s) in the model. Specifically, the hypotheses for the two main effects are:

- **Main effect test for A:**
  - ***H0:*** No differences in means across levels of A in population, given B in the model

    ⇔ All τ_j's = 0 in additive model.
  - ***HA:*** Some difference in means across levels A in population, given B in the model

    ⇔ At least one $\tau_j \neq 0$, in additive model.
- **Main effect test for B:**
  - ***H0:*** No differences in means across levels of B in population, given A in the model

    ⇔ All γ_k's = 0 in additive model.
  - ***HA:*** Some difference in means across levels B in population, given A in the model

    ⇔ At least one $\gamma_k \neq 0$, in additive model.

In order to test these effects (interaction in the interaction model and main effects in the additive model), F-tests are developed using Sums of Squares, Mean Squares, and degrees of freedom similar to those in Chapter 3. We won't worry about the details of the sums of squares formulas but you should remember the sums of squares decomposition, which still applies[5]. Table 4.1 summarizes the ANOVA results you will obtain for the interaction model and Table 4.2 provides the similar general results for the additive model. As we saw in Chapter 3, the degrees of freedom are the amount of information that is free to vary at a particular level and that rule generally holds here. For example, for factor A with J levels, there are J - 1 parameters that are free since the baseline is fixed. The residual degrees of freedom for both models are not as easily explained but have a simple formula. Note that the sum of the degrees of freedom from the main effects, (interaction if present), and error need to equal N - 1, just like in the One-Way ANOVA table.

Table 4.1: Interaction Model ANOVA Table.

Source	DF	SS	MS	F-statistics
A B	J - 1	SS_A	MSA = SSA/dfA	MSA/MSE
A:B (interaction)	K - 1	SS_B	MSB = SSB /dfB	MSB /MSE
	(J - 1)(K - 1)	SS_{AB}	MSAB = SSAB /dfAB	MSAB /MSE
Error	N - JK	SS_E	MSE = SSE /dfE	
Total	N - 1	SS_{Total}		

Table 4.2: Additive Model ANOVA Table.

Source	DF	SS	MS	F-statistics
A	J - 1	SSA	MSA = SSA/dfA	MSA/MSE
B	K - 1	SSB	MSB = SSB /dfB	MSB /MSE
Error	N - J - K + 1	SSE	MSE = SSE /dfE	
Total	N - 1	SSTotal		

The mean squares are formed by taking the sums of squares (we'll let R find those for us) and dividing by the df in the row. The F -ratios are found by taking the mean squares from the row and dividing by the mean squared error (MSE). They follow F -distributions with numerator degrees of freedom from the row and denominator degrees of freedom from the Error row (in R output this the Residuals row). It is possible to develop permutation tests for these methods but some technical issues arise in doing permutation tests for interaction model components so we will not use them here. This means we will have to place even more emphasis on the data not presenting clear violations of assumptions since we only have the parametric method available.

With some basic expectations about the ANOVA tables and F -statistic construction in mind, we can get to actually estimating the models and exploring the results. The first example involves the fake paper towel data displayed in Figure 4.1 and 4.2. It appeared that Scenario 5 was the correct story since the lines appeared to be non-parallel, but we need to know whether there is sufficient evidence to suggest that the interaction is "real" and we get that through the interaction hypothesis test. To fit the interaction model using lm, the general formulation is lm(y ~ x1 * x2, data = ...). The order of the variables doesn't matter as the most important part of the model, to start with, relates to the interaction of the variables.

The ANOVA table output shows the results for the interaction model obtained by running the anova function on the model called m1. Specifically, the test that ***H0:*** All ωjk 's = 0 has a test statistic of F (2, 24) = 1.92 (in the output from the row with brands: drops) and a p-value of 0.17. So there is weak evidence against the null hypothesis of no interaction, with a 17% chance we would observe a difference in the ωjk 's like we did or more extreme if the ωjk 's really were all 0. So we would conclude that the interaction is probably not needed6. Note that for the interaction model components, R presents them with a colon, :,between the variable names.

```
m1 <- lm(responses ~ brand * drops, data = pt)
anova(m1)
```

```
## Analysis of Variance Table
##
## Response: responses
##             Df Sum Sq Mean Sq F value   Pr(>F)
## brand        1 4.3322  4.3322 10.5192 0.003458
## drops        2 4.8581  2.4290  5.8981 0.008251
## brand:drops  2 1.5801  0.7901  1.9184 0.168695
## Residuals   24 9.8840  0.4118
```

It is useful to display the estimates from this model and we can utilize plot(all Effects(MODELNAME)) to visualize the results for the terms in our models. If we turn on the options for grid = T, multiline = T, and ci.style = "bars" we get a useful version of the basic "effect plot" for Two-Way ANOVA models with interaction. I also added lty = c(1:2) to change the line type for the two lines (replace 2 with the number of levels in the variable driving the different lines. The results of the estimated interaction model are displayed in Figure 4.7, which looks very similar to our previous interaction plot. The only difference is that this comes from model that assumes equal variance and these plots show 95% confidence intervals for the means instead of the ± 1 SE used in the intplot where each SE is calculated using the variance of the observations at each combination of levels. Note that other than the lines connecting the means, this plot also is similar to the pirate-plot in Figure 4.1 that also displayed the original responses for each of the six combinations of the two explanatory variables. That plot then provides a place to assess assumptions of the equal variance and distributions for each group as well as explore differences in the group means.

```
library(effects)
plot(allEffects(m1), grid = T, multiline = T, lty = c(1:2), ci.style = "bars")
```

In the absence of sufficient evidence to include the interaction, the model should be simplified to the additive model and the interpretation focused on each main effect, conditional on having the other variable in the model. To fit an additive model and not include an interaction, the model formula involves a "+" instead of a "*" between the explanatory variables.

```
m2 <- lm(responses ~ brand + drops, data = pt)
anova(m2)
```

```
## Analysis of Variance Table
##
## Response: responses
##           Df  Sum Sq Mean Sq F value   Pr(>F)
## brand      1  4.3322  4.3322  9.8251 0.004236
## drops      2  4.8581  2.4290  5.5089 0.010123
## Residuals 26 11.4641  0.4409
```

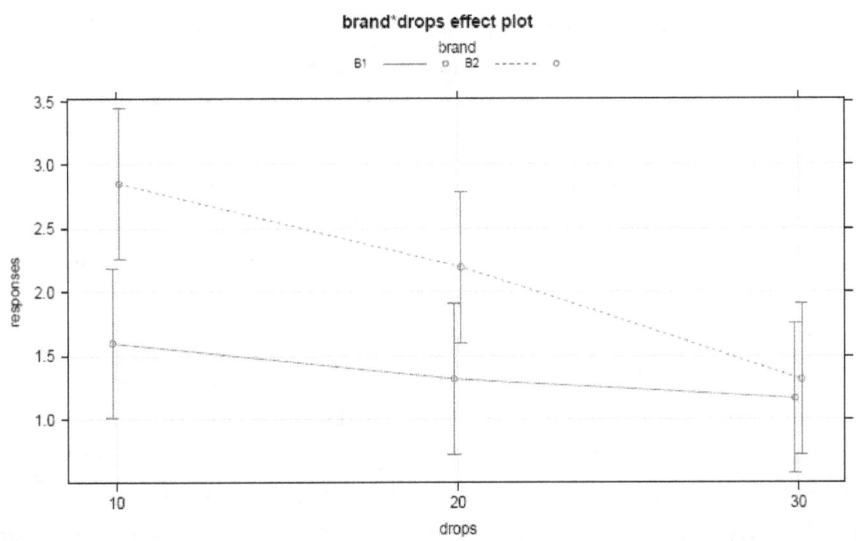

Figure 4.7: Plot of estimated results of interaction model for the paper towel performance data.

The p-values for the main effects of brand and drops change slightly from the results in the interaction model due to changes in the MSE from 0.4118 to 0.4409 (more variability is left over in the simpler model) and the DFerror that increases from 24 to 26. In both models, the SSTotal is the same (20.6544). In the interaction model,

$$\begin{aligned} SS_{\text{Total}} &= SS_{\text{brand}} + SS_{\text{drops}} + SS_{\text{brand:drops}} + SS_{\text{E}} \\ &= 4.3322 + 4.8581 + 1.5801 + 9.8840 \\ &= 20.6544. \end{aligned}$$

In the additive model, the variability that was attributed to the interaction term in the interaction model (SSbrand:drops = 1.5801) is pushed into the SSE, which increases from 9.884 to 11.4641. The sums of squares decomposition in the additive model is

$$\begin{aligned} SS_{\text{Total}} &= SS_{\text{brand}} + SS_{\text{drops}} + SS_{\text{E}} \\ &= 4.3322 + 4.8581 + 11.4641 \\ &= 20.6544. \end{aligned}$$

This shows that the sums of squares decomposition applies in these more complicated models as it did in the One-Way ANOVA. It also shows that if the interaction is removed from the model, that variability is lumped in with the other unexplained variability that goes in the SSE in any model.

The fact that the sums of squares decomposition can be applied here is useful, except that there is a small issue with the main effect tests in the ANOVA table results that follow this decomposition when the design is not balanced. It ends up that the tests in a typical ANOVA table are only conditional on the tests higher up in the table. For example, in the additive model ANOVA table, the Brand test is not conditional on the Drops effect, but the Drops effect is conditional on the Brand effect. In balanced designs, conditioning on the other variable does not change the results but in unbalanced designs, the order does matter. To get both results to be similarly conditional on the other variable, we have to use another type of sums of squares, called Type II sums of squares. These sums of squares will no longer always follow the rules of the sums of squares decomposition but they will test the desired hypotheses. Specifically, they provide each test conditional on any other terms at the same level of the model and match the hypotheses written out earlier in this section. To get the "correct" ANOVA results, the car package (Fox et al. [2022a], Fox and Weisberg [2011]) is required. We use the Anova function on our linear models from here forward to get the "right" tests in our ANOVA tables7. Note how the case-sensitive nature of R code shows up in the use of the capital "A" Anova function instead of the lower-case "a" anova function used previously. In this situation, because the design was balanced, the results are the same using either function. Observational studies rarely generate balanced designs (some designed studies can result in unbalanced designs too) so we will generally just use the Type II version of the sums of squares to give us the desired results across different data sets we might analyze. The Anova results using the Type II sums of squares are slightly more conservative than the results from anova, which are called Type I sums of squares. The sums of squares decomposition no longerapplies, but it is a small sacrifice to get each test after adjusting for all other variables8.

```
library(car)
Anova(m2)
```

```
## Anova Table (Type II tests)
##
## Response: responses
##             Sum Sq Df F value   Pr(>F)
## brand       4.3322  1  9.8251 0.004236
## drops       4.8581  2  5.5089 0.010123
## Residuals 11.4641 26
```

The new output switches the columns around and doesn't show you the mean squares, but gives the most critical parts of the output. Here, there is no change in results because it is a balanced design with equal counts of responses in each combination of the two explanatory variables.

The additive model, when appropriate, provides simpler interpretations for each explanatory variable compared to models with interactions because the effect of one variable is the same regardless of the levels of the other variable and vice versa. There are two tools to aid in understanding the impacts of the two variables in the additive model. First, the model summary provides estimated coefficients with interpretations like those seen in Chapter 3 (deviation of group j or k from the baseline group's mean), except with the additional wording of "controlling for" the other variable added to any of the discussion. Second, the term-plots now show each main effect and how the groups differ with one panel for each of the two explanatory variables in the model. These term-plots are created by holding the other variable constant at one of its levels (the most frequently occurring or first if the there are multiple groups tied for being most frequent) and presenting the estimated means across the levels of the variable in the plot.

```
summary(m2)
```

```
##
## Call:
## lm(formula = responses ~ brand + drops, data = pt)
##
## Residuals:
##     Min      1Q  Median      3Q     Max
## -1.4561 -0.4587  0.1297  0.4434  0.9695
##
## Coefficients:
##             Estimate Std. Error t value Pr(>|t|)
## (Intercept)   1.8454     0.2425   7.611 4.45e-08
## brandB2       0.7600     0.2425   3.134  0.00424
## drops20      -0.4680     0.2970  -1.576  0.12715
## drops30      -0.9853     0.2970  -3.318  0.00269
```

```
##
## Residual standard error: 0.664 on 26 degrees of freedom
## Multiple R-squared:  0.445,  Adjusted R-squared:  0.3809
## F-statistic: 6.948 on 3 and 26 DF,  p-value: 0.001381
```

In the model summary, the baseline combination estimated in the (Intercept) row is for Brand B1 and Drops 10 and estimates the mean failure time as 1.85 seconds for this combination. As before, the group labels that do not show up are the baseline but there are two variables' baselines to identify. Now the "simple" aspects of the additive model show up. The interpretation of the Brands B2 coefficient is as a deviation from the baseline but it applies regardless of the level of Drops. Any difference between B1 and B2 involves ashift up of 0.76 seconds in the estimated mean failure time. Similarly, going from 10 (baseline) to 20 drops results in a drop in the estimated failure mean of 0.47 seconds and going from 10 to 30 drops results in a drop of almost 1 second in the average time to failure, both estimated changes are the same regardless of the brand of paper towel being considered. Sometimes, especially in observational studies, we use the terminology "controlled for" to remind the reader that the other variable was present in the model9 and also explained some of the variability in the responses. The term-plots for the additive model (Figure 4.8) help us visualize the impacts of changes brand and changing water levels, holding the other variable constant. The differences in

```
library(effects)
plot(allEffects(m2))
```

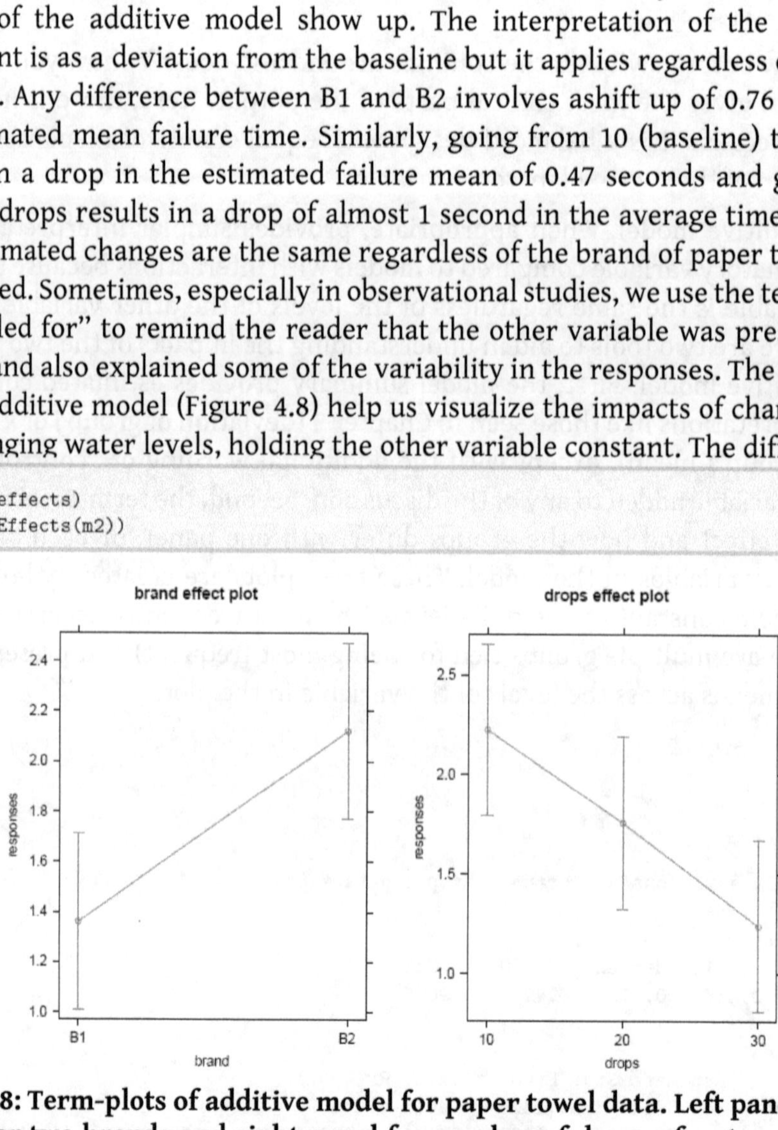

Figure 4.8: Term-plots of additive model for paper towel data. Left panel displays results for two brands and right panel for number of drops of water, each after controlling for the other.

With the first additive model we have considered, it is now the first time where we are working with a model where we can't display the observations together with the means that the model is producing because the results for each predictor are averaged across the levels of the other predictor. To visualize some aspects of the original observations with the estimates from each group, we can turn on an option in the term-plots(residuals = T) to obtain the partial residuals that show the residuals as a function of one variable after adjusting for the effects/impacts of other variables. We will avoid the specifics of the calculations for now, but you can use these to explore the residuals at different levels of each predictor. They will be most useful in the Chapters 7 and 8 but give us some insights in unexplained variation in each level of the predictors once we remove the impacts of other predictors in the model. Use plots like Figure 4.9 to look for different variability at different levels of the predictors and locations of possible outliers in these models. Note that the points (open circles) are jittered to aid in seeing all of them, the means of each group of residuals are indicated by a filled large circle, and the smaller circles in the center of the bars for the 95% confidence intervals are the means from the model. Term-plots with partial residuals accompany our regular diagnostic plots for assessing equal variance assumptions in these models – in some cases adding the residuals will clutter the term-plots so much that reporting them is not useful since one of the main purposes of the term-plots is to visualize the model estimates. So use the residuals = T option judiciously.

```
rary(effects)
t(allEffects(m2, residuals = T))
```

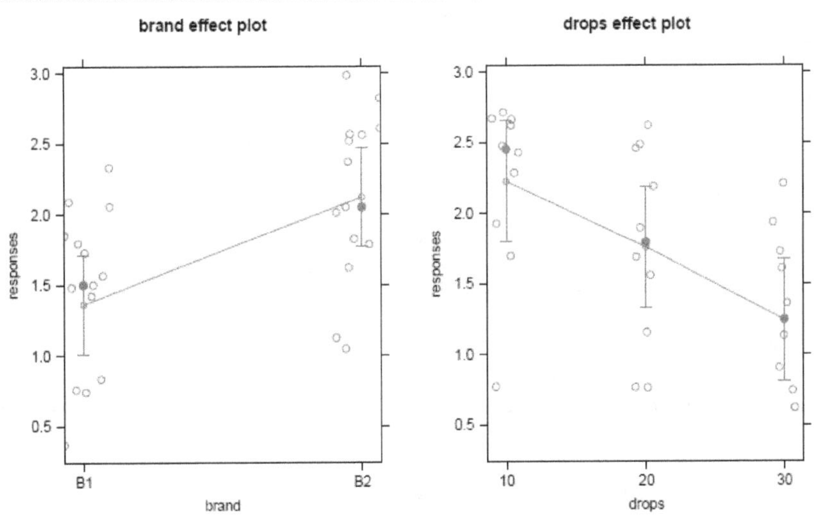

Figure 4.9: Term-plots of additive model for paper towel data with partial residuals added. Relatively similar variability seems to be present in each of the groups of residuals after adjusting for the other variable except for the residuals for the 10 drops where the variability is smaller, especially if one small outlier is ignored.

For the One-Way and Two-Way interaction models, the partial residuals are just the original observations so present similar information as the pirate-plots but do show the model estimated 95% confidence intervals. With interaction models, you can use the default settings in effects when adding in the partial residuals as seen below in Figure 4.12.

4.4 GUINEA PIG TOOTH GROWTH ANALYSIS WITH TWO-WAY ANOVA

The effects of dosage and delivery method of ascorbic acid on Guinea Pig odontoblast growth was analyzed as a One-Way ANOVA in Section 3.5 by assessing evidence of any difference in the means of any of the six combinations of dosage method (Vit C capsule vs Orange Juice) and three dosage amounts (0.5, 1, and 2 mg/day). Now we will consider the dosage and delivery methods as two separate variables and explore their potential interaction. A pirate-plot and interaction plot are provided in Figure 4.10.

```
data(ToothGrowth)
library(tibble)
ToothGrowth <- as_tibble(ToothGrowth)
```

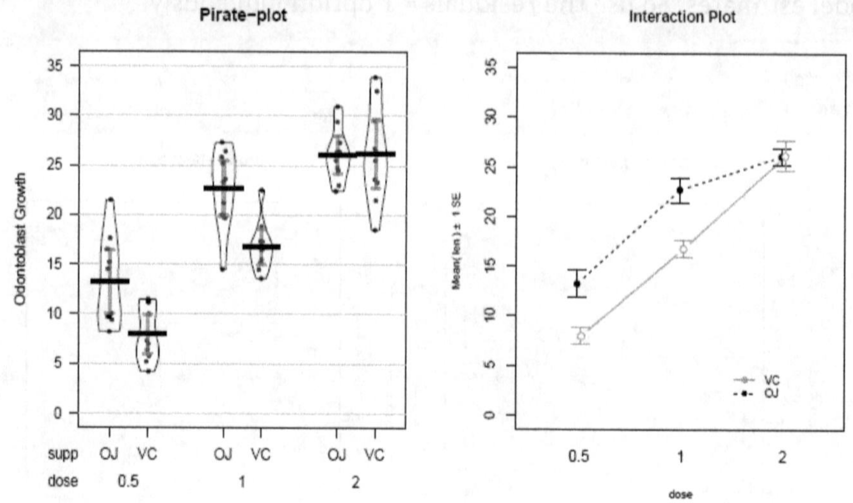

Figure 4.10: Pirate-plot and interaction plot of the odontoblast growth data set.

```
par(mfrow = c(1,2))
pirateplot(len ~ supp * dose, data = ToothGrowth, ylim = c(0,35),
        main = "Pirate-plot",  xlab = "Dosage", ylab = "Odontoblast Growth",
        inf.method = "ci", inf.disp = "line", theme = 2)
intplot(len ~ supp * dose, data = ToothGrowth, col = c(1,2),
       main = "Interaction Plot", ylim = c(0,35))
```

It appears that the effect of method changes based on the dosage as the interaction plot seems to show some evidence of non-parallel lines. Actually, it appears that the effect of delivery method is the same (parallel lines) for doses 0.5 and 1.0 mg/day but that the effect of delivery method changes for 2 mg/day.

We can use the ANOVA *F*-test for an interaction to assess whether we think the interaction is "real" relative to the variability in the responses. That is, is it larger than we would expect due to natural variation in the data? If yes, then we think it is a real effect and we should account for it. The following code fits the interaction model and provides an ANOVA table.

```
TG1 <- lm(len ~ supp * dose, data = ToothGrowth)
Anova(TG1)
```

```
## Anova Table (Type II tests)
##
## Response: len
##              Sum Sq Df  F value    Pr(>F)
## supp         205.35  1  12.3170 0.0008936
## dose        2224.30  1 133.4151 < 2.2e-16
## supp:dose     88.92  1   5.3335 0.0246314
## Residuals    933.63 56
```

The R output is reporting an interaction test result of $F(1, 56) = 5.3$ with a p-value of 0.025. But this should raise a red flag since the numerator degrees of freedom are not what we should expect based on Table 4.1 of $(K - 1) * (J - 1) = (2 - 1) * (3 - 1) = 2$. This brings up an issue in R when working with categorical variables. If the levels of a categorical variable are entered numerically, R will treat them as quantitative variables and not split out the different levels of the categorical variable. To make sure that R treats categorical variables the correct way, we should use the factor function on any variables[10] that are categorical in meaning but are coded numerically in the data set. The following code creates a new variable called dosef using mutate and the factor function to help us obtain correct results from the linear model. The re-run of the ANOVA table provides the correct analysis and the expected *df* for the two rows of output involving dosef:

```
ToothGrowth <- ToothGrowth %>% mutate(dosef = factor(dose))
TG2 <- lm(len ~ supp * dosef, data = ToothGrowth)
Anova(TG2)
```

```
## Anova Table (Type II tests)
##
## Response: len
##             Sum Sq Df F value    Pr(>F)
## supp         205.35  1  15.572 0.0002312
## dosef       2426.43  2  92.000 < 2.2e-16
## supp:dosef   108.32  2   4.107 0.0218603
## Residuals    712.11 54
```

The ANOVA F -test for an interaction between supplement type and dosage level is $F(2, 54) = 4.107$ with a p-value of 0.022. So there is moderate to strong evidence against the null hypothesis of no interaction between *Dosage* and *Delivery method*, so we would likely conclude that there is an interaction present that we should discuss and this supports a changing effect on odontoblast growth of dosage based on the delivery method in these guinea pigs.

Any similarities between this correct result and the previous WRONG result are coincidence. I once attended a Master's thesis defense where the results from a similar model were not as expected (small p-values in places they didn't expect and large p-values in places where they thought differences existed based on past results and plots of the data). During the presentation, the student showed some ANOVA tables and the four level categorical variable had 1 numerator *df* in all ANOVA tables. The student passed with *major* revisions but had to re-run **all** the results and re-write **all** the conclusions. . . So be careful to check the ANOVA results (*df* and for the right number of expected model coefficients) to make sure they match your expectations. This is one reason why you will be learning to fill in ANOVA tables based on information about the study so that you can be prepared to detect when your code has let you down[11]. It is also a great reason to explore term-plots and coefficient interpretations as that can also help diagnose errors in model construction.

Getting back to the previous results, we now have enough background information to more formally write up a focused interpretation of these results. The 6+ hypothesis testing steps in this situation would be focused on first identifying that the best analysis here is as a Two-Way ANOVA situation (these data were analyzed in Chapter 3 as a One-Way ANOVA but this version is likely better because it can explore whether there is an interaction between delivery method and dosage). We will focus on assessing the interaction. If the interaction had been dropped, we would have reported the test results for the interaction, then re-fit the additive model and used it to explore the main effect tests and estimates for *Dose* and *Delivery method*. But since we are inclined to retain the interaction component in the model, the steps focus on the interaction.

```
par(mfrow = c(2,2))
plot(TG2, pch = 16)
```

The RQ is whether there is an interaction of dosage and delivery method on odontoblast growth. Data were collected at all combinations of these predictor variables on the size of the cells, so they can address the size of the cells in these condition combinations. The interaction F -test will be used to assess the research question.

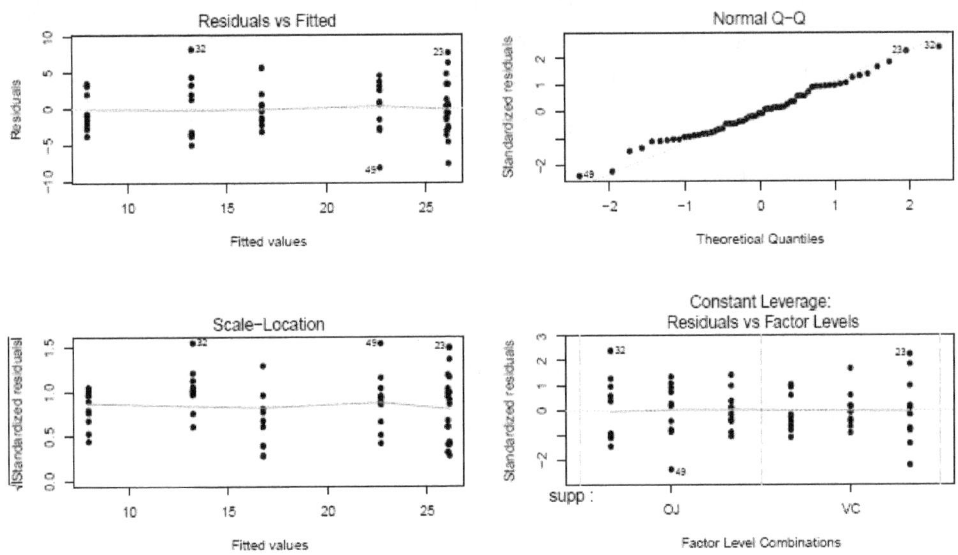

Figure 4.11: Diagnostic plots for the interaction model for odontoblast growth interaction model.

1. Hypotheses:

- H_0: No interaction between *Delivery method* and *Dose* on odontoblast growth in population of guinea pigs
  ⇔ All ωjk 's = 0.

- H_A: Interaction between *Delivery method* and *Dose* on odontoblast growth in population of guinea pigs
  ⇔ At least one ωjk /= 0.

2. Plot the data and assess validity conditions:

- Independence:
  - There is no indication of an issue with this assumption because we don't know of a reason why the independence of the measurements of odontoblast growth of across the guinea pigs as studied might be violated.

- Constant variance:
  - To assess this assumption, we can use the pirate-plot in Figure 4.10, the diagnostic plots in Figure 4.11, and by adding the partial residuals to the term-plot12 as shown in 4.12.
  - In the Residuals vs Fitted and the Scale-Location plots, the differences in variability among the groups (see the different x-axis positions for each group's fitted values) is minor, so there is not strong evidence of a problem with the equal variance assumption. Similarly, the original pirate-plots and adding the partial residuals to the term-plot do not highlight big differences in variability at any of the combinations of the predictors, so do not suggest clear issues with this assumption.

```
plot(allEffects(TG2, residuals = T, x.var = "dosef"))
```

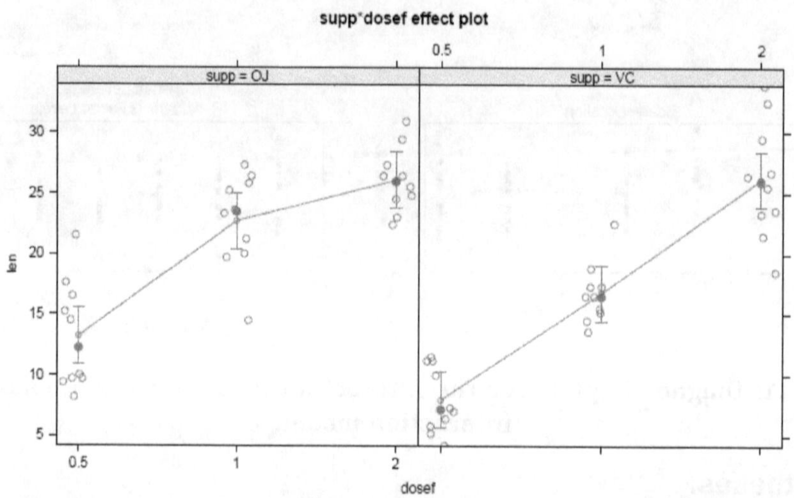

Figure 4.12: Term-plot for odontoblast growth interaction model with partial residuals added.

- Normality of residuals:
  - The QQ-Plot in Figure 4.11 does not suggest a problem with this assumption.

Note that these diagnostics and conclusions are the same as in Section 3.5 because the interaction model and the One-Way ANOVA model with all six combinations of the levels of the two variables fit exactly the same. But the RQ that we can address differs due to the different model parameterizations.

3. Calculate the test statistic and p-value for the interaction test.

```
TG2 <- lm(len ~ supp * dosef, data = ToothGrowth)
Anova(TG2)
```

```
## Anova Table (Type II tests)
##
## Response: len
##              Sum Sq Df F value    Pr(>F)
## supp         205.35  1  15.572 0.0002312
## dosef       2426.43  2  92.000 < 2.2e-16
## supp:dosef   108.32  2   4.107 0.0218603
## Residuals    712.11 54
```

- The test statistic is $F(2, 54) = 4.107$ with a p-value of 0.0219
- To find this p-value directly in R from the test statistic value and F-distribution, we can use thepf function.

```
pf(4.107, df1 = 2, df2 = 54, lower.tail = F)
```

```
## [1] 0.0218601
```

4. Conclusion based on p-value:

- With a p-value of 0.0219 (from $F(2, 54) = 4.107$), there is about a 2.19% chance we would observe an interaction like we did (or more extreme) if none were truly present. This provides moderate to strong evidence against the null hypothesis of no interaction between delivery method and dosage on odontoblast growth in the population so we would conclude that there is likely an interaction and would retain the interaction in the model.

5. Size of differences:

- See discussion below.

6. Scope of Inference:

- Based on the random assignment of treatment levels, causal inference is possible (the changes due to dosage in the differences based on supplement type caused the differences in growth) but because the guinea pigs were not randomly selected, the inferences only pertain to these guinea pigs.

In a Two-Way ANOVA, we need to go a little further to get to the final "size" interpretations since the models are more complicated. When there is an interaction present, we should focus on the term-plot of the interaction model for an interpretation of the form and pattern of the interaction. If the interaction were unimportant, then the hypotheses and results should focus on the additive model results, especially the estimated model coefficients. To see why we don't usually discuss all the estimated model coefficients in an interaction model, the six coefficients for this model are provided:

```
summary(TG2)$coefficients
```

```
##                  Estimate Std. Error    t value     Pr(>|t|)
## (Intercept)         13.23   1.148353 11.5208468 3.602548e-16
## suppVC              -5.25   1.624017 -3.2327258 2.092470e-03
## dosef1               9.47   1.624017  5.8312215 3.175641e-07
## dosef2              12.83   1.624017  7.9001660 1.429712e-10
## suppVC:dosef1       -0.68   2.296706 -0.2960762 7.683076e-01
## suppVC:dosef2        5.33   2.296706  2.3207148 2.410826e-02
```

There are two$\widehat{\omega}_{jk}$'s in the results, related to modifying the estimates for doses of 1 (-0.68) and 2 (5.33)for the Vitamin C group. If you want to re-construct the fitted values from the model that are displayed in Figure 4.13, you have to look for any coefficients that are "turned on" for a combination of levels of interest. For example, for the OJ group (solid line), the dosage of 0.5 mg/day has an estimate of an average growth of approximately 13 mm. This is the baseline group, so the model estimate for an observation in the OJ andyi,OJ,0.5mg0.5 mg/day dosage is simply $\widehat{y}_{i,OJ,0.5mg} = \widehat{\alpha}$ = 13.23 microns. For the OJ and 2 mg/day dosage estimate that has a value over 25 microns in the plot, the model incorporates the deviation for the 2 mg/day dosage: $\widehat{y}_{i,OJ,2mg} = \widehat{\alpha} + \widehat{\tau}_{2mg}$= 13.23 + 12.83 = 26.06 microns. For the Vitamin C group, another coefficient becomesyi,OJ,2mgα τ2mginvolved from its "main effect". For the VC and 0.5 mg dosage level, the estimate is approximately 8 microns. The pertinent model components are $\widehat{y}_{i,VC,0.5mg} = \widehat{\alpha} + \widehat{\gamma}_{VC}$= 13.23 + (-5.25) = 7.98 microns. Finally, whenyi,VC,0.5mgα γVCwe consider non-baseline results for both groups, three coefficients are required to reconstruct the results in the plot. For example, the estimate for the VC, 1 mg dosage is $\widehat{y}_{i,VC,1mg} = \widehat{\alpha} + \widehat{\tau}_{1mg} + \widehat{\gamma}_{VC} + \widehat{\omega}_{VC,1mg}$ =yi,VC,1mgα τ1mgγVCωVC,1mg13.23 + 9.47 + (-5.25) + (-0.68) = 16.77 microns. We usually will by-pass all this fun(!) with the coefficients in an interaction model and go from the ANOVA interaction test to focusing on the pattern of the responses in the interaction plot or going to the simpler additive model, but it is good to know that there are still model coefficients driving our results even if there are too many to be easily interpreted.

```
plot(allEffects(TG2), grid = T, multiline = T, lty = c(1:2), ci.style = "bars")
```

Given the presence of an important interaction, then the final step in the interpretation here is to interpret the results in the interaction plot or term-plot of the interaction model, supported by the p-value suggesting a different effect of supplement type based on the dosage level. To supplement this even more, knowing which combinations of levels differ can enhance our discussion. Tukey's HSD results (specifically the CLD) can be added to the original interaction plot by turning on the cld = T option in the intplot function as seen in Figure 4.14. Sometimes it is hard to see the letters and so there is also a cldshift = ... option to move the letters up or down; here a value of 1 seemed to work.

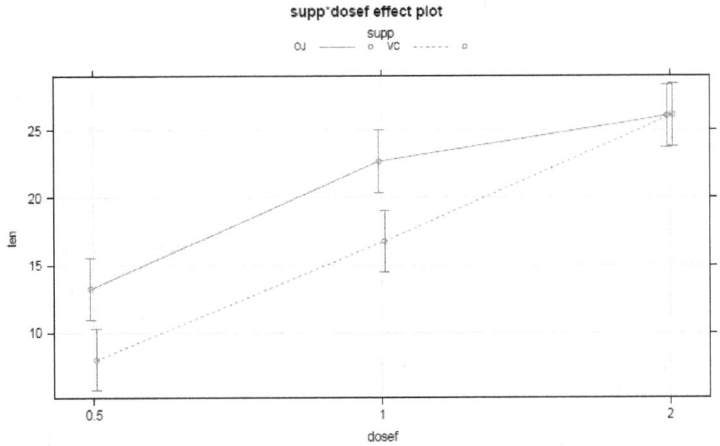

Figure 4.13: Term-plot for the estimated interaction for the Odontoblast Growth data using the multiline= T and ci.style = "bars" options.

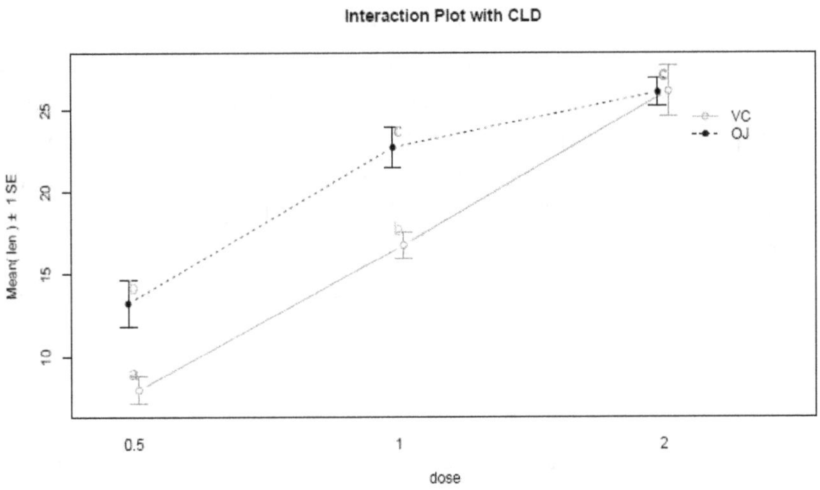

Figure 4.14: Interaction plot for Odontoblast data with added CLD from Tukey's HSD.

```
intplot(len ~ supp * dose, data = ToothGrowth, col = c(1,2), cldshift = 1,
        cld = T, main = "Interaction Plot with CLD")
```

The "size" interpretation of the previous hypothesis test result could be something like the following: Generally increasing the dosage increases the mean growth except for the 2 mg/day dosage level where the increase levels off in the OJ group (OJ 1 and 2 mg/day are not detectably different) and the differences between the two delivery methods disappear at the highest dosage level. But for 0.5 and 1 mg/day dosages, OJ is clearly better than VC by about 10 microns of growth on average.

4.5 OBSERVATIONAL STUDY EXAMPLE: THE PSYCHOLOGY OF DEBT

In this section, the analysis of a survey of N = 464 randomly sampled adults will be analyzed from a survey conducted by Lea et al. [1995] and available in the debt data set from the faraway package [Faraway, 2016]. The subjects responded to a variety of questions including whether they buy cigarettes (cigbuy: 0 if no, 1 if yes), their housing situation (house: 1 = rent, 2 = mortgage, and 3 = owned outright), their income group (incomegp: 1 = lowest, 5 = highest), and their score on a continuous scale of attitudes about debt (prodebt: 1 = least favorable, 6 = most favorable). The variable prodebt was derived as the average of a series of questions about debt with each question measured on an ordinal 1 to 6 scale, with higher values corresponding to more positive responses about going into debt of various kinds. The ordered scale on surveys that try to elicit your opinions on topics with scales from 1 to 5, 1 to 6, 1 to 7 or even, sometimes, 1 to 10 is called a Likert scale [Likert, 1932]. It is not a quantitative scale and really should be handled more carefully than taking an average of a set responses as was done here. That said, it is extremely common practice in social science research to treat ordinal responses as if they are quantitative and take the average of many of them to create a more continuous response variable like the one we are using here. If you continue your statistics explorations, you will see some better techniques for analyzing ordinal responses. That said, the scale of the response is relatively easy to understand as an amount of willingness to go into debt on a scale from 1 to 6 with higher values corresponding to more willingness to be in debt.

These data are typical of survey data where respondents were not required to answer all questions and there are some missing responses. We could clean out any individuals that failed to respond to all questions(called "complete cases") using the drop_na function, which will return responses only for subjects that responded to every question in the data set, debt. The change in sample size is available by running the dim function on the two data sets – there were 464 observations (rows) initially along with 13 variables (columns) and once observations with any missing values were dropped there are $N = 304$ for us to analyze. Losing35% of the observations is a pretty noticeable loss.

```
library(faraway)
data(debt)
library(tibble)
debt <- as_tibble(debt) %>% mutate(incomegp = factor(incomegp),
                                   cigbuy = factor(cigbuy)
                                   )
debtc <- debt %>% drop_na()

dim(debt)
```

```
## [1] 464  13
```

```
dim(debtc)
```

```
## [1] 304  13
```

Using drop_na() with a list a variable names, we can focus on the three variables we are using in this model and whether the responses are missing on them, only cleaning out rows that are missing on incomegp, cigbuy, and/or prodebt[13]. The missingness is less dramatic, retaining N = 388 observations in debtRc forour analysis using these three variables.

```
# Remove rows with missing values based on just three variables.
debtRc <- debt %>% drop_na(incomegp, cigbuy, prodebt)
dim(debtRc)
```

```
## [1] 388  13
```

The second approach seems better as it drops fewer observations so we will use that below. But suppose that people did not want to provide their income levels if they were in the lowest or highest income groups and that is why they are missing. Then we would be missing responses systematically and conclusions could be biased because of ignoring particular types of subjects. We don't have particular statistical tools to easily handle this problem but every researcher should worry about non-response when selected subjects do not respond at all or fail to answer some questions. When the missing values are systematic in some fashion and not just missing randomly (missing randomly might be thought of as caused by "demonic intrusion" [Hurlbert, 1984] that can't be easily explained or related to the types of responses), then we worry about non-response bias that is systematically biasing our results because of the missing responses. This also ties back into our discussion of who was sampled. We need to think carefully about who was part of the sample but refused to participate and how that might impact our inferences. And whether we can even address the research question of interest based on what was measured given those that refused/failed to respond. For example, suppose we are studying river flows and are interested in the height of a river. Missingness in these responses could arise because a battery fails or the data logger "crashes" (not related to the responses and so not definitely problematic) or because of something about the measurements to be taken that causes the missingness (suppose the gage can only can measure between one and three feet deep and the river is running at four feet deep during a flood or below 1 foot during a drought). The first machine failures are very different from the height-based missing responses; the height-based missingness clearly leads to bias in estimating mean river height because of what cannot be observed. In Chapter 5, we introduce the tableplotas another tool to visualize data that can also show missing data patterns to help you think about these sorts of issues further14. If you delete observations and the missing data are not random/non-systematic, your scope of inference is restricted to just those subjects that provided responses and were analyzed. If the missingness is random and not related to aspects of the measurements taken, then some missingness can be tolerated and still retain some comfort that inferences can be extended to the population a random sample of subjects was taken from.

Ignoring this potential for bias in the results for the moment, we are first interested in whether buying cigarettes/not and income groups interact in their explanation of the respondent's mean opinions on being in debt4.15) may suggest an interaction between cigbuy and incomegp where the lines cross, switching which of the cigbuy levels is higher (income levels 2, 3, and 5) or even almost not different (income levels 1 and 4). But it is not as clear as the previous examples, especially with how large the SEs are relative the variation in the means. The interaction F -test helps us objectively assess evidence against the null hypothesis of no interaction. Based on the plot, there do not appear to be differences based on cigarette purchasing but there might be some differences between the income groups if we drop the interaction from the model. If we drop the interaction, then this suggests that we might be in Scenario 2 or 3 where a single main effect of interest is present.

```
intplotarray(prodebt ~ cigbuy * incomegp, data = debtRc, col = c(1,3,4,5,6), lwd = 2)
```

As in other situations, and especially with observational studies where a single large sample is collected and then the levels of the factor variables are observed, it is important to check for balance – whether all the combinations of the two predictor variables are similarly represented. Even more critically, we need to check whether all the combinations of levels of factors are measured. If a combination is not measured, then we lose the ability to estimate the mean for that combination and the ability to test for an interaction. A solution to that problem would be to collapse the categories of one of the variables, changing the definitions of the levels but if you fail to obtain information for all combinations, you can't work with the interaction model. In this situation, we barely have enough information to proceed (the smallest n_{jk} is 13 for income group 4 that buys cigarettes). We have a very unbalanced design with counts between 13 and 60 in the different combinations, so lose some resistance to violation of assumptions but can proceed to explore the model with a critical eye on how the diagnostic plots look.

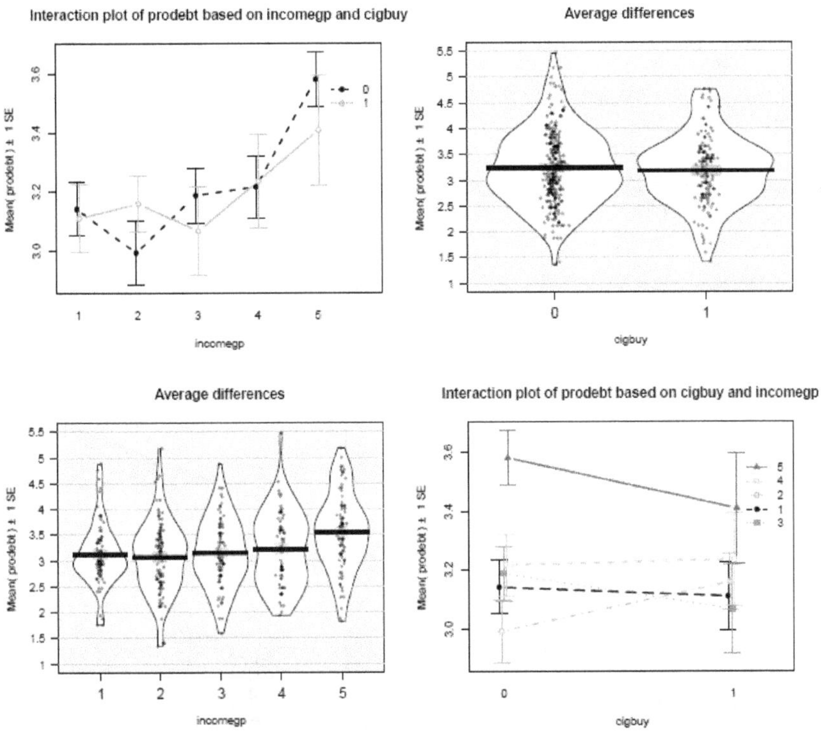

Figure 4.15: Interaction plot array of prodebt by income group (1 to 5) and whether they buy cigarettes (0= no, 1 = yes).

```
tally(cigbuy ~ incomegp, data = debtRc)
```

```
##         incomegp
## cigbuy  1  2  3  4  5
##       0 36 49 54 53 60
##       1 37 45 20 13 21
```

The test for the interaction is always how we start our modelling in Two-Way ANOVA situations. The ANOVA table suggests that there is little evidence against the null hypothesis of no interaction between the income level and buying cigarettes on the opinions of the respondents towards debt ($F(4, 378) = 0.686$,p-value = 0.6022), so we would conclude that there is likely not an interaction present here and we can drop the interaction from the model. This suggests that the initial assessment that the interaction wasn't too prominent was correct. We should move to the additive model here but first need to check the assumptions to make sure we can trust this initial test.

```
library(car)
debt1 <- lm(prodebt ~ incomegp * cigbuy, data = debtRc)
Anova(debt1)
```

```
## Anova Table (Type II tests)
##
## Response: prodebt
##                  Sum Sq  Df F value    Pr(>
## incomegp         10.742   4  5.5246 0.00024
## cigbuy            0.010   1  0.0201 0.88742
## incomegp:cigbuy   1.333   4  0.6857 0.60220
## Residuals       183.746 378
```

```
par(mfrow = c(2,2))
plot(debt1, pch = 16)
```

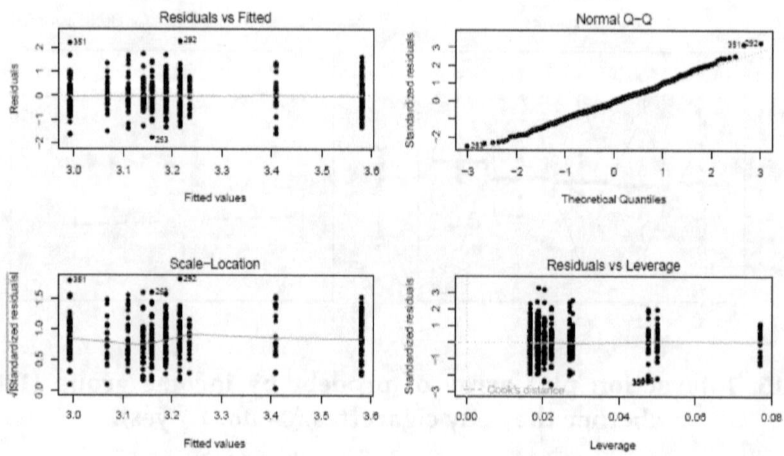

Figure 4.16: Diagnostic plot for prodebt by income group and buy cigarettes/not interaction model.

The diagnostic plots (Figure 4.16) seem to be pretty well-behaved with no apparent violations of the normality assumption and no clear evidence of a violation of the constant variance assumption. There is no indication of a problem with the independence assumption because there is no indication of structure to the measurements of the survey respondents that might create dependencies. In observational studies, violations of the independence assumption might come from repeated measures of the same person over time or multiple measurements within the same family/household or samples that are clustered geographically, none of which are part of the survey information we have. The random sampling from a population should allow inferences to a larger population except for that issue of removing partially missing responses so we can't safely generalize results beyond the complete observations we are using without worry that the missing subjects are systematically different from those we are able to analyze. We also don't have much information on the exact population sampled, so will just leave this vague here but know that there would be a population these conclusions apply since it was random sample (at least those that would answer the questions). All of this suggests proceeding to fitting and exploring the additive model is reasonable here. No causal inferences are possible because this is an observational study.

After ruling out the interaction of income and cigarette status on opinions about debt, we can focus on the additive model.

- Hypotheses (Two sets apply when the additive model is the focus!):
  - H_0: No difference in means for prodebt for income groups in population, given cigarette buying in model
  - ⇔ All τj's = 0 in additive model.
  - H_A: Some difference in means for prodebt for income group in population, given cigarette buying in model
  - ⇔ Not all τj's = 0 in additive model.
  - H_0: No difference in means for prodebt for cigarette buying/not in population, given income group in model
  - ⇔ All γk's = 0 in additive model.
  - H_A: Some difference in means for prodebt for cigarette buying/not in population, given income group in model
  - ⇔ Not all γk's = 0 in additive model.

2. Validity conditions – discussed above but with new plots for the additive model:

```
debt1r <- lm(prodebt ~ incomegp + cigbuy, data = debtRc)
par(mfrow = c(2,2))
plot(debt1r, pch = 16)
```

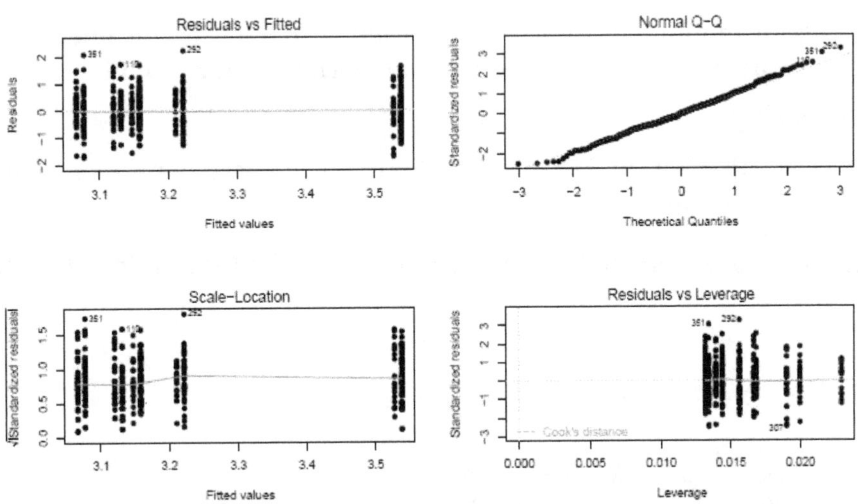

Figure 4.17: Diagnostic plot of additive model for "prodebt" by income group and whether they buy cigarettes/not.

⦿ Constant Variance:

- In the Residuals vs Fitted and the Scale-Location plots in Figure 4.17, the differences in variability among groups is minor and nothing suggests a violation. If you change models, you should always revisit the diagnostic plots to make sure you didn't create problems that were not present in more complicated models.
- We can also explore the partial residuals here as provided in Figure 4.18. The variability in the partial residuals appears to be similar across the different levels of each predictor, controlled for the other variable, and so does suggest any issues that were missed by just looking at the overall residuals versus fitted values in our regular diagnostic plots. Note how hard it is to see differences in the mean for levels of cigbuy in this plot relative to the variability in the partial residuals but that the differences in the means in incomegp are at least somewhat obvious.

```
plot(allEffects(debt1r, residuals = T))
```

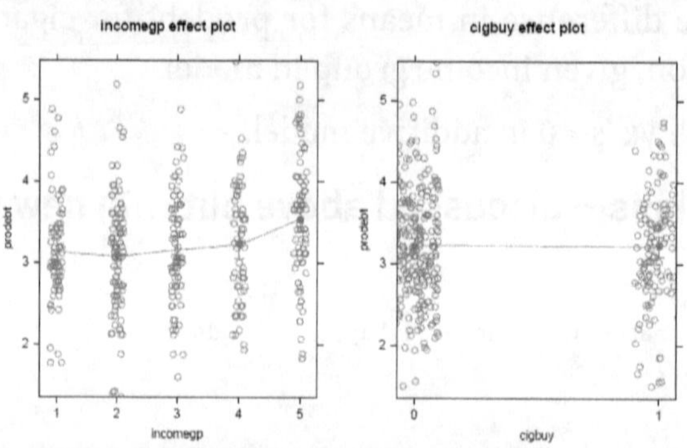

Figure 4.18: Term-plot for additive model for "prodebt" by income group and whether they buy cigarettes/not with partial residuals.

⦿ Normality of residuals:

- The QQ-Plot in Figure 4.17 does not suggest a problem with this assumption.

3. Calculate the test statistics and p-values for the two main effect tests.

```
Anova(debt1r)
```

```
## Anova Table (Type II tests)
##
## Response: prodebt
##            Sum Sq  Df F value    Pr(>F)
## incomegp   10.742   4  5.5428 0.0002399
## cigbuy      0.010   1  0.0201 0.8872394
## Residuals 185.079 382
```

- The test statistics are $F(4, 382) = 5.54$ and $F(1, 382) = 0.0201$ with p-values of 0.00024 and 0.887.

4. Conclusions (including for the initial work with the interaction test):

- There was initially little to no evidence against the null hypothesis of no interaction between income group and cigarette buying on pro-debt feelings ($F(4, 378) = 0.686$, p-value = 0.6022) so we would conclude that there is likely not an interaction in the population and the interaction was dropped from the model. There is strong evidence against the null hypothesis of no difference in the mean pro-debt feelings in the population across the income groups, after adjusting for cigarette buying ($F(4, 382) = 5.54$, p-value = 0.00024), so we would conclude that there is some difference in them. There is little evidence against the null hypothesis of no difference in the mean pro-debt feelings in the population based on cigarette buying/not, after adjusting for income group ($F(1, 382) = 0.0201$, p-value = 0.887), so we would conclude that there is probably not a difference across cigarette buying/not and could consider dropping this term from the model.
- So we learned that the additive model was more appropriate for these responses and that the results resemble Scenario 2 or 3 with only one main effect being important. In the additive model, the coefficients can be interpreted as shifts from the baseline after controlling for the other variable in the model.

5. Size:

- Figure 4.19 shows the increasing average comfort with being in debt as the income groups go up except between groups 1 and 2 where 1 is a little higher than two. Being a cigarette buyer was related to a lower comfort level with debt but is really no different from those that did not report buying cigarettes. It would be possible to consider follow-up tests akin to the Tukey's HSD comparisons for the levels of incomegp here but that is a bit beyond the scope of this course –focus on the estimated mean for the 5th income group being over 3.5 and none of the others over3.2. That seems like an interesting although modest difference in mean responses across income groups after controlling for cigarette purchasing or not.

(ref:fig4-19) Term-plots for the prodebt response additive model with left panel for income group and the right panel for buying cigarettes or not (0 for no, 1 for yes).

```
plot(allEffects(debt1r))
```

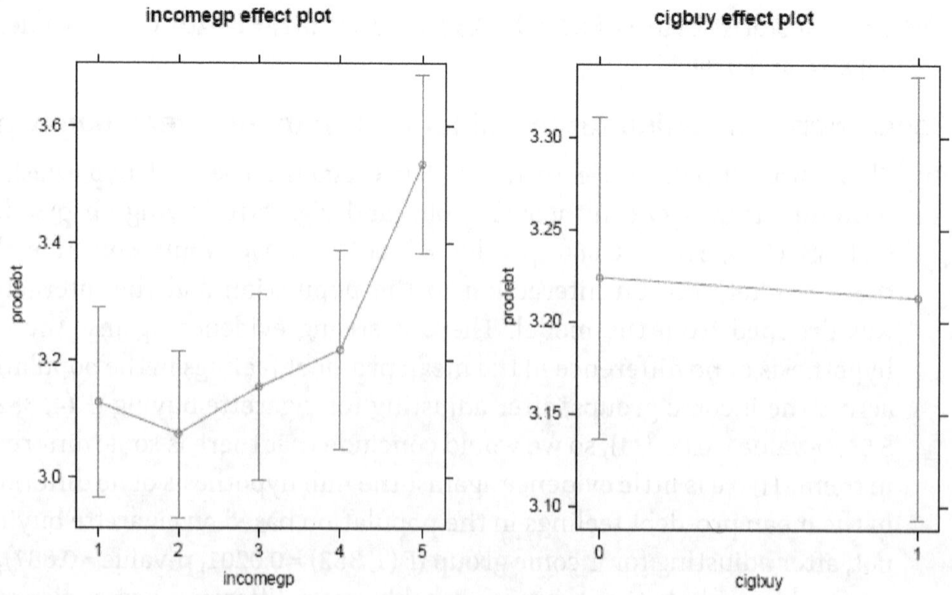

Figure 4.19:

6. Scope of inference:

- Because the income group and cigarette purchasing were not (and really could not) be randomly assigned, causal inference is not possible here. The data set came from a random sample but from an unspecified population and then there were missing observations. At best we can make inferences to those in that population that would answer these questions and it would be nice to know more about the population to really understand who this actually applies to. There would certainly be concerns about non-response bias in doing inference to the entire population that these data were sampled from.

The estimated coefficients can also be interesting to interpret for the additive model. Here are the model summary coefficients:

```
summary(debt1r)$coefficients
```

```
##                Estimate Std. Error    t value      Pr(>|t|)
## (Intercept)  3.13127172 0.09027437 34.6861672 4.283917e-120
## incomegp2   -0.05371924 0.10860898 -0.4946114  6.211588e-01
## incomegp3    0.02680595 0.11624894  0.2305909  8.177561e-01
## incomegp4    0.09072124 0.12059542  0.7522777  4.523474e-01
## incomegp5    0.40760033 0.11392712  3.5777288  3.911633e-04
## cigbuy1     -0.01088742 0.07672982 -0.1418929  8.872394e-01
```

In the model, the baseline group is for non-cigarette buyers (cigbuy = 0) and income group 1 with $\widehat{\alpha}$ = 3.131points. Regardless of the cigbuy level, the difference between income groups 2 and 1 is estimated to be $\widehat{\tau}_2$ = -0.054, an decrease in the mean score of 0.054 points. The difference between income groups 3 and 1 is $\widehat{\tau}_3$ = 0.027 points, regardless of cigarette smoking status. The estimated difference between cigarette buyers and non-buyers was estimated as $\widehat{\gamma}_2$ = -0.011 points for any income group, remember that this variable had a lar ge p-value in this model. The additive model-based estimates for all six combinations can be found in Table 4.3.

Table 4.3: Calculations to construct the estimates for all combinations of variables for the prodebt additive model.

Cig Buy	Income Group 1	Income Group 2	Income Group 3	Income Group 4	Income Group 5
0:No	$\hat{a}$ = 3.131 α	$\hat{a} + \hat{\tau}$ α τ2 = 3.131 – 0.016 = 3.115	$\hat{a} + \hat{\tau}$ α τ3 = 3.131 + 0.027 = 3.158	$\hat{a} + \hat{\tau}$ α τ4 = 3.131 + 0.091 = 3.222	$\hat{a} + \hat{\tau}$ α τ5 = 3.131 + = 3.539
1:Yes	$\hat{a} + \hat{\tau}$ α γ2 = 3 131.	$\hat{a} + \hat{\tau} + \gamma_2$ α τ2 γ2 = 3.131 – 0.016	$\hat{a} + \hat{\tau} + \gamma_2$ α τ3 γ2 = 3.131 + 0.027	$\hat{a} + \hat{\tau} + \gamma_2$ α τ4 γ2 = 3.131 + 0.091	$\hat{a} + \hat{\tau} + \gamma_2$ α τ5 γ2 = 3.131 +

One final plot of the fitted values from this additive model in Figure 4.20 hopefully crystallizes the implications of an additive model and reinforces that this model creates and assumes that the differences across levels of one variable are the same regardless of the level of the other variable and that this creates parallel lines. The difference between cigbuy levels across all income groups is a drop in -0.011 points. The income groups have the same differences regardless of cigarette buying or not, with income group 5 much higher than the other four groups. The minor differences in cigarette purchasing and large p-value for it controlled for income group suggest that we could also refine the model further and drop the cigbuy additive term and just focus on the income groups as a predictor – and this takes us right back to a One-Way ANOVA model so is not repeated here.

In general, we proceed through the following steps in any 2-WAY ANOVA situation:

- Make a pirate-plot and an interaction plot.
- Fit the interaction model; examine the test for the interaction.

- Check the residual diagnostic plots for the interaction model (especially normality and equal variance).
  - If there is a problem with normality or equal variance, consider a "transformation" of the response as discussed in Chapter 7. This can help make the responses have similar variances or responses (and the model residuals) to be more normal, but sometimes not both.

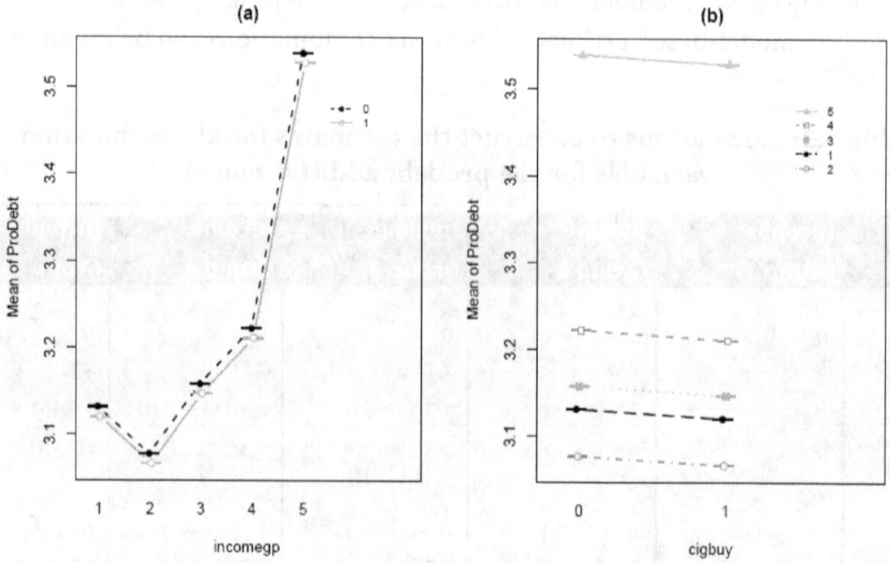

Figure 4.20: Illustration of the results from Table 4.3 showing the combined impacts of the components of the additive model for prodebt. Panel (a) uses income groups on the x-axis and different lines for cigarette buyers (1) or not (0). Panel (b) displays the different income groups as lines with the cigarette buying status on the x-axis.

- If the interaction test has a small p-value, that is your main result. Focus on the term-plot and the interaction plot from (1) to fully understand the results, adding Tukey's HSD results to intplot to see which means of the combinations of levels are detected as being different. Discuss the sizes of differences and the pattern of the estimated interaction.
- If the interaction is not considered important, then re-fit the model without the interaction (additive model) and re-check the diagnostic plots. If the diagnostics are reasonable to proceed:
  - Focus on the results for each explanatory variable, using Type II tests especially if the design is not balanced. Possibly consider further model refinement to only retain one of the two variables (the one with the smaller p-value) if a p-value is large. Follow One-Way ANOVA recommendations from this point on.

- Report the initial interaction test results and the results for the test for each variable from the model that is re-fit without the interaction.
- Model coefficients in the additive model are interesting as they are shifts from baseline for each level of each variable, controlling for the other variable – interpret those differences if the number of levels is not too great.

Whether you end up favouring an additive or interaction model or do further model refinement, all steps of the hypothesis testing protocol should be engaged and a story based on the final results should be compiled, supported by the graphical displays such as the term-plots and interaction plots.

4.6 PUSHING TWO-WAY ANOVA TO THE LIMIT: UN-REPLICATED DESIGNS AND ESTIMABILITY

In some situations, it is too expensive or impossible to replicate combinations of treatments and only one observation at each combination of the two explanatory variables, A and B, is possible. In these situations, even though we have information about all combinations of A and B, it is no longer possible to test for an interaction. Our regular rules for degrees of freedom show that we have nothing left for the error degrees of freedom and so we have to drop the interaction and call that potential interaction variability "error".

Without replication we can still perform an analysis of the responses and estimate all the coefficients in the interaction model but an issue occurs with trying to calculate the interaction F-test statistic – we run out of degrees of freedom for the error. To illustrate these methods, the paper towel example is revisited except that only one response for each combination is used. Now the entire data set can be easily printed out:

```
ptR <- read_csv("http://www.math.montana.edu/courses/s217/documents/ptR.csv")
ptR <- ptR %>% mutate(dropsf = factor(drops),
                      brand = factor(brand))
ptR
```

```
## # A tibble: 6 x 4
##   brand drops responses dropsf
##   <fct> <dbl>     <dbl> <fct>
## 1 B1       10     1.91  10
## 2 B2       10     3.05  10
## 3 B1       20     0.774 20
## 4 B2       20     2.84  20
## 5 B1       30     1.56  30
## 6 B2       30     0.547 30
```

Upon first inspection the interaction plot in Figure 4.21 looks like there might be some interesting interactions present with lines that look to be non-parallel.

But remember now that there is only a single observation at each combination of the brands and water levels so there is not much power to detect differences in this sort of situation and no replicates at any combinations of levels that allow estimation of SEs so no bands are produced in the plot.

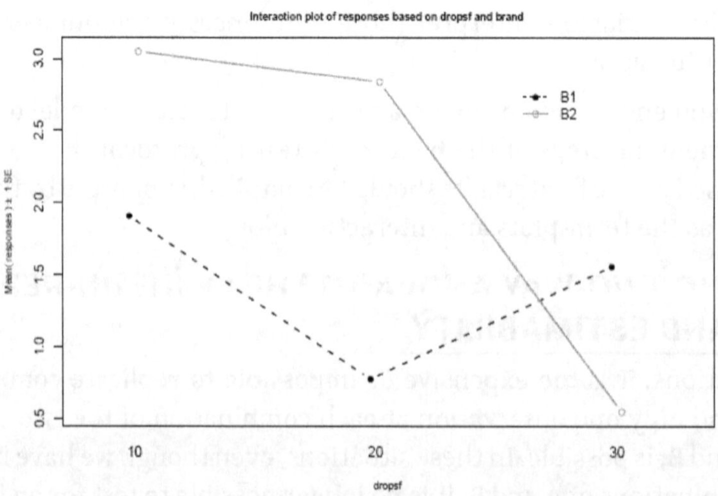

Figure 4.21: Interaction plot in paper towel data set with no replication.

```
intplot(responses ~ brand * dropsf, data = ptR, lwd = 2)
```

The next step would be to assess evidence related to the null hypothesis of no interaction between Brandand Drops. A problem will arise in trying to form the ANOVA table as you would see this when you run theanova15 function on the interaction model:

```
anova(lm(responses ~ dropsf * brand, data = ptR))
```

```
## Analysis of Variance Table
## Response: responses
##               Df  Sum Sq Mean Sq F value Pr(>F)
## dropsf         2 2.03872 1.01936
## brand          1 0.80663 0.80663
## dropsf:brand   2 2.48773 1.24386
## Residuals      0 0.00000
## Warning message:
## In anova.lm(lm(responses ~ dropsf * brand, data = ptR)) :
##   ANOVA F-tests on an essentially perfect fit are unreliable
```

Warning messages in R output show up after you run functions that contain problems and are generally not a good thing, but can sometimes be ignored. In this case, the warning message is not needed – there are no *F*-statistics or p-values in the results so we know there are some issues with the results. The **Residuals** line is key here – Residuals with 0 *df* and sums of squares of 0. Without replication, there are no degrees of freedom left to estimate the residual error. My first statistics professor, Dr. Gordon Bril at Luther College, used to refer to this as "shooting your load" by fitting too many terms in the model given the number of observations available. Maybe this is a bit graphic but hopefully will help you remember the need for replication if you want to test for interactions – it did for me. Without replication of observations, we run out of information to test all the desired model components.

So what can we do if we can't afford replication but want to study two variables in the same study? We can *assume* that the interaction does not exist and use those degrees of freedom and variability as the error variability. When we drop the interaction from Two-Way models, the interaction variability is added into the SSE so we assume that the interaction variability is really just "noise", which may not actually be true. We are not able to test for an interaction so must rely on the interaction plot to assess whether an interaction might be present. Figure 4.20 suggests there might be an interaction in these data (the two brands' lines suggesting non-parallel lines). So in this case, assuming no interaction is present is hard to justify. But if we proceed under this dangerous and untestable assumption, tests for the main effects can be developed.

```
norep1 <- lm(responses ~ dropsf + brand, data = ptR)
Anova(norep1)
```

```
## Anova Table (Type II tests)
##
## Response: responses
##            Sum Sq Df F value Pr(>F)
## dropsf    2.03872  2  0.8195 0.5496
## brand     0.80663  1  0.6485 0.5052
## Residuals 2.48773  2
```

In the additive model, the last row of the ANOVA table that is called the Residuals row is really the interaction row from the interaction model ANOVA table. Neither main effect had a small p-value (Drops:F (2, 2) = 0.82, p-value = 0.55 and Brand: F (1, 2) = 0.65, p-value = 0.51) in the additive model. To getsmall p-values with the small sample sizes that unreplicated designs would generate, the differences would need to be very large because the residual degrees of freedom have become very small. The term-plots in Figure 4.22 show that the differences among the levels are small relative to the residual variability as seen in the error bars around each point estimate.

```
plot(allEffects(norep1))
```

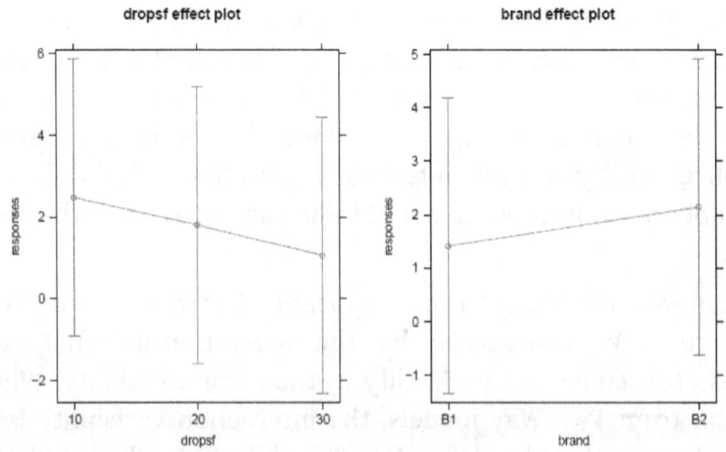

Figure 4.22: Term-plots for the additive model in paper towel data set with no replication.

In the extreme unreplicated situation it is possible to estimate all model coefficients in the interaction model but we can't do inferences for those estimates since there is no residual variability. Another issue in really any model with categorical predictors but especially noticeable in the Two-Way ANOVA situation is estimability issues. Instead of having issues with running out of degrees of freedom for tests we can run into situations where we do not have information to estimate some of the model coefficients. This happens any time you fail to have observations at either a level of a main effect or at a combination of levels in an interaction model.

To illustrate estimability issues, we will revisit the overtake data. Each of the seven levels of outfits was made up of a combination of different characteristics of the outfits, such as which helmet and pants were chosen, whether reflective leg clips were worn or not, etc. To see all these additional variables, we will introduce a new plot that will feature more prominently in Chapter 5 that allows us to explore relationships among a suite of categorical variables – the tableplot from the tabplot16 package [Tennekes and de Jonge,2019]. If this does not work, please contact your instructor or me for more information on possibly additional steps to get it working. The tabplot package allows us to sort the variables based on a single variable (think about how you might sort a spreadsheet based on one column and look at the results in other columns). The tableplot function displays bars for each response in a row17 based on the category of responses or as a barwith the height corresponding the value of quantitative variables18. It also plots a red cell if the observations were missing for a categorical variable and in grey for missing values on quantitative variables. The plotcan be obtained simply as tableplot (DATASETNAME) which will sort the data set based on the first variable. To use our previous work with the sorted levels of Condition2, the code dd[,-1] is used to specify the data set without Condition and then sort = Condition2 is used to sort based on the Condition2 variable. The pals = list("BrBG") option specifies a colour palette for the plot that is colour-blind friendly from the RColorBrewer package [Neuwirth, 2022].

```
dd <- read_csv("http://www.math.montana.edu/courses/s217/documents/Walker2014_mod.csv")
```

```
dd <- dd %>% mutate(Condition = factor(Condition),
                    Condition2 = reorder(Condition, Distance, FUN = mean),
                    Shirt = factor(Shirt),
                    Helmet = factor(Helmet),
                    Pants = factor(Pants),
                    Gloves = factor(Gloves),
                    ReflectClips = factor(ReflectClips),
                    Backpack = factor(Backpack)
                    )
```

```
library(remotes);
remotes::install_github("mtennekes/tabplot") # Only do this once on your computer
```

```
library(tabplot)
library(RColorBrewer)
# Options (sometimes) needed to prevent errors on PC
# options(ffbatchbytes = 1024^2 * 128); options(ffmaxbytes = 1024^2 * 128 * 32)
tableplot(dd[,-1], sort = Condition2, pals = list("BrBG"), sample = F,
          colorNA_num = "pink", numMode = "MB-ML")
```

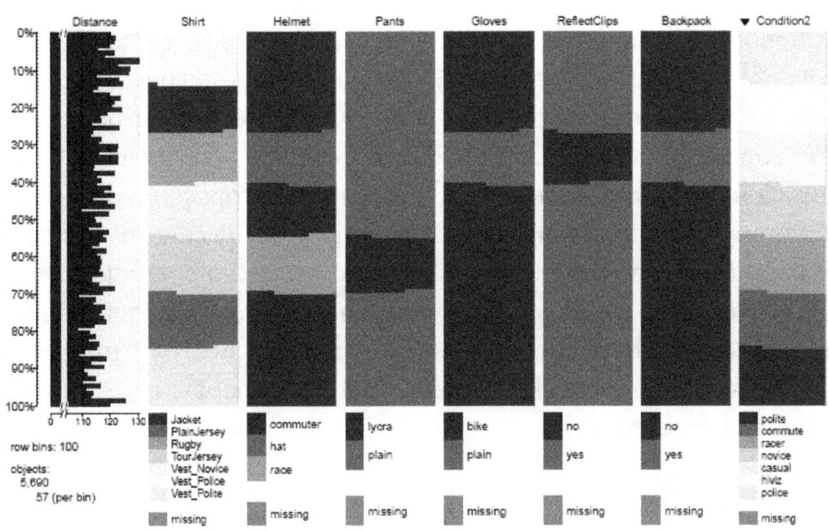

Figure 4.23: Tableplot of the full overtake data set sorted by outfit worn (Condition2).

In the tableplot in Figure 4.23, we can now explore the six variables created related to aspects of each outfit. For example, the commuter helmet (darkest shade in Helmet column) was worn with all outfits except for the racer and casual. So maybe we would like to explore differences in overtake distances based on the type of helmet worn. Similarly, it might be nice to explore whether wearing reflective pant clips is useful and maybe there is an interaction between helmet type and leg clips on impacts on overtake distance (should we wear both or just one, for example). So instead of using the seven level Condition2 in the model to assess differences based on all combinations of these outfits delineated in the other variables, we can try to fit a model with Helmet and ReflectClips and their interaction for overtake distances:

```
overtake_int <- lm(Distance ~ Helmet * ReflectClips, data = dd)
summary(overtake_int)
```

```
##
## Call:
## lm(formula = Distance ~ Helmet * ReflectClips, data = dd)
##
## Residuals:
##      Min       1Q   Median       3Q      Max
## -115.111  -17.756   -0.611   16.889  156.889
##
## Coefficients: (3 not defined because of singularities)
##                            Estimate Std. Error t value Pr(>|t|)
## (Intercept)                117.1106     0.4710 248.641   <2e-16
## Helmethat                    0.5004     1.1738   0.426    0.670
## Helmetrace                  -0.3547     1.1308  -0.314    0.754
## ReflectClipsyes                  NA         NA      NA       NA
## Helmethat:ReflectClipsyes        NA         NA      NA       NA
## Helmetrace:ReflectClipsyes       NA         NA      NA       NA
##
## Residual standard error: 30.01 on 5687 degrees of freedom
## Multiple R-squared:  5.877e-05,  Adjusted R-squared:  -0.0002929
## F-statistic: 0.1671 on 2 and 5687 DF,  p-value: 0.8461
```

The full model summary shows some odd things. First there is a warning after Coefficients of (3 not defined because of singularities). And then in the coefficient table, there are NAs for everything in the rows for ReflectClipsyes and the two interaction components. When lm encounters models where the data measured are not sufficient to estimate the model, it essentially drops parts of the model that you were hoping to estimate and only estimates what it can. In this case, it just estimates coefficients for the intercept and two deviation coefficients for Helmet types; the other three coefficients ($\gamma 2$ and the two ωs) are notestimable. This reinforces the need to check coefficients in any model you are fitting. A tally of the counts of observations across the two explanatory variables helps to understand the situation and problem:

```
tally(Helmet ~ ReflectClips, data = dd)
```

```
##           ReflectClips
## Helmet       no  yes
##   commuter    0 4059
##   hat       779    0
##   race        0  852
```

There are three combinations that have njk = 0 observations (for example for the commuter helmet, clips were always worn so no observations were made with this helmet without clips). So we have no hope of estimating a mean for the combinations with 0 observations and these are needed to consider interactions. If we revisit the tableplot, we can see how some of these needed combinations do not occur together. So this is an unbalanced design but also lacks necessary information to explore the potential research question of interest. In order to study just these two variables and their interaction, the researchers would have had to do rides with all six combinations of these variables. This could be quite informative because it could help someone tailor their outfit choice for optimal safety but also would have created many more than seven different outfit combinations to wear.

Hopefully by pushing the limits there are three conclusions available from this section. First, replication is important, both in being able to perform tests for interactions and for having enough power to detect differences for the main effects. Second, when dropping from the interaction model to additive model, the variability explained by the interaction term is pushed into the error term, whether replication is available or not. Third, we need to make sure we have observations at all combinations of variables if we want to be able to estimate models using them and their interaction.

4.7 CONCISE SUMMARY

In this chapter, methods for handling two different categorical predictors in the same model with a continuous response were developed. The methods build on techniques from Chapter 3 for the One-Way ANOVA and there are connections between the two models. This was most clearly seen in the Guinea Pig data set that was analyzed in both chapters. When two factors are available, it is better to start with the methods developed in this chapter because the interaction between the factors can, potentially, be separated from their main effects. The additive model is easier to interpret but should only be used when you are not convinced that there is an interaction is present. When an interaction is determined to be present, the main effects should not be interpreted and the interaction plot in combination with Tukey's HSD provides information on the important aspects of the results.

- If the interaction is retained in the model, there are two things you want to do with interpreting the interaction:
    - Describe the interaction, going through the changes from left to right in the interaction plot or term-plot for each level of the other variable.
    - Suggest optimal and worst combinations of the two variables to describe the highest and lowest possible estimated mean responses.

   - For example, you might want to identify a dosage and delivery method for the guinea pigs to recommend and one to avoid if you want to optimize odontoblast growth.
- If there is no interaction, then the additive model provides information on each of the variables and the differences across levels of each variable are the same regardless of the levels of the other variable.
   - You can describe the deviations from baseline as in Chapter 3, but for each variable, noting that you are controlling for the other variable.

Some statisticians might have different recommendations for dealing with interactions and main effects, especially in the context of models with interactions. We have chosen to focus on tests for interactions to screen for "real" interactions and then interpret the interaction plots aided by the Tukey's HSD for determining which combinations of levels are detectably different. Some suggest exploring the main effects tests even with interactions present. In some cases, those results are interesting but in others the results can be misleading and we wanted to avoid trying to parse out the scenarios when it might be safe to focus on the main effects in the presence of important interactions. Consider two scenarios, one where the main effects have large p-values but the interaction has a small p-value and the other where the main effects and the interaction all have small p-values. The methods discussed in this chapter allow us to effectively arrive at the interpretation of the differences in the results across the combinations of the treatments due to the interaction having a small p-value in both cases. The main effects results are secondary results at best when the interaction is important because we know that impacts of one explanatory variable is changing based on the levels of the other variable.

Chapter 5 presents a bit of a different set of statistical methods that allow analyses of data sets similar to those considered in the last two chapters but with a categorical response variable. The methods are very different in application but are quite similar in overall goals to those in Chapter 3 where differences in responses where explored across groups. After Chapter 5, the rest of the book will return to fitting models using the lm function as used here, but incorporating quantitative predictor variables and then eventually incorporating both categorical and quantitative predictor variables. The methods in Chapter 8 are actually quite similar to those considered here, so the better you understand these models, the easier that material will be to master.

4.8 SUMMARY OF IMPORTANT R CODE

The main components of R code used in this chapter follow with components to modify in lighter and/or ALL CAPS text, remembering that any R packages mentioned need to be installed and loaded for this code to have a chance of working:

- tally(A ~ B, data = DATASETNAME)
  - Requires the mosaic package be loaded.
  - Provides the counts of observations in each combination of categorical predictor variables A and B, used to check for balance and understand sample sizes in each combination.
- DATASETNAME <- DATASETNAME %>% mutate(VARIABLENAME = factor(VARIABLENAME))
  - Use the factor function on any numerically coded explanatory variable where the numerical codes represent levels of a categorical variable.
- intplot(Y ~ A*B, data = DATASETNAME)
  - Available in the catstats package or download and install using:source("http://www.math.montana.edu/courses/s217/documents/intplotfunctions_v3.R")
  - Provides interaction plot.
- intplotarray(Y ~ A*B, data = DATASETNAME)
  - Available in catstats or download and install using:
  - source("http://www.math.montana.edu/courses/s217/documents/intplotfunctions_v3.R")
  - Provides interaction plot array that makes interaction plots switching explanatory variable roles and makes pirate-plots of the main effects.
- INTERACTIONMODELNAME <- lm(Y ~ A*B, data = DATASETNAME)
  - Fits the interaction model with main effects for A and B and an interaction between them.
  - This is the first model that should be fit in Two-Way ANOVA modeling situations.
- ADDITIVEMODELNAME <- lm(Y ~ A + B, data = DATASETNAME)
  - Fits the additive model with only main effects for A and B but no interaction between them.
  - Should only be used if the interaction has been decided to be unimportant using a test for the interaction.
- summary(MODELNAME)
  - Generates model summary information including the estimated model coefficients, SEs, *t*-tests, and p-values.

- Anova(MODELNAME)
  - Requires the car package to be loaded.
  - Generates a Type II Sums of Squares ANOVA table that is useful for both additive and interaction models, but it is most important to use when working with the additive model as it provides inferences for each term conditional on the other one.
- par(mfrow = c(2,2)); plot(MODELNAME)
  - Generates four diagnostic plots including the Residuals vs Fitted and Normal Q-Q plot.
- plot(allEffects(MODELNAME))
  - Requires the effects package be loaded.
  - Plots the results from the estimated model.
- plot(allEffects(MODELNAME, residuals = T))
  - Plots the results from the estimated model with partial residuals.

4.9 PRACTICE PROBLEMS

4.1. Mathematics Usage Test Scores Analysis To practice the Two-Way ANOVA, consider a data set on N = 861 ACT Mathematics Usage Test scores from 1987. The test was given to a sample of high school seniors who met one of three profiles of high school mathematics course work: (a) Algebra I only;(b) two Algebra courses and Geometry; and (c) two Algebra courses, Geometry, Trigonometry, Advanced Mathematics, and Beginning Calculus. These data were generated from summary statistics for one particular form of the test as reported by Doolittle and Welch [1989]. The source of this version of the data set is Ramsey and Schafer [2012] and the Sleuth3 package [by F.L. Ramsey et al., 2019]. First install and then load that package.

```
library(Sleuth3)
library(mosaic)
library(tibble)
math <- as_tibble(ex1320)
math
names(math)
favstats(Score ~ Sex + Background, data = math)
```

Use the favstats summary to discuss whether the design was balanced or not.

4.1.2Make a pirate-plot and interaction plot array of the results and discuss the relationship between Sex, Background, and ACT Score.

4.1.3Write out the interaction model in terms of the Greek letters, making sure to define all the terms and don't forget the error terms in the model.

4.1.4Fit the interaction plot and find the ANOVA table. For the test you should consider first (the interaction), write out the hypotheses, report the test statistic, p-value, distribution of the test statistic under the null, and write a conclusion related to the results of this test.

4.1.5Re-fit the model as an additive model (why is this reasonable here?) and use Anova to find the Type II sums of squares ANOVA. Write out the hypothesis for the Background variable, report the test statistic, p-value, distribution of the test statistic under the null, and write a conclusion related to the results of this test. Make sure to discuss the scope of inference for this result.

4.1.6 Use the effects package to make a term-plot from the additive model from 4.5 and discuss the results. Specifically, discuss what you can conclude about the average relationship across both sexes, between Background and average ACT score?

4.1.7Add partial residuals to the term-plot and make our standard diagnostic plots and assess the assumptions using these plots. Can you assess independence using these plots? Discuss this assumption in this situation.

4.1.8 Use the term-plot and the estimated model coefficients to determine which of the combinations of levels provides the highest estimated average score.

4.2 Sleep Quality Analysis As a second example, consider data based on Figure 3 from Puhan et al. [2006], which is available at http://www.bmj.com/content/332/7536/266. In this study, the researchers were interested in whether didgeridoo playing might impact sleep quality (and therefore daytime sleepiness). They obtained volunteers and they randomized the subjects to either get a lesson or be placed on a waiting list for lessons. They constrained the randomization based on the high/low apnoea and high/low on the Epworth scale of the subjects in their initial observations to make sure they balanced the types of subjects going into the treatment and control groups. They measured the subjects' Epworth value (daytime sleepiness, higher is more sleepy) initially and after four months, where only the treated subjects (those who took lessons) had any intervention. We are interested in whether the mean Epworth scale values changed differently over the four months in the group that got didgeridoo lessons than it did in the control group (that got no lessons). Each subject was measured twice in the data set provided that is available at http://www.math.montana.edu/courses/s217/documents/epworthdata.csv.

```
library(readr)
epworthdata <- read_csv("http://www.math.montana.edu/courses/s217/documents/epworthdata.csv")
epworthdata <- epworthdata %>% mutate(Time = factor(Time),
                                      Group = factor(Group)
                                      )
levels(epworthdata$Time) <- c("Pre" , "Post")
levels(epworthdata$Group) <- c("Control" , "Didgeridoo")
```

4.2.1 Make a pirate-plot and an interaction plot array to graphically explore the potential interaction of Time and Group on the Epworth responses.

4.2.2 Fit the interaction model and find the ANOVA table. For the test you should consider first (the interaction), write out the hypotheses, report the test statistic, p-value, distribution of the test statistic under the null, and write a conclusion related to the results of this test.

4.2.3 Discuss the independence assumption for the previous model. The researchers used an analysis based on matched pairs. Discuss how using ideas from matched pairs might be applicable to the scenario discussed here.

4.2.4 Refine the model based on the previous test result and continue refining the model as the results might suggest. This should lead to retaining just a single variable. Make term-plot plot for this model and discuss this result related to the intent of the original research. If you read the original paper, they did find evidence of an effect of learning to play the didgeridoo (that there was a different change over time in the treated control when compared to the control group) – why might they have gotten a different result (hint: think about the previous question).

Note that the didgeridoo example is revisited in the case-studies in Chapter 9 with some information on an even better way to analyze these data.

CHAPTER-5

CHI-SQUARE TESTS

INTRODUCTION: AN OVERVIEW

The chi-square test is a fundamental statistical method that plays a crucial role in analyzing categorical data and uncovering patterns or associations within it. Developed by Karl Pearson in the late 19th century, the chi-square test has since become a staple in statistical analyses across various disciplines, from social sciences to biology and market research. This non-parametric test assesses the independence or association between two categorical variables by comparing observed frequencies with expected frequencies. By quantifying the degree of deviation between the observed and expected values, the chi-square test allows researchers to make informed inferences about the relationships within categorical data sets. Whether exploring the distribution of traits among different groups or examining the independence of variables in contingency tables, the chi-square test provides a versatile and powerful tool for researchers seeking to discern meaningful insights from categorical data. In this exploration of chi-square tests, we delve into the principles and applications that underpin this statistical technique, illuminating its significance in drawing conclusions from categorical data analyses.

5.1 SITUATION, CONTINGENCY TABLES, AND TABLEPLOTS

In this chapter, the focus shifts briefly from analyzing quantitative response variables to methods for handling categorical response variables. This is important because in some situations it is not possible to measure the response variable quantitatively. For example, we will analyze the results from a clinical trial where the results for the subjects were measured as one of three categories: no improvement, some improvement, and marked improvement. While that type of response could be treated as numerical, coded possibly as 1, 2, and 3, it would be difficult to assume that the responses such as those follow a normal distribution since they are discrete (not continuous, measured at whole number values only) and, more importantly, the difference between no improvement and some improvement is not necessarily the same as the difference between some and marked improvement. If it is treated numerically, then the differences between levels are assumed to be the same unless a different coding scheme is used (say 1, 2, and 5). It is better to treat this type of responses as being in one of the three categories and use statistical methods that don't make unreasonable and arbitrary assumptions about what the numerical coding might mean. The study being performed here involved subjects randomly assigned to either a treatment or a placebo (control) group and we want to address research questions similar to those considered in Chapters 2 and 3 – assessing differences in a response variable among two or more groups. With quantitative responses, the differences in the distributions are parameterized via the means of the groups and we used linear models. With categorical responses, the focus is on the probabilities of getting responses in each category and whether they differ among the groups.

We start with some useful summary techniques, both numerical and graphical, applied to some examples of studies these methods can be used to analyze. Graphical techniques provide opportunities for assessing specific patterns in variables, relationships between variables, and for generally understanding the responses obtained. There are many different types of plots and each can elucidate certain features of data. The tableplot, briefly introduced1 in Chapter 4, is a great and often fun starting point for working with data sets that contain categorical variables. We will start here with using it to help us understand some aspects of the results from a double-blind randomized clinical trial investigating a treatment for rheumatoid arthritis. These data are available in the Arthritis data set available in the vcd package [Meyer et al., 2022]. There were n = 84 subjects, with some demographic information recorded along with the Treatment status (Treated, Placebo) and whether the patients' arthritis symptoms Improved (with levels of None, Some, and Marked). When using tableplot, we may not want to display everything in the tibble and can just select some of the variables. We use Treatment, Improved, Gender, and Age in the select = ... option with a c() and commas between the names of the variables we want to display as shown below. The first one in the list is also the one that the data are sorted on and is what we want here – to start with sorting observations based on Treatment status.

```
library(vcd)
data(Arthritis) #Double-blind clinical trial with treatment and control groups
library(tibble)
Arthritis <- as_tibble(Arthritis)
# Homogeneity example
library(tabplot)
library(RColorBrewer)
# Options needed to (sometimes) prevent errors on PC
# options(ffbatchbytes = 1024^2 * 128); options(ffmaxbytes = 1024^2 * 128 * 32)
tableplot(Arthritis, select = c(Treatment, Improved, Sex, Age), pals = list("BrBG"),
          sample = F, colorNA_num = "orange", numMode = "MB-ML")
```

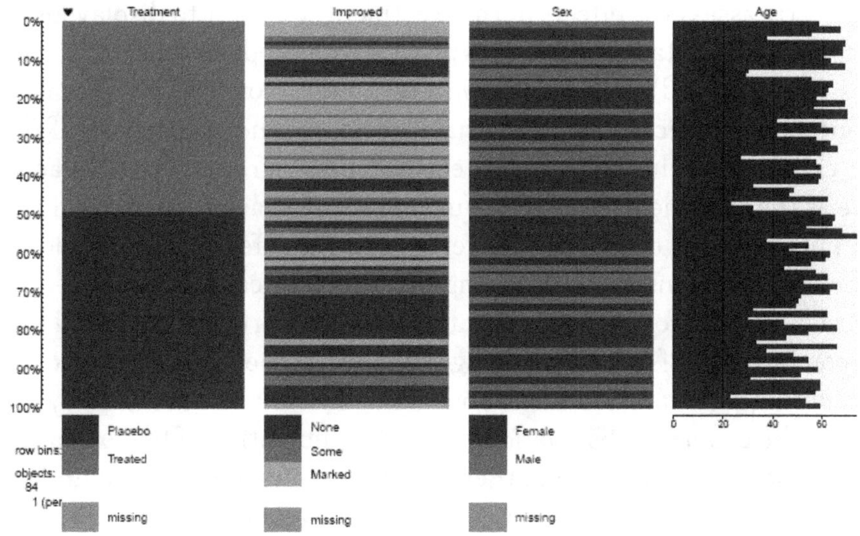

Figure 5.1: Tableplot of the arthritis data set.

The first thing we can gather from Figure 5.1 is that there are no red cells so there were no missing observations in the data set. Missing observations regularly arise in real studies when observations are not obtained for many different reasons and it is always good to check for missing data issues – this plot provides a quick visual method for doing that check. Primarily we are interested in whether the treatment led to a different pattern (or rates) of improvement responses. There seems to be more light (Marked) improvement responses in the treatment group and more dark (None) responses in the placebo group. This sort of plot also helps us to simultaneously consider the role of other variables in the observed responses. You can see the sex of each subject in the vertical panel for Sex and it seems that there is a relatively balanced mix of males and females in the treatment/placebo groups. Quantitative variables are also displayed with horizontal bars corresponding to the responses (the x-axis provides the units of the responses, here in years). From the panel for Age, we can see that the ages of subjects ranged from the 20s to 70s and that there is no clear difference in the ages between the treated and placebo groups. If, for example, all the male subjects had ended up being randomized into the treatment group, then we might have worried about whether sex and treatment were confounded and whether any differences in the responses might be due to sex instead of the treatment. The random assignment of treatment/placebo to the subjects appears to have been successful here in generating a mix of ages and sexes among the two treatment groups2. The main benefit of this sort of plot is the ability tovisualize more than two categorical variables simultaneously. But now we want to focus more directly on the researchers' main question – does the treatment lead to different improvement outcomes than the placebo?

To directly assess the effects of the treatment, we want to display just the two variables of interest. Stacked bar charts provide a method of displaying the response patterns (in Improved) across the levels of a predictor variable (Treatment) by displaying a bar for each predictor variable level and the proportions of responses in each category of the response in each of those groups. If the placebo is as effective as the treatment, then we would expect similar proportions of responses in each improvement category. A difference in the effectiveness would manifest in different proportions in the different improvement categories between Treated and Placebo. To get information in this direction, we start with obtaining the counts in each combination of categories using the tally function to generate contingency tables. Contingency tables with R rows and C columns (called R by C tables) summarize the counts of observations in each combination of the explanatory and response variables. In these data, there are R = 2 rows and C = 3 columns making a 2 × 3 table – note that you do not count the row and column for the "Totals" in defining the size of the table. In the table, there seems to be many more Marked improvement responses (21 vs 7) and fewer None responses (13 vs 29) in the treated group compared to the placebo group.

```
library(mosaic)
tally(~ Treatment + Improved, data = Arthritis, margins = T)
```

```
##            Improved
## Treatment None Some Marked Total
##   Placebo   29    7      7    43
##   Treated   13    7     21    41
##   Total     42   14     28    84
```

Using the tally function with ~ x + y provides a contingency table with the x variable on the rows and the y variable on the columns, with margins = T as an option so we can obtain the totals along the rows, columns, and table total of N = 84. In general, contingency tables contain the counts nrc in the rth row and cth column where r = 1, . . . , R and c = 1, . . . , C. We can also define the row totals as the sum across the columns of the counts in row r as

$$\mathbf{n_{r\bullet}} = \Sigma_{c=1}^{C} n_{rc},$$

the column totals as the sum across the rows for the counts in column c as

$$\mathbf{n_{\bullet c}} = \Sigma_{r=1}^{R} n_{rc},$$

and the table total as

$$\mathrm{N} = \Sigma_{r=1}^{R}\mathbf{n_{r\bullet}} = \Sigma_{c=1}^{C}\mathbf{n_{\bullet c}} = \Sigma_{r=1}^{R}\Sigma_{c=1}^{C}\mathbf{n_{rc}}.$$

We'll need these quantities to do some calculations in a bit. A generic contingency table with added row, column, and table totals just like the previous result from the tally function is provided in Table 5.1.

Table 5.1: General notation for counts in an R by C contingency table.

	Response Level 1	Response Level 2	Response Level 3	...	Response Level C	Totals
Group 1	n11	n12	n13	...	n1C	n1•
Group 2	n21	n22	n23	...	n2C	n2•
...	...	...	...	...	...	...
Group R	nR1	nR2	nR3	...	nRC	nR•
Totals	n•1	n•2	n•3	...	n•C	N

Comparing counts from the contingency table is useful, but comparing proportions in each category is better, especially when the sample sizes in the levels of the explanatory variable differ. Switching the formula used in the tally function formula to ~ y | x and adding the format = "proportion" option provides the proportions in the response categories conditional on the category of the predictor (these are called conditional proportions or the conditional distribution of, here, Improved on Treatment)3. Note that they sum to 1.0 in each level of x, placebo or treated:

```
tally(~ Improved | Treatment, data = Arthritis, format = "proportion", margins = T)
```

```
##           Treatment
## Improved    Placebo   Treated
##    None   0.6744186 0.3170732
##    Some   0.1627907 0.1707317
##    Marked 0.1627907 0.5121951
##    Total  1.0000000 1.0000000
```

This version of the tally result switches the variables between the rows and columns from the first summary of the data but the single "Total" row makes it clear to read the proportions down the columns in this version of the table. In this application, it shows how the proportions seem to be different among categories of Improvement between the placebo and treatment groups. This matches the previous thoughts on these data, but now a difference of marked improvement of 16% vs 51% is more clearly a big difference. We can also display this result using a stacked bar chart4 that displays the same information using the plot function with a y ~ x formula:

```
par(mai = c(1.5,1.5,0.82,0.42), #Adds extra space to bottom and left margin,
    las = 2, #Rotates text labels, optional code
    mgp = c(6,1,0)) #Adds space to labels, order is axis label, tick label, tick mark
plot(Improved ~ Treatment, data = Arthritis,
     main = "Stacked Bar Chart of Arthritis Data")
```

The stacked bar chart in Figure 5.2 displays the previous conditional proportions for the groups, with the same relatively clear difference between the groups persisting. If you run the plot function with variables that are coded numerically, it will make a very different looking graph (R is smart!) so again be careful that you are instructing R to treat your variables as categorical if they really are categorical. R is powerful but can't read your mind!

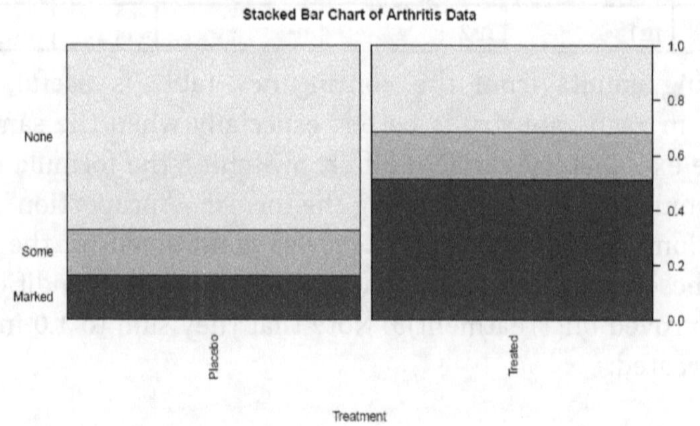

Figure 5.2: Stacked bar chart of Arthritis data. The left bar is for the Placebo group and the right bar is for the Treated group. The width of the bars is based on relative size of each group and the portion of the total height of each shaded area is the proportion of that group in each category. The lightest shading is for "none", medium shading for "some", and the darkest shading for "marked", as labelled on the y-axis.

In this chapter, we analyze data collected in two different fashions and modify the hypotheses to reflect the differences in the data collection processes, choosing either between what are called Homogeneity and Independence tests. The previous situation where levels of a treatment are randomly assigned to the subjects in a study describes the situation for what is called a Homogeneity Test. Homogeneity also applies when random samples are taken from each population of interest to generate the observations in each group of the explanatory variable based on the population groups. These sorts of situations resemble many of the examples from Chapter 3 where treatments were assigned to subjects. The other situation considered is where a single sample is collected to represent a population and then a contingency table is formed based on responses on two categorical variables. When one sample is collected and analyzed using a contingency table, the appropriate analysis is called a Chi-square test of Independence or Association. In this situation, it is not necessary to have variables that are clearly classified as explanatory or response although it is certainly possible. Data that often align with Independence testing are collected using surveys of subjects randomly selected from a single, large population. An example, analyzed below, involves a survey of voters and whether their party affiliation is related to who they voted for – the Republican, Democrat, or other candidate. There is clearly an explanatory variable of the Party affiliation but a single large sample was taken from the population of all likely voters so the Independence test needs to be applied. Another example where Independence is appropriate involves a study of student cheating behaviour. Again, a single sample was taken from the population of students at a university and this determines that it will be an Independence test. Students responded to questions about lying to get out of turning in a paper and/or taking an exam (none, either, or both) and copying on an exam and/or turning in a paper written by someone else (neither, either, or both). In this situation, it is not clear which variable is response or explanatory (which should explain the other) and it does not matter with the Independence testing framework. Figure 5.3 contains a diagram of the data collection processes and can help you to identify the appropriate analysis situation.

You will discover that the test statistics are the same for both methods, which can create some desire to assume that the differences in the data collection don't matter. In Homogeneity designs, the sample size in each group (n1•, n2•, . . . , nR•) is fixed (researcher chooses the size of each group). In Independence situations, the total sample size N is fixed but all the nr•'s are random (we need the data set to know how many are in each group). These differences impact the graphs, hypotheses, and conclusions used even though the test statistics and p-values are calculated the same way – so we only need to learn one test statistic to handle the two situations, but we need to make sure we know which we're doing!

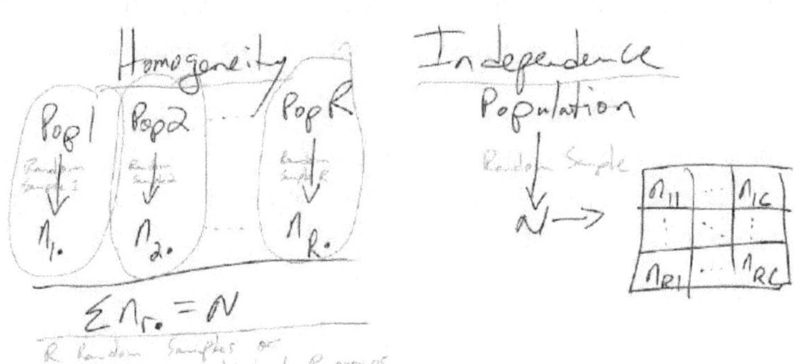

Figure 5.3: Diagram of the scenarios involved in Homogeneity and Independence tests. Homogeneity testing involves R random samples or subjects assigned to R groups. Independence testing involves a single random sample and measurements on two categorical variables.

5.2 HOMOGENEITY TEST HYPOTHESES

If we define some additional notation, we can then define hypotheses that allow us to assess evidence related to whether the treatment "matters" in Homogeneity situations. This situation is similar to what we did in the One-Way ANOVA (Chapter 3) situation with quantitative responses but the parameters now relate to proportions in the response variable categories across the groups. First we can define the conditional population proportions in level *c* (column $c = 1, \ldots, C$) of group *r* (row $r = 1, \ldots, R$) as *prc*. Table 5.2 shows the proportions, noting that the proportions in each row sum to 1 since they are conditional on the group of interest. A ***transposed*** (rows and columns flipped) version of this table is produced by the tally function if you use the formula ~ y | x.

Table 5.2: Table of conditional proportions in the Homogeneity testing scenario.

	Response Level 1	Response Level 2	Response Level 3	...	Response Level C	Totals
Group 1	p11	p12	p13	...	p1C	1.0
Group 2	p21	p22	p23	...	p2C	1.0
...	...	...	...	...	...	...
Group R	pR1	pR2	pR3	...	pRC	1.0
Totals	p•1	n•2	p•3	...	p•C	1.0

In the Homogeneity situation, the null hypothesis is that the distributions are the same in all the *R* populations. This means that the null hypothesis is:

$\mathbf{H_0 : p_{11} = p_{21} = \ldots = p_{R1}}$ **and** $\mathbf{p_{12} = p_{22} = \ldots = p_{R2}}$ **and** $\mathbf{p_{13} = p_{23} = \ldots = p_{R3}}$ **and ... and** $\mathbf{p_{1C} = p_{2C} = \ldots = p_{RC}}$.

If all the groups are the same, then they all have the same conditional proportions and we can more simply write the null hypothesis as:

$\mathbf{H_0 : (p_{r1}, p_{r2}, \ldots, p_{rC}) = (p_1, p_2, \ldots, p_C)}$ **for all r.**

In other words, the pattern of proportions across the columns are the same for all the R groups. The alternative is that there is some difference in the proportions of at least one response category for at least one group. In slightly more gentle and easier to reproduce words, equivalently, we can say:

H0 : The population distributions of the responses for variable y are the same across the R groups.

The alternative hypothesis is then:

HA : The population distributions of the responses for variable y are NOT ALL the same across the R groups.

To make this concrete, consider what the proportions could look like if they satisfied the null hypothesis for the Arthritis example, as displayed in Figure 5.4. Stacked bar charts provide a natural way to visualize the null hypothesis (equal distributions) to compare to the observed proportions in the observed data. Stacked bar charts are the appropriate visual display to present the summarized data in homogeneity test situations.

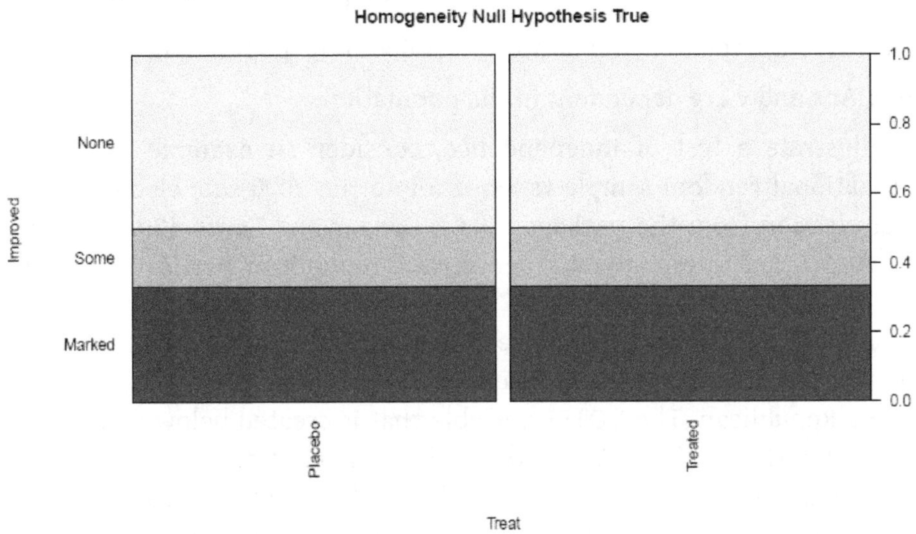

Figure 5.4: Stacked bar chart of one way that the Arthritis proportions could have been if the null hypothesis had been true.

Note that the proportions in the different response categories do not need to be the same just that the distribution needs to be the same across the groups. The null hypothesis does *not* require that all three response categories (*none, some, marked*) be equally likely. It assumes that whatever the distribution of proportions is across these three levels of the response that there is no difference in that distribution between the explanatory variable (here treated/placebo) groups. Figure 5.4 shows an example of a situation where the null hypothesis is true and the distributions of responses across the groups look the same but the proportions for *none, some* and *marked* are not all equally likely. That situation satisfies the null hypothesis. Compare this plot to the one for the real data set in Figure 5.2. It looks like there might be some differences in the responses between the treated and placebo groups as that plot looks much different from this one, but we will need a test statistic and a p-value to fully address the evidence relative to the previous null hypothesis.

5.3 INDEPENDENCE TEST HYPOTHESES

When we take a single random sample of size *N* and make a contingency table, our inferences relate to whether there is a relationship or ***association*** (that they are not independent) between the variables. This is related to whether the distributions of proportions match across rows in the table but is a more general question since we do not need to determine a variable to condition on, one that takes on the role of an explanatory variable, from the two variables of interest. In general, the hypotheses for an Independence test for variables *x* and *y* are:

H0: There is no relationship between x and y in the population.

Or: H0: x and y are independent in the population.

HA: There is a relationship between x and y in the population.

Or: HA: x and y are dependent in the population.

To illustrate a test of independence, consider an example involving data from a national random sample taken prior to the 2000 U.S. elections from the data set election from the package poLCA (Linzer and Lewis. [2022], Linzer and Lewis [2011]). Each respondent's democratic-republican partisan identification was collected, provided in the PARTY variable for measurements on a seven-point scale from (1) Strong Democrat,(2) Weak Democrat, (3) Independent-Democrat, (4) Independent-Independent, (5) Independent-Republican, (6) Weak Republican, to (7) Strong Republican. The VOTEF variable that is created below will contain the candidate that the participants voted for (the data set was originally coded with 1, 2, and 3 for the candidates and we replaced those levels with the candidate names). The contingency table shows some expected results, that individuals with strong party affiliations tend to vote for the party nominee with strong support for Gore in the Democrats (PARTY = 1 and 2) and strong support for Bush in the Republicans (PARTY = 6 and 7). As always, we want to support our explorations with statistical inferences, here with the potential to extend inferences to the overall population of voters. The inferences in an independence test are related to whether there is a relationship between the two variables in the population. A relationship between variables occurs when knowing the level of one variable for a person, say that they voted for Gore, informs the types of responses that you would expect for that person, here that they are likely affiliated with the Democratic Party. When there is no relationship (the null hypothesis here), knowing the level of one variable is not informative about the level of the other variable.

```
library(poLCA)
# 2000 Survey - use package = "" because other data sets in R have same name
data(election, package = "poLCA")
election <- as_tibble(election)
# Subset variables and remove missing values
election2 <- election %>%
  select(PARTY, VOTE3) %>%
  mutate(VOTEF = factor(VOTE3)) %>%
  drop_na()

levels(election2$VOTEF) <- c("Gore", "Bush", "Other") #Replace 1,2,3 with meaningful names
levels(election2$VOTEF) #Check new names of levels in VOTEF
```

```
## [1] "Gore"  "Bush"  "Other"
```

```
electable <- tally(~ PARTY + VOTEF, data = election2) #Contingency table
```

```
electable
```

```
##        VOTEF
## PARTY Gore Bush Other
##     1  238    6     2
##     2  151   18     1
##     3  113   31    13
##     4   37   37    11
##     5   21  124    12
##     6   20  121     2
##     7    3  189     1
```

The hypotheses for an Independence/Association Test here are:

- ***H0:*** There is no relationship between party affiliation and voting status in the population.
  - **Or: *H0:*** Party affiliation and voting status are independent in the population.
- ***HA:*** There is a relationship between party affiliation and voting status in the population.
  - **Or: *HA:*** Party affiliation and voting status are dependent in the population.

You could also write these hypotheses with the variables switched and that is also perfectly acceptable. Because these hypotheses are ambivalent about the choice of a variable as an "x" or a "y", the summaries of results should be consistent with that idea. We should not calculate conditional proportions or make stacked bar charts since they imply a directional relationship from x to y (or results for y conditional on the levels ofx) that might be hard to justify. Our summaries in these situations are the contingency table (tally(~ var1+ var2, data = DATASETNAME)) and a new graph called a mosaic plot (using the mosaicplot function).

Mosaic plots display a box for each cell count whose area corresponds to the proportion of the total data set that is in that cell (nrc/N). In some cases, the bars can be short or narrow if proportions of the total are small and the labels can be hard to read but the same bars or a single line exist for each category of the variables in all rows and columns. The mosaic plot makes it easy to identify the most common combination of categories. For example, in Figure 5.5 the Gore and PARTY = 1 (Strong Democrat) box in the top segment under column 1 of the plot has the largest area so is the highest proportion of the total. Similarly, the middle segment on the right for the PARTY category 7s corresponds to the Bush voters who were a 7 (Strong Republican). Knowing that the middle box in each column is for Bush voters is a little difficult as "Other" and "Bush" overlap each other in the y-axis labelling but it is easy enough to sort out the story here if we

have briefly explored the contingency table. We can also get information about the variable used to make the columns as the width of the columns is proportional to the number of subjects in each PARTY category in thisplot. There were relatively few 4s (Independent-Independent responses) in total in the data set. Also, the Other category was the highest proportion of any vote-getter in the PARTY = 4 column but there were actually slightly more Other votes out of the total in the 3s (Independent-Democrat) party affiliation. Comparing the size of the 4s & Other segment with the 3s & Other segment, one should conclude that the 3s & Other segment is a slightly larger portion of the total data set. There is generally a gradient of decreasing/increasing voting rates for the two main party candidates across the party affiliations, but there are a few exceptions. For example, the proportion of Gore voters goes up slightly between the PARTY affiliations of 5s and 6s –as the voters become more strongly republican. To have evidence of a relationship, there just needs to be a pattern of variation across the plot of some sort but it does not need to follow such an easily described pattern, especially when the categorical variables do not contain natural ordering.

The mosaic plots are best made on the tables created by the tally function from a table that just contains the counts (**no totals**):

```
# Makes a mosaic plot where areas are related to the proportion of
# the total in the table
mosaicplot(electable, main = "Mosaic plot of observed results")
```

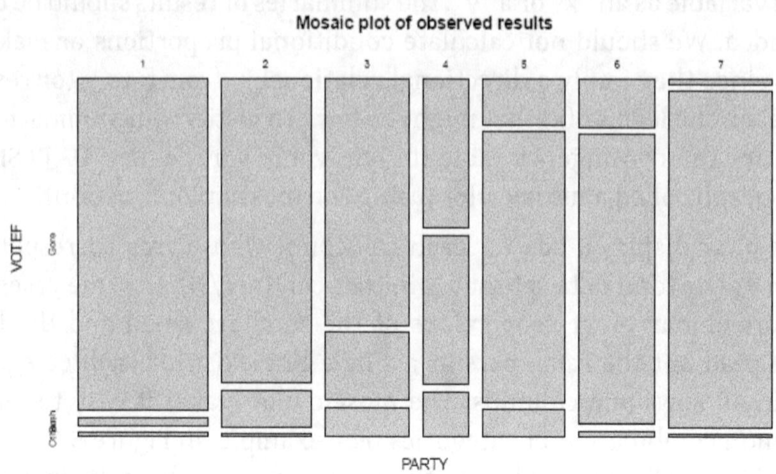

Figure 5.5: Mosaic plot of the 2000 election data comparing party affiliation and voting results.

In general, the results here are not too surprising as the respondents became more heavily republican, they voted for Bush and the same pattern occurs as you look at more democratic respondents. As the voters leaned towards being independent, the proportion voting for "Other" increased. So it certainly seems that there is some sort of relationship between party affiliation and voting status. As always, it is good to compare the observed results to what we would expect if the null hypothesis is true. Figure 5.6 assumes that the null hypothesis is true and shows the variation in the proportions in each category in the columns and variation in the proportions across the rows, but displays no relationship between PARTY and VOTEF. Essentially, the pattern5.6 but was in certain places in Figure 5.5.

5.4 MODELS FOR R BY C TABLES

This section is very short in this chapter because we really do not use any "models" in this Chapter. There are some complicated statistical models that can be employed in these situations, but they are beyond the scope of this book. What we do have in this situation is our original data summary in the form of a contingency table, graphs of the results like those seen above, a hypothesis test and p-value (presented below), and some post-test plots that we can use to understand the "source" of any evidence we found in the test.

5.5 PERMUTATION TESTS FOR THE X^2 STATISTIC

In order to assess the evidence against our null hypotheses of no difference in distributions or no relationship between the variables, we need to define a test statistic and find its distribution under the null hypothesis. The test statistic used with both types of tests is called the X2 statistic (we want to call the statistic X-square not Chi-square). The statistic compares the observed counts in the contingency table to the expected counts under the null hypothesis, with large differences between what we observed and what we expect under the null leading to evidence against the null hypothesis. To help this statistic to follow a named parametric distribution and provide some insights into sources of interesting differences from the null hypothesis, we standardize5 the difference between the observed and expected counts by the square-root of the expected count. The X2 statistic is based on the sum of squared standardized differences,

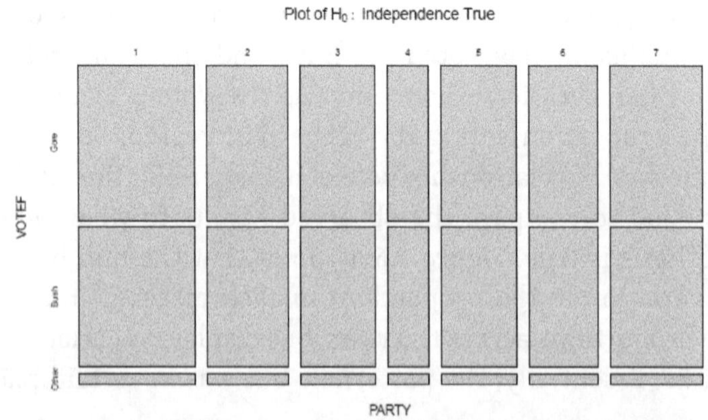

Figure 5.6: Mosaic plot of what the 2000 election data would look like if the null hypothesis of no relationship were true.

$$X^2 = \Sigma_{i=1}^{RC}\left(\frac{Observed_i - Expected_i}{\sqrt{Expected_i}}\right)^2,$$

which is the sum over all (R times C) cells in the contingency table of the square of the difference between observed and expected cell counts divided by the square root of the expected cell count. To calculate this test statistic, it useful to start with a table of expected cell counts to go with our contingency table of observed counts. The expected cell counts are easiest to understand in the homogeneity situation but are calculated the same in either scenario.

The idea underlying finding the expected cell counts is to find how many observations we would expect in category c given the sample size in that group, nr•, if the null hypothesis is true. Under the null hypothesis across all R groups the conditional probabilities in each response category must be the same. Consider Figure 5.7 where, under the null hypothesis, the probability of None, Some, and Marked are the same in both treatment groups. Specifically we have Pr(None) = 0.5, Pr(Some) = 0.167, and Pr(Marked) = 0.333. With nPlacebo• = 43 and Pr(None) = 0.50, we would expect 43 $*$ 0.50 = 21.5 subjects to be found in the Placebo, None combination if the null hypothesis were true. Similarly, with Pr(Some) = 0.167, we would expect 43 $*$ 0.167 = 7.18 in the Placebo, Some cell. And for the Treated group with nTreated• = 41, the expected count in the Marked improvement group would be 41 $*$ 0.333 = 13.65. Those conditional probabilities came from aggregating across the rows because, under the null, the row (Treatment) should not matter. So, the conditional probability was actually calculated as n•c/N = total number of responses in category c divided by the table total. Since each expected cell count was a conditional probability times the number of observations in the row, we can re-write the expected cell count formula for row r and column c as:

$$\text{Expected cell count}_{rc} = \frac{(\mathbf{n_{r\bullet}} * \mathbf{n_{\bullet c}})}{\mathrm{N}} = \frac{(\text{row } r \text{ total } * \text{ column } c \text{ total})}{\text{table total}}.$$

Table 5.3 demonstrates the calculations of the expected cell counts using this formula for all 6 cells in the2 × 3 table.

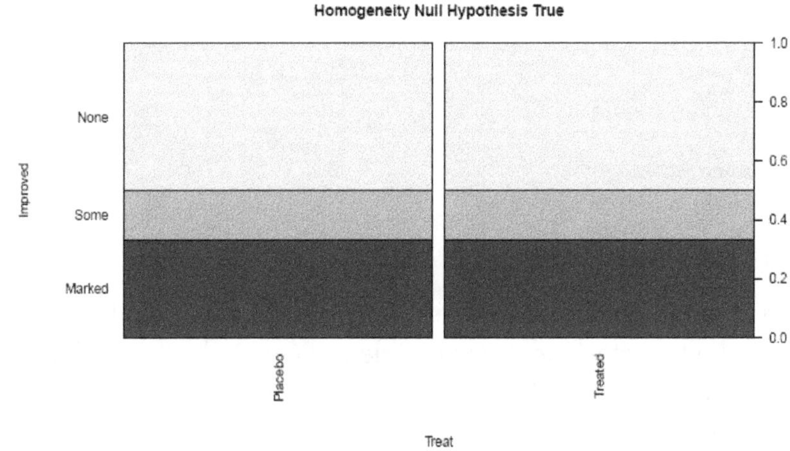

Figure 5.7: Stacked bar chart that could occur if the null hypothesis were true for the Arthritis study.

Table 5.3: Demonstration of calculation of expected cell counts for Arthritis data.

	None	Some	Marked	Totals
Placebo	$\frac{n_{\text{Placebo}\bullet} * n_{\bullet\text{None}}}{N} = \frac{43*42}{84} = 21.5$	$\frac{n_{\text{Placebo}\bullet} * n_{\bullet\text{Some}}}{N} = \frac{43*14}{84} = 7.167$	$\frac{n_{\text{Placebo}\bullet} * n_{\bullet\text{Marked}}}{N} = \frac{43*28}{84} = 14.33$	$n_{\text{Placebo}\bullet} = 43$
Treated	$\frac{n_{\text{Treated}\bullet} * n_{\bullet\text{None}}}{N} = \frac{41*42}{84} = 20.5$	$\frac{n_{\text{Treated}\bullet} * n_{\bullet\text{Some}}}{N} = \frac{41*14}{84} = 6.83$	$\frac{n_{\text{Treated}\bullet} * n_{\bullet\text{Marked}}}{N} = \frac{41*28}{84} = 13.67$	$n_{\text{Treated}\bullet} = 41$
Totals	$n_{\bullet\text{None}} = 42$	$n_{\bullet\text{Some}} = 14$	$n_{\bullet\text{Marked}} = 28$	$N = 84$

Of course, using R can help us avoid tedium like this. . . The main engine for results in this chapter is the chisq.test function. It operates on a table of counts that has been produced without row or columntotals.

For example, Arthtable below contains just the observed cell counts. Applying the chisq.test function6 to Arthtable provides a variety of useful output. For the moment, we are just going to extract the information in the "expected" attribute of the results from running this function (using chisq.test(TABLENAME)$expected). These are the expected cell counts which match the previous calculations except for some rounding in the hand-calculations.

```
Arthtable <- tally(~ Treatment + Improved, data = Arthritis)
Arthtable
```

```
##           Improved
## Treatment None Some Marked
##    Placebo   29    7      7
##    Treated   13    7     21
```

```
chisq.test(Arthtable)$expected
```

```
##          Improved
## Treatment None     Some   Marked
##   Placebo 21.5 7.166667 14.33333
##   Treated 20.5 6.833333 13.66667
```

With the observed and expected cell counts in hand, we can turn our attention to calculating the test statistic. It is possible to lay out the "contributions" to the X2 statistic in a table format, allowing a simple way to finally calculate the statistic without losing any information. For each cell we need to find square them, and then we need to add them all up. In the current example, there are 6 cells to add up (R = 2 times C = 3), shown in Table 5.4.

$$(\text{observed} - \text{expected})/\sqrt{\text{expected}},$$

Table 5.4: X2 contributions for the Arthritis data.

	None	Some	Marked
Placebo	$\left(\frac{29-21.5}{\sqrt{21.5}}\right)^2 = 2.616$	$\left(\frac{7-7.167}{\sqrt{7.167}}\right)^2 = 0.004$	$\left(\frac{7-14.33}{\sqrt{14.33}}\right)^2 = 3.752$
Treated	$\left(\frac{13-20.5}{\sqrt{20.5}}\right)^2 = 2.744$	$\left(\frac{7-6.833}{\sqrt{6.833}}\right)^2 = 0.004$	$\left(\frac{21-13.67}{\sqrt{13.67}}\right)^2 = 3.935$

Finally, the X2 statistic here is the sum of these six results = 2.616 + 0.004 + 3.752 + 2.744 + 0.004 + 3.935 = 13.055

Our favourite function in this chapter, chisq.test, does not provide the contributions to the X2 statistic directly. It provides a related quantity called the

$$\text{standardized residual} = \left(\frac{\text{Observed}_i - \text{Expected}_i}{\sqrt{\text{Expected}_i}}\right),$$

which, when squared (in R, squaring is accomplished using ^2), is the contribution of that particular cell to the X2 statistic that is displayed in Table 5.4.

```
(chisq.test(Arthtable)$residuals)^2
```

```
##          Improved
## Treatment        None        Some      Marked
##   Placebo 2.616279070 0.003875969 3.751937984
##   Treated 2.743902439 0.004065041 3.934959350
```

The most common error made in calculating the X2 statistic by hand involves having observed less than expected and then failing to make the X2 contribution positive for all cells (remember you are squaring the entire quantity in the parentheses and so the sign has to go positive!). In R, we can add up the cells using the sum function over the entire table of numbers:

```
(chisq.test(Arthtable)$residuals)^2)
```

```
## [1] 13.05502
```

Or we can let R do all this hard work for us and get straight to the good stuff:

```
chisq.test(Arthtable)
```

```
##
##  Pearson's Chi-squared test
##
## data:  Arthtable
## X-squared = 13.055, df = 2, p-value = 0.001463
```

The chisq.test function reports a p-value by default. Before we discover how it got that result, we can rely on our permutation methods to obtain a distribution for the X^2 statistic under the null hypothesis. As in Chapters 2 and 3, this will allow us to find a p-value while relaxing one of our assumptions[7]. In the One-WAY ANOVA in Chapter 3, we permuted the grouping variable relative to the responses, mimicking the null hypothesis that the groups are the same and so we can shuffle them around if the null is true. That same technique is useful here. If we randomly permute the grouping variable used to form the rows in the contingency table relative to the responses in the other variable and track the possibilities available for the X^2 statistic under permutations, we can find the probability of getting a result as extreme as or more extreme than what we observed assuming the null is true, our p-value. The observed statistic is the X^2 calculated using the formula above. Like the F-statistic, it ends up that only results in the right tail of this distribution are desirable for finding evidence against the null hypothesis because all the values showing deviation from the null in any direction going into the statistic have to be positive. You can see this by observing that values of the X^2 statistic close to 0 are generated when the observed values are close to the expected values and that sort of result should not be used to find evidence against the null. When the observed and expected values are "far apart", then we should find evidence against the null. It is helpful to work through some examples to be able to understand how the X^2 statistic "measures" differences between observed and expected.

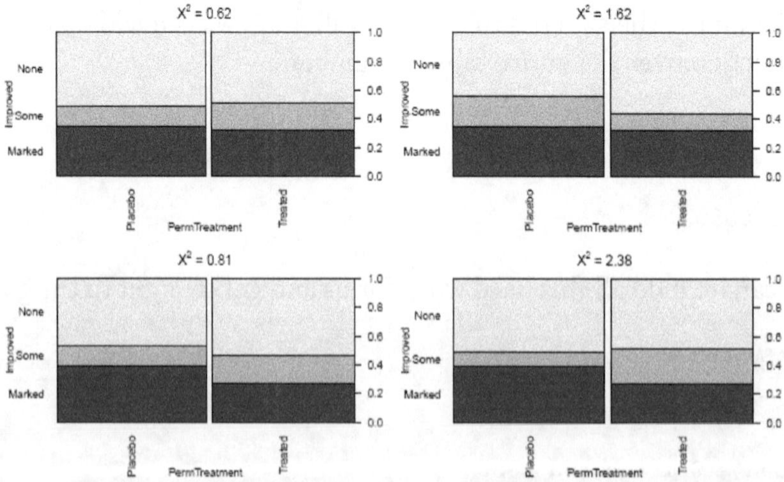

Figure 5.8: Stacked bar charts of four permuted Arthritis data sets that produced X^2 between 0.62 and 2.38.

To start, compare the previous observed X^2 of 13.055 to the sort of results we obtain in a single permutation of the treated/placebo labels – Figure 5.8 (top left panel) shows a permuted data set that produced X^{2*} = 0.62.Visually, you can only see minimal differences between the treatment and placebo groups showing up in the stacked bar chart. Three other permuted data sets are displayed in Figure 5.8 showing the variability in results in permutations but that none get close to showing the differences in the bars observed in the real data set in Figure 5.2.

```
Arthperm <- Arthritis
Arthperm <- Arthperm %>% mutate(PermTreatment = shuffle(Treatment))
```

```
plot(Improved ~ PermTreatment, data = Arthperm,
     main = "Stacked Bar Chart of Permuted Arthritis Data")
```

```
##              Improved
## PermTreatment None Some Marked
##       Placebo   22    6     15
##       Treated   20    8     13
```

```
chisq.test(Arthpermtable)
```

```
##
##  Pearson's Chi-squared test
##
## data:  Arthpermtable
## X-squared = 0.47646, df = 2, p-value = 0.788
```

To build the permutation-based null distribution for the X^2 statistic, we need to collect up the test statistics (X^{2*}) in many of these permuted results. The code is similar to permutation tests in Chapters 2and 3 except that each permutation generates a new contingency table that is summarized and provided tochisq.test to analyze. We extract the $statistic attribute of the results from running chisq.test.

```
- chisq.test(Arthtable)$statistic; Tobs
```

```
## X-squared
##   13.05502
```

```
par(mfrow = c(1,2))
B <- 1000
Tstar <- matrix(NA, nrow = B)
for (b in (1:B)){
  Tstar[b] <- chisq.test(tally(~ shuffle(Treatment) + Improved,
                         data = Arthritis))$statistic
}
pdata(Tstar, Tobs, lower.tail = F)[[1]]
```

```
## [1] 0.002
```

```
tibble(Tstar) %>% ggplot(aes(x = Tstar)) +
  geom_histogram(aes(y = ..ncount..), bins = 20, col = 1, fill = "khaki") +
  geom_density(aes(y = ..scaled..)) +
  theme_bw() +
  labs(y = "Density") +
  geom_vline(xintercept = Tobs, col = "red", lwd = 2) +
  stat_bin(aes(y = ..ncount.., label = ..count..), bins = 20,
           geom = "text", vjust = -0.75)
```

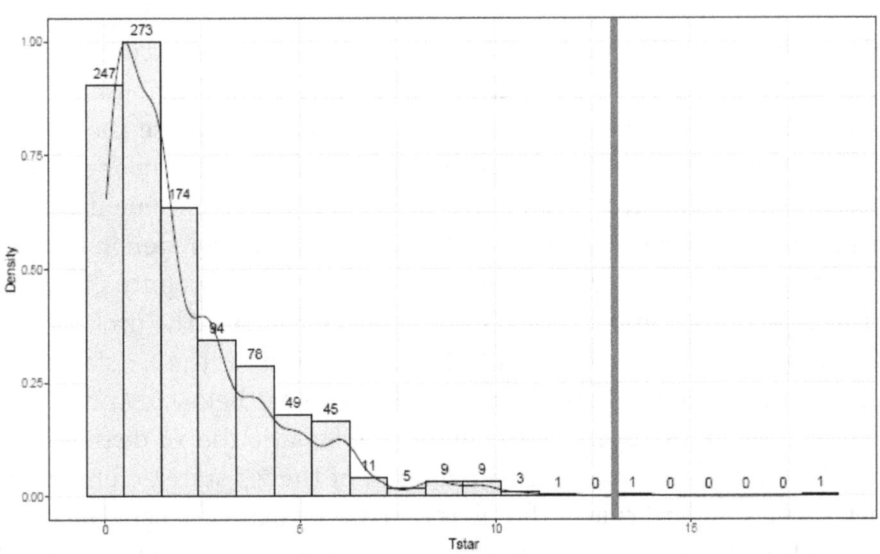

Figure 5.9: Permutation distribution for the X2 statistic for the Arthritis data with an observed X2 of 13.1 (bold, vertical line).

For an observed X2 statistic of 13.055, two out of 1,000 permutation results matched or exceeded this value (pdata returned a value of 0.002) as displayed in Figure 5.9. This suggests that our observed result is quite extreme relative to the null hypothesis and provides strong evidence against it.

Validity conditions for a permutation X2 test are:

- Independence of observations.
- Both variables are categorical.
- Expected cell counts > 0 (otherwise X^2 is not defined).

For the permutation approach described here to provide valid inferences we need to be working with observations that are independent of one another. One way that a violation of independence can sometimes occur in this situation is when a single subject shows up in the table more than once. For example, if a single individual completes a survey more than once and those results are reported as if they came from *N* independent individuals. Be careful about this as it is really easy to make tables of poorly collected or non-independent observations and then consider them for these analyses. Poor data still lead to poor conclusions even if you have fancy new statistical tools to use!

5.6 CHI-SQUARE DISTRIBUTION FOR THE X^2 STATISTIC

When one additional assumption beyond the previous assumptions for the permutation test is met, it is possible to avoid permutations to find the distribution of the X2 statistic under the null hypothesis and get a p-value using what is called the Chi-square or χ2-distribution. The name of our test statistic, X-squared, is meant to allude to the potential that this will follow a χ2-distribution in certain situations but may not do that all the time and we still can use the methods in Section 5.5. Along with the previous assumption regarding independence and all expected cell counts are greater than 0, we make a requirement that N (the total sample size) is "large enough" and this assumption is written in terms of the expected cell counts. If N is large, then all the expected cell counts should also be large because all those observations have to go somewhere. The problems for the χ2-distribution as an approximation to the distribution of the X2 statistic under the null hypothesis come when expected cell counts are below 5. And the smaller the expected cell counts become, the more problematic the χ2-distribution is as an approximation of the sampling distribution of the X2 statistic under the null hypothesis. The standard rule of thumb is that all the expected cell counts need to exceed 5 for the parametric approach to be valid. When this condition is violated, it is better to use the permutation approach. The chisq.test function will provide a warning message to help you notice this. But it is good practice to always explore the expected cell counts using chisq.test(...)$expected.

```
chisq.test(Arthtable)$expected
```

```
##           Improved
## Treatment None     Some   Marked
##    Placebo 21.5 7.166667 14.33333
##    Treated 20.5 6.833333 13.66667
```

In the Arthritis data set, the sample size was sufficiently large for the $\chi 2$-distribution to provide an accurate p-value since the smallest expected cell count is 6.833 (so all expected counts are larger than 5).

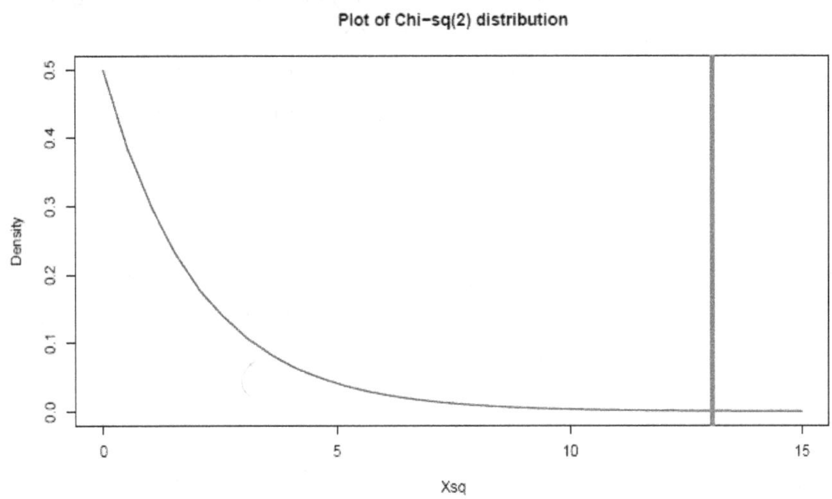

Figure 5.10: χ^2-distribution with two degrees of freedom with the observed statistic of 13.1 indicated with a vertical line.

The $\chi 2$-distribution is a right-skewed distribution that starts at 0 as shown in Figure 5.10. Its shape changes as a function of its degrees of freedom. In the contingency table analyses, the degrees of freedom for the Chi-square test are calculated as

DF= (R - 1) $*$ (C - 1) = (number of rows - 1) $*$ (number of columns - 1).

In the 2 × 3 table above, the DF = (2 - 1) $*$ (3 - 1) = 2 leading to a Chi-square distribution with 2 df for the distribution of X2 under the null hypothesis. The p-value is based on the area to the right of the observed X2 value of 13.055 and the pchisq function provides that area as 0.00146. Note that this is very similar to the permutation result found previously for these data.

```
pchisq(13.055, df = 2, lower.tail = F)
```

```
## [1] 0.001462658
```

We'll see more examples of the χ^2-distributions in each of the examples that follow.

A small side note about sample sizes is warranted here. In contingency tables, especially those based on survey data, it is common to have large overall sample sizes (N). With large sample sizes, it becomes easy to find strong evidence against the null hypothesis, even when the "distance" from the null is relatively minor and possibly unimportant. By this we mean that the observed proportions are a small practical distance from the situation described in the null. After obtaining a small p-value, we need to consider whether we have obtained practical significance (or maybe better described as practical importance) to accompany our discussion of strong evidence against the null hypothesis. Whether a result is large enough to be of practical importance can only be judged by knowing something about the situation we are studying and by providing a good summary of our results to allow experts to assess the size and importance of the result. Unfortunately, many researchers are so happy to see small p-values that this is their last step. We encountered a similar situation in the car overtake distance data set where a large sample size provided a data set that had a small p-value and possibly minor differences in the means driving it.

If we revisit our observed results, re-plotted in Figure 5.11 since it was quite a ways back that we saw the original data in Figure 5.2, knowing that we have strong evidence against the null hypothesis of no difference between Placebo and Treated groups, what can we say about the effectiveness of the arthritis medication? It seems that there is a real and important increase in the proportion of patients getting improvement (Some or Marked). If the differences "looked" smaller, even with a small p-value you8 might not recommend someone take the drug. . .

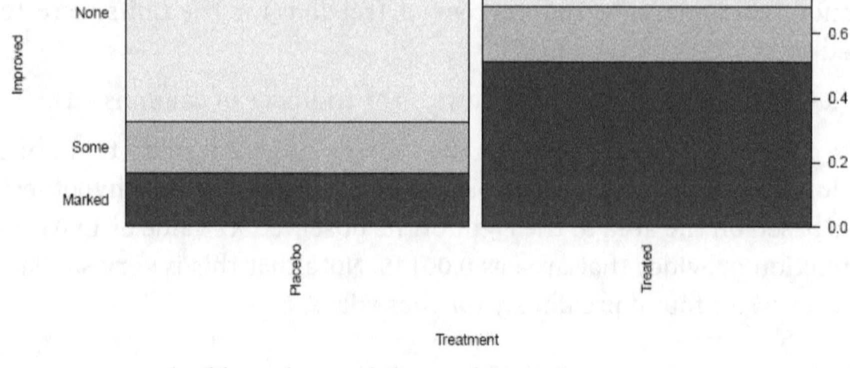

Figure 5.11: Stacked bar chart of the Arthritis data comparing Treated and Placebo.

5.7 EXAMINING RESIDUALS FOR THE SOURCE OF DIFFERENCES

Small p-values are generated by large X2 values. If we want to understand the source of a small p-value, we need to understand what made the test statistic large. To get a large X2 value, we either need many small contributions from lots of cells or a few large contributions. In most situations, there are just a few cells that show large deviations between the null hypothesis (expected cell counts) and what was observed (observed cell counts). It is possible to explore the "size" and direction of the differences between observed and expected counts to learn something about the behaviour of the relationship between the variables, especially as it relates to evidence against the null hypothesis of no difference or no relationship. The standardized residual, provides a measure of deviation of the observed from expected which retains the direction of deviation (whether observed was more or less than expected is interesting for interpretations) for each cell in the table. It is scaled much like a standard normal distribution providing a scale for "large" deviations for absolute values that are over 2 or 3. In other words, values with magnitude over 2 should be your focus in the standardized residuals, noting whether the observed counts were much more or less than expected. On the X2 scale, standardized residuals of 2 or more mean that the cells are contributing 4 or more units to the overall statistic, which is a pretty noticeable bump up in the size of the statistic. A few contributions at 4 or higher and you will likely end up with a small p-value.

$$\left(\frac{\text{Observed}_i - \text{Expected}_i}{\sqrt{\text{Expected}_i}}\right),$$

There are two ways to explore standardized residuals. First, we can obtain them via the chisq.test and manually identify the "big ones". Second, we can augment a mosaic plot of the table with the standardized results by turning on the shade = T option and have the plot help us find the big differences. This technique can be applied whether we are performing an Independence or Homogeneity test – both are evaluated with the same X^2 statistic so the large standardized residuals are of interest in both situations. Both types of results are shown for the Arthritis data table:

```
##            Improved
## Treatment         None         Some       Marked
##   Placebo   1.61749160 -0.06225728 -1.93699199
##   Treated  -1.65647289  0.06375767  1.98367320
```

```
.isq.test(Arthtable)$residuals
'saicplot(Arthtable, shade = T)
```

In these data, the standardized residuals are all less than 2 in magnitude so Figure 5.12 isn't too helpful but this type of plot is in other examples. The largest contributions to the X2 statistic come from the Placebo and Treated groups in the Marked improvement cells. Those standardized residuals are -1.94 and 1.98 (both really close to 2), showing that the placebo group had noticeably fewer Marked improvement results than expected and the Treated group had noticeably more Marked improvement responses than expected if the null hypothesis was true. Similarly but with smaller magnitudes, there were more None results than expected in the Placebo group and fewer None results than expected in the Treated group. The standardized residuals were very small in the two cells for the Some improvement category, showing that the treatment/placebo were similar in this response category and that the results were about what would be expected if the null hypothesis of no difference were true.

5.8 GENERAL PROTOCOL FOR X^2 TESTS

In any contingency table situation, there is a general protocol to completing an analysis.

- ◉ Identify the data collection method and whether the proper analysis is based on the Independence or Homogeneity hypotheses (Section 5.1).

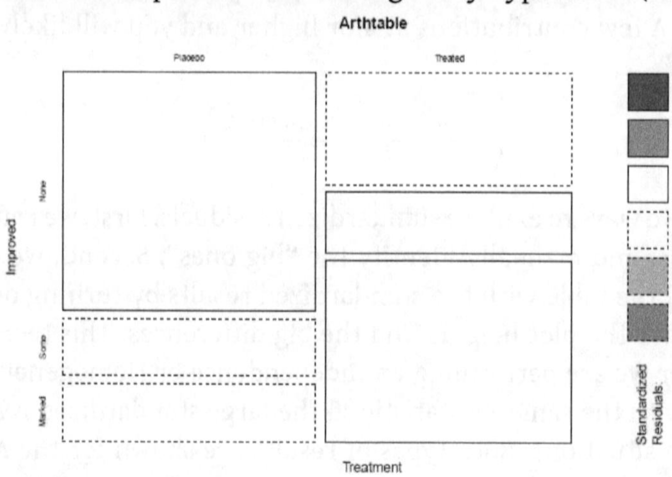

Figure 5.12: Mosaic plot of the Arthritis data with large standardized residuals indicated (actually, there were none that were indicated because all were less than 2). Note that dashed borders correspond to negative standardized residuals (observed less than expected) and solid borders are positive standardized residuals (observed more than expected).

- ◉ **2.** Make contingency table and get a general sense of response patterns. Pay attention to "small" counts, especially cells with 0 counts.
  - a) If there are many small count cells, consider combining categories on one or both variables to make a new variable with fewer categories that

has larger counts per cell to have more robust inferences (see Section 5.10 for a related example).

- Make the appropriate graphical display of results and generally describe the pattern of responses.
  - For Homogeneity, make a stacked bar chart.
  - For Independence, make a mosaic plot.
  - Consider a more general exploration using a tableplot if other variables were measured to check for confounding and other interesting multi-variable relationships. Also check for missing data if you have not done this before.
- Conduct the 6+ steps of the appropriate type of hypothesis test.
  - Use permutations if any expected cell counts are below 5.
  - If all expected cell counts greater than 5, either permutation or parametric approaches are acceptable.
- Explore the standardized residuals for the "source" of any evidence against the null – this can be the start of your "size" discussion.
  - Tie the interpretation of the "large" standardized residuals and their direction (above or below expected under the null) back into the original data display (this really gets to "size"). Work to find a story for the pattern of responses. If little evidence is found against the null, there is not much to do here.

5.9 POLITICAL PARTY AND VOTING RESULTS: COMPLETE ANALYSIS

As introduced in Section 5.3, a national random sample of voters was obtained related to the 2000 Presidential Election with the party affiliations and voting results recorded for each subject. The data are available in election in the poLCA package [Linzer and Lewis., 2022]. It is always good to start with a bit of data exploration with a tableplot, displayed in Figure 5.13. Many of the lines of code here are just for making sure that R is treating the categorical variables that were coded numerically as categorical variables.

```
election <- election %>% mutate(VOTEF = factor(VOTE3),
                                PARTY = factor(PARTY),
                                EDUC = factor(EDUC),
                                GENDER = factor(GENDER)
                                )

levels(election$VOTEF) <- c("Gore","Bush","Other")
# (Possibly) required options to avoid error when running on a PC,
# should have no impact on other platforms
# options(ffbatchbytes = 1024^2 * 128); options(ffmaxbytes = 1024^2 * 128 * 32)
tableplot(election, select = c(VOTEF, PARTY, EDUC, GENDER), pals = list("BrBG"),
          sample = F)
```

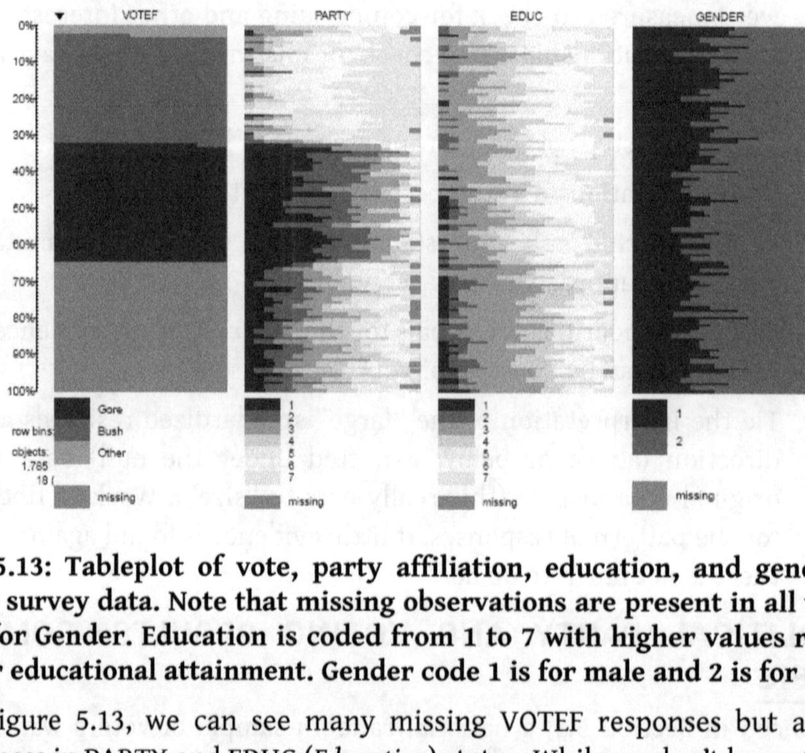

Figure 5.13: Tableplot of vote, party affiliation, education, and gender from election survey data. Note that missing observations are present in all variables except for Gender. Education is coded from 1 to 7 with higher values related to higher educational attainment. Gender code 1 is for male and 2 is for female.

In Figure 5.13, we can see many missing VOTEF responses but also some missingness in PARTY and EDUC (*Education*) status. While we don't know too much about why people didn't respond on the Vote question – they could have been unwilling to answer it or may not have voted. It looks like those subjects have more of the lower education level responses (more dark colours, especially level 2 of education) than in the responders to this question. There are many "middle" ratings in the party affiliation responses for the missing VOTEF responses, suggesting that independents were less likely to answer the question in the survey for whatever reason. Even though this comes with concerns about who these results actually apply to (likely not the population that was sampled from), we want to focus on those that did respond in VOTEF, so will again use drop_na to clean out any subjects with any missing responses after using select to focus just on these four variables. Then we remake the tableplot (Figure 5.14). The code also adds the sort option to the tableplot function call that provides an easy way to sort the data set based on other variables. It is interesting, for example, to sort the responses by *Education* level and explore the differences in other variables. These explorations are omitted here but easily available by changing the sorting column from 1 to sort = 3 or sort = EDUC. Figure 5.14 shows us that there are clear differences in party affiliation based on voting for *Bush*, *Gore*, or *Other*. It is harder to see if there are differences in education level or gender based on the voting status in this plot, but, as noted above, sorting on these other variables can sometimes help to see other relationships between variables.

```
election2 <- election %>%
  select(VOTEF, PARTY, EDUC, GENDER) %>%
  drop_na()
tableplot(election2, select = c(VOTEF, PARTY, EDUC, GENDER), sort = 1,
          pals = list("BrBG"), sample = F)
```

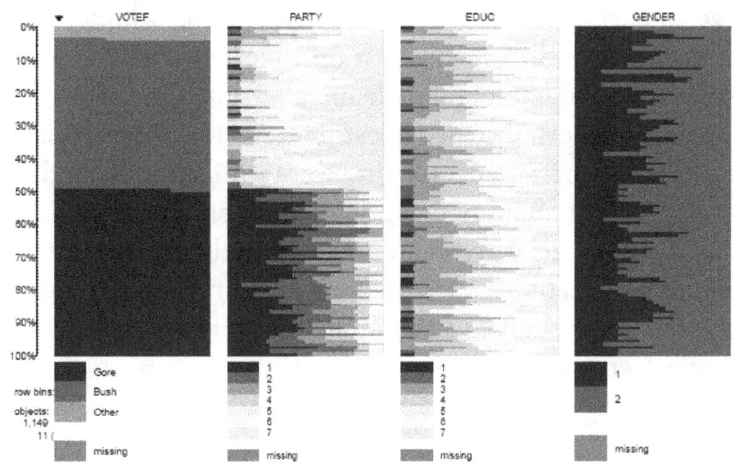

Figure 5.14: Tableplot of election data with subjects without any missing responses (complete cases).

Focusing on the party affiliation and voting results, the appropriate analysis is with an Independence test because a single random sample was obtained from the population. The total sample size for the complete responses was $N = 1{,}149$ (out of the original 1,785 subjects). Because this is an Independence test, the mosaic plot is the appropriate display of the results, which was provided in Figure 5.5.

```
electable <- tally(~ PARTY + VOTEF, data = election2)
electable
```

```
##       VOTEF
## PARTY Gore Bush Other
##     1  238    6     2
##     2  151   18     1
##     3  113   31    13
##     4   37   36    11
##     5   21  124    12
##     6   20  121     2
##     7    3  188     1
```

There is a potential for bias in some polls because of the methods used to find and contact people. As U.S. residents have transitioned from land-lines to cell phones, the early adopting cell phone users were often excluded from political polling. These policies are being reconsidered to adapt to the decline in residential phone lines and most polling organizations now include cell phone numbers in their list of potential respondents. This study may have some bias regarding who was considered as part of the population of interest and who was actually found that was willing to respond to their questions. We don't have much information here but biases arising from unobtainable members of populations are a potential issue in many studies, especially when questions tend toward more sensitive topics. We can make inferences here to people that were willing to respond to the request to answer the survey but should be cautious in extending it to all Americans or even voters in the year 2000. When we say "population" below, this nuanced discussion is what we mean. Because the political party is not randomly assigned to the subjects, we cannot make causal inferences for political affiliation causing different voting patterns 9.

Here are our 6+ steps applied to this example:

0. The desired RQ is about assessing the relationship between part affiliation and vote choice, but this is constrained by the large rate of non-response in this data set. This is an Independence test and so the tableplot and mosaic plot are good visualizations to consider and the X^2-statistic will be used.

⦿ Hypotheses:

- H_0: There is no relationship between the party affiliation (7 levels) and voting results (*Bush, Gore, Other*) in the population.
- H_A: There is a relationship between the party affiliation (7 levels) and voting results (*Bush, Gore, Other*) in the population.

⦿ Plot the data and assess validity conditions:

- **Independence:** There is no indication of an issue with this assumption since each subject is measured only once in the table. No other information suggests a potential issue since a random sample was taken from presumably a large national population and we have no information that could suggest dependencies among observations.

- All expected cell counts larger than 5 to use the parametric χ^2-distribution to find p-values:
- We need to generate a table of expected cell counts to be able to check this condition:

```
chisq.test(electable)$expected
```

- When we request the expected cell counts, R tries to help us with a warning message if the expected cell counts might be small, as in this situation.
- There is one expected cell count below 5 for Party = 4 who voted Other with an expected cell count of 3.07, so the condition is violated and the permutation approach should be used to obtain more trustworthy p-values. The conditions are met for performing a permutation test.

```
## Warning in chisq.test(electable): Chi-squared approximation may be incorrect

##       VOTEF
## PARTY       Gore      Bush    Other
##     1 124.81984 112.18799 8.992167
##     2  86.25762  77.52829 6.214099
##     3  79.66144  71.59965 5.738903
##     4  42.62141  38.30809 3.070496
##     5  79.66144  71.59965 5.738903
##     6  72.55788  65.21497 5.227154
##     7  97.42037  87.56136 7.018277
```

Calculate the test statistic and p-value:

- The test statistic is best calculated by the chisq.test function since there are 21 cells and many potential places for a calculation error if performed by hand.

```
chisq.test(electable)
```

```
##
##  Pearson's Chi-squared test
##
## data:  electable
## X-squared = 762.81, df = 12, p-value < 2.2e-16
```

- The observed X^2 statistic is 762.81.
- The parametric p-value is < 2.2e-16 from the R output which would be reported as < 0.0001. This was based on a χ^2-distribution with $(7 - 1) * (3 - 1) = 12$ degrees of freedom displayed in Figure 5.15.

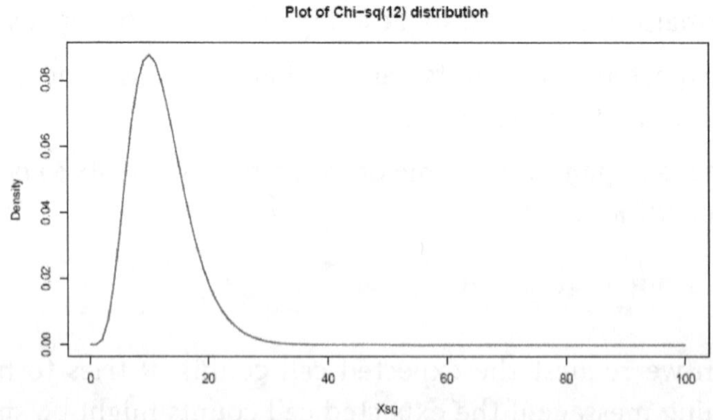

Figure 5.15: Plot of χ^2-distribution with 12 degrees of freedom.

- If you want to repeat this calculation directly you get a similarly tiny value that R reports as 1.5e-155. Again, reporting less than 0.0001 is just fine.

```
pchisq(762.81, df = 12, lower.tail = F)
```

```
## [1] 1.553744e-155
```

- But since the expected cell count condition is violated, we should use permutations as implemented in the following code to provide a more trustworthy p-value:

```
Tobs <- chisq.test(electable)$statistic; Tobs
```

```
## X-squared
##  762.8095
```

```
par(mfrow = c(1,2))
B <- 1000
Tstar <- matrix(NA, nrow = B)
for (b in (1:B)){
  Tstar[b] <- chisq.test(tally(~ shuffle(PARTY) + VOTEF, data = election2,
                         margins = F))$statistic
}
pdata(Tstar, Tobs, lower.tail = F)[[1]]
```

```
## [1] 0
```

```
tibble(Tstar) %>% ggplot(aes(x = Tstar)) +
  geom_histogram(aes(y = ..ncount..), bins = 30, col = 1, fill = "khaki") +
  geom_density(aes(y = ..scaled..)) +
  theme_bw() +
  labs(y = "Density") +
  geom_vline(xintercept = Tobs, col = "red", lwd = 2) +
  stat_bin(aes(y = ..ncount.., label = ..count..), bins = 30,
           geom = "text", vjust = -0.75)
```

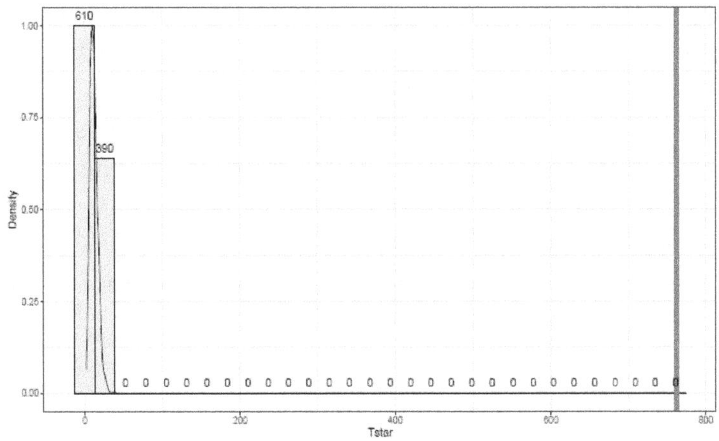

Figure 5.16: Permutation distribution of X^2 for the election data.

- The last results tells us that there were no permuted data sets that produced larger X^2's than the observed X^2 in 1,000 permutations, so we report that the **p-value was less than 0.001** using the permutation approach. The permutation distribution in Figure 5.16 contains no results over40, so the observed configuration was really far from the null hypothesis of no relationship between party status and voting.

Conclusion

- There is strong evidence against the null hypothesis of no relationship between party affiliation and voting results in the population ($X^2 = 762.81$, p-value<0.001), so we would conclude that there is a relationship between party affiliation and voting results.

Size

We can add insight into the results by exploring the standardized residuals. The numerical results are obtained using chisq.test(electable)$residuals and visually using mosaicplot(electable, shade = T) in Figure 5.17. The standardized residuals show some clear sources of the differences from the results expected if there were no relationship present. The largest contributions are found in the highest democrat category (PARTY = 1) where the standardized residual for *Gore* is 10.13 and for *Bush* is -10.03, showing much higher than expected (under *H*0) counts for Gore voters and much lower than expected (under *H*0) for Bush.

Similar results in the opposite direction are found in the strong republicans (PARTY = 7). Note how the brightest shade of blue in Figure 5.17 shows up for much higher than expected results and the brighter red for results in the other direction, where observed counts were much lower than expected. When there are many large standardized residuals, it is OK to focus on the largest results but remember that some of the intermediate deviations, or lack thereof, could also be interesting. For example, the Gore voters from PARTY = 3 had a standardized residual of 3.75 but the PARTY = 5 voters for Bush had a standardized residual of 6.17. So maybe Gore didn't have as strong of support from his center-leaning supporters as Bush was able to obtain from the same voters on the other side of the middle? Exploring the relative proportion of each vertical bar in the response categories is also interesting to see the proportions of each level of party affiliation and how they voted. A political scientist would easily obtain many more (useful) theories based on this combination of results.

```
chisq.test(electable)$residuals #(Obs - expected)/sqrt(expected)
```

```
##        VOTEF
## PARTY        Gore        Bush      Other
##     1  10.1304439 -10.0254117 -2.3317373
##     2   6.9709179  -6.7607252 -2.0916557
##     3   3.7352759  -4.7980730  3.0310127
##     4  -0.8610559  -0.3729136  4.5252413
##     5  -6.5724708   6.1926811  2.6135809
##     6  -6.1701472   6.9078679 -1.4115200
##     7  -9.5662296  10.7335798 -2.2717310
```

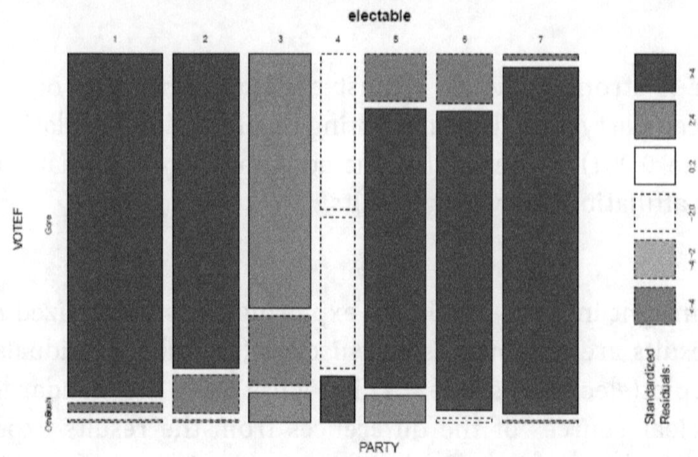

Figure 5.17: Mosaic plot with shading based on standardized residuals for the election data.

```
#Adds information on the size of the residuals
mosaicplot(electable, shade = T)
```

Scope of inference

- The results are not causal since no random assignment was present but they do apply to the population of voters in the 2000 election that were able to be contacted by those running the poll and who would be willing to answer all the questions and actually voted.

5.10 IS CHEATING AND LYING RELATED IN STUDENTS?

A study of student behaviour was performed at a university with a survey of N = 319 undergraduate students (cheating data set from the poLCA package originally published by Dayton [1998]). They were asked to answer four questions about their various academic frauds that involved cheating and lying. Specifically, they were asked if they had ever lied to avoid taking an exam (LIEEXAM with 1 for no and 2 for yes), if they had lied to avoid handing in a term paper on time (LIEPAPER with 1 for no, 2 for yes), if they had purchased a term paper to hand in as their own or obtained a copy of an exam prior to taking the exam (FRAUD with 1 for no, 2 for yes), and if they had copied answers during an exam from someone near them (COPYEXAM with 1 for no, 2 for yes). Additionally, their GPAs were obtained and put into categories: (<2.99, 3.0 to 3.25, 3.26 to3.50, 3.51 to 3.75, and 3.76 to 4.0). These categories were coded from 1 to 5, respectively. Again, the code starts with making sure the variables are treated categorically by applying the factor function.

```
library(poLCA)
data(cheating) #Survey of students
cheating <- as_tibble(cheating)

cheating <- cheating %>% mutate(LIEEXAM = factor(LIEEXAM),
                                LIEPAPER = factor(LIEPAPER),
                                FRAUD = factor(FRAUD),
                                COPYEXAM = factor(COPYEXAM),
                                GPA = factor(GPA)
                                )

tableplot(cheating, sort = GPA, pals = list("BrBG"))
```

We can explore some interesting questions about the relationships between these variables. The tableplot in Figure 5.18 again helps us to get a general idea of the data set and to assess some complicated aspects of the relationships between variables. For example, the rates of different unethical behaviours seem to decrease with higher GPA students (but do not completely disappear!). This data set also has a few missing GPAs that we would want to carefully consider – which sorts of students might not be willing to reveal their GPAs?It ends up that these students did not *admit* to any of the unethical behaviours. Note that we used the sort= GPA option in the tableplot function to sort the responses based on GPA to see how GPA might relate to patterns of unethical behaviour.

While the relationship between GPA and presence/absence of the different behaviours is of interest, we want to explore the types of behaviours. It is possible to group the lying behaviours as being a different type (less extreme?) of unethical behaviour than obtaining an exam prior to taking it, buying a paper, or copying someone else's answers. We want to explore whether there is some sort of relationship between the lying and copying behaviours - are those that engage in one type of behaviour more likely to do the other? Or are they independent of each other? This is a hard story to elicit from the previous plot because there are so many variables involved.

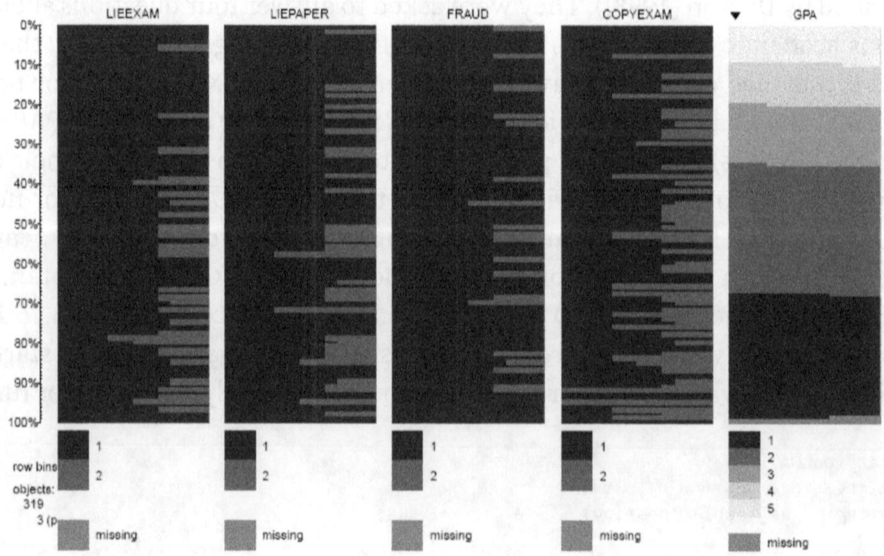

Figure 5.18: Tableplot of initial cheating and lying data set. Note that a few GPAs were missing in the data set.

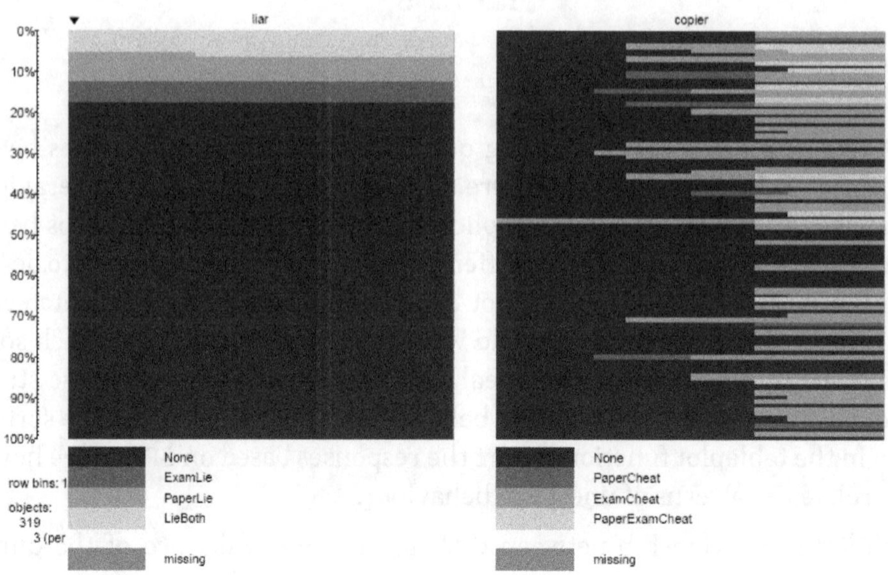

Figure 5.19: Tableplot of new variables liar and copier that allow exploration of relationships between different types of lying and cheating behaviours.

To simplify the results, combining the two groups of variables into the four possible combinations on each has the potential to simplify the results – or at least allow exploration of additional research questions. The interaction function is used to create two new variables that have four levels that are combinations of the different options from none to both of each type (copier and liar). In the tableplot in Figure 5.19, you can see the four categories for each, starting with no bad behaviour of either type (which is fortunately the most popular response on both variables!). For each variable, there are students who admitted to one of the two violations and some that did both. The liar variable has categories of None, ExamLie, PaperLie, and LieBoth. The copier variable has categories of None, PaperCheat, ExamCheat, and PaperExamCheat (for doing both). The last category for copier seems to mostly occur at the top of the plot which is where the students who had lied to get out of things reside, so maybe there is a relationship between those two types of behaviours? On the other hand, for the students who have never lied, quite a few had cheated on exams. The contingency table can help us dig further into the hypotheses related to the Chi-square test of Independence that is appropriate in this situation.

```
cheating <- cheating %>% mutate(liar = interaction(LIEEXAM, LIEPAPER),
                                copier = interaction(FRAUD, COPYEXAM)
                                )
levels(cheating$liar) <- c("None", "ExamLie", "PaperLie", "LieBoth")
levels(cheating$copier) <- c("None", "PaperCheat", "ExamCheat", "PaperExamCheat")

tableplot(cheating, sort = liar, select = c(liar, copier), pals = list("BrBG"))
```

```
cheatlietable <- tally(~ liar + copier, data = cheating)
cheatlietable
```

```
##             copier
## liar         None PaperCheat ExamCheat PaperExamCheat
##   None        207          7        46              5
##   ExamLie      10          1         3              2
##   PaperLie     13          1         4              2
##   LieBoth      11          1         4              2
```

Unfortunately for our statistic, there were very few responses in some combinations of categories even with $N = 319$. For example, there was only one response each in the combinations for students that copied on papers and lied to get out of exams, papers, and both. Some other categories were pretty small as well in the groups that only had one behaviour present. To get a higher number of counts in the combinations, we combined the single behaviour only levels into "*either* " categories and left the *none* and *both* categories for each variable. This creates two new variables called liar2 and copier2 (tableplot in Figure 5.20). The code to create these variables and make the plot is below which employs the levels function to assign the same label to two different levels from the original list.

```
# Collapse the middle categories of both variables by making both have the same level name
cheating <- cheating %>% mutate(liar2 = liar,
                                copier2 = copier
                                )
levels(cheating$liar2) <- c("None", "ExamorPaper", "ExamorPaper", "LieBoth")
levels(cheating$copier2) <- c("None", "ExamorPaper", "ExamorPaper", "CopyBoth")
tableplot(cheating, sort = liar2, select = c(liar2, copier2), pals = list("BrBG"))
```

```
cheatlietable <- tally(~ liar2 + copier2, data = cheating)
cheatlietable
```

```
##              copier2
## liar2         None ExamorPaper CopyBoth
##   None         207          53        5
##   ExamorPaper   23           9        4
##   LieBoth       11           5        2
```

This 3 × 3 table is more manageable and has few really small cells so we will proceed with the 6+ steps of hypothesis testing applied to these data using the Independence testing methods (again a single sample was taken from the population so that is the appropriate procedure to employ):

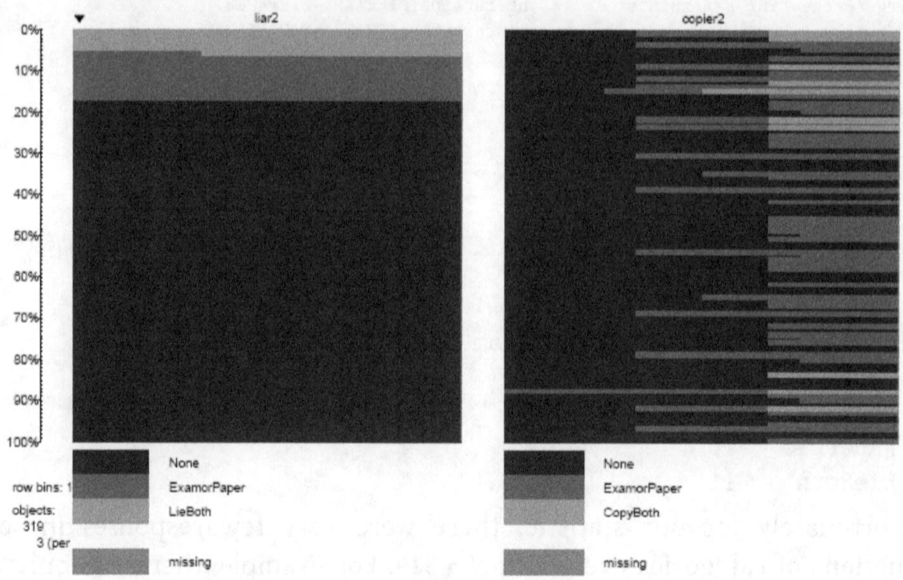

Figure 5.20: Tableplot of lying and copying variables after combining categories.

0. The RQ is about relationships between lying to instructors and cheating and these questions, after some work and simplifications, allow us to address a version of that RQ even though it might not be the one that we started with. The tableplots help to visualize the results and the X^2-statistic will be used to do the hypothesis test.

Hypotheses:

- H_0: Lying and copying behaviour are independent in the population of students at this university.
- H_A: Lying and copying behaviour are dependent in the population of students at this university.

Validity conditions:

- **Independence:** There is no indication of a violation of this assumption since each subject is measured only once in the table. No other information suggests a potential issue but we don't have much information on how these subjects were obtained. What happens if we had sampled from students in different sections of a multi-section course and one of the sections had recently had a cheating scandal that impacted many students in that section?
- All expected cell counts larger than 5 (required to use χ^2-distribution to find p-values):
- We need to generate a table of expected cell counts to check this condition:

```
chisq.test(cheatlietable)$expected
```

```
##                copier2
## liar2                None ExamorPaper  CopyBoth
##   None          200.20376   55.658307 9.1379310
##   ExamorPaper    27.19749    7.561129 1.2413793
##   LieBoth        13.59875    3.780564 0.6206897
```

- When we request the expected cell counts, there is a warning message (not shown).
- There are three expected cell counts below 5, so the condition is violated and a permutation approach should be used to obtain more trustworthy p-values.

Calculate the test statistic and p-value:

- Use chisq.test to obtain the test statistic, although this table is small enough to do by hand if you want the practice – see if you can find a similar answer to what the function provides:

```
chisq.test(cheatlietable)
```

```
##
##  Pearson's Chi-squared test
##
## data:  cheatlietable
## X-squared = 13.238, df = 4, p-value = 0.01017
```

- The X^2 statistic is 13.24.
- The parametric p-value is 0.0102 from the R output. This was based on a $\chi 2$-distribution with $(3 - 1) * (3 - 1) = 4$ degrees of freedom that is displayed in Figure 5.21. Remember that this isn't quite the right distribution for the test statistic since our expected cell count condition was violated.

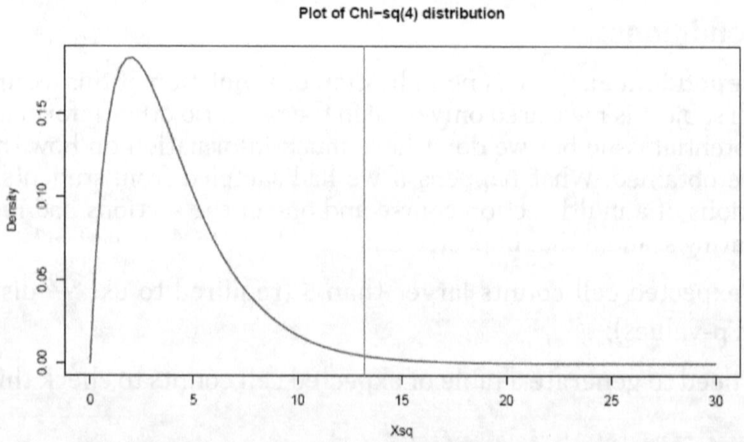

Figure 5.21: Plot of χ^2-distribution with 4 degrees of freedom.

- If you want to repeat the p-value calculation directly:

```
pchisq(13.2384, df = 4, lower.tail = F)
```

```
## [1] 0.01016781
```

- But since the expected cell condition is violated, we should use permutations as implemented in the following code with the number of permutations increased to 10,000 to help get a better estimate of the p-value since it is possibly close to 0.05:

```
Tobs <- chisq.test(tally(~ liar2 + copier2, data = cheating))$statistic
Tobs
```

```
## X-squared
##  13.23844
```

```
par(mfrow = c(1,2))
B <- 10000 # Now performing 10,000 permutations
Tstar <- matrix(NA,nrow = B)
for (b in (1:B)){
  Tstar[b] <- chisq.test(tally(~ shuffle(liar2) + copier2,
                               data = cheating))$statistic
}
pdata(Tstar, Tobs, lower.tail = F)[[1]]
```

```
## [1] 0.0174
```

```
tibble(Tstar) %>% ggplot(aes(x = Tstar)) +
  geom_histogram(aes(y = ..ncount..), bins = 20, col = 1, fill = "khaki") +
  geom_density(aes(y = ..scaled..)) +
  theme_bw() +
  labs(y = "Density") +
  geom_vline(xintercept = Tobs, col = "red", lwd = 2) +
  stat_bin(aes(y = ..ncount.., label = ..count..), bins = 20,
      geom = "text", vjust = -0.75)
```

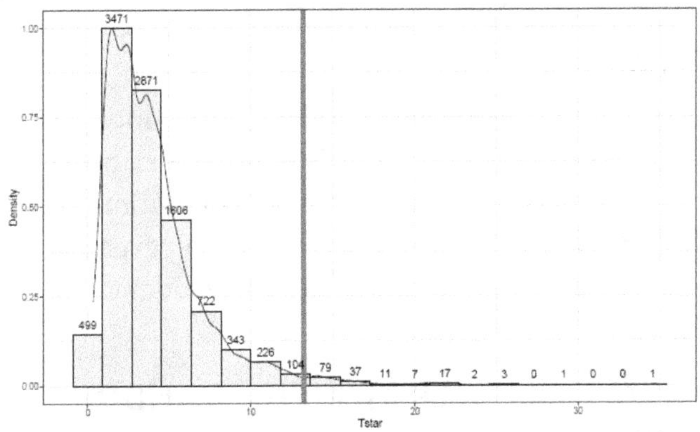

Figure 5.22: Plot of permutation distributions for cheat/lie results with observed value of 13.24 (bold, vertical line).

- There were 174 of B = 10,000 permuted data sets that produced as large or larger X^2*'s than the observed as displayed in Figure 5.22, so we report that the p-value was 0.0174 using the permutation approach, which was slightly larger than the result provided by the parametric method.

Conclusion:

- There is strong evidence against the null hypothesis of no relationship between lying and copying behaviour in the population of students (X^2-statistic = 13.24, permutation p-value of 0.0174), so conclude that there is a relationship between lying and copying behaviour at the university in the population of students studied.

Size:

- The standardized residuals can help us more fully understand this result – the mosaic plot only had one cell shaded and so wasn't needed here.

```
chisq.test(cheatlietable)$residuals
```

```
##              copier2
## liar2               None ExamorPaper   CopyBoth
##   None         0.4803220  -0.3563200 -1.3688609
##   ExamorPaper -0.8048695   0.5232734  2.4759378
##   LieBoth     -0.7047165   0.6271633  1.7507524
```

- There is really only one large standardized residual for the *ExamorPaper* liars and the *CopyBoth* copiers, with a much larger observed value than expected of 2.48. The only other medium-sized standardized residuals came from the *CopyBoth* copiers column with fewer than expected students in the *None* category and more than expected in the *LieBoth* type of lying category. So we are seeing more than expected that lied somehow and copied – we can say this suggests that the students who lie tend to copy too!

Scope of inference:

- There is no causal inference possible here since neither variable was randomly assigned (really neither is explanatory or response here either) but we can extend the inferences to the population of students that these were selected from that would be willing to reveal their GPA (see initial discussion related to some differences in students that wouldn't answer that question).

5.11 ANALYZING A STRATIFIED RANDOM SAMPLE OF CALIFORNIA SCHOOLS

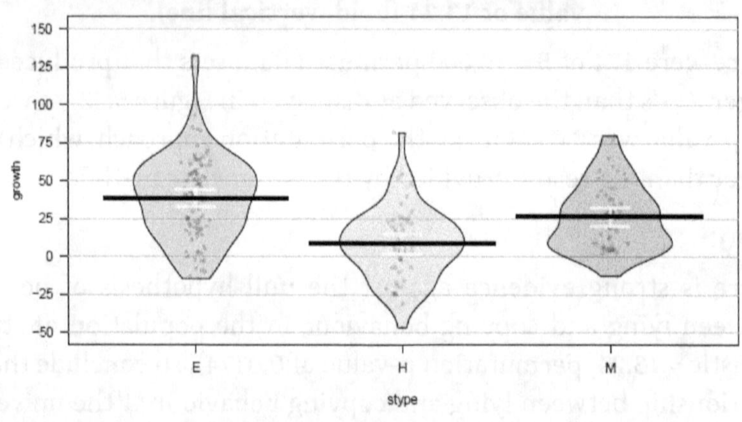

Figure 5.23: Pirate-plot of the API growth scores by level of school in the stype variable (coded E for elementary, M for Middle, and H for High school).

In recent decades, there has been a push for quantification of school performance and tying financial punishment and rewards to growth in these metrics both for schools and for teachers. One example is the API (Academic Performance Index) in California that is based mainly on student scores on standardized tests. It ranges between 200 and 1000 and year to year changes are of interest to assess "performance" of schools – calculated as one year minus the previous year (negative "growth" is also possible!). Suppose that a researcher is interested in whether the growth metric might differ between different levels of schools. Maybe it is easier or harder for elementary, middle, or high schools to attain growth? The researcher has a list of most of the schools in the state of each level that are using a database that the researcher has access to. In order to assess this question, the researcher takes a stratified random sample10, selecting nelementary = 100 schools from the population of 4421 elementary schools, nmiddle = 50 from the population of 1018 middle schools, and nhigh = 50 from the population of 755 high schools. These data are available in the survey package [Lumley, 2021] and the api data object that loads both apipop (population) and apistrat (stratified random sample) data sets. The growth (change!) in API scores for the schools between 1999 and 2000 (taken as the year 2000score minus 1999 score) is used as the response variable. The pirate-plot of the growth scores are displayed in Figure 5.23. They suggest some differences in the growth rates among the different levels. There are also a few schools flagged as being possible outliers.

```
library(survey)
data(api)
apistrat <- as_tibble(apistrat)
apipop <- as_tibble(apipop)
tally(~ stype, data = apipop) #Population counts
```

```
## stype
##    E    H    M
## 4421  755 1018
```

```
tally(~ stype, data = apistrat) #Sample counts
```

```
## stype
##   E   H   M
## 100  50  50
```

```
pirateplot(growth ~ stype, data = apistrat, inf.method = "ci", inf.disp = "line")
```

The One-Way ANOVA F -test, provided below, suggests strong evidence against the null hypothesis of no difference in the true mean growth scores among the different types of schools (F (2, 197) = 23.56, p-value < 0.0001). But the residuals from this model displayed in the QQ-Plot in Figure 5.24 contain a slightly long right tail and short left tail, suggesting a right skewed distribution for the residuals. In a high-stakes situation such as this, reporting results with violations of the assumptions probably would not be desirable, so another approach is needed. The permutation methods would be justified here but there is another "simpler" option available using our new Chi-square analysis methods.

```
m1 <- lm(growth ~ stype, data = apistrat)
library(car)
Anova(m1)
```

```
## Anova Table (Type II tests)
##
## Response: growth
##            Sum Sq  Df F value    Pr(>F)
## stype       30370   2  23.563 6.685e-10
## Residuals 126957 197
```

```
plot(m1, which = 2, pch = 16)
```

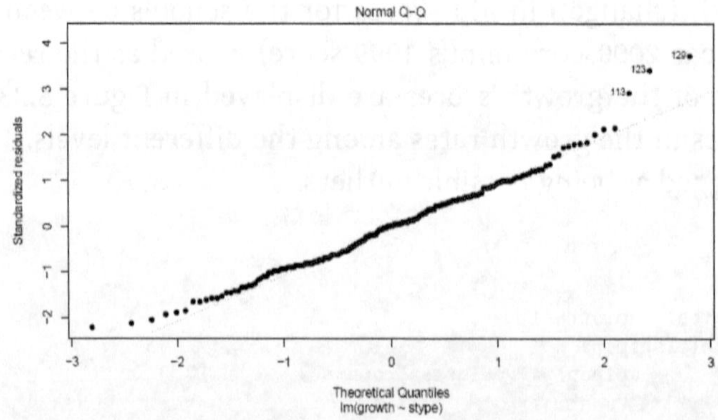

Figure 5.24: QQ-plot of standardized residuals from the One-Way ANOVA linear model.

One way to get around the normality assumption is to use a method that does not assume the responses follow a normal distribution. If we ***bin*** or cut the quantitative response variable into a set of ordered categories and apply a Chi-square test, we can proceed without concern about the lack of normality in the residuals of the ANOVA model. To create these bins, a simple idea would be to use the quartiles to generate the response variable categories, binning the quantitative responses into groups for the lowest 25%, second 25%, third 25%, and highest 25% by splitting the data at Q1, the Median, and Q3. In R, the cut function is available to turn a quantitative variable into a categorical variable. First, we can use the information from favstats to find the cut-points:

```
favstats(~ growth, data = apistrat)
```

```
## min   Q1 median Q3 max   mean      sd   n missing
## -47 6.75     25 48 133 27.995 28.1174 200       0
```

The cut function can provide the binned variable if it is provided with the end-points of the desired intervals (breaks = ...) to create new categories with those names in a new variable called growth cut.

```
apistrat <- apistrat %>% mutate(growthcut = cut(growth, breaks = c(-47,6.75,25,48,133),
                                                include.lowest = T))
```

```
tally(~ growthcut, data = apistrat)
```

```
## growthcut
## [-47,6.75]  (6.75,25]    (25,48]   (48,133]
##         50         52         49         49
```

Now that we have a categorical response variable, we need to decide which sort of Chi-square analysis to perform. The sampling design determines the correct analysis as always in these situations. The stratified random sample involved samples from each of the three populations so a Homogeneity test should be employed. In these situations, the stacked bar chart provides the appropriate summary of the data. It also shows us the labels of the categories that the cut function created in the new growth cut variable:

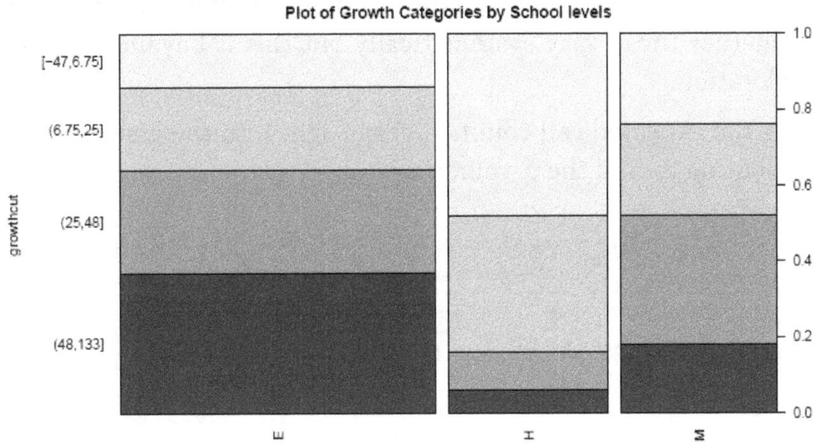

Figure 5.25: Stacked bar chart of the growth category responses by level of school.

```
plot(growthcut ~ stype, data = apistra,
     main = "Plot of Growth Categories by School levels")
```

Figure 5.25 suggests that the distributions of growth scores may not be the same across the levels of the schools with many more high growth *Elementary* schools than in either the *Middle* or *High* school groups (the "high" growth category is labelled as (48, 133] providing the interval of growth scores placed in this category). Similarly, the proportion of the low or negative growth (category of (-47.6, 6.75] for "growth" between-47.6 and 6.75) is least frequently occurring in *Elementary* schools and most frequent in the *High* schools. Statisticians often work across many disciplines and so may not always have the subject area knowledge to know why these differences exist (just like you might not), but an education researcher could take this sort of information – because it is a useful summary of interesting school-level data – and generate further insights into why growth in the API metric may or may not be a good or fair measure of school performance.

Of course, we want to consider whether these results can extend to the population of all California schools. The homogeneity hypotheses for assessing the growth rate categories across the types of schools would be:

- **H0:** There is no difference in the distribution of growth categories across the three levels of schools in the population of California schools.
- **HA:** There is some difference in the distribution of growth categories across the three levels of schools in the population of California schools.

There might be an issue with the independence assumption in that schools within the same district might be more similar to one another and different between one another. Sometimes districts are accounted for in education research to account for differences in policies and demographics among the districts. We could explore this issue by finding district-level average growth rates and exploring whether those vary systematically but this is beyond the scope of the current exploration.

Checking the expected cell counts gives insight into the assumption for using the χ^2-distribution to find the p-value:

```
growthtable <- tally(~ stype + growthcut, data = apistrat)
growthtable
```

```
##       growthcut
## stype [-47,6.75] (6.75,25] (25,48] (48,133]
##     E         14        22      27       37
##     H         24        18       5        3
##     M         12        12      17        9
```

```
chisq.test(growthtable)$expected
```

```
##       growthcut
## stype [-47,6.75] (6.75,25] (25,48] (48,133]
##     E       25.0        26   24.50    24.50
##     H       12.5        13   12.25    12.25
##     M       12.5        13   12.25    12.25
```

The smallest expected count is 12.25, occurring in four different cells, so we can use the parametric approach.

```
chisq.test(growthtable)
```

```
##
##  Pearson's Chi-squared test
##
## data:  growthtable
## X-squared = 38.668, df = 6, p-value = 8.315e-07
```

The observed test statistic is $X^2 = 38.67$ and, based on a $\chi^2(6)$ distribution, the p-value is 0.0000008. This p-value suggests that there is very strong evidence against the null hypothesis of no difference in the distribution of API growth of schools among *Elementary*, *Middle* and *High School* in the population of schools in California between 1999 and 2000, and we can conclude that there is some difference in the population (California schools). Because the schools were randomly selected from all the California schools we can make valid inferences to all the schools but because the level of schools, obviously, cannot be randomly assigned, we cannot say that level of school causes these differences.

The standardized residuals can enhance this interpretation, displayed in Figure 5.26. The *Elementary* schools have fewer low/negative growth schools and more high growth schools than expected under the null hypothesis. The *High* schools have more low growth and fewer higher growth (growth over 25 points) schools than expected if there were no difference in patterns of response across the school levels. The *Middle* school results were closer to the results expected if there were no differences across the school levels.

```
chisq.test(growthtable)$residuals
```

```
##        growthcut
## stype [-47,6.75]  (6.75,25]     (25,48]    (48,133]
##      E -2.2000000 -0.7844645  0.5050763  2.5253814
##      H  3.2526912  1.3867505 -2.0714286 -2.6428571
##      M -0.1414214 -0.2773501  1.3571429 -0.9285714
```

```
mosaicplot(growthcut ~ stype, data = apistrat, shade = T)
```

The binning of quantitative variables is not a first step in analyses – the quantitative version is almost always preferable. However, this analysis avoided the violation of the normality assumption that was somewhat problematic for the ANOVA and still provided useful inferences to the differences in the types of schools. When one goes from a quantitative to categorical version of a variable, one loses information (the specific details of the quantitative responses within each level created) and this almost always will result in a loss of statistical power of the procedure. In this situation, the p-value from the ANOVA was of the order 10-10 while the Chi-square test had a p-value of order 10-7. This larger p-value is typical of the loss of power in going to a categorical response when more information was available. In many cases, there are no options but to use contingency table analyses. This example shows that there might be some situations where "going categorical" could be an acceptable method for handing situations where an assumption is violated.

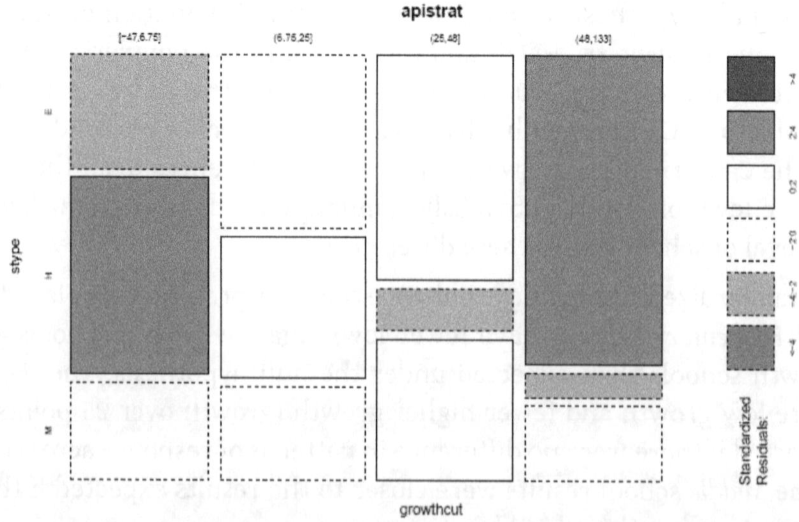

Figure 5.26: Mosaic plot of the API Growth rate categories versus level of the school with shading for size of standardized residuals.

5.12 CONCISE SUMMARY

Chi-square tests can be generally used to perform two types of tests, the Independence and Homogeneity tests. The appropriate analysis is determined based on the data collection methodology. The parametric Chi-square distribution for which these tests are named is appropriate when the expected cell counts are large enough (related to having a large enough overall sample). When the expected cell count condition is violated, the permutation approach can provide valuable inferences in these situations in most situations.

Data displays of the stacked bar chart (Homogeneity) and mosaic plots (Independence) provide a visual summary of the results that can also be found in contingency tables. You should have learned how to calculate the X^2 (X-squared) test statistic based on first finding the expected cell counts. Under certain assumptions, it will follow a Chi-Square distribution with $(R - 1)(C - 1)$ degrees of freedom. When those assumptions are not met, it is better to use a permutation approach to find p-values. Either way, the same statistic is used to test either kind of hypothesis, independence or homogeneity. After assessing evidence against the null hypothesis, it is interesting to see which cells in the table contributed to the deviations from the null hypothesis. The standardized residuals provide that information. Graphing them in a mosaic plot makes for a fun display to identify the large residuals and often allows you to better understand the results. This should tie back into the original data display (tableplot, stacked bar chart or mosaic plot) and contingency table where you identified initial patterns and help to tell the story of the results.

5.13 SUMMARY OF IMPORTANT R CODE

The main components of R code used in this chapter follow with components to modify in lighter and/or ALL CAPS text where y is a response variable and x is a predictor are easily identified:

- TABLENAME <- tally(~ x + y, data = DATASETNAME)
  - This function requires that the mosaic package has been loaded.
  - This provides a table of the counts in the variable called TABLENAME.
  - margins = T is used if you want to display row, column, and table totals.
- plot(y ~ x, data = DATASETNAME)
  - Makes a stacked bar chart useful for homogeneity test situations.
- mosaicplot(TABLENAME)
  - Makes a mosaic plot useful for finding patterns in the table in independence test situations.
- tableplot(data = DATASETNAME, sortCol = VARIABLENAME, pals = list("BrBG"))
  - Makes a tableplot sorted by VARIABLENAME, requires that the tabplot and RColor Brewer packages have been loaded.
  - The pals = list("BrBG") option provides a colourblind friendly colour palette, although other options are possible, such as pals = list("RdBu").
- chisq. test(TABLENAME)
  - Provides X^2 and p-values based on the χ^2-distribution with $(R - 1)(C - 1)$ degrees of freedom.
- chisq. test (TABLENAME) $ expected
  - Provides expected cell counts.
- pchisq (X-SQUARED, df = (R - 1)*(C - 1), lower. tail = F)
  - Provides p-value from χ^2-distribution with $(R - 1)(C - 1)$ degrees of freedom for observed test statistic.
  - See Section 5.5 for code related to finding a permutation-based p-value.
  - chisq.test(TABLENAME)$residuals^2
  - Provides X^2 contributions from each cell in table.
- chisq.test(TABLENAME)$residuals
  - Provides standardized residuals.

- mosaicplot(TABLENAME, shade = T)
  - Provides a mosaic plot with shading based on standardized residuals.

5.14 PRACTICE PROBLEMS

5.1 Determine type of Chi-Square test Determine which type of test is appropriate in each situation Independence or Homogeneity?

5.1.1.Concerns over diseases being transmitted between birds and humans have led to many areas developing monitoring plans for the birds that are in their regions. The duck pond on campus at MSU-Bozeman is a bit like a night club for the birds that pass through Bozeman.

- Suppose that a researcher randomly samples 20 ducks at the duck pond on campus on 4 different occasions and records the number ducks that are healthy and number that are sick on each day. The variables in this study are the day of measurement and sick/healthy.
- In another monitoring study, a researcher goes to a wetland area and collects a random sample from all birds present on a single day, classifies them by type of bird (ducks, swans, etc.) and then assesses whether each is sick or healthy. The variables in this study are type of bird and sick/healthy.

5.1.2 Psychologists performed an experiment on 48 male bank supervisors attending a management institute to investigate biases against women in personnel decisions. The supervisors were asked to make a decision on whether to promote a hypothetical applicant based on a personnel file. For half of them, the application file described a female candidate; for the others it described a male.

5.1.3. Researchers collected data on death penalty sentencing in Georgia. For 243 crimes, they categorized the crime by severity from 1 to 6 with Category 1 comprising barroom brawls, liquor-induced arguments, lovers' quarrels, and similar crimes and Category 6 including the most vicious, cruel, cold-blooded, unprovoked crimes. They also recorded the perpetrator's race. They wanted to know if there was a relationship between race and type of crime.

5.1.4. Epidemiologists want to see if Vitamin C helped people with colds. They would like to give some patients Vitamin C and some a placebo then compare the two groups. However, they are worried that the placebo might not be working. Since vitamin C has such a distinct taste, they are worried the participants will know which group they are in. To test if the placebo was working, they collected 200 subjects and randomly assigned half to take a placebo and the other half to take Vitamin C. 30 minutes later, they asked the subjects which supplement they received (hoping that the patients would not know which group they were assigned to).

5.1.5. Is zodiac sign related to GPA? 300 randomly selected students from MSU were asked their birthday and their current GPA. GPA was then categorized as < 1.50 = F, 1.51-2.50 = D, 2.51 - 3.25 = C, 3.26-3.75= B, 3.76-4.0 = A and their birthday was used to find their zodiac sign.

5.1.6. In 1935, the statistician R. A. Fisher famously had a colleague claim that she could distinguish whether milk or tea was added to a cup first. Fisher presented her, in a random order, 4 cups that were filled with milk first and 4 cups that were filled with tea first.

5.1.7. Researchers wanted to see if people from Rural and Urban areas aged differently. They contacted 200 people from Rural areas and 200 people from Urban areas and asked the participants their age (<40, 41-50, 51-60, >60).

5.2 Data is/are analysis The Five Thirty Eight Blog often shows up with interesting data summaries that have general public appeal. Their staff includes a bunch of quants with various backgrounds. When starting their blog, they had to decide on the data is/are question that we introduced in Section 2.1. To help them think about this, they collected a nationally representative sample that contained three questions about this. Based on their survey, they concluded that Relevant to the interests of Five Thirty Eight in particular, we also asked whether people preferred using "data" as a singular or plural noun. To those who prefer the plural, I'll put this in your terms: The data are pretty conclusive that the vast majority of respondents think we should say "data is." The singular crowd won by a 58 percentage-point margin, with 79 percent of respondents liking "data is" to 21 percent preferring "data are." But only half of respondents had put any thought to the usage prior to our survey, so it seems that it's not a pressing issue for most.

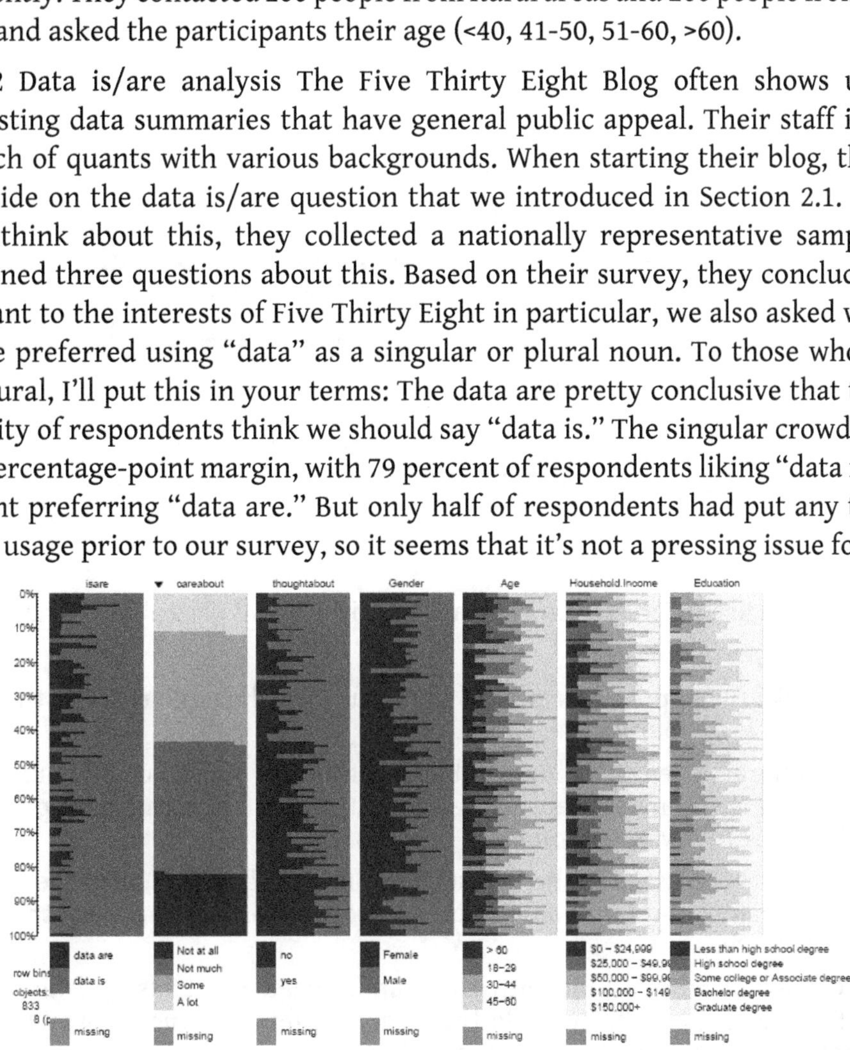

Figure 5.27: Tableplot of data from "data-is-vs-data-are" survey, sorted by "CareAbout" responses.

This came from a survey that contained questions about *which is the correct usage*, (isare), *have you thought about this issue* (thought about) with levels Yes/No, and *do you care about this issue* (careabout) with four levels from *Not at all* to *A lot*. The following code loads their data set after missing responses were removed, does a little re-ordering of factor levels using the fct_relevel function to help make the results easier to understand, and makes a tableplot (Figure 5.27) to get a general sense of the results including information on the respondents' gender, age, income, and education.

```
library(readr)
csd <- read_csv("http://www.math.montana.edu/courses/s217/documents/csd.csv")
```

```
library(tabplot)
# Need to make it explicit that these are factor variables and reorder
# factor levels to be in "correct" order using fct_relevel:
csd <- csd %>% mutate(careabout = factor(careabout),
                      careabout = fct_relevel(careabout,"Not at all", "Not much",
                                              "Some", "A lot"),
                      Education = factor(Education),
                      Education = fct_relevel(Education,
                                          levels(Education)[c(4,3,5,1,2)]),
                      Household.Income = factor(Household.Income),
                      Household.Income = fct_relevel(Household.Income,
                                          levels(Household.Income)[c(1,4,5,6,2,3)])
                      )

# Sorts plot by careabout responses
tableplot(csd, select = c(isare, careabout, thoughtabout, Gender,
                  Age, Household.Income, Education), sortCol = careabout,
          pals = list("BrBG"))
```

5.2.1. If we are interested in the variables is are and care about, what sort of test should we perform?

5.2.2. Make the appropriate plot of the results for the table relating those two variables relative to your answer to 5.8.

5.2.3. Generate the contingency table and find the expected cell counts, first "by hand" and then check them using the output. Is the parametric procedure appropriate here? Why or why not?

5.2.4. Report the value of the test statistic, its distribution under the null, the parametric p-value, and write a decision and conclusion, making sure to address scope of inference.

5.2.5. Make a mosaic plot with the standardized residuals and discuss the results. Specifically, in what way do the is/are preferences move away from the null hypothesis for people that care more about this?

We might be fighting a losing battle on "data is a plural word", but since we are in the group that cares a lot about this, we are going to keep trying. . .

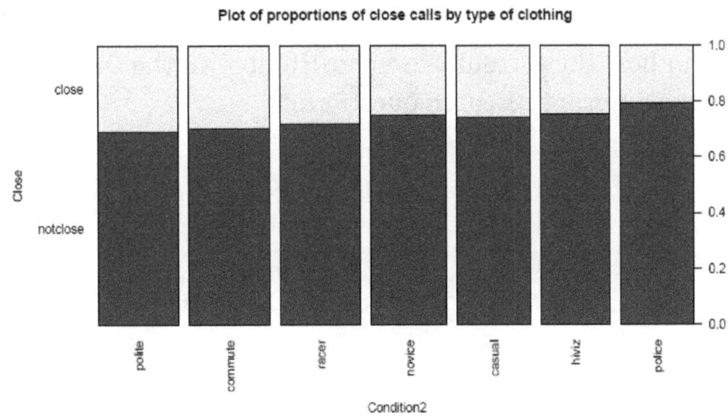

Figure 5.28: Stacked bar chart of the close calls/not (overtakes less than or equal to 100 cm or not) by outfit.

5.3 Overtake close calls by outfit analysis We can revisit the car overtake passing distance data from Chapter 3 and to focus in on the "close calls". The following code uses the if else function to create the close call/not response variable. It works to create a two-category variable where the first category (close) is encountered when the condition is true (Distance <= 100, so the passing distance was less than or equal to100 cm) from the "if" part of the function (if Distance is less than or equal to 100 cm, then "close") and the "else" is the second category (when the Distance was over 100 cm) and gets the category of notclose. The factor function is applied to the results from if else to make this a categorical variable for later use. Some useful code and a stacked bar chart in Figure 5.28 is provided.

```
dd <- read_csv("http://www.math.montana.edu/courses/s217/documents/Walker2014_mod.csv")
```

```
dd <- dd %>% mutate(Condition = factor(Condition),
                    Condition2 = reorder(Condition, Distance, FUN = mean),
                    Close = factor(ifelse(Distance <= 100, "close", "notclose"))
                    )

plot(Close ~ Condition2, data = dd)
```

```
table1 <- tally(Close ~ Condition2, data = dd)

chisq.test(table1)
```

```
##
##  Pearson's Chi-squared test
##
## data:  table1
## X-squared = 30.861, df = 6, p-value = 2.695e-05
```

5.3.1. This is a Homogeneity test situation. Why?

5.3.2. Perform the 6+ steps of the hypothesis test using the provided results.

5.3.3. Explain how these results are consistent with the One-Way ANOVA test but also address a different research question.

CHAPTER-6

CORRELATION AND SIMPLE LINEAR REGRESSION

INTRODUCTION: AN OVERVIEW

Correlation and simple linear regression are two foundational statistical techniques that facilitate the exploration and quantification of relationships between variables. In the realm of data analysis, understanding the associations between variables is paramount, and these methods offer valuable insights into the strength and direction of such connections. Correlation assesses the degree of linear relationship between two continuous variables, providing a numerical measure of their association. On the other hand, simple linear regression takes this exploration a step further by modelling the relationship between a dependent variable and an independent variable, enabling predictions and uncovering patterns in the data. Both methods are employed across diverse fields such as economics, psychology, and natural sciences, offering researchers a means to interpret and leverage the interconnectedness of variables. In this exploration of correlation and simple linear regression, we embark on a journey to unravel the intricacies of these statistical tools, unveiling their applications and implications in deciphering the complex web of relationships within data sets.

6.1 RELATIONSHIPS BETWEEN TWO QUANTITATIVE VARIABLES

The independence test in Chapter 5 provided a technique for assessing evidence of a relationship between two categorical variables. The terms relationship and association are synonyms that, in statistics, imply that particular values on one variable tend to occur more often with some other values of the other variable or that knowing something about the level of one variable provides information about the patterns of values on the other variable. These terms are not specific to the "form" of the relationship – any pattern (strong or weak, negative or positive, easily described or complicated) satisfy the definition. There are two other aspects to using these terms in a statistical context. First, they are not directional – an association between x and y is the same as saying there is an association between y and x. Second, they are not causal unless the levels of one of the variables are randomly assigned in an experimental context. We add to this terminology the idea of correlation between variables x and y. Correlation, in most statistical contexts, is a measure of the specific type of relationship between the variables: the linear relationship between two quantitative variables1. So as we start to review these ideas from your previous statistics course, remember that associations and relationships are more general than correlations and it is possible to have no correlation where there is a strong relationship between variables. "Correlation" is used colloquially as a synonym for relationship but we will work to reserve it for its more specialized usage here to refer specifically to the linear relationship.

Assessing and then modelling relationships between quantitative variables drives the rest of the chapters, so we should get started with some motivating examples to start to think about what relationships between quantitative variables "look like"... To motivate these methods, we will start with a study of the effects of beer consumption on blood alcohol levels (BAC, in grams of alcohol per deciliter of blood). A group of n = 16 student volunteers at The Ohio State University drank a randomly assigned number of beers2. Thirty minutes later, a police officer measured their BAC. Your instincts, especially as well-educated college students with some chemistry knowledge, should inform you about the direction of this relationship – that there i sa positive relationship between Beers and BAC. In other words, higher values of one variable are associated with higher values of the other. Similarly, lower values of one are associated with lowervalues of the other. In fact there are online calculators that tell you how much your BAC increases for each extra beer consumed (for example: http://www.craftbeer.com/beer-studies/blood-alcohol-content-calculatorif you plug in 1 beer). The increase in y (BAC) for a 1 unit increase in x (here, 1 more beer) is an example of a slope coefficient that is applicable if the relationship between the variables is linear and something that will be fundamental in what is called a simple linear regression model. In a simple linear regression model (simple means that there is only one explanatory variable) the slope is the expected change in the mean response for a one unit increase in the explanatory variable. You could also use the BAC calculator and the models that we are going to develop to pick a total number of beers you will consume and get a predicted BAC, which employs the entire equation we will estimate.

Before we get to the specifics of this model and how we measure correlation, we should graphically explore the relationship between Beers and BAC in a scatterplot. Figure 6.1 shows a scatterplot of the results that display the expected positive relationship. Scatterplots display the response pairs for the two quantitative variables with the explanatory variable on the x-axis and the response variable on the y-axis. The relationship between Beers and BAC appears to be relatively linear but there is possibly more variability than one might expect. For example, for students consuming 5 beers, their BAC s range from 0.05 to 0.10. If you look at the online BAC calculators, you will see that other factors such as weight, sex, and beer percent alcohol can impact the results. We might also be interested in previous alcohol consumption. In Chapter 8, we will learn how to estimate the relationship between Beers and BAC after correcting or controlling for those "other variables" using multiple linear regression, where we incorporate more than one quantitative explanatory variable into the linear model (somewhat like in the 2-Way ANOVA). Some of this variability might be hard or impossible to explain regardless of the other variables available and is considered unexplained variation and goes into the residual errors in our models, just like in the ANOVA models. To make scatterplots as in Figure 6.1, you could use the base R function plot, but we will want to again access the power of ggplot2 so will use geom_point to add the points to the plot at the "x" and "y" coordinates that you provide in aes(x= ..., y = ...).

```
library(readr)
BB <- read_csv("http://www.math.montana.edu/courses/s217/documents/beersbac.csv")
```

```
BB %>% ggplot(mapping = aes(x = Beers, y = BAC)) +
  geom_point() +
  theme_bw()
```

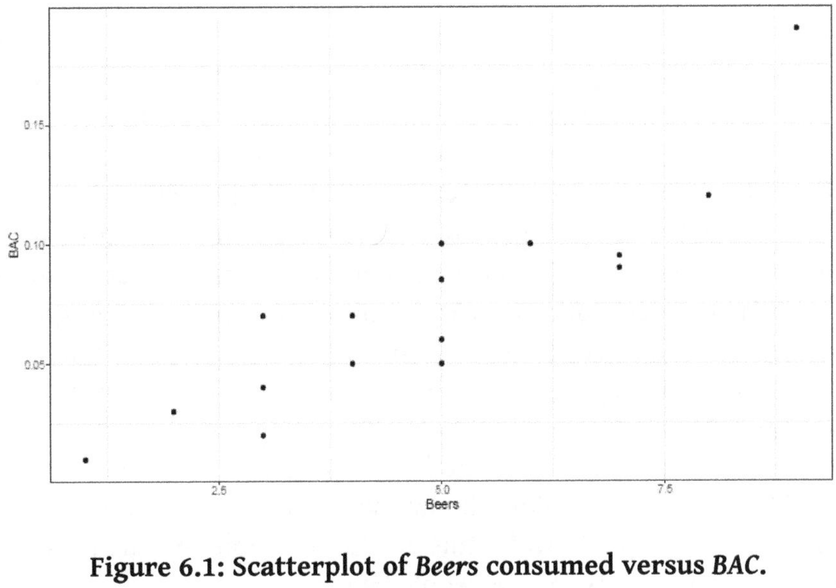

Figure 6.1: Scatterplot of *Beers* consumed versus *BAC*.

There are a few general things to look for in scatterplots:

- **Assess the direction of the relationship** – is it positive or negative?

- **Consider the strength of the relationship.** The general idea of assessing strength visually is about how hard or easy it is to see the pattern. If it is hard to see a pattern, then it is weak. If it is easy to see, then it is strong.
- **Consider the linearity of the relationship.** Does it appear to curve or does it follow a relatively straight line? Curving relationships are called curvilinear or nonlinear and can be strong or weak just like linear relationships – it is all about how tightly the points follow the pattern you identify.
- **Check for unusual observations** – outliers – by looking for points that don't follow the overall pattern. Being large in x or y doesn't mean that the point is an outlier. Being unusual relative to the overall pattern makes a point an outlier in this setting.
- **Check for changing variability in one variable based on values of the other variable.** This will tie into a constant variance assumption later in the regression models.
- **Finally, look for distinct groups in the scatterplot.** This might suggest that observations from two populations, say males and females, were combined but the relationship between the two quantitative variables might be different for the two groups.

Going back to Figure 6.1 it appears that there is a moderately strong linear relationship between Beers and BAC – not weak but with some variability around what appears to be a fairly clear to see straight-line relationship. There might even be a hint of a nonlinear relationship in the higher beer values. There are no clear outliers because the observation at 9 beers seems to be following the overall pattern fairly closely. There is little evidence of non-constant variance mainly because of the limited size of the data set – we'll check this with better plots later. And there are no clearly distinct groups in this plot, possibly because the# of beers was randomly assigned. These data have one more interesting feature to be noted – that subjects managed to consume 8 or 9 beers. This seems to be a large number. I have never been able to trace this data set to the original study so it is hard to know if (1) they had this study approved by a human subjects research review board to make sure it was "safe", (2) every subject in the study was able to consume their randomly assigned amount, and (3) whether subjects were asked to show up to the study with *BAC* s of 0. We also don't know the exact alcohol concentration of the beer consumed or volume. So while this is a fun example to start these methods with, a better version of this data set would be nice...

In making scatterplots, there is always a choice of a variable for the *x*-axis and the *y*-axis. It is our convention to put explanatory or independent variables (the ones used to explain or predict the responses) on the x-axis. In studies where the subjects are randomly assigned to levels of a variable, this is very clearly an explanatory variable, and we can go as far as making causal inferences with it. In observational studies, it can be less clear which variable explains which. In these cases, make the most reasonable choice based on the observed variables but remember that, when the direction of relationship is unclear, you could have switched the axes and thus the implication of which variable is explanatory.

6.2 ESTIMATING THE CORRELATION COEFFICIENT

In terms of quantifying relationships between variables, we start with the correlation coefficient, a measure that is the same regardless of your choice of variables as explanatory or response. We measure the strength and direction of linear relationships between two quantitative variables using Pearson's r or Pearson's Product Moment Correlation Coefficient. For those who really like acronyms, Wikipedia even suggests calling it the PPMCC. However, its use is so ubiquitous that the lower case r or just "correlation coefficient" are often sufficient to identify that you have used the PPMCC. Some of the extra distinctions arise because there are other ways of measuring correlations in other situations (for example between two categorical variables), but we will not consider them here.

The correlation coefficient, *r*, is calculated as

$$r = \frac{1}{n-1}\sum_{i=1}^{n}\left(\frac{x_i - \bar{x}}{s_x}\right)\left(\frac{y_i - \bar{y}}{s_y}\right),$$

where *sx* and *sy* are the standard deviations of *x* and *y*. This formula can also be written as

$$r = \frac{1}{n-1}\sum_{i=1}^{n} z_{x_i} z_{y_i}$$

where *zxi* is the z-score (observation minus mean divided by standard deviation) for the ith observation *onx* and *zyi* is the z-score for the i observation on y. We won't directly use this formula, but its contentsin form the behaviour of r . First, because it is a sum divided by (n - 1) it is a bit like an average - it combines information across all observations and, like the mean, is sensitive to outliers. Second, it is a dimension-less measure, meaning that it has no units attached to it. It is based on z-scores which have units of standard deviations of x or y so the original units of measurement are cancelled out going into this calculation. This also means that changing the original units of measurement, say from Fahrenheit to Celsius or from miles to km for one or the other variable will have no impact on the correlation. Less obviously, the formula guarantees that r is between -1 and 1. It will attain -1 for a perfect negative linear relationship, 1 for a perfect positive linear relationship, and 0 for no linear relationship. We are being careful here to say linear relationship because you can have a strong nonlinear relationship with a correlation of 0. For example, consider Figure 6.2.

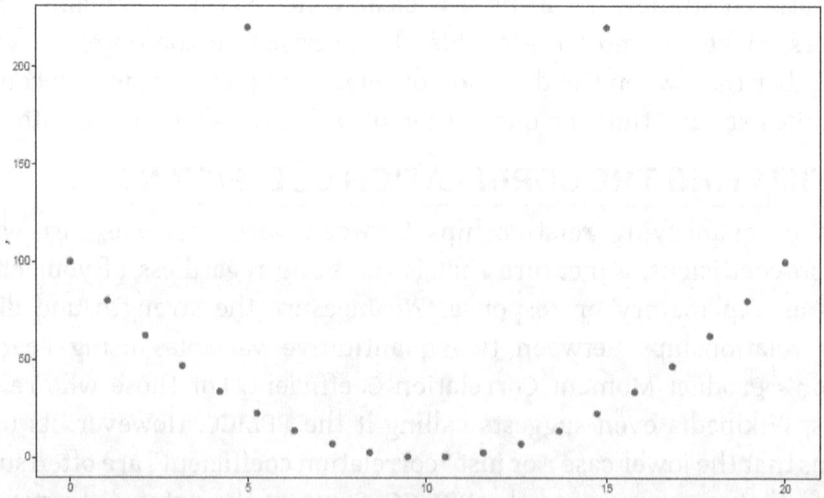

Figure 6.2: Scatterplot of an amusing (and strong) relationship that has *r* = 0.

There are some conditions for trusting the results that the correlation coefficient provides:

- Two quantitative variables measured.
    - This might seem silly, but categorical variables can be coded numerically and a meaningless correlation can be estimated if you are not careful what you correlate.
- The relationship between the variables is relatively linear.
    - If the relationship is nonlinear, the correlation is meaningless since it only measures linear relationships and can be misleading if applied to a nonlinear relationship.
- There should be no outliers.
    - The correlation is very sensitive (technically not resistant) to the impacts of certain types of outliers and you should generally avoid reporting the correlation when they are present.
    - One option in the presence of outliers is to report the correlation with and without outliers to see how they influence the estimated correlation.

The correlation coefficient is dimensionless but larger magnitude values (closer to -1 OR 1) mean stronger linear relationships. A rough interpretation scale based on experiences working with correlations follows, but this varies between fields and types of research and variables measured. It depends on the levels of correlation researchers become used to obtaining, so can even vary within fields. Use this scale for the discussing the strength of the linear relationship until you develop your own experience with typical results in a particular field and what is expected:

- $|r| < 0.3$: weak linear relationship,
- $0.3 < |r| < 0.7$: moderate linear relationship,
- $0.7 < |r| < 0.9$: strong linear relationship, and
- $0.9 < |r| < 1.0$: very strong linear relationship.

And again note that this scale only relates to the **linear** aspect of the relationship between the variables.

When we have linear relationships between two quantitative variables, x and y, we can obtain estimated correlations from the cor function either using y ~ x or by running the cor function3 on the entire data set. When you run the cor function on a data set it produces a correlation matrix which contains a matrix of correlations where you can triangulate the variables being correlated by the row and column names, noting that the correlation between a variable and itself is 1. A matrix of correlations is useful for comparing more than two variables, discussed below.

```
library(mosaic)
cor(BAC ~ Beers, data = BB)
```

```
## [1] 0.8943381
```

```
cor(BB)
```

```
##           Beers       BAC
## Beers 1.0000000 0.8943381
## BAC   0.8943381 1.0000000
```

Based on either version of using the function, we find that the correlation between Beers and BAC is estimated to be 0.89. This suggests a strong linear relationship between the two variables. Examples are about the only way to build up enough experience to become skilful in using the correlation coefficient. Some additional complications arise in more complicated studies as the next example demonstrates.

Gude et al. [2009] explored the relationship between average summer temperature (degrees F) and area burned (natural log of hectares4 = log(hectares)) by wildfires in Montana from 1985 to 2007. The log-transformation is often used to reduce the impacts of really large observations with non-negative (strictly greater than 0) variables (more on transformations and their impacts on regression models in Chapter 7). Based on your experiences with the wildfire "season" and before analyzing the data, I'm sure you would assume that summer temperature explains the area burned by wildfires. But could it be that more fires are related to having warmer summers? That second direction is unlikely on a state-wide scale but could apply at a particular weather station that is near a fire. There is another option – some other variable is affecting both variables. For example, drier summers might be the real explanatory variable that is related to having both warm summers and lots of fires. These variables are also being measured over time making them examples of time series. In this situation, if there are changes over time, they might be attributed to climate change. So there are really three relationships to explore with the variables measured here (remembering that the full story might require measuring even more!): log-area burned versus temperature, temperature versus year, and log-area burned versus year.

As demonstrated in the following code, with more than two variables, we can use the cor function on all the variables and end up getting a matrix of correlations or, simply, the correlation matrix. If you triangulate the row and column labels, that cell provides the correlation between that pair of variables. For example, in the first row (Year) and the last column (log hectares), you can find that the correlation coefficient is r = 0.362. Note the symmetry in the matrix around the diagonal of 1's – this further illustrates that correlation between x and y does not depend on which variable is viewed as the "response". The estimated correlation between Temperature and Year is -0.004 and the correlation between log hectares (log-hectares burned) and Temperature is 0.81. So Temperature has almost no linear change over time. And there is a strong linear relationship between log hectares and Temperature. So it appears that temperatures may be related to log-area burned but that the trend over time in both is less clear (at least the linear trends).

```
mtfires <- read_csv("http://www.math.montana.edu/courses/s217/documents/climateR2.csv")
```

```
# natural log transformation of area burned
mtfires <- mtfires %>% mutate(loghectares = log(hectares))

# Cuts the original hectares data so only log-scale version in tibble
mtfiresR <- mtfires %>%
  select(-hectares)
cor(mtfiresR)
```

```
##                   Year Temperature loghectares
## Year         1.0000000  -0.0037991   0.3617789
## Temperature -0.0037991   1.0000000   0.8135947
## loghectares  0.3617789   0.8135947   1.0000000
```

The correlation matrix alone is misleading – we need to explore scatterplots to check for nonlinear relationships, outliers, and clustering of observations that may be distorting the numerical measure of the linear relationship. The ggpairs function from the GGally package [Schloerke et al., 2021] combinesthe numerical correlation information and scatterplots in one display5. As in the correlation matrix, you triangulate the variables for the pair wise relationship. The upper right panel of Figure 6.3 displays acorrelation of 0.362 for Year and log hectares and the lower left panel contains the scatterplot with Year on the *x*-axis and log hectares on the *y*-axis. The correlation between Year and Temperature is really small, both in magnitude and in display, but appears to be nonlinear (it goes down between 1985 and 1995 and then goes back up), so the correlation coefficient doesn't mean much here since it just measures the overall linear relationship. We might say that this is a moderate strength (moderately "clear") curvilinear relationship. In terms of the underlying climate process, it suggests a decrease in summer temperatures between 1985 and 1995 and then an increase in the second half of the data set.

```
library(GGally)
mtfiresR %>% ggpairs() + theme_bw()
```

As one more example, the Australian Institute of Sport collected data on 102 male and 100 female athletes that are available in the ais data set from the alr4 package (Weisberg [2018], Weisberg [2014]). They measured a variety of variables including the athlete's Hematocrit (Hc, units of percentage of red blood cells in the blood), Body Fat Percentage (Bfat, units of percentage of total body weight), and height (Ht, units of cm). Eventually we might be interested in predicting Hc based on the other associations are of interest.

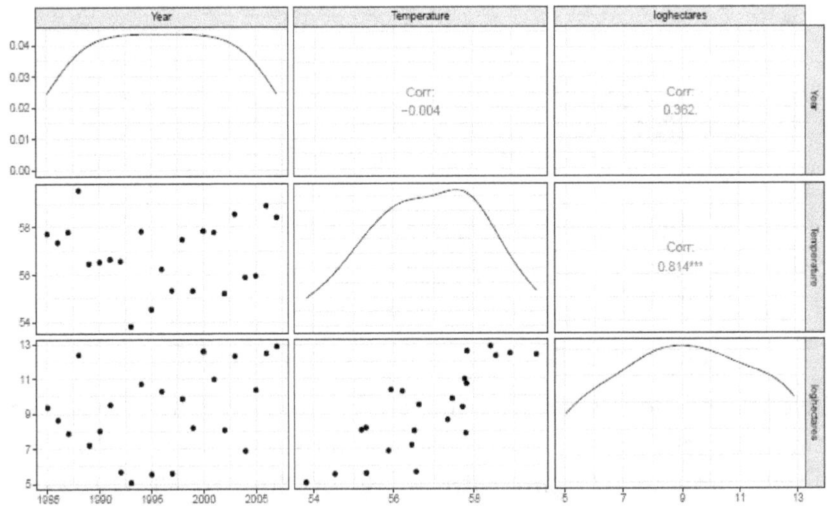

Figure 6.3: Scatterplot matrix of Montana fires data.

```
library(alr4)
data(ais)
library(tibble)
ais <- as_tibble(ais)
aisR <- ais %>%
  select(Ht, Hc, Bfat)
summary(aisR)
```

```
##        Ht              Hc             Bfat       
##  Min.   :148.9   Min.   :35.90   Min.   : 5.630  
##  1st Qu.:174.0   1st Qu.:40.60   1st Qu.: 8.545  
##  Median :179.7   Median :43.50   Median :11.650  
##  Mean   :180.1   Mean   :43.09   Mean   :13.507  
##  3rd Qu.:186.2   3rd Qu.:45.58   3rd Qu.:18.080  
##  Max.   :209.4   Max.   :59.70   Max.   :35.520  
```

```
aisR %>% ggpairs() + theme_bw()
```

```
cor(aisR)
```

```
##              Ht         Hc       Bfat
## Ht    1.0000000  0.3711915 -0.1880217
## Hc    0.3711915  1.0000000 -0.5324491
## Bfat -0.1880217 -0.5324491  1.0000000
```

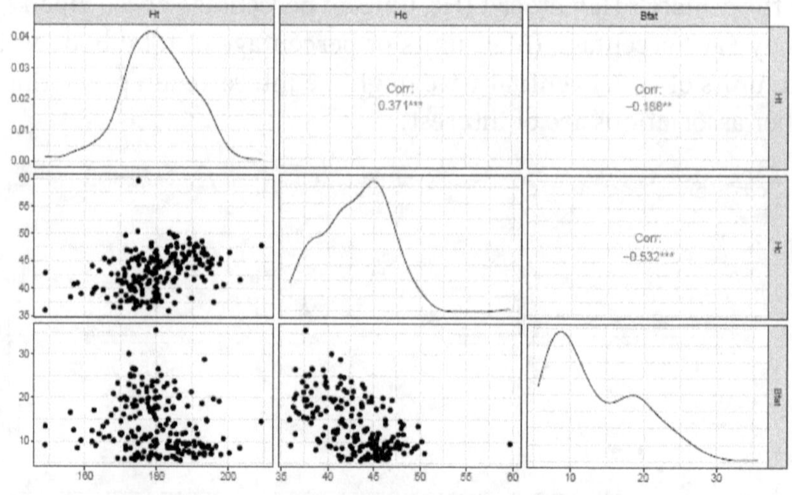

Figure 6.4: Scatterplot matrix of athlete data.

Ht (Height) and Hc (Hematocrit) have a moderate positive relationship that may contain a slight nonlinearity. It also contains one clear outlier for a middle height athlete (around 175 cm) with an Hc of close to 60% (a result that is extremely high). One might wonder about whether this athlete has been doping or if that measurement involved a recording error. We should consider removing that observation to see how our results might change without it impacting the results. For the relationship between Bfat (*body fat*) and Hc (*hematocrit*), that same high Hc value is a clear outlier. There is also a high Bfat (*body fat*) athlete (35%) with a somewhat low Hc value. This also might be influencing our impressions so we will remove both "unusual" values and remake the plot. The two offending observations were found for individuals numbered 56 and 166 in the data set. To access those observations (and then remove them), we introduce the slice function that we can apply to a tibble as a way to use the row number to either select (as used here) or remove those rows:

```
aisR %>% slice(56, 166)
```

```
## # A tibble: 2 x 3
##        Ht     Hc  Bfat
##     <dbl> <dbl> <dbl>
## 1   180.   37.6 35.5
## 2   175.   59.7  9.56
```

We can create a reduced version of the data (aisR2) using the slice function to slice "out" the rows we don't want by passing a vector of the rows we don't want to retain with a minus sign in front of each of them, slice(-56, -166), or as vector of rows with a minus in front of the concatenated (c(...)) vector (slice(-c(56, 166))), and then remake the plot:

```
aisR2 <- aisR %>% slice(-56, -166) #Removes observations in rows 56 and 166
aisR2 %>% ggpairs() + theme_bw()
```

After removing these two unusual observations, the relationships between the variables are more obvious (Figure 6.5). There is a moderate strength, relatively linear relationship between Height and Hematocrit. There is almost no relationship between Height and Body Fat % (r = -0.20). There is a negative, moderate strength, somewhat curvilinear relationship between Hematocrit and Body Fat % (r = -0.54). As hematocrit increases initially, the body fat percentage decreases but at a certain level (around 45% for Hc), the body fat percentage seems to level off. Interestingly, it ended up that removing those two outliers had only minor impacts on the estimated correlations – this will not always be the case.

Sometimes we want to just be able to focus on the correlations, assuming we trust that the correlation isa reasonable description of the results between the variables. To make it easier to see patterns of positive and negative correlations, we can employ a different version of the same display from the corrplot package [Wei and Simko, 2021] with the corrplot. mixed function. In this case (Figure 6.6), it tells much the same story but also allows the viewer to easily distinguish both size and direction and read off the numerical correlations if desired.

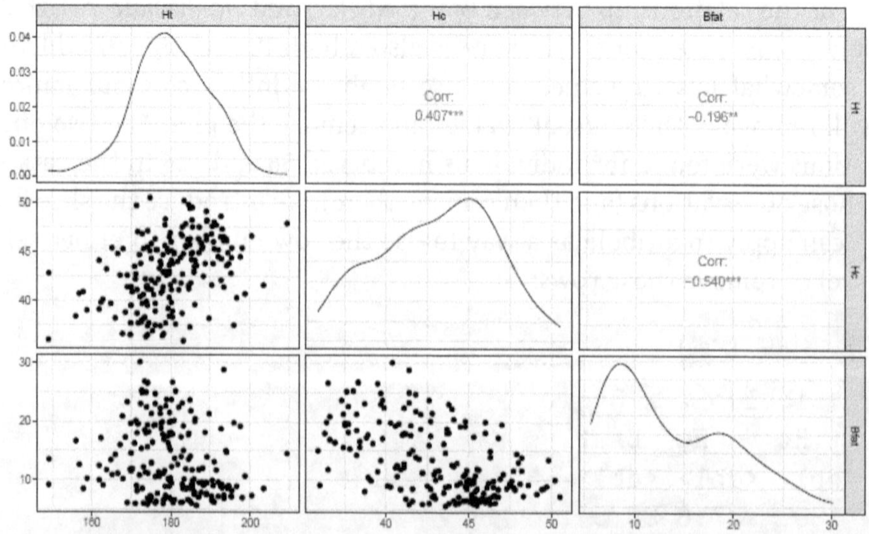

Figure 6.5: Scatterplot matrix of athlete data with two potential outliers removed.

```
library(corrplot)
corrplot.mixed(cor(aisR2), upper.col = c("black", "orange"),
        lower.col = c("black", "orange"))
```

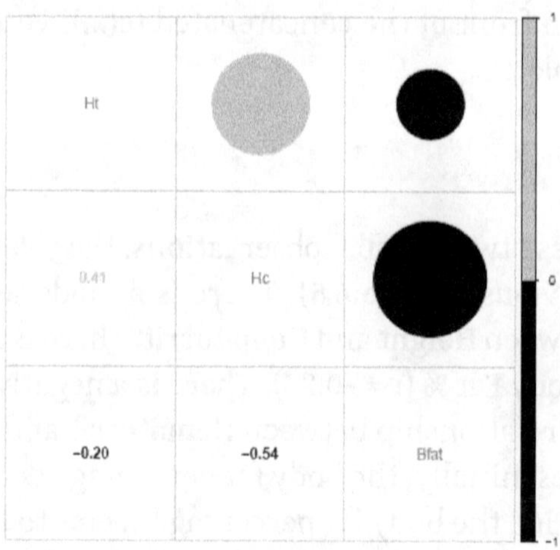

Figure 6.6: Correlation plot of the athlete data with two potential outliers removed. Lighter (orange) circle for positive correlations and black for negative correlations.

6.3 RELATIONSHIPS BETWEEN VARIABLES BY GROUPS

In assessing the relationship between variables, incorporating information from a third variable can often enhance the information gathered by either showing that the relationship between the first two variables is the same across levels of the other variable or showing that it differs. When the other variable is categorical (or just can be made categorical), it can be added to scatterplots, changing the symbols and colours for the points based on the different groups. These techniques are especially useful if the categorical variable corresponds to potentially distinct groups in the responses. In the previous example, the data set was built with male and female athletes. For some characteristics, the relationships might be the same for both sexes but for others, there are likely some physiological differences to consider.

This set of material is where the ggplot2 methods will really pay off for us, providing you with an extensive set of tools for visualizing relationships between two quantitative variables and incorporating information from other variables. There are three ways to add a categorical variable to a scatterplot that we will use. The first is to modify the colours, the second is modify the plotting symbol, and the third is to split the graph into panels or facets based on the groups of the variable. We usually combine the first two options to give the reader the best chance of detecting the group differences using both colours and symbols by groups; we will save faceting for a little later in the material. In these modifications, we can modify the colours and symbols based on the levels of categorical variable (say groupfactor) by adding colour =groupfactor, shape = groupfactor to the aes() definition in the initial ggplot part of the function or within an aesthetic inside geom_point. Defining the colours and shape within the geom_point only is useful ifyou want to change colours or symbols for the points in a way that might differ from the colours and groupings you use for other layers in the plot. The addition of grouping information in the initial ggplot aesthetic iscalled a "global" aesthetic and will apply to all the following geom's. Defining the colours or symbols within geom_point is called a "local" aesthetic and only applies to that layer of the plot. To enhance visibility of the points in the scatterplot, we often engage different colour palettes, using a version6 of the viridis colours with scale_color_viridis_d(end = 0.7). Using these ggplot additions, Figure 6.7 displays the Height and Hematocrit relationship with information on the sex of the athletes where sex was coded 0 for males and 1 for females, changing both the symbol and colour for the groups – with a legend to help to understand the plot.

Adding the grouping information really changes the impressions of the relationship between Height and Hematocrit – within each sex, there is little relationship between the two variables. The overall relationship is of moderate strength and positive but the subgroup relationships are weak at best. The overall relationship is created by inappropriately combining two groups that had different means in both the x and y directions. Men have higher mean heights and hematocrit values than women and putting them together in one large group creates the misleading overall relationship

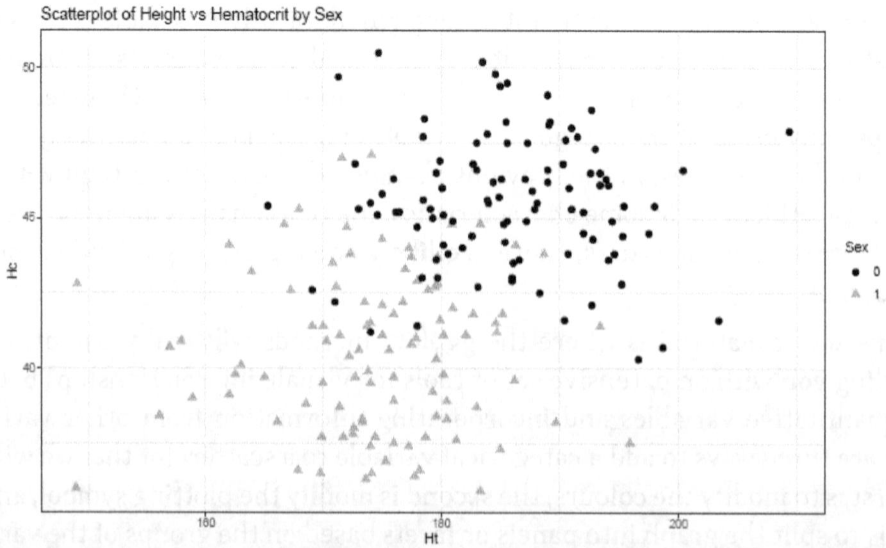

Figure 6.7: Scatterplot of athlete's height and hematocrit by sex of athletes. Males were coded as 0s and females as 1s.

```
aisR2 <- ais %>%
  slice(-c(56, 166)) %>%
  select(Ht, Hc, Bfat, Sex) %>%
  mutate(Sex = factor(Sex))


aisR2 %>% ggplot(mapping = aes(x = Ht, y = Hc)) +
  geom_point(aes(shape = Sex, color = Sex), size = 2.5) +
  theme_bw() +
  scale_color_viridis_d(end = 0.7) +
  labs(title = "Scatterplot of Height vs Hematocrit by Sex")
```

To get the correlation coefficients by groups, we can subset the data set using a logical inquiry on the Sex variable in the updated aisR2 data set, using Sex == 0 in the filter function to get a tibble with male subjects only and Sex == 1 for the female subjects, then running the cor function on each version of the data set:

```
cor(Hc ~ Ht, data = aisR2 %>% filter(Sex == 0)) #Males only
```

```
## [1] -0.04756589
```

```
cor(Hc ~ Ht, data = aisR2 %>% filter(Sex == 1)) #Females only
```

```
## [1] 0.02795272
```

These results show that r = -0.05 for Height and Hematocrit for males and r = 0.03 for females. The first suggests a very weak negative linear relationship and the second suggests a very weak positive linear relationship. The correlation when the two groups were combined (and group information was ignored!)was that r = 0.37. So one conclusion here is that correlations on data sets that contain groups can be very misleading (if the groups are ignored). It also emphasizes the importance of exploring for potential subgroups in the data set – these two groups were not obvious in the initial plot, but with added information the real story became clear.

For the Body Fat vs Hematocrit results in Figure 6.8, with an overall correlation of r = -0.54, the subgroup correlations show weaker relationships that also appear to be in different directions (r = 0.13 for men and r = -0.17 for women). This doubly reinforces the dangers of aggregating different groups and ignoring the group information.

```
cor(Hc ~ Bfat, data = aisR2 %>% filter(Sex == 0)) #Males only
```

```
## [1] 0.1269418
```

```
cor(Hc ~ Bfat, data = aisR2 %>% filter(Sex == 1)) #Females only
```

```
## [1] -0.1679751
```

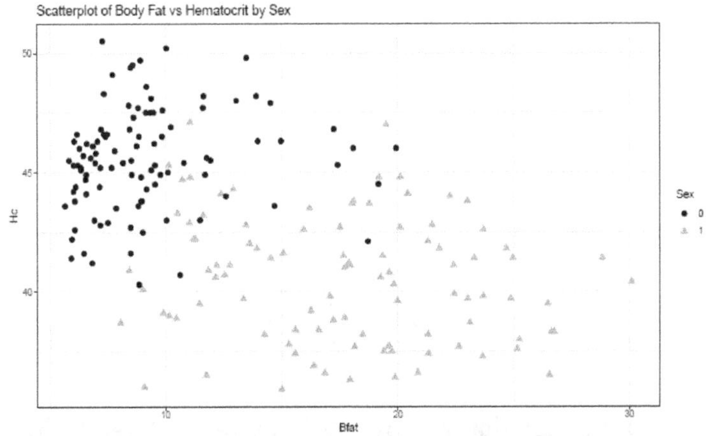

Figure 6.8: Scatterplot of athlete's body fat and hematocrit by sex of athletes. Males were coded as 0s and females as 1s.

```
aisR2 %>% ggplot(mapping = aes(x = Bfat, y = Hc)) +
  geom_point(aes(shape = Sex, color = Sex), size = 2.5) +
  theme_bw() +
  scale_color_viridis_d(end = 0.7) +
  labs(title = "Scatterplot of Body Fat vs Hematocrit by Sex")
```

One final exploration for these data involves the *body fat* and *height* relationship displayed in Figure **6.9** This relationship shows an even greater disparity between overall and subgroup results. The overall relationship is characterized as a weak negative relationship (r = -0.20) that is not clearly linear or nonlinear. The subgroup relationships are both clearly positive with a stronger relationship for men that might also be nonlinear (for the linear relationships r = 0.45 for women and r = 0.20 for men). Especially for female athletes, those that are taller seem to have higher body fat percentages. This might be related to the types of sports they compete in (there were 10 in the data set) – that would be another categorical variable we could incorporate. . . Both groups also seem to demonstrate slightly more variability in *Body Fat* associated with taller athletes (each sort of "fans out").

```
cor(Bfat ~ Ht, data = aisR2 %>% filter(Sex == 0)) #Males only
```

```
## [1] 0.1954609
```

```
cor(Bfat ~ Ht, data = aisR2 %>% filter(Sex == 1)) #Females only
```

```
## [1] 0.4476962
```

```
aisR2 %>% ggplot(mapping = aes(x = Ht, y = Bfat)) +
```

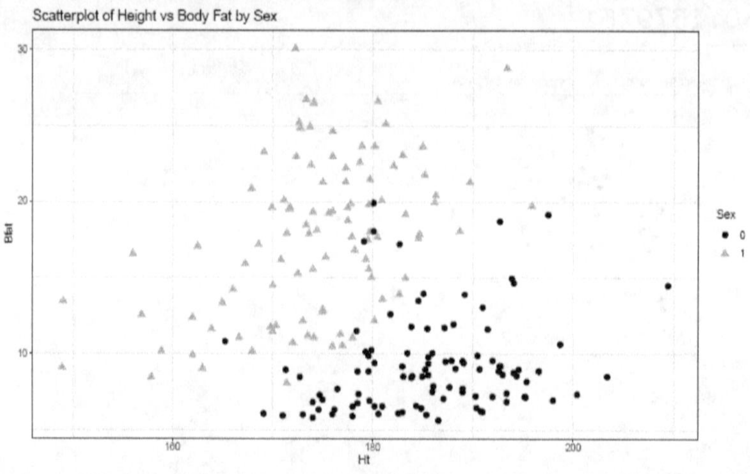

Figure 6.9: Scatterplot of athlete's body fat and height by sex.

```
geom_point(aes(shape = Sex, color = Sex), size = 2.5) +
theme_bw() +
scale_color_viridis_d(end = 0.7) +
labs(title = "Scatterplot of Height vs Body Fat by Sex")
```

In each of these situations, the sex of the athletes has the potential to cause misleading conclusions if ignored. There are two ways that this could occur – if we did not measure it then we would have no hope to account for it OR we could have measured it but not adjusted for it in our results, as was done initially. We distinguish between these two situations by defining the impacts of this additional variable as either a confounding or lurking variable:

- **Confounding variable:** affects the response variable and is related to the explanatory variable. The impacts of a confounding variable on the response variable cannot be separated from the impacts of the explanatory variable.
- **Lurking variable:** a potential confounding variable that is not measured and is not considered in the interpretation of the study.

Lurking variables show up in studies sometimes due to lack of knowledge of the system being studied or a lack of resources to measure these variables. Note that there may be no satisfying resolution to the confounding variable problem but that it is better to have measured it and know about it than to have it remain a lurking variable.

To help think about confounding and lurking variables, consider the following situation. On many highways, such as Highway 93 in Montana and north into Canada, recent construction efforts have been involved in creating safe passages for animals by adding fencing and animal crossing structures. These structures both can improve driver safety, save money from costs associated with animal-vehicle collisions, and increase connectivity of animal populations. Researchers (such as Clevenger and Waltho [2005]) involved in these projects are interested in which characteristics of underpasses lead to the most successful structures, mainly measured by rates of animal usage (number of times they cross under the road). Crossing structures are typically made using culverts and those tend to be cylindrical. Researchers are interested in studying the effect of height and width of crossing structures on animal usage. Unfortunately, all the tallest structures are also the widest structures. If animals prefer the tall and wide structures, then there is no way to know if it is due to the height or width of the structure since they are confounded. If the researchers had only measured width, then they might assume that it is the important characteristic of the structures but height could be a lurking variable that really was the factor related to animal usage of the structures. This is an example where it may not be possible to design a study that prevents confounding of the two variables *height* and *width*. If the researchers could control the height and width of the structures independently, then they could randomly assign both variables to make sure that some narrow structures are installed that are tall and some that are short. Additionally, they would also want to have some wide structures that are short and some are tall. Careful design of studies can prevent confounding of variables if they are known in advance and it is possible to control them, but in observational studies the observed combinations of variables are uncontrollable. This is why we need to employ additional caution in interpreting results from observational studies. Here that would mean that even if width was found to be a predictor of animal usage, we would likely want to avoid saying that width of the structures caused differences in animal usage.

6.4 INFERENCE FOR THE CORRELATION COEFFICIENT

We used bootstrapping briefly in Chapter 2 to generate nonparametric confidence intervals based on the middle 95% of the bootstrapped version of the statistic. Remember that bootstrapping involves sampling *with replacement* from the data set and creates a distribution centered near the statistic from the real data set. This also mimics sampling under the alternative as opposed to sampling under the null as in our permutation approaches. Bootstrapping is particularly useful for making confidence intervals where the distribution of the statistic may not follow a named distribution. This is the case for the correlation coefficient which we will see shortly.

The correlation is an interesting summary but it is also an estimator of a population parameter called ρ (the symbol rho), which is the ***population correlation coefficient***. When ρ = 1, we have a perfect positive linear relationship in the population; when ρ = -1, there is a perfect negative linear relationship in the population; and when ρ = 0, there is no linear relationship in the population. Therefore, to test if there is a linear relationship between two quantitative variables, we use the null hypothesis *H*0 : ρ = 0 (testsif the true correlation, ρ, is 0 – no linear relationship). The alternative hypothesis is that there is some (positive or negative) relationship between the variables in the population, *HA* : ρ /= 0. The distribution of the Pearson correlation coefficient can be complicated in some situations, so we will use bootstrapping methods to generate confidence intervals for ρ based on repeated random samples with replacement from the original data set. If the C% confidence interval contains 0, then we would find little to no evidence against the null hypothesis since 0 is in the interval of our likely values for ρ. If the C% confidence interval does not contain 0, then we would find strong evidence against the null hypothesis. Along with its use in testing, it is also interesting to be able to generate a confidence interval for ρ to provide an interval where we are C% confident that the true parameter lies.

The beers and BAC example seemed to provide a strong relationship with r = 0.89. As correlations approach -1 or 1, the sampling distribution becomes more and more skewed. This certainly shows up in the bootstrap distribution that the following code produces (Figure 6.10). Remember that bootstrapping utilizes the resample function applied to the data set to create new realizations of the data set by re-sampling with replacement from those observations. The bold vertical line in Figure 6.10 corresponds to the estimated correlation r = 0.89 and the distribution contains a noticeable left skew with a few much smaller T*'s possible in bootstrap samples. The C% confidence interval is found based on the middle C% of the distribution or by finding the values that put (100 - C)/2 into each tail of the distribution with the qdata function.

```
Tobs <- cor(BAC ~ Beers, data = BB); Tobs
```

```
## [1] 0.8943381
```

```
set.seed(614)
B <- 1000
Tstar <- matrix(NA, nrow = B)
```

```
for (b in (1:B)){
  Tstar[b] <- cor(BAC ~ Beers, data = resample(BB))
}
quantiles <- qdata(Tstar, c(0.025, 0.975)) #95% Confidence Interval
```

```
quantiles
```

```
##       2.5%     97.5%
## 0.7633606 0.9541518
```

```
tibble(Tstar) %>% ggplot(aes(x = Tstar)) +
  geom_histogram(aes(y = ..ncount..), bins = 15, col = 1, fill = "skyblue", center = 0) +
  geom_density(aes(y = ..scaled..)) +
  theme_bw() +
  labs(y = "Density") +
  geom_vline(xintercept = quantiles, col = "blue", lwd = 2, lty = 3) +
  geom_vline(xintercept = Tobs, col = "red", lwd = 2) +
  stat_bin(aes(y = ..ncount.., label = ..count..), bins = 15,
           geom = "text", vjust = -0.75)
```

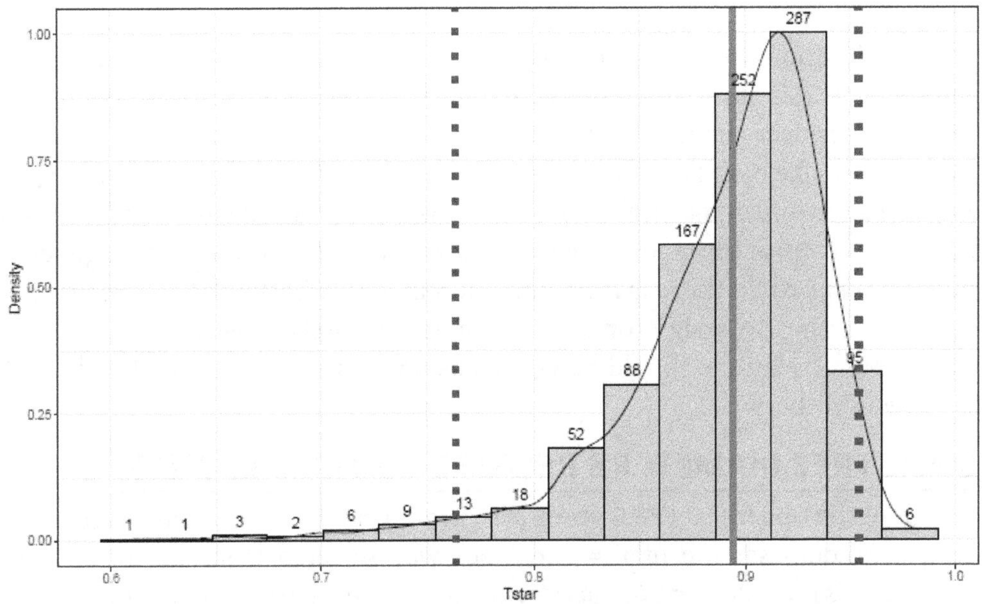

Figure 6.10: Histogram and density curve of the bootstrap distribution of the correlation coefficient with bold vertical line for observed correlation and dashed lines for bounds for the 95% bootstrap confidence interval.

These results tell us that the bootstrap 95% CI is from 0.76 to 0.95 – we are 95% confident that the true correlation between *Beers* and *BAC* in all OSU students like those that volunteered for this study is between and 0.95. Note that there are no units on the correlation coefficient or in this interpretation of it.

We can also use this confidence interval to test for a linear relationship between these variables.

- **H0 :** $\rho = 0$ **:** There is no linear relationship between Beers and BAC in the population.
- **HA :** $\rho \neq 0$ **:** There is a linear relationship between Beers and BAC in the population.

The 95% confidence level corresponds to a 5% significance level test and if the 95% CI does not contain 0, you know that the p-value would be less than 0.05 and if it does contain 0 that the p-value would be more than0.05. The 95% CI is from 0.76 to 0.95, which does not contain 0, so we find strong evidence[8] against the null hypothesis and conclude that there is a linear relationship between *Beers* and *BAC* in OSU students. We'll revisit this example using the upcoming regression tools to explore the potential for more specific conclusions about this relationship. Note that for these inferences to be accurate, we need to be able to trust that the sample correlation is reasonable for characterizing the relationship between these variables along with the assumptions we will discuss below.

In this situation with randomly assigned levels of *x* and strong evidence against the null hypothesis of no relationship, we can further conclude that changing beer consumption **causes** changes in the *BAC*. This is a much stronger conclusion than we can typically make based on correlation coefficients. Correlations and scatterplots are enticing for infusing causal interpretations in non-causal situations. Statistics teachers often repeat the mantra that ***correlation is not causation*** and that generally applies – except when there is randomization involved in the study. It is rarer for researchers either to assign, or even to be able to assign, levels of quantitative variables so correlations should be viewed as non-causal unless the details of the study suggest otherwise.

6.5 ARE TREE DIAMETERS RELATED TO TREE HEIGHTS?

In a study at the Upper Flat Creek study area in the University of Idaho Experimental Forest, a random sample of n = 336 trees was selected from the forest, with measurements recorded on Douglas Fir, Grand Fir, Western Red Cedar, and Western Larch trees. The data set called ufc is available from the spuRs package [Jones et al., 2018] and contains dbh.cm (tree diameter at 1.37 m from the ground, measured in cm) and height.m (tree height in meters). The relationship displayed in Figure 6.11

is positive, moderately strong with some curvature and increasing variability as the diameter increases. There do not appear to be groups in the data set but since this contains four different types of trees, we would want to revisit this plot by type of tree. To assist in the linearity assessment, we also add the geom_smooth to the plot with an option of method = "lm", which provides a straight line to best describe the relationship (more on that line in the coming sections and chapters). The bands around the line are based on the 95% confidence intervals we can generate for any x-value and relate to pinning down the true mean value of the y-variable at that value of the x-variable – but only apply if the linear relationship is a good description of the relationship between the variables (which it is not here!).

```
library(spuRs) #install.packages("spuRs")
data(ufc)
ufc <- as_tibble(ufc)

ufc %>% ggplot(mapping = aes(x = dbh.cm, y = height.m)) +
  geom_point() +
  geom_smooth(method = "lm") +
  theme_bw()
```

Of particular interest is an observation with a diameter around 58 cm and a height of less than 5 m. Observing a tree with a diameter around 60 cm is not unusual in the data set, but none of the other trees with this diameter had heights under 15 m. It ends up that the likely outlier is in observation number 168 and because it is so unusual it likely corresponds to either a damaged tree or a recording error.

```
ufc %>% slice(168)
```

```
## # A tibble: 1 x 5
```

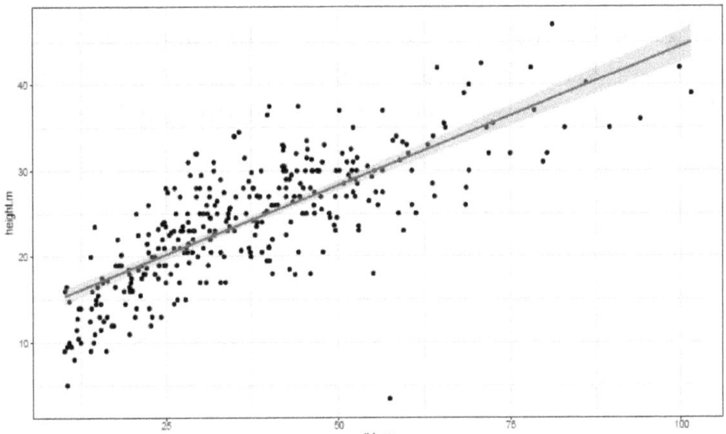

Figure 6.11: Scatterplot of tree heights (m) vs tree diameters (cm) with estimated straight line relationship (blue line) and 95% confidence interval (grey band).

```
##     plot  tree species dbh.cm height.m
##    <int> <int> <fct>     <dbl>    <dbl>
## 1     67     6 WL         57.5      3.4
```

With the outlier in the data set, the correlation is 0.77 and without it, the correlation increases to 0.79. The removal does not create a big change because the data set is relatively large and the diameter value is close to the mean of the x's9 but it has some impact on the strength of the correlation.

```
cor(dbh.cm ~ height.m, data = ufc)
```

```
## [1] 0.7699552
```

```
cor(dbh.cm ~ height.m, data = ufc %>% slice(-168))
```

```
## [1] 0.7912053
```

With the outlier included, the bootstrap 95% confidence interval goes from 0.702 to 0.820 – we are 95% confident that the true correlation between *diameter* and *height* in the population of trees is between 0.708 and 0.819. When the outlier is dropped from the data set, the 95% bootstrap CI is 0.753 to 0.826, which shifts the lower endpoint of the interval up, reducing the width of the interval from 0.111 to 0.073 (Figure 6.12). In other words, the uncertainty regarding the value of the population correlation coefficient is reduced. The reason to remove the observation is that it is unusual based on the observed pattern, which implies an error in data collection or sampling from a population other than the one used for the other observations and, if the removal is justified, it helps us refine our inferences for the population parameter. But measuring the linear relationship in these data where there is a clear curve violates one of our assumptions of using these methods – we'll see some other ways of detecting this issue in Section 6.10 and we'll try to "fix" this example using transformations in Chapter 7.

```
Tobs <- cor(dbh.cm ~ height.m, data = ufc); Tobs
```

```
## [1] 0.7699552
```

```
set.seed(208)
B <- 1000
Tstar <- matrix(NA, nrow = B)
for (b in (1:B)){
  Tstar[b] <- cor(dbh.cm ~ height.m, data = resample(ufc))
}
quantiles <- qdata(Tstar, c(.025, .975)) #95% Confidence Interval
quantiles
```

```
##       2.5%     97.5%
## 0.7075771 0.8190283
```

```
p1 <- tibble(Tstar) %>% ggplot(aes(x = Tstar)) +
  geom_histogram(aes(y = ..ncount..), bins = 25, col = 1, fill = "skyblue", center = 0) +
  geom_density(aes(y = ..scaled..)) +
  theme_bw() +
  labs(y = "Density", title = "Bootstrap distribution of correlation with all data") +
  geom_vline(xintercept = quantiles, col = "blue", lwd = 2, lty = 3) +
  geom_vline(xintercept = Tobs, col = "red", lwd = 2) +
  stat_bin(aes(y = ..ncount.., label = ..count..), bins = 25,
           geom = "text", vjust = -0.75) +
  xlim(0.6, 0.85) +
  ylim(0, 1.1)

Tobs <- cor(dbh.cm ~ height.m, data = ufc %>% slice(-168)); Tobs
```

```
## [1] 0.7912053
```

```
Tstar <- matrix(NA, nrow = B)
for (b in (1:B)){
  Tstar[b] <- cor(dbh.cm ~ height.m, data = resample(ufc %>% slice(-168)))
}
quantiles <- qdata(Tstar, c(.025, .975)) #95% Confidence Interval
quantiles
```

```
##       2.5%     97.5%
## 0.7532338 0.8259416
```

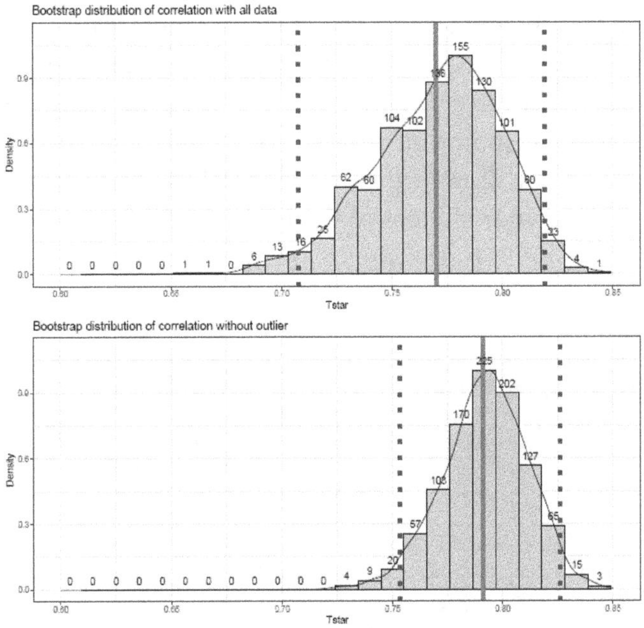

Figure 6.12: Bootstrap distributions of the correlation coefficient for the full data set (top) and without potential outlier included (bottom) with observed correlation (bold line) and bounds for the 95% confidence interval (dashed lines). Notice the change in spread of the bootstrap distributions as well as the different centers.

```
p2 <- tibble(Tstar) %>% ggplot(aes(x = Tstar)) +
  geom_histogram(aes(y = ..ncount..), bins = 25, col = 1, fill = "skyblue", center = 0) +
  geom_density(aes(y = ..scaled..)) +
  theme_bw() +
  labs(y = "Density", title = "Bootstrap distribution of correlation without outlier") +
  geom_vline(xintercept = quantiles, col = "blue", lwd = 2, lty = 3) +
  geom_vline(xintercept = Tobs, col = "red", lwd = 2) +
  stat_bin(aes(y = ..ncount.., label = ..count..), bins = 25,
           geom = "text", vjust = -0.75) +
```

```
  xlim(0.6, 0.85) +
  ylim(0, 1.1)

grid.arrange(p1, p2, ncol = 1)
```

6.6 DESCRIBING RELATIONSHIPS WITH A REGRESSION MODEL

When the relationship appears to be relatively linear, it makes sense to estimate and then interpret a line to represent the relationship between the variables. This line is called a regression line and involves finding a line that best fits (explains variation in) the response variable for the given values of the explanatory variable. For regression, it matters which variable you choose for x and which you choose for y – for correlation it did not matter. This regression line describes the "effect" of x on y and also provides an equation for predicting values of y for given values of x. The Beers and BAC data provide a nice example to start our exploration of regression models. The beer consumption is a clear explanatory variable, detectable in the story because it was randomly assigned to subjects and (2) basic science supports beer consumption amount being an explanatory variable for BAC. In some situations, this will not be so clear, but look for random assignment or scientific logic to guide your choices of variables as explanatory or response10.

```
BB %>% ggplot(mapping = aes(x = Beers, y = BAC)) +
  geom_smooth(method = "lm", col = "cyan4") +
  geom_point() +
  theme_bw() +
  geom_segment(aes(y = 0.05914, yend = 0.05914, x = 4, xend = 0), col = "blue",
               lty = 2, arrow = arrow(length = unit(.3, "cm"))) +
  geom_segment(aes(x = 4, xend = 4, y = 0, yend = 0.05914),
               arrow = arrow(length = unit(.3, "cm")), col = "blue") +
  geom_segment(aes(y = 0.0771, yend = 0.0771, x = 5, xend = 0), col = "forestgreen",
               lty = 2, arrow = arrow(length = unit(.3, "cm"))) +
  geom_segment(aes(x = 5, xend = 5, y = 0, yend = 0.0771),
               arrow = arrow(length = unit(.3, "cm")), col = "forestgreen")
```

The equation for a line is y = a + bx, or maybe y = mx + b. In the version mx + b you learned that m is a slope coefficient that relates a change in x to changes in y and that b is a y-intercept (the value of y when x is 0). In Figure 6.13, extra lines are added to help you see the defining characteristics of the line. The slope, whatever letter you use, is the change in y for a one-unit increase in x. Here, the slope is the change in BAC for a 1 beer increase in Beers, such as the change from 4 to 5 beers. The y-values (dashed lines with arrows) for Beers = 4 and 5 go from 0.059 to 0.077. This means that for a 1 beer increase (+1 unit change in x), the BAC goes up by 0.077 - 0.059 = 0.018 (+0.018 unit change in y). We can also try to find the y-intercept on the graph by looking for the BAC level for 0 Beers consumed. The y-value (BAC) ends up being around -0.01 if you extend the regression line to Beers = 0. You might assume that the BAC should be 0 for Beers = 0 but the researchers did not observe any students at 0 Beers, so we don't really know what the BAC might be at this value. We have to use our line to predict this value. This ends up providing a prediction below 0 – an impossible value for BAC. If the y-intercept were positive, it would suggest that the students have a BAC over 0 even without drinking.

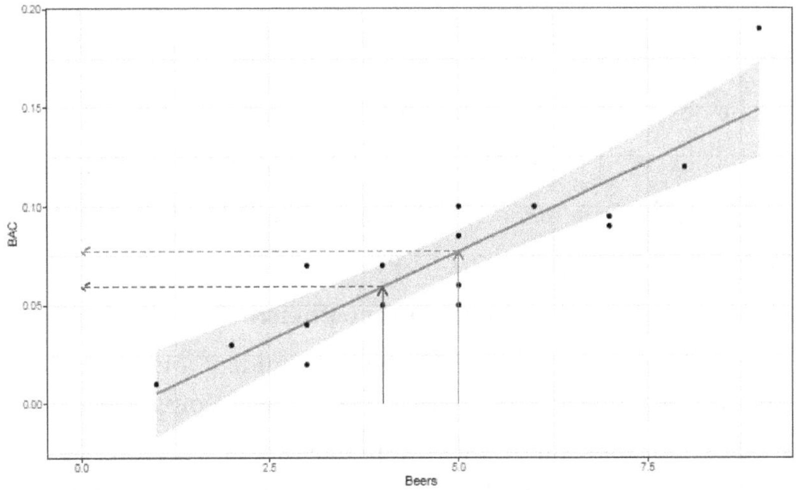

Figure 6.13: Scatterplot with estimated regression line (solid line) for the *Beers* and *BAC* data. The solid arrows indicate the predictor variable values of 4 and 5 beers and the dashed lines illustrate the predicted mean BAC for 4 and 5 beers consumed based on the SLR model.

The numbers reported were very accurate because we weren't using the plot alone to generate the values – we were using a linear model to estimate the equation to describe the relationship between Beers and BAC. In statistics, we estimate "m" and "b". We also write the equation starting with the y-intercept and use slightly different notation that allows us to extend to more complicated models with more variables. Specifically, the estimated regression equation is $\hat{y}$ = b0 + b1x, wherey

- $\hat{y}$ is the estimated value of *y* for a given *x*,
- b_0 is the estimated *y*-intercept (predicted value of *y* when *x* is 0),
- b_1 is the estimated slope coefficient, and
- *x* is the explanatory variable.

One of the differences between when you learned equations in algebra classes and our situation is that the line is not a perfect description of the relationship between x and y – it is an "on average" description and the will usually leave differences between the line and the observations, which we call residuals (e = y–$\widehat{y}$). We worked with residuals in the ANOVA11material. The residuals describe the vertical distance in scatterplot between our model (regression line) and the actual observed data point. The lack of a perfect fit of the line to the observations distinguishes statistical equations from those you learned in math classes. The equations work the same, but we have to modify interpretations of the coefficients to reflect this. We also tie this estimated model to a theoretical or population regression model:

$$y_i = \beta_0 + \beta_1 x_i + \varepsilon_i$$

where:

- yi is the observed response for the ith observation,
- xi is the observed value of the explanatory variable for the ith observation,
- β0 + β1xi is the true mean function evaluated at xi,
- β0 is the true (or population) y-intercept,
- β1 is the true (or population) slope coefficient, and
- the deviations, εi, are assumed to be independent and normally distributed with mean 0 and standard deviation σ or, more compactly, εi ~ N (0, σ2).

This presents another version of the linear model from Chapters 2, 3, and 4, now with a quantitative explanatory variable instead of categorical explanatory variable(s). This chapter focuses mostly on the estimated regression coefficients, but remember that we are doing statistics and our desire is to make inferences to a larger population. So, estimated coefficients, *b*0 and *b*1, are approximations to theoretical coefficients, β0 and β1. In other words, *b*0 and *b*1 are the statistics that try to estimate the true population parameters β0 and β1, respectively.

To get estimated regression coefficients, we use the lm function and our standard lm(y ~ x, data = ...) setup. This is the same function used to estimate our ANOVA models and much of this will look familiar. In fact, the ties between ANOVA and regression are deep and fundamental but not the topic of this section. For the Beers and BAC example, the estimated regression coefficients can be found from:

```
m1 <- lm(BAC ~ Beers, data = BB)
m1
```

```
##
## Call:
```

```
## lm(formula = BAC ~ Beers, data = BB)
##
## Coefficients:
## (Intercept)        Beers
##    -0.01270      0.01796
```

More often, we will extract these from the coefficient table produced by a model summary:

```
summary(m1)
```

```
##
## Call:
## lm(formula = BAC ~ Beers, data = BB)
##
## Residuals:
##       Min        1Q    Median        3Q       Max
## -0.027118 -0.017350  0.001773  0.008623  0.041027
##
## Coefficients:
##              Estimate Std. Error t value Pr(>|t|)
## (Intercept) -0.012701   0.012638  -1.005    0.332
## Beers        0.017964   0.002402   7.480 2.97e-06
##
## Residual standard error: 0.02044 on 14 degrees of freedom
## Multiple R-squared:  0.7998, Adjusted R-squared:  0.7855
## F-statistic: 55.94 on 1 and 14 DF,  p-value: 2.969e-06
```

From either version of the output, you can find the estimated y-intercept in the (Intercept) part of the output and the slope coefficient in the Beers part of the output. So b0 = -0.0127, b1 = 0.01796, and the estimated regression equation is

$$\widehat{\text{BAC}}_i = -0.0127 + 0.01796 \cdot \text{Beers}_i.$$

This is the equation that was plotted in Figure 6.13. In writing out the equation, it is good to replace x and y with the variable names to make the predictor and response variables clear. If you prefer to write all equations with x and y, you need to define x and y or else these equations are not clearly defined.

There is a general interpretation for the slope coefficient that you will need to master. In general, we interpret the slope coefficient as:

- **Slope interpretation (general):** For a 1 [unit of X] increase in X , we expect, on average, a b1 [unit of Y] change in Y .

Figure 6.14 can help you think about the different sorts of slope coefficients we might need to interpret, both providing changes in the response variable for 1 unit increases in the predictor variable.

Applied to this problem, for each additional 1 beer consumed, we expect a 0.018 gram per dL change in the *BAC on average*. Using "change" in the interpretation for what happened in the response allows you to use the same template for the interpretation even with negative slopes – be careful about saying "decrease" when the slope is negative as you can create a double-negative and end up implying an increase. . . Note also that you need to carefully incorporate the units of *x* and the units of *y* to make the interpretation clear. For example, if the change in *BAC* for 1 beer increase is 0.018, then we could also modify the size of the change in *x* to be a 10 beer increase and then the estimated change in *BAC* is 10 ∗ 0.018 = 0.18 g/dL. Bothare correct as long as you are clear about the change in *x* you are talking about. Typically, we will just use the units used in the original variables and only change the scale of "change in *x*" when it provides an interpretation we are particularly interested in.

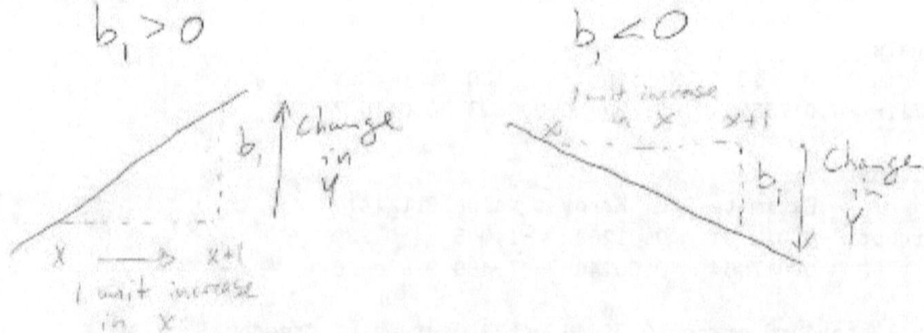

Figure 6.14: Diagram of interpretation of slope coefficients.

Similarly, the general interpretation for a *y*-intercept is:

- **Y -intercept interpretation (general):** For X = 0 [units of X], we expect, on average, b0 [units of Y] in Y .

Again, applied to the BAC data set: For 0 beers for Beers consumed, we expect, on average, -0.012 g/dL BAC. The y-intercept interpretation is often less interesting than the slope interpretation but can be interesting in some situations. Here, it is predicting average BAC for Beers = 0, which is a value outside the scope of the x's (Beers was observed between 1 and 9). Prediction outside the scope of the predictor values is called extrapolation. Extrapolation is dangerous at best and misleading at worst. That said, if you are asked to interpret the y-intercept you should still interpret it, but it is also good to note if it is outside of the region where we had observations on the explanatory variable. Another example is useful for practicing how to do these interpretations.

In the Australian Athlete data, we saw a weak negative relationship between Body Fat (% body weight that is fat) and Hematocrit (% red blood cells in the blood). The scatterplot in Figure 6.15 shows just the results for the female athletes along with the regression line which has a negative slope coefficient. The estimated regression coefficients are found using the lm function:

```
m2 <- lm(Hc ~ Bfat, data = aisR2 %>% filter(Sex == 1)) #Results for Females
summary(m2)
```

```
##
## Call:
## lm(formula = Hc ~ Bfat, data = aisR2 %>% filter(Sex == 1))
##
## Residuals:
##     Min      1Q  Median      3Q     Max
## -5.2399 -2.2132 -0.1061  1.8917  6.6453
##
## Coefficients:
##             Estimate Std. Error t value Pr(>|t|)
## (Intercept) 42.01378    0.93269  45.046   <2e-16
## Bfat        -0.08504    0.05067  -1.678   0.0965
##
## Residual standard error: 2.598 on 97 degrees of freedom
## Multiple R-squared:  0.02822,    Adjusted R-squared:  0.0182
## F-statistic: 2.816 on 1 and 97 DF,  p-value: 0.09653
```

```
aisR2 %>% filter(Sex == 1) %>% ggplot(mapping = aes(x = Bfat, y = Hc)) +
  geom_point() +
  geom_smooth(method = "lm") +
  theme_bw() +
  labs(title = "Scatterplot of Body Fat vs Hematocrit for Female Athletes",
       y = "Hc (% blood)", x = "Body fat (% weight)")
```

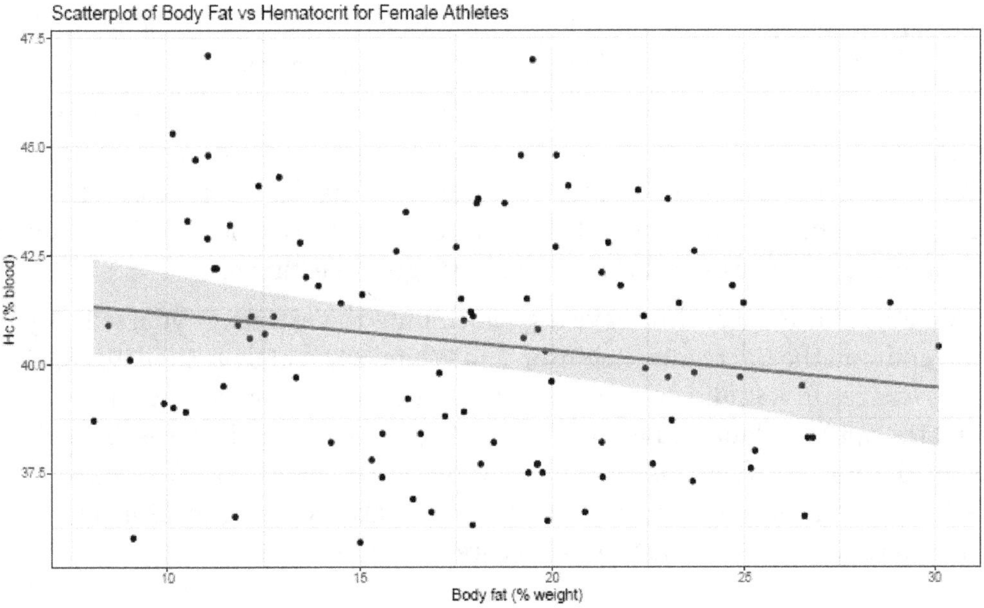

Figure 6.15: Scatterplot of Hematocrit versus Body Fat for female athletes. Note how the filter was used to pipe the subset of the data set to the plot.

Based on these results, the estimated regression equation is H--ci = 42.014 - 0.085 · BodyFat$_i$ with $b0$ = 42.014and $b1$ = 0.085. The slope coefficient interpretation is: For a one percent increase in body fat, we expect, on average, a -0.085% (blood) change in Hematocrit for Australian female athletes. For the y-intercept, the interpretation is: For a 0% body fat female athlete, we expect a Hematocrit of 42.014% on average. Again, this y-intercept involves extrapolation to a region of x's that we did not observed. None of the athletes had body fat below 5% so we don't know what would happen to the hematocrit of an athlete that had no body fat except that it probably would not continue to follow a linear relationship.

6.7 LEAST SQUARES ESTIMATION

The previous results used the lm function as a "black box" to generate the estimated coefficients. The lines produced probably look reasonable but you could imagine drawing other lines that might look equally plausible. Because we are interested in explaining variation in the response variable, we want a model that in yi some sense minimizes the residuals (ei = yi - $\widehat{y_i}$) and explains the responses as well as possible, in other words as small as possible. We can't just add these 's up because it would always be 0 (remember why eyi has yi - $\widehat{y_i}$ i we use the variance to measure spread from introductory statistics?). We use a similar technique in regression, we find the regression line that minimizes the squared residuals $e_i^2 = (y_i - \widehat{y}_i)^2$ over all the observations, minimizing the Sum of Squared Residuals= Σe_i^2. Finding regression coefficients that2 the estimated-minimize the sum of squared residuals is called least squares estimation and provides us a reasonable method for finding the "best" estimated regression line of all the possible choices.

For the Beers vs BAC data, Figure 6.16 shows the result of a search for the optimal slope coefficient between values of 0 and 0.03. The plot shows how the sum of the squared residuals was minimized for the value that lm returned at 0.018. The main point of this is that if any other slope coefficient was tried, it did not do as good on the least squares criterion as the least squares estimates.

Sometimes it is helpful to have a go at finding the estimates yourself. If you install and load the tiger stats [Robinson and White, 2020] and manipulate [Allaire, 2014] packages in RStudio and then run Find Reg Line (), you get a chance to try to find the optimal slope and intercept for a fake data set. Click on the "sprocket" icon in the upper left of the plot and you will see something like Figure 6.17. This interaction can help you see how the residuals are being measuring in the y-direction and appreciate that lm takes care of this for us.

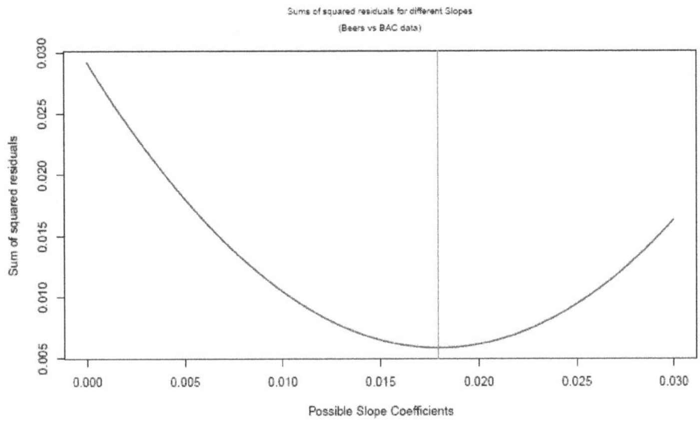

Figure 6.16: Plot of sum of squared residuals vs possible slope coefficients for *Beers* vs *BAC* data, with vertical line for the least squares estimate that minimizes the sum of squared residuals.

```
> library(tigerstats)
> library(manipulate)
> FindRegLine()

Equation of the regression line is:
y = 4.34 + -0.02x

Your final score is 13143.99
Thanks for playing!
```

It ends up that the least squares criterion does not require a search across coefficients or trial and error – there are some "simple" equations available for calculating the estimates of the *y*-intercept and slope:

$$b_1 = \frac{\Sigma_i(x_i - \bar{x})(y_i - \bar{y})}{\Sigma_i(x_i - \bar{x})^2} = r\frac{s_y}{s_x} \text{ and } b_0 = \bar{y} - b_1\bar{x}.$$

You will never need to use these equations but they do inform some properties of the regression line. The slope coefficient, b1, is based on the variability in x and y and the correlation between them. If r = 0, then the slope coefficient will also be 0. The intercept is a function of the means of x and y and what the estimated slope coefficient is. If the slope coefficient, b1, is 0, then b0 = y¯ (which is just the mean of the response variable for all observed values of x – this is a very boring model!). The slope is 0 when the correlation is 0. So when there is no linear relationship between x and y (r = 0), the least squares regression line is a horizontal line with height y¯, and the line produces the same fitted values for all x values. You can also think about this as when there is no relationship between x and y, the best prediction of y is the mean of the y-values and it doesn't change based on the values of x. It is less obvious in these equations, but they also imply that the regression line ALWAYS goes through the point (x¯, y¯). It provides a sort of anchor point for all regression lines.

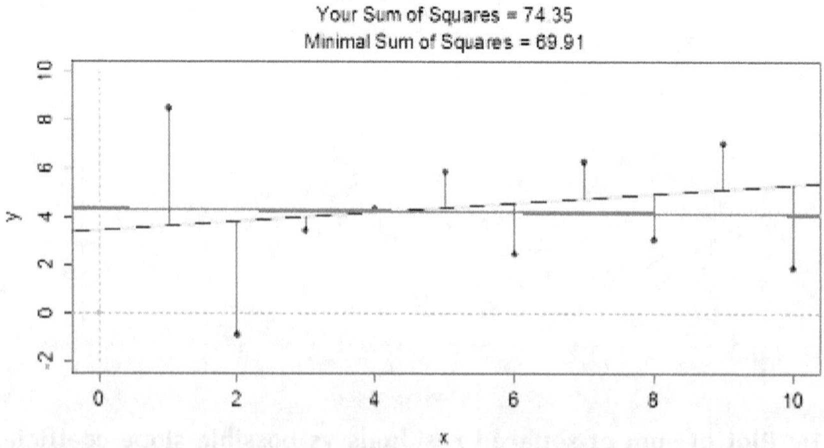

Figure 6.17: Results of running FindRegLine() where I didn't quite find the least squares line. The correct line is the bold (red) line and produced a smaller sum of squared residuals than the guessed thinner (black) line.

For one more example, we can revisit the Montana wildfire areas burned (log-hectares) and the average summer temperature (degrees F), which had r = 0.81. The interpretations of the different parts of the regression model follow the least squares estimation provided by lm:

```
fire1 <- lm(loghectares ~ Temperature, data = mtfires)
summary(fire1)
```

```
##
## Call:
## lm(formula = loghectares ~ Temperature, data = mtfires)
##
## Residuals:
##     Min      1Q  Median      3Q     Max
## -3.0822 -0.9549  0.1210  1.0007  2.4728
##
## Coefficients:
##             Estimate Std. Error t value Pr(>|t|)
## (Intercept) -69.7845    12.3132  -5.667 1.26e-05
## Temperature   1.3884     0.2165   6.412 2.35e-06
##
## Residual standard error: 1.476 on 21 degrees of freedom
## Multiple R-squared:  0.6619, Adjusted R-squared:  0.6458
## F-statistic: 41.12 on 1 and 21 DF,  p-value: 2.347e-06
```

- **Regression Equation (Completely Specified):**
  - **Estimated model:** log(Ha) = -69.78 + 1.39 · Temp
  - Or $\widehat{y}_i$ = -69.78 + 1.39x with **Y = log(Ha) and X = Temperature**
- **Response Variable:** Yearly *log* Hectares burned by wildfires
  - **Explanatory Variable:** Average Summer Temperature

- **Estimated *y*-Intercept (*b*0):** -69.78
- **Estimated slope (*b*1):** 1.39
- **Slope Interpretation:** For a 1 degree Fahrenheit increase in Average Summer Temperature we would expect, on average, a 1.39 log(Hectares) change in log(Hectares) burned in Montana.
- **Y -intercept Interpretation:** If temperature were 0 degrees F, we would expect -69.78 log(Hectares) burned on average in Montana.

One other use of regression equations is for prediction. It is a trivial exercise (or maybe not – we'll see when you try it!) to plug an x-value of interest into the regression equation and get an estimate for y at that x. Basically, the regression lines displayed in the scatterplots show the predictions from the regression line across the range of x's. Formally, prediction involves estimating the response for a particular value of x. We know that it won't be perfect but it is our best guess. Suppose that we are interested in predicting the log-area burned for a summer that had an average temperature of 59°F. If we plug 59°F into the regression equation, log(Ha) = -69.78 + 1.39 • Temp, we get

$$\begin{aligned}\log(\text{Ha}) &= -69.78 \text{ log-hectares } + 1.39 \text{ log-hectares}/^{\circ}\text{F} \bullet 59^{\circ}\text{F} \\ &= -69.78 \text{ log-hectares } + 1.39 \text{ log-hectares}/^{\circ}\text{F} \bullet 59^{\circ}\text{F} \\ &= 12.23 \text{ log-hectares}\end{aligned}$$

We did not observe any summers at exactly x = 59 but did observe some nearby and this result seems relatively reasonable.

Now suppose someone asks you to use this equation for predicting Temperature = 65°F . We can run that through the equation: -69.78 + 1.39 $*$ 65 = 20.57 log-hectares. But can we trust this prediction? We did not observe any summers over 60 degrees F so we are now predicting outside the scope of our observations–performing extrapolation. Having a scatterplot in hand helps us to assess the range of values where we can reasonably use the equation–here between 54 and 60 degrees F seems reasonable.

```
mtfires %>% ggplot(mapping = aes(x = Temperature, y = loghectares)) +
  geom_point(aes(color = Year), size = 2.5) +
  geom_smooth(method = "lm") +
  theme_bw() +
  scale_color_viridis() +
  labs(title = "Scatterplot with regression line for Area burned vs
      Temperature, colored by year")
```

6.8 MEASURING THE STRENGTH OF REGRESSIONS: R^2

At the beginning of the chapter, we used the correlation coefficient to measure the strength and direction of the linear relationship. The regression line provides an even more detailed description of the direction of the linear relationship than the correlation provided; in regression we addressed the question of "for a unit change in x, what sort of change in y do we expect, on average?" whereas the correlation just addressed whether the relationship was positive or negative. However, the regression line tells us nothing about the strength of the relationship. Consider the three scatterplots in Figure 6.19: the left panel is the original BAC data and the two right panels have fake data that generated exactly the same estimated regression model with a weaker (middle panel) and then a stronger (right panel) linear relationship between Beers and BAC. This suggests that the regression line is a useful but incomplete characterization of relationships between variables– we need a measure of strength of the relationship to go with the equation.

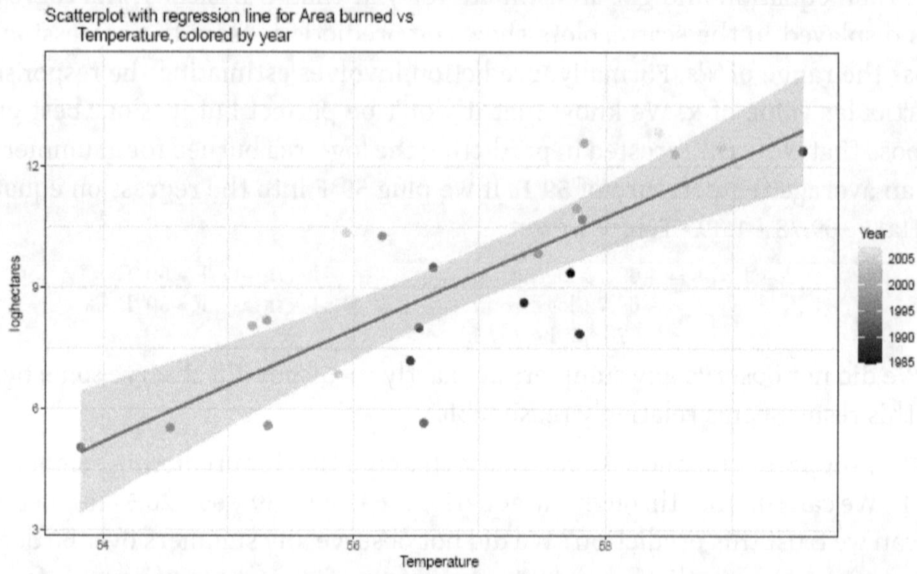

Figure 6.18: Scatterplot of log-hectares burned versus temperature with estimated regression line. Information on the year of each observation is added using a local aesthetic inside geom_point to colour the points on a colour gradient based on Year.

We could use the correlation coefficient, *r*, again to characterize strength but it is somewhat redundant to report a measure that contains direction information. It also will not extend to multiple regression models where we have more than one predictor variable in the same model.

In regression models, we use the coefficient of determination (symbol: R2) to accompany our regression line and describe the strength of the relationship and assess the quality of the model fit. It can either be scaled between 0 and 1 or 0 to 100% and has "units" of the proportion or percentage of the variation in y that is explained by the model that includes x (and later more than one x). For example, an R2 of 0% corresponds to explaining 0% of the variation in the response with our model (worst possible fit) and R2 = 100% means that all the variation in the response was explained by the model (best possible fit). In between, it provides a nice summary of how much of the total variability in the response we can account for with our model including x (and, in Chapter 8, including multiple predictor variables).

The $\mathbf{R^2}$ is calculated using the sums of squares we encountered in the ANOVA methods. We once again have some total amount of variability that is attributed to the variation based on the model fit, here we call it SSregression, and the residual variability, still $SS_{error} = \Sigma(y - \widehat{y})^2$ is most easily calculated as $SS_{regression} = SS_{Total} - SS_{error}$, the difference between the total variability and the variability not explained by the model under consideration. Using these quantities, we calculate the portion of the total variability that the model explains as

$$R^2 = \frac{\text{SS}_{\text{regression}}}{\text{SS}_{\text{Total}}} = 1 - \frac{\text{SS}_{\text{error}}}{\text{SS}_{\text{Total}}}.$$

It also ends up that the coefficient of determination for models with one predictor is the correlation coefficient (r) squared ($R^2 = r^2$). So we can quickly find coefficients of determination if we know correlations in simple linear regression models. In the real Beers and BAC data, r = 0.8943. So $R^2 = 0.79998$ or approximately 0.80. So 80% of the variation in BAC is explained by Beer consumption. That leaves 20% of the variation in the responses to be unexplained by our model. In this case much of the unexplained variation is likely attributable to differences in physical characteristics (that were not measured) but the statistical model places that unexplained variation into the category of "random errors". We don't actually have to find r to get coefficients of determination – the result is part of the regular summary of a regression model that we have not discussed. We repeat the full lm model summary below – note that a number is reported for the

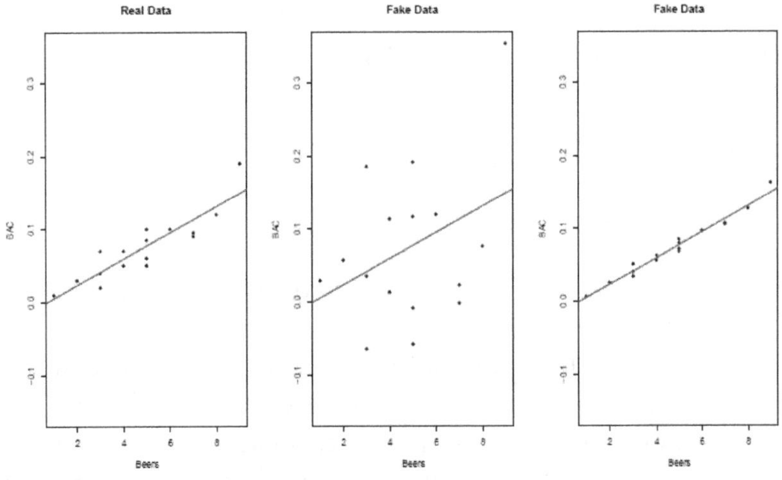

Figure 6.19: Three scatterplots with the same estimated regression line.

```
m1 <- lm(BAC ~ Beers, data = BB)
summary(m1)
```

```
##
## Call:
## lm(formula = BAC ~ Beers, data = BB)
##
## Residuals:
##       Min        1Q    Median        3Q       Max
## -0.027118 -0.017350  0.001773  0.008623  0.041027
##
## Coefficients:
##              Estimate Std. Error t value Pr(>|t|)
## (Intercept) -0.012701   0.012638  -1.005    0.332
## Beers        0.017964   0.002402   7.480 2.97e-06
##
## Residual standard error: 0.02044 on 14 degrees of freedom
## Multiple R-squared:  0.7998, Adjusted R-squared:  0.7855
## F-statistic: 55.94 on 1 and 14 DF,  p-value: 2.969e-06
```

In this output, be careful because there is another related quantity called Adjusted R-squared that we will discuss later. This other quantity is not a measure of the strength of the relationship but will be useful.

We could also revisit the ANOVA table for this model to verify the source of the R2 of 0.80 based on $S_{Sregression}$ = 0.02337 and S_{STotal} = 0.02337 + 0.00585. This provides 0.80 from 0.02337/0.02922.

```
anova(m1)
```

```
## Analysis of Variance Table
##
## Response: BAC
##           Df    Sum Sq   Mean Sq F value    Pr(>F)
## Beers      1 0.0233753 0.0233753  55.944 2.969e-06
## Residuals 14 0.0058497 0.0004178
```

```
SStotal <- 0.0233753 + 0.0058497
SSregression <- 0.0233753
SSregression/SStotal
```

```
## [1] 0.7998392
```

In Figure 6.19, there are three examples with the same regression model, but different strengths of relationships. In the real data set R_2 = 80%. For the first fake data set (middle panel), the R_2 drops to 13.8% and for the second fake data set (right panel), R_2 is 97.3%. As a summary, R_2 provides a natural scale to understand "how good" each model is at explaining the responses. We can revisit some of our previous models to get a little more practice with using this summary of strength or quality of regression models.

For the Montana fire data, R_2 = 66.2%. So the proportion of variation of log-area burned that is explained by average summer temperature is 0.662. This is "good" but also leaves quite a bit of unexplained variation in the responses. There is a long list of reasons why this explanatory variable leaves a lot of variation in the response unexplained. Note that we were careful about using the scaling of the response variable (log(area burned)) in the interpretation – this is because we would get a much different answer if area burned vs temperature was considered.

```
fire1 <- lm(loghectares ~ Temperature, data = mtfires)
summary(fire1)
```

```
##
## Call:
## lm(formula = loghectares ~ Temperature, data = mtfires)
##
## Residuals:
##      Min       1Q  Median      3Q     Max
## -3.0822 -0.9549  0.1210  1.0007  2.4728
##
## Coefficients:
##             Estimate Std. Error t value Pr(>|t|)
## (Intercept) -69.7845    12.3132  -5.667 1.26e-05
## Temperature   1.3884     0.2165   6.412 2.35e-06
##
## Residual standard error: 1.476 on 21 degrees of freedom
## Multiple R-squared:  0.6619, Adjusted R-squared:  0.6458
## F-statistic: 41.12 on 1 and 21 DF,  p-value: 2.347e-06
```

For the model for female Australian athletes that used Body fat to explain Hematocrit, the estimated regression model was $\widehat{Hc}_i$ = 42.014 - 0.085 · Body Fat i and r = -0.168. The coefficient of determination is R2 = (-0.168)2 = 0.0282. So body fat explains 2.8% of the variation in Hematocrit in these women. That is not a very good regression model with over 97% of the variation in Hematocrit unexplained by this model. The scatterplot showed a fairly weak relationship but this provides numerical and interpretable information that drives that point home.

```
2 <- lm(Hc ~ Bfat, data = aisR2 %>% filter(Sex == 1)) #Results for Females
ummary(m2)
```

```
##
## Call:
## lm(formula = Hc ~ Bfat, data = aisR2 %>% filter(Sex == 1))
##
## Residuals:
##     Min      1Q  Median      3Q     Max
## -5.2399 -2.2132 -0.1061  1.8917  6.6453
##
## Coefficients:
##             Estimate Std. Error t value Pr(>|t|)
## (Intercept) 42.01378    0.93269  45.046   <2e-16
## Bfat        -0.08504    0.05067  -1.678   0.0965
##
## Residual standard error: 2.598 on 97 degrees of freedom
## Multiple R-squared:  0.02822,    Adjusted R-squared:  0.0182
## F-statistic: 2.816 on 1 and 97 DF,  p-value: 0.09653
```

6.9 OUTLIERS: LEVERAGE AND INFLUENCE

In the review of correlation, we loosely considered the impacts of outliers on the correlation. We removed unusual points to see both the visual changes (in the scatterplot) as well as changes in the correlation coefficient in Figures 6.4 and 6.5. In this section, we formalize these ideas in the context of impacts of unusual points on our regression equation. In regression, it is possible for a single point to have a big impact on the overall regression results but it is also possible to have a clear outlier that has little impact on the results. We call an observation influential if its removal causes a "big" change in the regression line, specifically in terms of impacting the slope coefficient. Points that are on the edges of the x's (far from the mean of the x's) have the potential for more impact on the line as we will see in some examples shortly. You can think of the regression line being balanced at x¯ and the further from that location a point is, the more a single point can move the line. We can measure the distance of points from x¯ to quantify each observation's potential for impact on the line using what is called the leverage of a point. Leverage is a positive numerical measure with larger values corresponding to more leverage. The scale changes depending on the sample size (n) and the complexity of the model so all that matters is which observations have more or less relative leverage in a particular data set. The observations with x-values that provide higher leverage have increased potential to influence the estimated regression line. Along with measuring the leverage, we can also measure the influence that each point has on the regression line using Cook's Distance or Cook'sD. It also is a positive measure with higher values suggesting more influence. The rule of thumb is that Cook's D values over 1.0 correspond to clearly influential points, values over 0.5 have some influence and values lower than 0.5 indicate points that are not influential on the regression model slope coefficients. One part of the regular diagnostic plots we will use for regression models displays the leverages on the x-axis, the standardized residuals on the y-axis, and adds contour lines for Cook's Distances in a panel that is labelled "Residuals vs Leverage". This allows us to see the potential for impact of a point (leverage), how far it's observation was from the regression line (residual), and to see a measure of that point's influence (Cook's D).

To extract the level of Cook's D on the "Residuals vs Leverage" plot, look for contours to show up on the upper and lower right of the plot. They show increasing levels of influence going to the upper and lower right corners as you combine higher leverage (*x*-axis) and larger residuals (*y*-axis) – the two ingredients required to be influential on the line. The contours are displayed for Cook's D values of 0.5 and 1.0 if there are points near or over those levels. The Cook's D values come from a topographical surface of values that is a sort of U-shaped valley in the middle of the plot centered at $y = 0$ with the lowest contour corresponding to Cook's D values below 0.5 (no influence). As you move to the upper right or lower right corners, the influence increases and the edges of the valley get steeper. If you do not see any contours in the plot, then no points were even close to being influential based on Cook's D.

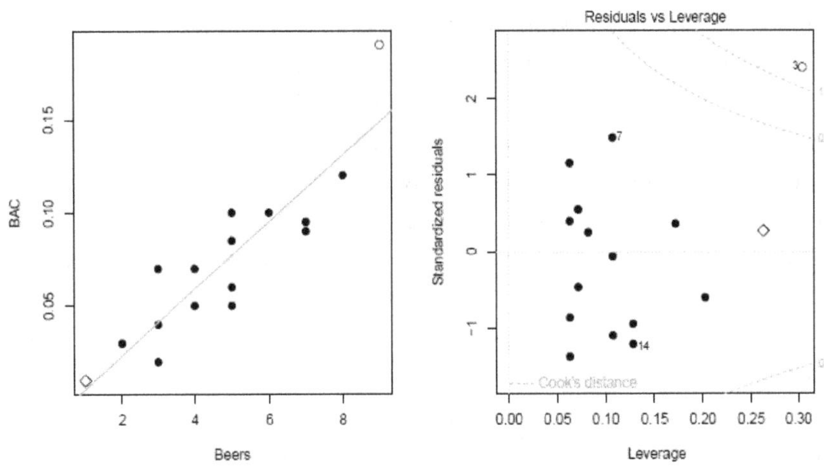

Figure 6.20: Scatterplot and Residuals vs Leverage plot for the real BAC data. Two high leverage points are flagged, with only one that has a Cook's D value over 1 (“o ”) and is indicated as influential.

To illustrate these concepts, the original Beers and BAC data are used again. In the scatter plot in Figure6.20, two points are plotted with different characters. The point for 1 Beer and BAC of 0.010 is displayed as a “<” and the 9 Beer and BAC 0.19 observation is displayed with a “o”. These two points are the furthest from the mean of the of the x's (Beers = 4.8) but show two different levels of influence on the line. The “<” point has a leverage of 0.27 and the 9 Beer observation (“o”) had a leverage of 0.30. The 1 Beer observation was close to the pattern defined by the other points, had a small residual, and a Cook's D value below 0.5 (it did not exceed the first of the contours). So even though it had high leverage, it was not an influential point. The 9 Beer observation had the highest leverage in the data set and was quite a bit above the pattern defined by the other points and ends up being an influential point with a Cook's D over 1. We might want to consider fitting this model without that observation to get a better estimate of the effects of beer consumption on BAC or revisit our assumption that the relationship is really linear here.

To further explore influence, we will add a point to the original data set and move it around so you can see how those changes impact the results. For each scatterplot in Figure 6.21, the Residuals vs Leverage plot is displayed to its right. The original data are “•” and the original regression line is the dashed line in Figure 6.21. First, a fake observation at 11 Beers and 0.1 BAC is added, at (11, 0.1), in the top panels of the figure. This observation is clearly an outlier and heavily impacts the slope of the regression line (so is clearly influential). This added point drops the R2 from 0.80 in the original data to 0.24. The accompanying Residuals vs Leverage plot shows that this point has extremely high leverage and a Cook's D over 1 – it is a clearly influential point. However, having high leverage does not always make points influential. Consider the second row of plots with an added point of (11, 0.19). The regression line barely changes and R2 increases a little. This point has the same leverage as in the first example since it is the same set of x's and the distance to the mean of the x's is unchanged. But it is not influential since its Cook's D value is less than 0.5. This occurred because it followed the overall pattern of observations even though it was "far away" from the other observations in the x-direction. The last two rows of plots show what happens when low leverage outliers are encountered. If observations are near the center of the x's, it ends up that to be influential the points have to be very far from the pattern of the other observations. The (5, 0.19) example almost attains a Cook's D of 0.5 but has little impact on the regression line, especially the slope coefficient. It does impact the y-intercept and drops the R-squared value to 0.57. The same result occurs if the observation is noticeably lower than the other points.

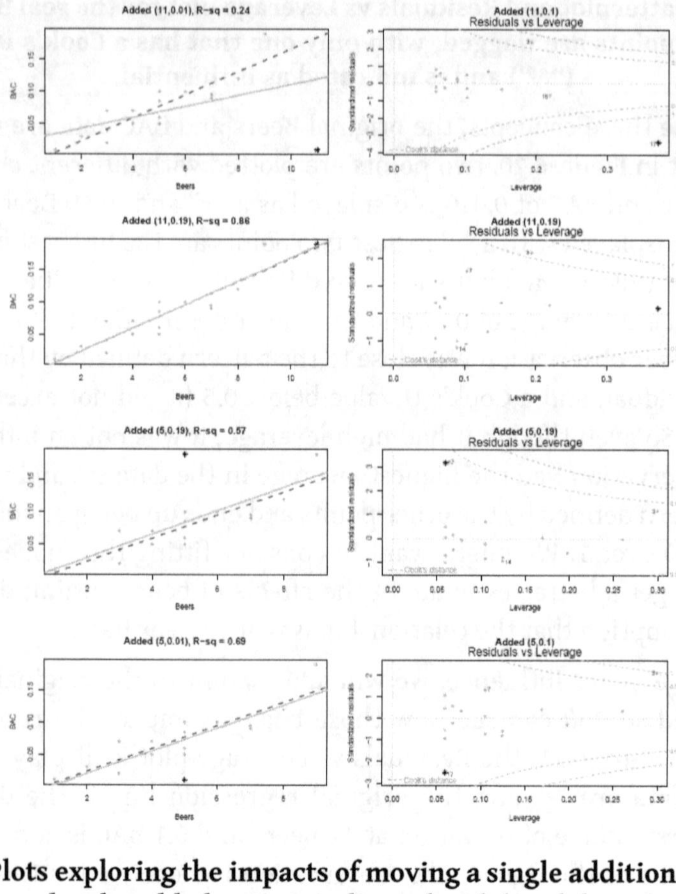

Figure 6.21: Plots exploring the impacts of moving a single additional observation in the BAC example. The added point is indicated with * and the original regression line is the dashed line in the left column.

When we are doing regressions, we get very worried about points "at the edges" having an undue influence on the results. When we start using multiple predictors, say if we had body weight data on these subjects as well as beer consumption, it becomes harder to "see" if the points are "far away" from the other observations and we will trust the Residuals vs Leverage plots to help us identify the influential points. These techniques work the same in the multiple regression models in Chapter 8 as they do in these simpler, single predictor regression models.

6.10 RESIDUAL DIAGNOSTICS – SETTING THE STAGE FOR INFERENCE

Influential points are not the only potential issue that can cause us to have concerns about our regression model. There are two levels to these considerations. The first is related to issues that directly impact the least squares regression line and cause concerns about whether a line is a reasonable representation of the relationship between the two variables. These issues for regression model estimation have been discussed previously (the same concerns in estimating correlation apply to regression models). The second level is whether the line we have will be useful for making inferences for the population that our data were collected *yi*from and whether the data follow our assumed model. Our window into problems of both types is the residuals ($e_i = y_i - \hat{y}_i$). By exploring patterns in how the line "misses" the responses we can gain information about the reasonableness of using the estimated regression line and sometimes information about how we might fix problems. The validity conditions for doing inference in a regression setting (Chapter 7) involve two sets of considerations, those that are assessed based on the data collection and measurement process and those that can be assessed using diagnostic plots. The first set is:

- **Quantitative variables condition**
    - We'll discuss using categorical predictor variables later – to use simple linear regression both the explanatory and response variables need to quantitative.
- **Independence of observations**
    - As in the ANOVA models, linear regression models assume that the observations are collected in a fashion that makes them independent.
    - This will be based on the "story" of the data. Consult a statistician if your data violate this assumption as there are more advanced methods that adjust for dependency in observations but they are beyond the scope of this material.

The remaining assumptions for getting valid inferences from regression models can be assessed using diagnostic plots:

- **Linearity of relationship**
  - We should not report a linear regression model if the data show a curve (curvilinear relationship between *x* and *y*).
  - Examine the initial scatterplot to assess the potential for a curving relationship.
  - Examine the Residuals vs Fitted (top left panel of diagnostic plot display) plot:
  - If the model missed a curve in the relationship, the residuals often will highlight that missed pattern and a curve will show up in this plot.
  - Try to explain or understand the pattern in what is left over. If we have a good model, there shouldn't be much left to "explain" in the residuals (i.e., no patterns left over after accounting for x).
- **Equal (constant) variance**
  - We assume that the variation is the same for all the observations and especially that the variability does not change in the responses as a function of our predictor variables or the fitted values.
  - There are three plots to help with this:
    - Examine the original scatterplot and look at the variation around the line and whether it looks constant across values of x.
    - Examine the Residuals vs Fitted plot and look for evidence of changing spread in the residuals, being careful to try to separate curving patterns from non-constant variance (and look for situations where both are present as you can violate both conditions simultaneously).
    - Examine the "Scale-Location" plot and look for changing spread as a function of the fitted values.
    - The y-axis in this plot is the square-root of the absolute value of the standardized residual. This scale flips the negative residuals on top of the positive ones to help you better assess changing variability without being distracted by whether the residuals are above or below 0.
    - Because of the absolute value, curves in the Residuals vs Fitted plot can present as sort of looking like non-constant variance in the Scale-Location plot – check for nonlinearity in the residuals vs fitted values before using this plot. If nonlinearity is present, just

use the Residuals vs Fitted and original scatterplot for assessing constant variance around the curving pattern.

- If there are patterns of increasing or decreasing variation (often described as funnel or cone shapes), then it might be possible to use a transformation to fix this problem (more later). It is possible to have decreasing and then increasing variability and this also is a violation of this condition.

◉ **Normality of residuals**

- Examine the Normal QQ-plot for violations of the normality assumption as in Chapters 3 and 4.
  - Specifically review the discussion of identifying skews in different directions and heavy vs light tailed distributions.
  - Skewed and heavy-tailed distributions are the main problems for our inferences, especially since both kinds of distributions can contain outliers that can wreak havoc on the estimated regression line.
  - Light-tailed distributions cause us no real inference issues except that the results are conservative so you should note when you observe these situations but feel free to proceed with using your model results.
  - Remember that clear outliers are an example of a violation of the normality assumption but some outliers may just influence the regression line and make it fit poorly and this issue will be more clearly observed in the residuals vs fitted than in the QQ-plot.

◉ No influential points

- Examine the Residuals vs Leverage plot as discussed in the previous section.
- Consider removing influential points (one at a time) and focusing on results without those points in the data set.

To assess these later assumptions, we will use the four residual diagnostic plots that R provides from lm fitted models. They are similar to the results from ANOVA models but the Residuals vs Leverage plot is now interesting as was discussed in Section 6.9. Now we can fully assess the potential for trusting the estimated regression models in a couple of our examples:

◉ Beers vs BAC:

- Quantitative variables condition:

- Both variables are quantitative.
- Independence of observations:
- We can assume that all the subjects are independent of each other. There is only one measurement per student and it is unlikely that one subject's beer consumption would impact another's BAC. Unless the students were trading blood it isn't possible for one person's beer consumption to change someone else's BAC.

```
m1 <- lm(BAC ~ Beers, data = BB)
par(mfrow = c(2,2))
plot(m1, add.smooth = F, main = "Beers vs BAC", pch = 16)
```

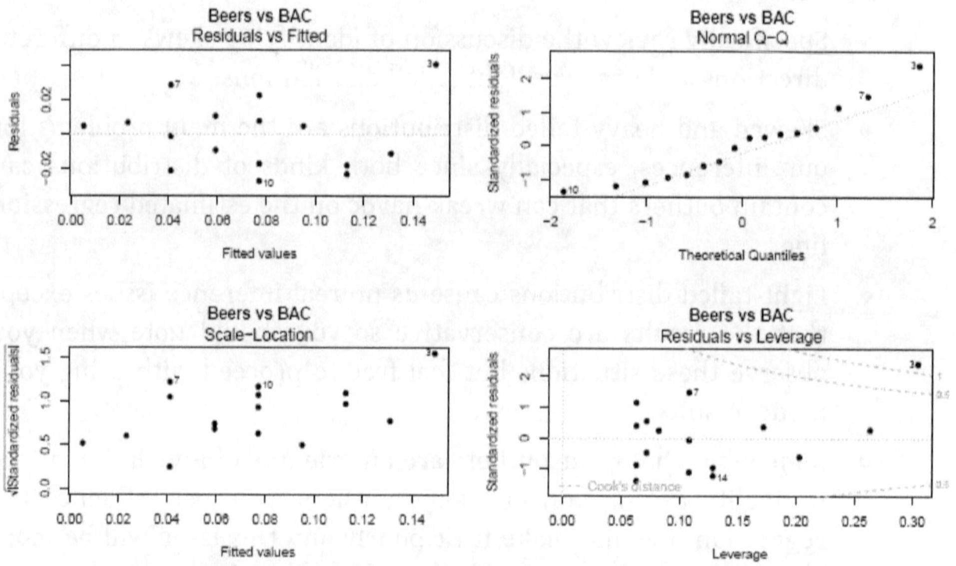

Figure 6.22: Full suite of diagnostics plots for Beer vs BAC data.

⊙ Linearity, constant variance from Residuals vs Fitted:

- We previously have identified a potentially influential outlier point in these data. Consulting the Residuals vs Fitted plot in Figure 6.22, if you trust that influential point, shows some curvature with a pattern of decreasing residuals as a function of the fitted values and then an increase at the right. Or, if you do not trust that highest BAC observation, then there is a mostly linear relationship with an outlier identified. We would probably suggest that it is an outlier, should be removed from the analysis, and inferences constrained to the region of beer consumption from 1 to 8 beers since we don't know what might happen at higher values.

⊙ Constant variance from Scale-Location:

- There is some evidence of increasing variability in this plot as the spread of the results increases from left to right, however this is just an artifact of the pattern in the original residuals and not real evidence of non-constant variance. Note that there is little to no evidence of non-constant variance in the Residuals vs Fitted.

⦿ Normality from Normal QQ Plot:

- The left tail is a little short and the right tail is a little long, suggesting a slightly right skewed distribution in the residuals. This also corresponds to having a large positive outlying value. But we would conclude that there is a minor issue with normality in the residuals here.
- Influential points from Residuals vs Leverage:
- Previously discussed, this plot shows one influential point with a Cook's D value over 1 that is distorting the fitted model and is likely the biggest issue here.

⦿ Tree height and tree diameter (suspicious observation already removed):

- Quantitative variables: Met
- Independence of observations:
- There are multiple trees that were measured in each plot. One problem might be that once a tree is established in an area, the other trees might not grow as tall. The other problem is that some sites might have better soil conditions than others. Then, all the trees in those rich soil areas might be systematically taller than the trees in other areas. Again, there are statistical methods to account for this sort of "clustering" of measurements but this technically violates the assumption that the trees are independent of each other. So this assumption is violated, but we will proceed with that caveat on our results – the precision of our inferences might be slightly over-stated due to some potential dependency in the measurements.
- Linearity, constant variance from Residuals vs Fitted in Figure 6.23.
    - There is evidence of a curve that was missed by the linear model.
    - There is also evidence of increasing variability AROUND the curve in the residuals.
- Constant variance from Scale-Location:
    - This plot actually shows relatively constant variance but this plot is misleading when curves are present in the data set. **Focus on**

the Residuals vs Fitted to diagnose non-constant variance in situations where a curve was missed.

- Normality in Normal QQ plot:
  - There is no indication of any problem with the normality assumption.
- Influential points?
  - The Cook's D contours do not show up in this plot so none of the points are influential.

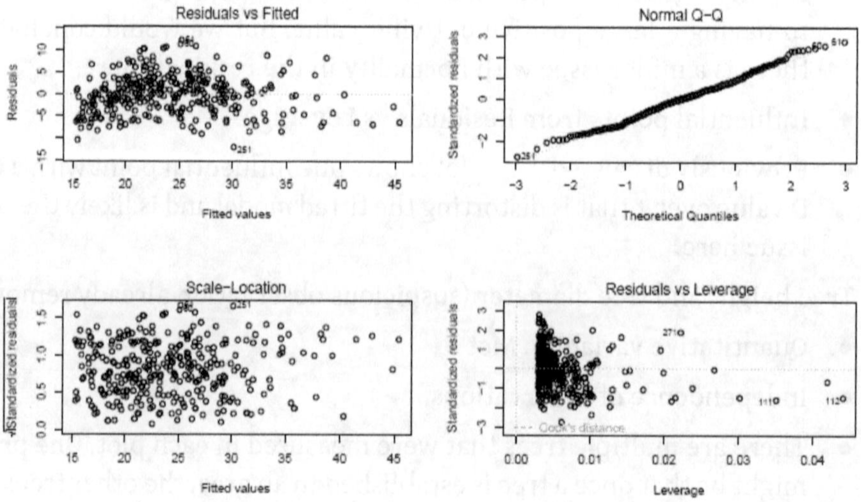

Figure 6.23: Diagnostics plots for tree height and diameter simple linear regression model.

So the main issues with this model are the curving relationship and non-constant variance. We'll revisit this example later to see if we can find a model on transformed variables that has better diagnostics. Reporting the following regression model that has a decent R^2 of 62.6% would be misleading since it does not accurately represent the relationship between tree diameter and tree height.

```
tree1 <- lm(height.m ~ dbh.cm, data = ufc %>% slice(-168))
summary(tree1)
```

```
##
## Call:
## lm(formula = height.m ~ dbh.cm, data = ufc %>% slice(-168))
##
## Residuals:
##      Min       1Q   Median       3Q      Max
## -12.1333  -3.1154   0.0711   2.7548  12.3076
##
## Coefficients:
##             Estimate Std. Error t value Pr(>|t|)
## (Intercept) 11.98364    0.57422   20.87   <2e-16
## dbh.cm       0.32939    0.01395   23.61   <2e-16
##
## Residual standard error: 4.413 on 333 degrees of freedom
## Multiple R-squared:  0.626,  Adjusted R-squared:  0.6249
## F-statistic: 557.4 on 1 and 333 DF,  p-value: < 2.2e-16
```

```
par(mfrow = c(2,2))
plot(tree1, add.smooth = F)
```

6.11 OLD FAITHFUL DISCHARGE AND WAITING TIMES

A study in August 1985 considered time for Old Faithful and how that might relate to *waiting time* for the next eruption (Ripley [2022], Azzalini and Bowman [1990]). This sort of research provides the Yellowstone National Park (YNP) staff a way to show tourists a predicted time to next eruption so they can quickly see it erupt and then get back in their cars, not wasting too much time in the outdoors. Or, less cynically, the opportunity to study the behaviour of the eruption of a geyser. Both variables are measured in minutes and the scatterplot in Figure 6.24 shows a moderate to strong positive and relatively linear relationship. We added a ***smoothing line*** (dashed line) using geom_smooth to this plot – this is actually the default choice in geom_smooth and we have to use geom_smooth(method = "lm") to get the regression (straight)line. Smoothing lines provide regression-like fits but are performed on local areas of the relationship between the two variables and so can highlight where the relationships change, especially highlighting curvilinear relationships. They can also return straight lines just like the regression line if that is reasonable. The technical details of regression smoothing are not covered here but they are a useful graphical addition to help visualize nonlinearity in relationships and a topic you can explore further based on the sources related to themgcv R package [Wood, 2022], which is being used by geom_smooth.

In these data, there appear to be two groups of eruptions (shorter length, shorter wait and longer length, longer wait) – but we don't know enough about these data to assume that there are two groups. The smoothing line does help us to see if the relationship appears to change or stay the same across different values of the explanatory variable, Duration. The smoothing line suggests that the upper group might have a less steep slope than the lower group as it sort of levels off for observations with Duration of over 4 minutes. It also indicates that there is one point for an eruption under 1 minute in Duration that might be causing some problems for both the linear fit and the smoothing line. The story of these data involve some measurements during the night that were just noted as being short, medium, and long – and they were re-coded as 2, 3, or 4 minute duration eruptions. You can see responses stacking up at 2 and 4 minute durations and this is obviously a problematic aspect of these data. We'll see if our diagnostics detect some of these issues when we fit a simple linear regression to try to explain waiting time based on duration of prior eruption.

```
data(geyser, package = "MASS")
geyser <- as_tibble(geyser)
# Aligns the duration with time to next eruption
G2 <- tibble(Waiting = geyser$waiting[-1], Duration = geyser$duration[-299])

G2 %>% ggplot(mapping = aes(x = Duration, y = Waiting)) +
  geom_point() +
  geom_smooth(method = "lm") +
  geom_smooth(lty = 2, col = "red", lwd = 1.5, se = F) + #Add smoothing line
  theme_bw() +
  labs(title = "Scatterplot with regression and smoothing line,
       Waiting Time vs Duration")
```

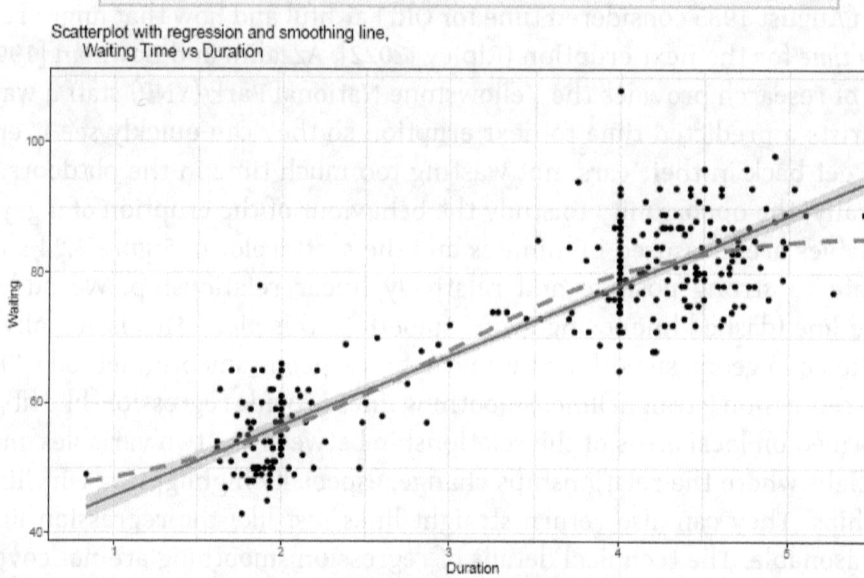

Figure 6.24: Scatterplot of Old Faithful waiting times to next eruption (minutes) and duration of prior eruption (minutes) with smoothing line (dashed) and regression line (solid).

An initial concern with these data is that the observations are likely not independent. Since they were taken consecutively, one waiting time might be related to the next waiting time – violating the independence assumption. As noted above, there might be two groups (types) of eruptions – short ones and long ones. The Normal QQ-Plot in Figure 6.25 also suggests a few observations creating a slightly long right tail. Those observations might warrant further exploration as they also show up as unusual in the Residuals vs Fitted plot. There are no highly influential points in the data set with all points having Cook's D smaller than 0.5 (contours are not displayed because no points are near or over them), so these outliers are not necessarily moving the regression line around. There are two distinct groups of observations but the variability is not clearly changing so we do not have to worry about non-constant variance here. So these results might be relatively trustworthy if we assume that the same relationship holds for all levels of duration of eruptions.

```
OF1 <- lm(Waiting ~ Duration, data = G2)
summary(OF1)
```

```
##
## Call:
## lm(formula = Waiting ~ Duration, data = G2)
##
## Residuals:
##      Min       1Q   Median       3Q      Max
## -14.6940  -4.4954  -0.0966   3.9544  29.9544
##
## Coefficients:
##             Estimate Std. Error t value Pr(>|t|)
## (Intercept)  34.9452     1.1807   29.60   <2e-16
## Duration     10.7751     0.3235   33.31   <2e-16
##
## Residual standard error: 6.392 on 296 degrees of freedom
## Multiple R-squared:  0.7894, Adjusted R-squared:  0.7887
## F-statistic:  1110 on 1 and 296 DF,  p-value: < 2.2e-16
```

```
par(mfrow = c(2,2))
plot(OF1)
```

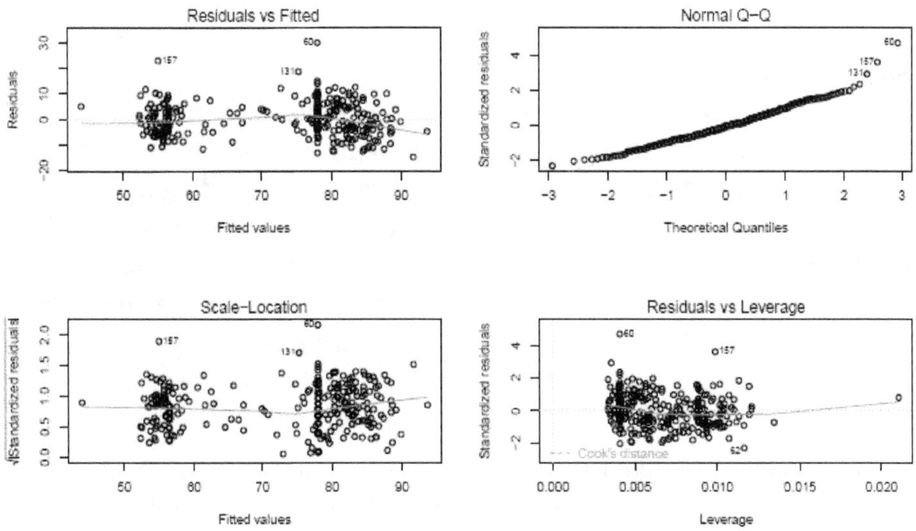

Figure 6.25: Diagnostic plots for Old Faithful waiting time model.

The estimated regression equation is Waiting Timei = 34.95 + 10.78 · Durationi, suggesting that for a 1 minute increase in eruption Duration we would expect, on average, a 10.78 minute change in the Waiting Time. This equation might provide a useful tool for the YNP staff to predict waiting times. The R^2 is fairly large: 78.9% of the variation in waiting time is explained by the duration of the previous eruption. But maybe this is more about two types of eruptions/waiting times? We could consider the relationship within the shorter and longer eruptions but since there are observations residing between the two groups, it is difficult to know where to split the explanatory variable into two groups. Maybe we really need to measure additional information that might explain why there are two groups in the responses.

6.12 CONCISE SUMMARY

The correlation coefficient (***r*** or Pearson's Product Moment Correlation Coefficient) measures the strength and direction of the linear relationship between two quantitative variables. Regression models estimate the impacts of changes in x on the mean of the response variable y. Direction of the assumed relationship (which variable explains or causes the other) matters for regression models but does not matter for correlation. Regression lines only describe the pattern of the relationship; in regression, we use the coefficient of determination to describe the strength of the relationship between the variables as a percentage of the response variable that is explained by the model. If we are choosing between models, we prefer them to have higher R^2 values for obvious reasons, but we will discover in Chapter 8 that maximizing the coefficient of determination is not a good way to pick a model when we have multiple candidate options.

In this chapter, a wide variety of potential problems were explored when using regression models. This included a discussion of the conditions that will be required for using the models to perform trustworthy inferences in the remaining chapters. It is important to remember that correlation and regression models only measure the **linear** association between variables and that can be misleading if a nonlinear relationship is present. Similarly, influential observations can completely distort the apparent relationship between variables and should be assessed before trusting any regression output. It is also important to remember that regression lines should not be used outside the scope of the original observations – extrapolation should be checked for and avoided whenever possible or at least acknowledged when it is being performed.

Regression models look like they estimate the changes in y that are caused by changes in x, especially when you use x to predict y. This is not true unless the levels of x are randomly assigned and only then we can make causal inferences. Since this is not generally true, you should initially always assume that any regression equation describes the relationship – if you observe two subjects that are 1 unit of *x* apart, you can expect their mean to differ by *b*1 – you should not, however, say that changing *x* causes a change in the mean of the responses. Despite all these cautions, regression models are very popular statistical methods. They provide detailed descriptions of relationships between variables and can be extended to situations where we are interested in multiple predictor variables. They also share ties to the ANOVA models discussed previously. When you are running R code, you will note that all the ANOVAs and the regression models are estimated using lm. The assumptions and diagnostic plots are quite similar. And in the next chapter, we will see that inference techniques look similar. People still like to distinguish among the different types of situations, but the underlying ***linear models*** are actually exactly the same...

6.13 SUMMARY OF IMPORTANT R CODE

The main components of the R code used in this chapter follow with the components to modify in lighter and/or ALL CAPS text where y is a response variable, x is an explanatory variable, and the data are in DATASETNAME.

- DATASETNAME %>% ggpairs()
  - Requires the GGally package.
  - Makes a scatterplot matrix that also displays the correlation coefficients.
- cor(y ~ x, data = DATASETNAME)
  - Provides the estimated correlation coefficient between *x* and *y*.
- plot(y ~ x, data = DATASETNAME)
  - Provides a base R scatter plot.
- DATASETNAME %>% ggplot(mapping = aes(x = x, y = y) + geom_point() +geom_smooth(method = "lm")
  - Provides a scatter plot with a regression line.
  - Add **color = groupfactor** to the **aes()** to colour points and get regression lines based on a grouping (categorical) variable.
  - Add + **geom_smooth(se = F, lty = 2)** to add a smoothing line to the scatterplot as a dashed line.
- MODELNAME <- lm(y ~ x, data = DATASETNAME)
  - Estimates a regression model using least squares.
- summary(MODELNAME)
  - Provides parameter estimates and R-squared (used heavily in Chapter 7 and 8 as well).

- par(mfrow = c(2, 2)); plot(MODELNAME)
  - Provides four regression diagnostic plots in one plot.

6.14 PRACTICE PROBLEMS

6.1 Treadmill data analysis These questions revisit the treadmill data set from Chapter 1. Researchers were interested in whether the run test variable could be used to replace the treadmill oxygen consumption variable that is expensive to measure. The following code loads the data set and provides a scatterplot matrix using ggpairs for all variables except for the subject identifier variable that was in the first column and was removed by select(-1).

```
treadmill <- read_csv("http://www.math.montana.edu/courses/s217/documents/treadmill.csv")
library(psych)
treadmill %>% select(-1) %>% ggpairs()
```

6.1.1. First, we should get a sense of the strength of the correlation between the variable of primary interest, TreadMillOx, and the other variables and consider whether outliers or nonlinearity are going to be major issues here. Which variable is it most strongly correlated with? Which variables are next most strongly correlated with this variable?

6.1.2. Fit the SLR using RunTime as explanatory variable for TreadMillOx. Report the estimated model.

6.1.3. Predict the treadmill oxygen value for a subject with a run time of 14 minutes. Repeat for a subject with a run time of 16 minutes. Is there something different about these two predictions?

6.1.4. Interpret the slope coefficient from the estimated model, remembering the units on the variables.

6.1.5. Report and interpret the y-intercept from the SLR.

6.1.6. Report and interpret the R2 value from the output. Show how you can find this value from the original correlation matrix result.

6.1.7. Produce the diagnostic plots and discuss any potential issues. What is the approximate leverage of the highest leverage observation and how large is its Cook's D? What does that tell you about its potential influence in this model?

CHAPTER-7

SIMPLE LINEAR REGRESSION INFERENCE

INTRODUCTION: AN OVERVIEW

Simple linear regression is a fundamental statistical method used to analyze the relationship between two variables. It provides a straightforward approach to model and understand the linear association between a dependent variable (response) and an independent variable (predictor). This method assumes that the relationship between the variables can be adequately represented by a straight line. By fitting the data to this line, simple linear regression allows us to make predictions, draw inferences, and quantify the strength and direction of the linear relationship. The simplicity of this regression model makes it an essential tool in various fields, such as economics, finance, biology, and social sciences, enabling researchers and analysts to gain valuable insights into the connections between variables and make informed decisions based on their findings.

7.1 MODEL

In Chapter 6, we learned how to estimate and interpret correlations and regression equations with a single predictor variable (simple linear regression or SLR). We carefully explored the variety of things that could go wrong and how to check for problems in regression situations. In this chapter, that work provides the basis for performing statistical inference that mainly focuses on the population slope coefficient based on the sample slope coefficient. As a reminder, the estimated regression model is $yi = b_0 + b_1xi$. The population regression equation is $yi = \beta 0 + \beta 1xi + \varepsilon i$ where $\beta 0$ is the population (or true) y-intercept and $\beta 1$ is the population (or true) slope coefficient. These are population parameters (fixed but typically unknown).This model can be re-written to think about different components and their roles. The mean of a random variable is statistically denoted as $E(yi)$, the expected value of yi, or as μyi and the mean of the response variable in a simple linear model is specified by $E(yi) = \mu yi = \beta 0 + \beta 1xi$. This uses the true regression line todefine the model for the mean of the responses as a function of the value of the explanatory variable1.

The other part of any statistical model is specifying a model for the variability around the mean. There are two aspects to the variability to specify here – the shape of the distribution and the spread of the distribution. This is where the normal distribution and our "normality assumption" re-appears. And for normal distributions, we need to define a variance parameter, σ^2. Combined, the complete regression model is which can be read as "y follows a normal distribution with mean mu-y and variance sigma-squared" and that "mu-y is equal to beta-0 plus beta-1 times x". This also implies that the random variability around the true mean, the errors, follow a normal distribution with mean 0 and that same variance, $\varepsilon_i \sim N(0, \sigma^2)$. The true deviations ($\varepsilon_i$) are once again estimated by the residuals, $e_i = y_i - \hat{y}_i$ = observed response –predicted response. We can use the residuals to estimate σ, which is also called the residual standard error, $\hat{\sigma} = \sqrt{\Sigma e_i^2/(n-2)}$. We will find this quantity near the end of the regression output as discussed below so the formula is not heavily used here. This provides us with the three parameters that are estimated as part of our SLR model: β0, β1, and σ.

$$y_i \sim N(\mu_{y_i}, \sigma^2), \text{ with } \mu_{y_i} = \beta_0 + \beta_1 x_i,$$

These definitions also formalize the assumptions implicit in the regression model:

- The errors follow a normal distribution (Normality assumption).
- The errors have the same variance (Constant variance assumption).
- The observations are independent (Independence assumption).
- The model for the mean is "correct" (Linearity, No Influential points, Only one group).

The diagnostics described at the end of Chapter 6 provide techniques for checking these assumptions – at least not having clear issues with those assumptions is fundamental to having a regression line that we trust and inferences from it that we also can trust.

To make this clearer, suppose that in the Beers and BAC study that they had randomly assigned 20 students to consume each number of beers. We would expect some variation in the BAC for each group of 20 at each level of Beers but that each group of observations will be centered at the true mean BAC for each number of Beers. The regression model assumes that the BAC values are normally distributed around the mean for each Beer level, $BAC_i \sim N(\beta_0 + \beta_1 Beers_i, \sigma^2)$, with the mean defined by the regression equation.

We actually do not need to obtain more than one observation at each x value to make this assumption or assess it, but the plots below show you what this could look like. The sketch in Figure 7.1 attempts to show the idea of normal distributions that are centered at the true regression line, all with the same shape and variance that is an assumption of the regression model. Figure 7.2 contains simulated realizations from a normal distribution of 20 subjects at each Beer level around the assumed true regression line with two different residual SEs of 0.02 and 0.06. The original BAC model has a residual SE of 0.02 but had many fewer observations at each Beer value.

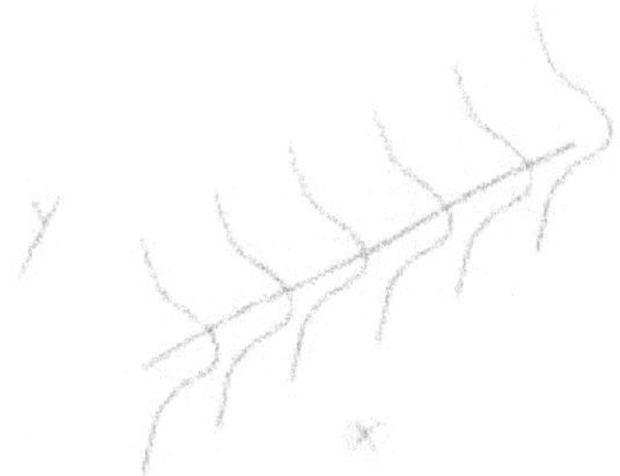

Figure 7.1: Sketch of assumed normal distributions for the responses centered at the regression line.

```
BB <- read_csv("http://www.math.montana.edu/courses/s217/documents/beersbac.csv")
```

Along with getting the idea that regression models define normal distributions in the y-direction that are centered at the regression line, you can also get a sense of how variable samples from a normal distribution can appear. Each distribution of 20 subjects at each *x* value came from a normal distribution but there are some of those distributions that might appear to generate small outliers and have slightly different variances. This can help us to remember to not be too particular when assessing assumptions and allow for some variability in spreads and a few observations from the tails of the distribution to occasionally arise.

In sampling from the population, we expect some amount of variability of each estimator around its true value. This variability leads to the potential variability in estimated regression lines (think of a suite of potential estimated regression lines that would be created by different random samples from the same population). Figure 7.3 contains the true regression line (bold, red) and realizations of the estimated regression line in simulated data based on results similar to the real data set. This variability due to random sampling is something that needs to be properly accounted for to use the **single** estimated regression line to make inferences about the true line and parameters based on the sample-based estimates. The next sections develop those inferential tools.

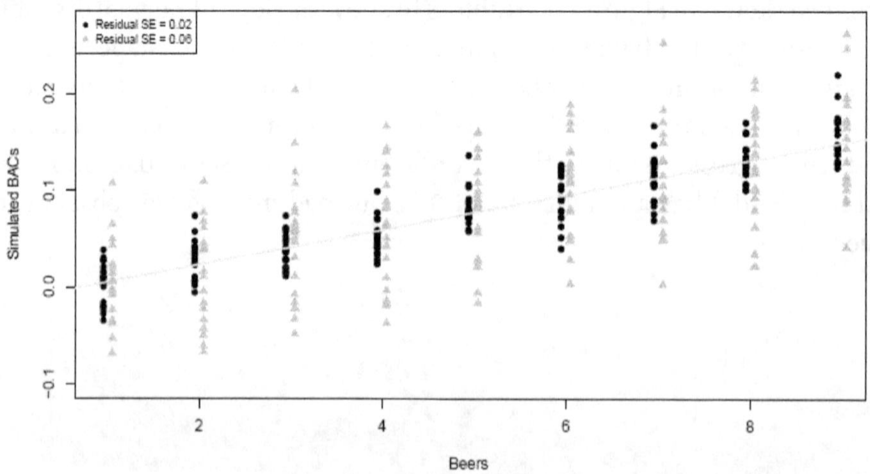

Figure 7.2: Simulated data for Beers and BAC assuming two different residual standard errors (0.02 and 0.06).

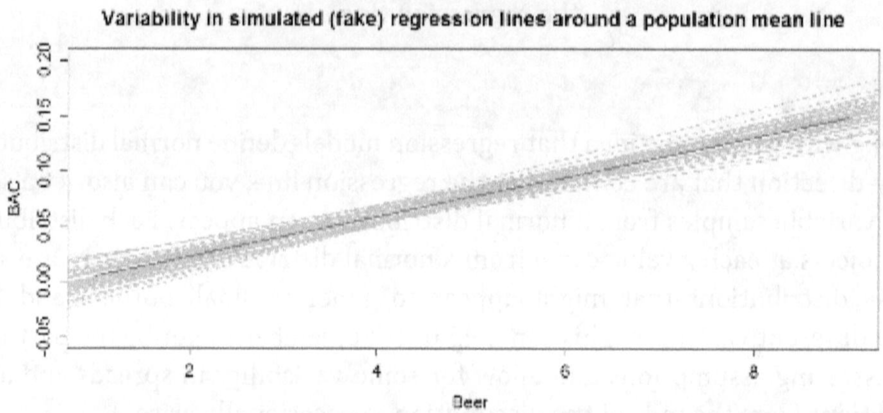

Figure 7.3: Variability in realized regression lines based on sampling variation. Light grey lines are simulated realizations assuming the bold (red) line is the true SLR model and variability is similar to the original BACdata set. Simulated observations from the estimated models using the simulate function as was used in Chapter 2 were used to create this plot.

7.2 CONFIDENCE INTERVAL AND HYPOTHESIS TESTS FOR THE SLOPE AND INTERCEPT

Our inference techniques will resemble previous material with an interest in forming confidence intervals and doing hypothesis testing, although the interpretation of confidence intervals for slope coefficients take some extra care. Remember that the general form of any parametric confidence interval is

$$\text{estimate} \mp t^* \text{SE}_{estimate},$$

so we need to obtain the appropriate standard error for regression model coefficients and the degrees of freedom to define the t-distribution to look up t* multiplier. We will find the SEb0 and SEb1 in the model summary. The degrees of freedom for the t-distribution in simple linear regression are df = n - 2. Putting this together, the confidence interval for the true y-intercept, β0, is $b_0 \mp t^*_{n-2}SE_{b_0}$ although this confidence interval is rarely of interest. The confidence interval that is almost always of interest is for the true slopen-2coefficient, β1, that is $b_1 \mp t^*_{n-2}SE_{b_1}$. The slope confidence interval is used to do two things: (1) inference for the amount of change in the mean of y for a unit change in x in the population and (2) to potentially do hypothesis testing by checking whether 0 is in the CI or not. The sketch in Figure 7.4 illustrates the roles of the CI for the slope in terms of determining where the population slope, β1, coefficient might be – centered at the sample slope coefficient – our best guess for the true slope. This sketch also informs an interpretation of the slope coefficient confidence interval:

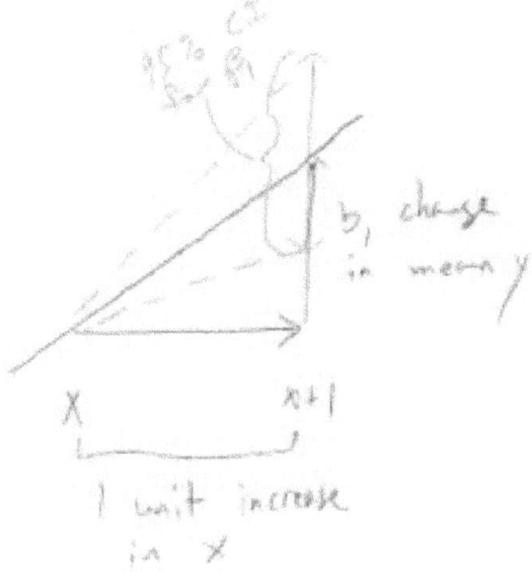

Figure 7.4: Graphic illustrating the confidence interval for a slope coefficient for a 1 unit increase in x.

For a 1 [units of X] increase in X, we are___% confident that the true change in the mean of Y will be between LL and UL [units of Y].

In this interpretation, LL and UL are the calculated lower and upper limits of the confidence interval. This builds on our previous interpretation of the slope coefficient, adding in the information about pinning down the true change (population change) in the mean of the response variable for a difference of 1 unit in the *x*-direction. The interpretation of the y-intercept CI is:

For an x of 0 [units of X], we are 95% confident that the true mean of Y will be between LL and **UL [*units of Y*].**

This is really only interesting if the value of x = 0 is interesting – we'll see a method for generating CIs for the true mean at potentially more interesting values of x in Section 7.7. To trust the results from these confidence intervals, it is critical that any issues with the regression validity conditions are minor.

The only hypothesis test of interest in this situation is for the slope coefficient. To develop the hypotheses of interest in SLR, note the effect of having β1 = 0 in the mean of the regression equation, μyi = β0 + β1xi = β0 + 0xi = β0. This is the "intercept-only" or "mean-only" model that suggests that the mean of y does not vary with different values of x as it is always β0. We saw this model in the ANOVA material as the reduced model when the null hypothesis of no difference in the true means across the groups was true. Here, this is the same as saying that there is no linear relationship between x and y, or that x is of no use in predicting y, or that we make the same prediction for y for every value of x. Thus is a test for no linear relationship between x and y in the population. The alternative of HA : β1 /= 0, that there is some linear relationship between x and y in the population, is our main test of interest in these situations. It is also possible to test greater than or less than alternatives in certain situations.

$$H_0 : \beta_1 = 0$$

Test statistics for regression coefficients are developed, if we can trust our assumptions, using the t-distribution with *n* - 2 degrees of freedom. The *t*-test statistic is generally

$$t = \frac{b_i}{\text{SE}_{b_i}}$$

with the main interest in the test for β1 based on *b*1 initially. The p-value would be calculated using the two-tailed area from the *tn*-2 distribution calculated using the pt function. The p-value to test these hypotheses is also provided in the model summary as we will see below.

The greater than or less than alternatives can have interesting interpretations in certain situations. For example, the greater than alternative (HA : β1 > 0) tests an alternative of a positive linear relationship, with the p-value extracted just from the right tail of the same t-distribution. This could be used when a researcher would only find a result "interesting" if a positive relationship is detected, such as in the study of tree height and tree diameter where a researcher might be justified in deciding to test only for a positive linear relationship. Similarly, the left-tailed alternative is also possible, HA : β1 < 0. To get one-tailed p-values from two-tailed results (the default), first check that the observed test statistic is in the direction of the alternative (t > 0 for HA : β1 > 0 or t < 0 for HA : β1 < 0). If these conditions are met, then the p-value for the one-sided test from the two-sided version is found by dividing the reported p-value by 2. If t > 0 for HA : β1 > 0 or t < 0 for HA : β1 < 0 are not met, then the p-value would be greater than 0.5 and it would be easiest to look it up directly using pt using the tail area direction in the direction of the alternative.

We can revisit a couple of examples for a last time with these ideas in hand to complete the analyses. For the *Beers, BAC* data, the 95% confidence for the true slope coefficient, β1, is

$$\begin{aligned} b_1 \mp t^*_{n-2}\mathrm{SE}_{b_1} &= 0.01796 \mp 2.144787 * 0.002402 \\ &= 0.01796 \mp 0.00515 \\ &\rightarrow (0.0128, 0.0231). \end{aligned}$$

You can find the components of this calculation in the model summary and from qt(0.975, df = n-2)

which was 2.145 for the *t**-multiplier. Be careful not to use the *t*-value of 7.48 in the model summary to make confidence intervals – that is the test statistic used below. The related calculations are shown at the bottom of the following code:

```
m1 <- lm(BAC ~ Beers, data = BB)
summary(m1)
```

```
##
## Call:
## lm(formula = BAC ~ Beers, data = BB)
##
## Residuals:
##        Min         1Q    Median        3Q       Max
## -0.027118 -0.017350  0.001773  0.008623  0.041027
##
## Coefficients:
##              Estimate Std. Error t value Pr(>|t|)
## (Intercept) -0.012701   0.012638  -1.005    0.332
## Beers        0.017964   0.002402   7.480 2.97e-06
##
## Residual standard error: 0.02044 on 14 degrees of freedom
## Multiple R-squared:  0.7998, Adjusted R-squared:  0.7855
## F-statistic: 55.94 on 1 and 14 DF,  p-value: 2.969e-06
```

```
qt(0.975, df = 14) #t* multiplier for 95% CI
```

```
## [1] 2.144787
```

```
0.017964 + c(-1,1)*qt(0.975, df = 14)*0.002402
```

```
## [1] 0.01281222 0.02311578
```

```
qt(0.975, df = 14)*0.002402
```

```
## [1] 0.005151778
```

We can also get the confidence interval directly from the confint function run on our regression model, saving some calculation effort and providing both the CI for the y-intercept and the slope coefficient.

```
confint(m1)
```

```
##                   2.5 %     97.5 %
## (Intercept) -0.03980535 0.01440414
## Beers        0.01281262 0.02311490
```

We interpret the 95% CI for the slope coefficient as follows: For a 1 beer increase in number of beers consumed, we are 95% confident that the true change in the mean BAC will be between 0.0128 and 0.0231 g/dL. While the estimated slope is our best guess of the impacts of an extra beer consumed based on our sample, this CI provides information about the likely range of potential impacts on the mean in the population. It also could be used to test the two-sided hypothesis test and would suggest strong evidence against the null hypothesis since the confidence interval does not contain 0, but its main use is to quantify where we think the true slope coefficient resides.

The width of the CI, interpreted loosely as the precision of the estimated slope, is impacted by the variability of the observations around the estimated regression line, the overall sample size, and the positioning of the x-observations. Basically all those aspects relate to how "clearly" known the regression line is and that determines the estimated precision in the slope. For example, the more variability around the line that is present, the more uncertainty there is about the correct line to use (Least Squares (LS) can still find an estimated line but there are other lines that might be "close" to its optimizing choice). Similarly, more observations help us get a better estimate of the mean – an idea that permeates all statistical methods. Finally, the location of x-values can impact the precision in a slope coefficient. We'll revisit this in the context of multicollinearity in the next chapter, and often we have no control of x-values, but just note that different patterns of x-values can lead to different precision of estimated slope coefficients2.

For hypothesis testing, we will almost always stick with two-sided tests in regression modelling as it is a more conservative approach and does not require us to have an expectation of a direction for relationships a priori. In this example, the null hypothesis for the slope coefficient is that there is no linear relationship between Beers and BAC in the population. The alternative hypothesis is that there is some linear relationship between Beers and BAC in the population. The test statistic is $t = 0.01796/0.002402 = 7.48$ which, if model assumptions hold, follows a $t(14)$ distribution under the null hypothesis. The model summary provides the calculation of the test statistic and the two-sided test p-value of 2.97e-6 = 0.00000297. So we would just report "p-value < 0.0001". This suggests that there is very strong evidence against the null hypothesis of no linear relationship between Beers and BAC in the population, so we would conclude that there is a linear relationship between them. Because of the random assignment, we can also say that drinking beers causes changes in BAC but, because the sample was made up of volunteers, we cannot infer that these results would hold in the general population of OSU students or more generally.

There are also results for the y-intercept in the output. The 95% CI is from -0.0398 to 0.0144, that the true mean BAC for a 0 beer consuming subject is between -0.0398 to 0.01445. This is really not a big surprise but possibly is comforting to know that these results would not show much evidence against the null hypothesis that the true mean BAC for 0 Beers is 0. Finding little evidence of a difference from 0 makes sense and makes the estimated y-intercept of -0.013 not so problematic. In other situations, the results for the y-intercept may be more illogical but this will often be because the y-intercept is extrapolating far beyond the scope of observations. The y-intercept's main function in regression models is to be at the right level for the slope to "work" to make a line that describes the responses and thus is usually of lesser interest even though it plays an important role in the model.

As a second example, we can revisit modelling the Hematocrit of female Australian athletes as a function of body fat %. The sample size is n = 99 so the df are 97 in any t-distributions. In Chapter 6, the relationship between Hematocrit and body fat % for females appeared to be a weak negative linear association. The 95% confidence interval for the slope is -0.186 to 0.0155. For a 1% increase in body fat %, we are 95% confident that the change in the true mean Hematocrit is between -0.186 and 0.0155% of blood. This suggests that we would find little evidence against the null hypothesis of no linear relationship because this CI contains 0. In fact the p-value is 0.0965 which is larger than 0.05 and so provides a consistent conclusion with using the 95% confidence interval to perform a hypothesis test. Either way, we would conclude that there is not strong evidence against the null hypothesis but there is some evidence against it with a p-value of that size since more extreme results are somewhat common but still fairly rare if we assume the null is true. If you think p-values around 0.10 provide moderate evidence, you might have a different opinion about the evidence against the null hypothesis here. For this reason, we sometimes interpret this sort of marginal result as having some or marginal evidence against the null but certainly would never say that this presents strong evidence.

```
library(alr4)
data(ais)
library(tibble)
ais <- as_tibble(ais)
aisR <- ais %>% slice(-56, -166) #Removes observations in rows 56 and 166
m2 <- lm(Hc ~ Bfat, data = aisR %>% filter(Sex == 1)) #Results for Females
summary(m2)
```

```
##
## Call:
## lm(formula = Hc ~ Bfat, data = aisR %>% filter(Sex == 1))
##
## Residuals:
##     Min      1Q  Median      3Q     Max
## -5.2399 -2.2132 -0.1061  1.8917  6.6453
##
## Coefficients:
##             Estimate Std. Error t value Pr(>|t|)
## (Intercept) 42.01378    0.93269  45.046   <2e-16
## Bfat        -0.08504    0.05067  -1.678   0.0965
##
## Residual standard error: 2.598 on 97 degrees of freedom
## Multiple R-squared:  0.02822,    Adjusted R-squared:  0.0182
## F-statistic: 2.816 on 1 and 97 DF,  p-value: 0.09653
```

```
confint(m2)
```

```
##                  2.5 %      97.5 %
## (Intercept) 40.1626516 43.86490713
## Bfat        -0.1856071  0.01553165
```

One more worked example is provided from the Montana fire data. In this example pay particular attention to how we are handling the units of the response variable, log-hectares, and to the changes to doing inferences with a 99% confidence level CI, and where you can find the needed results in the following output:

```
mtfires <- read_csv("http://www.math.montana.edu/courses/s217/documents/climateR2.csv")
```

```
mtfires <- mtfires %>% mutate(loghectares = log(hectares))
fire1 <- lm(loghectares ~ Temperature, data = mtfires)
summary(fire1)
```

```
##
## Call:
## lm(formula = loghectares ~ Temperature, data = mtfires)
##
## Residuals:
##     Min      1Q  Median      3Q     Max
## -3.0822 -0.9549  0.1210  1.0007  2.4728
##
## Coefficients:
##             Estimate Std. Error t value Pr(>|t|)
## (Intercept) -69.7845    12.3132  -5.667 1.26e-05
## Temperature   1.3884     0.2165   6.412 2.35e-06
##
## Residual standard error: 1.476 on 21 degrees of freedom
## Multiple R-squared:  0.6619, Adjusted R-squared:  0.6458
## F-statistic: 41.12 on 1 and 21 DF,  p-value: 2.347e-06
```

```
confint(fire1, level = 0.99)
```

```
##                    0.5 %     99.5 %
## (Intercept) -104.6477287 -34.921286
## Temperature    0.7753784   2.001499
```

```
qt(0.995, df = 21)
```

```
## [1] 2.83136
```

- Based on the estimated regression model, we can say that if the average temperature is 0, we expect that, on average, the log-area burned would be -69.8 log-hectares.
- From the regression model summary, $b1$ = 1.39 with SEb_1 = 0.2165 and **t = 6.41.**
- There were n = 23 measurements taken, so df = n - 2 = 23 - 3 = 21.
- Suppose that we want to test for a linear relationship between temperature and log-hectares burned:

$$H_0 : \beta_1 = 0$$

- In words, the true slope coefficient between *Temperature* and *log-area burned* is 0 OR there is no linear relationship between *Temperature* and *log-area burned* in the population.

$$H_A : \beta_1 /= 0$$

- In words, the alternative states that the true slope coefficient between *Temperature* and *log-area burned* is not 0 OR there is a linear relationship between *Temperature* and *log-area burned* in the population.

Test statistic: t = 1.39/0.217 = 6.41

- Assuming the null hypothesis to be true (no linear relationship), the t-statistic follows a t-distribution with n - 2 = 23 - 2 = 21 degrees of freedom (or simply $t21$).

p-value:

- From the model summary, the p-value is 2.35 ∗ 10¯6
  - **Interpretation:** There is less than a 0.01% chance that we would observe slope coefficient like we did or something more extreme (greater than 1.39 log(hectares)/$^{\circ}F$) if there were in fact no linear relationship between temperature ($^{\circ}F$) and log-area burned (log-hectares) in the population.

Conclusion

There is very strong evidence against the null hypothesis of no linear relationship, so we would conclude that there is, in fact, a linear relationship between Temperature and log(Hectares) burned.

Scope of Inference: Since we have a time series of results, our inferences pertain to the results we could have observed for these years but not for years we did not observe – so just for the true slope for this sample of years. Because we can't randomly assign the amount of area burned, we cannot make causal inferences – there are many reasons why both the average temperature and area burned would vary together that would not involve a direct connection between them.

$$99\% \text{ CI for } \beta_1 : b_1 \mp t^*_{n-2}\text{SE}_{b_1} \rightarrow 1.39 \mp 2.831 \bullet 0.217 \rightarrow (0.78, 2.00)$$

Interpretation of 99% CI for slope coefficient:

- For a 1 degree F increase in *Temperature*, we are 99% confident that the change in the true mean log-area burned is between 0.78 and 2.00 log(Hectares).

Another way to interpret this is:

- For a 1 degree F increase in *Temperature*, we are 99% confident that the mean Area Burned will change by between 0.78 and 2.00 log(Hectares) in the population.

Also R2 is 66.2%, which tells us that Temperature explains 66.2% of the variation in log(Hectares) burned. Or that the linear regression model built using Temperature explains 66.2% of the variation in yearly *log(Hectares)* burned so this model explains quite a bit but not all the variation in the responses.

7.3 BOZEMAN TEMPERATURE TREND

For a new example, consider the yearly average maximum temperatures in Bozeman, MT. For over 100 years, daily measurements have been taken of the minimum and maximum temperatures at hundreds of weather stations across the US. In early years, this involved manual recording of the temperatures and resetting the thermometer to track the extremes for the following day. More recently, these measures have been replaced by digital temperature recording devices that continue to track this sort of information with much less human effort and, possibly, errors. This sort of information is often aggregated to monthly or yearly averages to be able to see "on average" changes from month-to-month or year-to-year as opposed to the day-to-day variation in the temperature3. Often the local information is aggregated further to provide regional, hemispheric, or even global average temperatures. Climate change research involves attempting to quantify the changes over time in these and other long-term temperature or temperature proxies.

-These data were extracted from the National Oceanic and Atmospheric Administration's National Centers for Environmental Information's database (http://www.ncdc.noaa.gov/cdo-web/) and we will focus on the yearly average of the monthly averages of the daily maximum temperature in Bozeman in degrees F from 1901 to 2014. We can call them yearly average maximum temperatures but note that it was a little more complicated than that to arrive at the response variable we are analyzing.

```
bozemantemps <- read_csv("http://www.math.montana.edu/courses/s217/documents/BozemanMeanMax.csv")
```

```
summary(bozemantemps)
```

```
##      meanmax           Year
##  Min.   :49.75   Min.   :1901
##  1st Qu.:53.97   1st Qu.:1930
##  Median :55.43   Median :1959
##  Mean   :55.34   Mean   :1958
##  3rd Qu.:57.02   3rd Qu.:1986
##  Max.   :60.05   Max.   :2014
```

```
length(bozemantemps$Year) #Some years are missing (1905, 1906, 1948, 1950, 1995)
```

```
## [1] 109
```

```
bozemantemps %>% ggplot(mapping = aes(x = Year, y = meanmax)) +
  geom_point() +
  geom_smooth(method = "lm") +
  geom_smooth(lty = 2, col = "red", lwd = 1.5, se = F) + #Add smoothing line
  theme_bw() +
  labs(title = "Scatterplot of Bozeman Yearly Average Max Temperatures",
       y = "Mean Maximum Temperature (degrees F)")
```

The scatterplot in Figure 7.5 shows the results between 1901 and 2014 based on a sample of n = 109 years because four years had too many missing months to fairly include in the responses. Missing values occur for many reasons and in this case were likely just machine or human error4. These are time series data and in time series analysis we assume that the population of interest for inference is all possible realizations from the underlying process over this time frame even though we only ever get to observe one realization. In terms of climate change research, we would want to (a) assess evidence for a trend over time (hopefully assessing whether any observed trend is clearly different from a result that could have been observed by chance if there really is no change over time in the true process) and (b) quantify the size of the change over time along with the uncertainty in that estimate relative to the underlying true mean change over time. The hypothesis test for the slope answers (a) and the confidence interval for the slope addresses (b). We also should be concerned about problematic (influential) points, changing variance, and potential nonlinearity in the trend over time causing problems for the SLR inferences. The scatterplot suggests that there is a moderate or strong positive linear relationship between temperatures and year. Both looking at the points and at the smoothing line does not suggest a clear curve in these responses over time and the variability seems similar across the years. There appears to be one potential large outlier in the late 1930s.We'll perform all 6+ steps of the hypothesis test for the slope coefficient and use the confidence interval

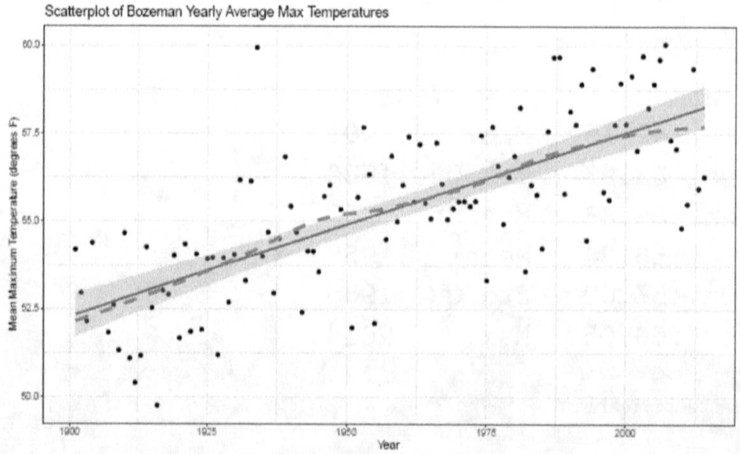

Figure 7.5: Scatterplot of average yearly maximum temperatures in Bozeman from 1900 to 2014 with SLR (solid) and smoothing (dashed) lines.

interpretation to discuss the size of the change. First, we need to select our hypotheses (the 2-sided test would be a conservative choice and no one that does climate change research wants to be accused of taking a liberal approach in their analyses5) and our test statistic, $t = \frac{b_1}{\text{SE}_{b_1}}$. The scatterplot is the perfect tool to illustrate the situation.

1. Hypotheses for the slope coefficient test:

$$H_0 : \beta_1 = 0 \text{ vs } H_A : \beta_1 \neq 0$$

2. Validity conditions:

- ⦿ Quantitative variables condition
    - Both Year and yearly average Temperature are quantitative variables so are suitable for an SLR analysis.
- ⦿ Independence of observations
    - There may be a lack of independence among years since a warm year might be followed by another warmer than average year. It would take more sophisticated models to account for this and the standard error on the slope coefficient could either get larger or smaller depending

on the type of ***autocorrelation*** (correlation between neighbouring time points or correlation with oneself at some time lag) present. This creates a caveat on these results but this model is often the first one researchers fit in these situations and often is reasonably correct even in the presence of some autocorrelation.

To assess the remaining conditions, we need to fit the regression model and use the diagnostic plots in Figure 7.6 to aid our assessment:

```
temp1 <- lm(meanmax ~ Year, data = bozemantemps)
par(mfrow = c(2,2))
plot(temp1, add.smooth = F, pch = 16)
```

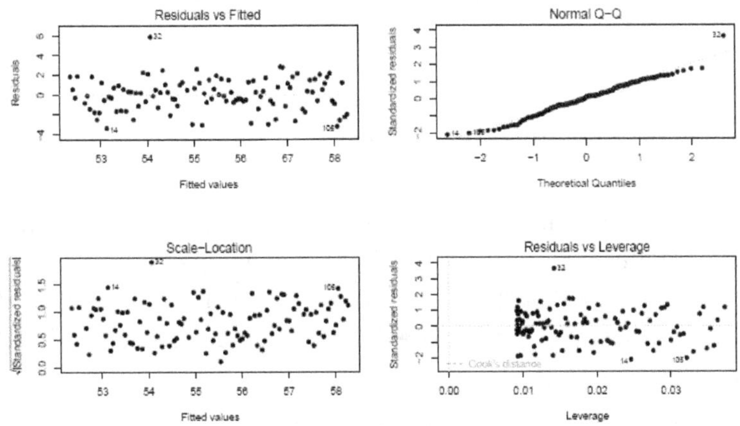

Figure 7.6: Diagnostic plots of the Bozeman yearly temperature simple linear regression model.

- ⦿ Linearity of relationship
  - Examine the Residuals vs Fitted plot:
    - There does not appear to be a clear curve remaining in the residuals so we should be able to proceed without worrying too much about missed nonlinearity.
  - Compare the smoothing line to the regression line in Figure 7.5:
    - There does not appear to be a big difference between the straight line and the smoothing line.
- ⦿ Equal (constant) variance
  - Examining the Residuals vs Fitted and the "Scale-Location" plots provide little to no evidence of changing variance. The variability does decrease slightly in the middle fitted values but those changes are really minor and present no real evidence of changing variability.

- Normality of residuals
  - Examining the Normal QQ-plot for violations of the normality assumption shows only one real problem in the outlier from the 32nd observation in the data set (the temperature observed in 1934) which was identified as a large outlier when examining the original scatterplot. We should be careful about inferences that assume normality and contain this point in the analysis. We might consider running the analysis with and without that point to see how much it impacts the results just to be sure it isn't creating evidence of a trend because of a violation of the normality assumption. The next check reassures us that re-running the model without this point would only result in slightly changing the SEs and not the slopes.
- No influential points:
  - There are no influential points displayed in the Residuals vs Leverage plot since the Cook's D contours are not displayed.
  - **Note:** by default this plot contains a smoothing line that is relatively meaningless, so ignore it if is displayed. We suppressed it using the add.smooth = F option in plot(temp1) but if you forget to do that, just ignore the smoothers in the diagnostic plots especially in the Residuals vs Leverage plot.
  - These results tells us that the outlier was not influential. If you look back at the scatterplot, it was located near the middle of the observed *x*'s so its potential leverage is low. You can find its leverage based on the plot to be around 0.12 when there are observations in the data set with leverages over 0.3. The high leverage points occur at the beginning and the end of the record because they are at the edges of the observed *x*'s and most of these points follow the overall pattern fairly well.

So the main issues are with the assumption of independence of observations and one non-influential outlier that might be compromising our normality assumption a bit.

3. Calculate the test statistic and p-value:

- $t = 0.05244/0.00476 = 11.02$

```
summary(temp1)
```

- From the model summary: p-value < 2e-16 or just < 0.0001
- The test statistic is assumed to follow a *t*-distribution with $n - 2 = 109 - 2 = 107$ degrees of freedom. The p-value can also be calculated as:

```
##
## Call:
## lm(formula = meanmax ~ Year, data = bozemantemps)
##
## Residuals:
##     Min      1Q  Median      3Q     Max
## -3.3779 -0.9300  0.1078  1.1960  5.8698
##
## Coefficients:
##              Estimate Std. Error t value Pr(>|t|)
## (Intercept) -47.35123    9.32184   -5.08 1.61e-06
## Year          0.05244    0.00476   11.02  < 2e-16
##
## Residual standard error: 1.624 on 107 degrees of freedom
## Multiple R-squared:  0.5315, Adjusted R-squared:  0.5271
## F-statistic: 121.4 on 1 and 107 DF,  p-value: < 2.2e-16
```

```
2*pt(11.02, df = 107, lower.tail = F)
```

```
## [1] 2.498481e-19
```

- Which is then reported as < 0.0001, which means that the chances of observing a slope coefficient as extreme or more extreme than 0.052 if the null hypothesis of no linear relationship is true is less than 0.01%.

4. Conclusion:

- There is very strong evidence ($t107$ = 11.02, p-value < 0.0001) against the null hypothesis of no linear relationship between *Year* and yearly mean *Temperature* so we can conclude that there is, in fact, some linear relationship between *Year* and yearly mean maximum *Temperature* in Bozeman.

5. Size:

- For a one year increase in Year, we estimate that, on average, the yearly average maximum temperature will change by 0.0524 $^{\circ}F$ (95% CI: 0.043 to 0.062). This suggests a modest but noticeable change in the mean temperature in Bozeman and the confidence suggests minimal variation around this estimate, going from 0.04 to 0.06 $^{\circ}F$. The "size" of this change is discussed more in Section 7.5.

```
confint(temp1)
```

```
##                     2.5 %       97.5 %
## (Intercept) -65.83068375 -28.87177785
## Year          0.04300681   0.06187746
```

6. Scope of inference:

- We can conclude that this detected trend pertains to the Bozeman area in the years 1901 to 2014 but not outside of either this area or time frame. We cannot say that time caused the observed changes since it was not randomly assigned and we cannot attribute the changes to any other factors because we did not consider them. But knowing that there was a trend toward increasing temperatures is an intriguing first step in a more complete analysis of changing climate in the area.

It is also good to report the percentage of variation that the model explains: *Year* explains 54.91% of the variation in yearly average maximum *Temperature.* If the coefficient of determination value had been very small, we might discount the previous result. Since it is moderately large with over 50% of the variation in the response explained, that suggests that just by using a linear trend over time we can account for quite a bit of the variation in yearly average maximum temperatures in Bozeman. Note that the percentage of variation explained would get much worse if we tried to analyze the monthly or original daily maximum temperature data even though we might find about the same estimated mean change over time.

Interpreting a confidence interval provides more useful information than the hypothesis test here – instead of just assessing evidence against the null hypothesis, we can actually provide our best guess at the true change in the mean of y for a change in x. Here, the 95% CI is (0.043, 0.062). This tells us that for a 1 year increase in *Year*, we are 95% confident that the change in the true mean of the yearly average maximum *Temperatures* in Bozeman is between 0.043 and 0.062 $^{\circ}F$.

Sometimes the scale of the x-variable makes interpretation a little difficult, so we can re-scale it to make the resulting slope coefficient more interpretable without changing how the model fits the responses. One option is to re-scale the variable and re-fit the regression model and the other (easier) option is to re-scale our interpretation. The idea here is that a 100-year change might be easier and more meaningful scale to interpret than a single year change. If we have a slope in the model of 0.052 (for a 1 year change), we can also say that a 100 year change in the mean is estimated to be 0.052*100 = 0.52$^{\circ}F$. Similarly, the 95% CI for the population mean 100-year change would be from 0.43$^{\circ}F$ to 0.62$^{\circ}F$. In 2007, the IPCC (Intergovernmental Panel on Climate Change; http://www.ipcc.ch/publications_and_data/ar4/wg1/en/tssts-3-1-1.html) estimated the global temperature change from 1906 to 2005 to be 0.74$^{\circ}C$ per decade or, scaled up, 7.4$^{\circ}C$ per century (1.33$^{\circ}F$). There are many reasons why our local temperature trend might differ, including that our analysis was of average maximum temperatures and the IPCC was considering the average temperature (which was not measured locally or in most places in a good way until digital instrumentation was installed) and that local trends are likely to vary around the global average change based on localized environmental conditions.

One issue that arises in studies of climate change is that researchers often consider these sorts of tests at many locations and on many response variables (if I did the maximum temperature, why not also do the same analysis of the minimum temperature time series as well? And if I did the analysis for Bozeman, what about Butte and Helena and. . . ?). Remember our discussion of multiple testing issues? This issue can arise when regression modelling is repeated in many similar data sets, say different sites or different response variables or both, in one study. In Moore et al. [2007], we considered the impacts on the assessment of evidence of trends of earlier spring onset timing in the Mountain West when the number of tests across many sites is accounted for. We found that the evidence for time trends decreases substantially but does not disappear. In a related study, Greenwood et al. [2011] found evidence for regional trends to earlier spring onset using more sophisticated statistical models. The main point here is to be careful when using simple statistical methods repeatedly if you are not accounting for the number of tests performed.

Along with the confidence interval, we can also plot the estimated model (Figure 7.7) using a term-plot from the effects package (Fox, 2003). This is the same function we used for visualizing results in the ANOVA models and in its basic application you just need plot(allEffects(MODELNAME)) although we from time to time, we will add some options. In regression models, we get to see the regression line along with bounds for 95% confidence intervals for the mean at every value of x that was observed (explained in the next section). Note that there is also a rug plot on the x-axis showing you where values of the explanatory variable were obtained, which is useful to understanding how much information is available for different aspects of the line. Here it provides gaps for missing years of observations as sort of broken teeth in a comb. Also not used here, we can also turn on the residuals = T option, which in SLR just plots the original points and adds a smoothing line to this plot to reinforce the previous assessment of assumptions.

```
library(effects)
plot(allEffects(temp1, xlevels = list(Year = bozemantemps$Year)),
     grid = T)
```

If we extended the plot for the model to Year = 0, we could see the reason that the y-intercept in this model is -47.4$^{\circ}F$. This is obviously a large extrapolation for these data and provides a silly result. However, in paleo-climate data that goes back thousands of years using tree rings, ice cores, or sea sediments, the estimated mean in year 0 might be interesting and within the scope of observed values or it might not. For example, in Santibáñez et al. [2018], the data were a time series from

27,000 to about 9,000 years before present extracted from Antarctic ice cores. It all depends on the application.

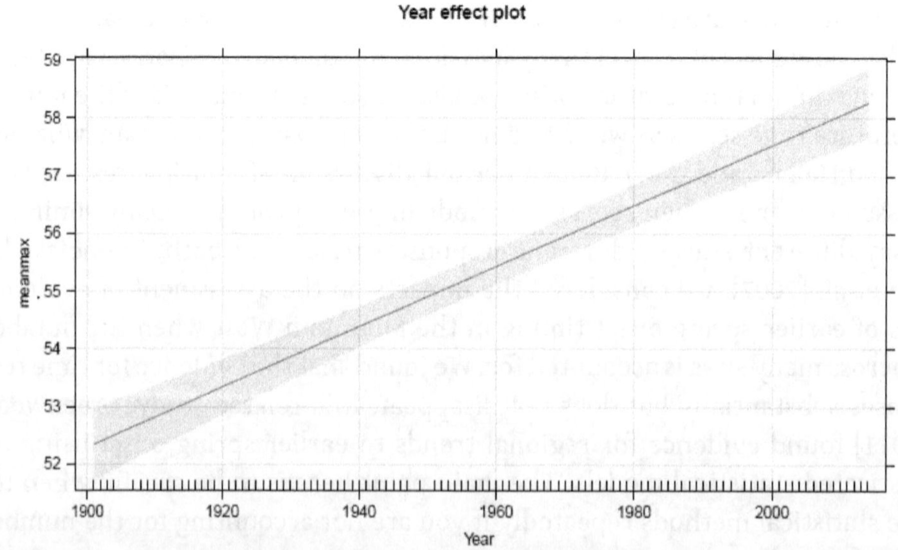

Figure 7.7: Term-plot for the Bozeman mean yearly maximum temperature linear regression model with 95% confidence interval bands for the mean in each year.

To make the y-intercept more interesting for our data set, we can re-scale the *x*'s using mutate before we fit the model to have the first year in the data set (1901) be "0". This is accomplished by calculating Year2 = Year - 1901.

```
bozemantemps <- bozemantemps %>% mutate(Year2 = Year - 1901)
summary(bozemantemps$Year2)
```

```
##    Min. 1st Qu.  Median    Mean 3rd Qu.    Max.
##    0.00   29.00   58.00   57.27   85.00  113.00
```

The new estimated regression equation is $\widehat{Temp}_i = 52.34 + 0.052 \cdot Year2_i$. The slope and its test statistic are the same as in the previous model. The y-intercept has changed dramatically with a 95% CI from 51.72°*F* to 52.96°*F* for Year2 = 0. But we know that Year2 has a 0 value for 1901 because of our subtraction. That means that this CI is for the true mean in 1901 and is now at least somewhat interesting. If you revisit Figure you will actually see that the displayed confidence intervals provide upper and lower bounds that match this result for 1901 – the y-intercept CI matches the 95% CI for the true mean in the first year of the data set.

```
temp2 <- lm(meanmax ~ Year2, data = bozemantemps)
summary(temp2)
```

```
##
## Call:
## lm(formula = meanmax ~ Year2, data = bozemantemps)
##
## Residuals:
##     Min      1Q  Median      3Q     Max
## -3.3779 -0.9300  0.1078  1.1960  5.8698
##
## Coefficients:
##             Estimate Std. Error t value Pr(>|t|)
## (Intercept) 52.34126    0.31383  166.78   <2e-16
## Year2        0.05244    0.00476   11.02   <2e-16
##
## Residual standard error: 1.624 on 107 degrees of freedom
## Multiple R-squared:  0.5315, Adjusted R-squared:  0.5271
## F-statistic: 121.4 on 1 and 107 DF,  p-value: < 2.2e-16
```

```
confint(temp2)
```

```
##                   2.5 %      97.5 %
## (Intercept) 51.71913822 52.96339150
## Year2        0.04300681  0.06187746
```

Ideally, we want to find a regression model that does not violate any assumptions, has a high R2 value, and a slope coefficient with a small p-value. If any of these are not the case, then we are not completely satisfied with the regression and should be suspicious of any inference we perform. We can sometimes resolve some of the systematic issues noted above using transformations, discussed in Sections 7.5 and 7.6.

7.4 RANDOMIZATION-BASED INFERENCES FOR THE SLOPE COEFFICIENT

Exploring permutation testing in SLR provides an opportunity to gauge the observed relationship against the sorts of relationships we would expect to see if there was no linear relationship between the variables. If the relationship is linear (not curvilinear) and the null hypothesis of $\beta 1 = 0$ is true, then any configuration of the responses relative to the predictor variable's values is as good as any other. Consider the four scatterplots of the Bozeman temperature data versus Year and permuted versions of Year in Figure 7.8. First, think about which of the panels you think present the most evidence of a linear relationship between Year and Temperature?

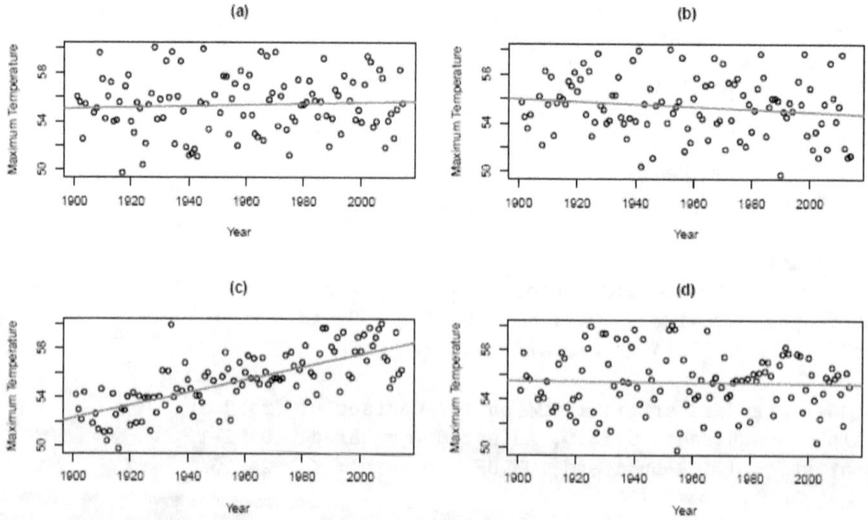

Figure 7.8: Plot of the Temperature responses versus four versions of Year, three of which are permutations of the Year variable relative to the Temperatures.

Hopefully you can see that panel (c) contains the most clear linear relationship among the choices. The plot in panel (c) is actually the real data set and pretty clearly presents itself as "different" from the other results. When we have small p-values, the real data set will be clearly different from the permuted results because it will be almost impossible to find a permuted data set that can attain as large a slope coefficient as was observed in the real data set[6]. This result ties back into our original interests in this climate change research situation – does our result look like it is different from what could have been observed just by chance if there were no linear relationship between x and y? It seems unlikely. . .

Repeating this permutation process and tracking the estimated slope coefficients, as T^*, provides another method to obtain a p-value in SLR applications. This could also be performed on the t-statistic for the slope coefficient and would provide the same p-values but the sampling distribution would have a different x-axis scaling. In this situation, the observed slope of 0.052 is really far from any possible values that can be obtained using permutations as shown in Figure 7.9. The p-value would be reported as < 0.001 for the two-sided permutation test.

```
Tobs <- lm(meanmax ~ Year, data = bozemantemps)$coef[2]
Tobs
```

```
##        Year
## 0.05244213
```

```
B <- 1000
Tstar <- matrix(NA, nrow = B)

for (b in (1:B)){
  Tstar[b] <- lm(meanmax ~ shuffle(Year), data = bozemantemps)$coef[2]
}
pdata(abs(Tstar), abs(Tobs), lower.tail = F)[[1]]
```

```
## [1] 0
```

```
tibble(Tstar) %>% ggplot(aes(x = Tstar)) +
  geom_histogram(aes(y = ..ncount..), bins = 20, col = 1, fill = "skyblue") +
  geom_density(aes(y = ..scaled..)) +
  theme_bw() +
  labs(y = "Density") +
  geom_vline(xintercept = c(-1,1)*Tobs, col = "red", lwd = 2) +
  stat_bin(aes(y = ..ncount.., label = ..count..), bins = 20,
           geom = "text", vjust = -0.75)
```

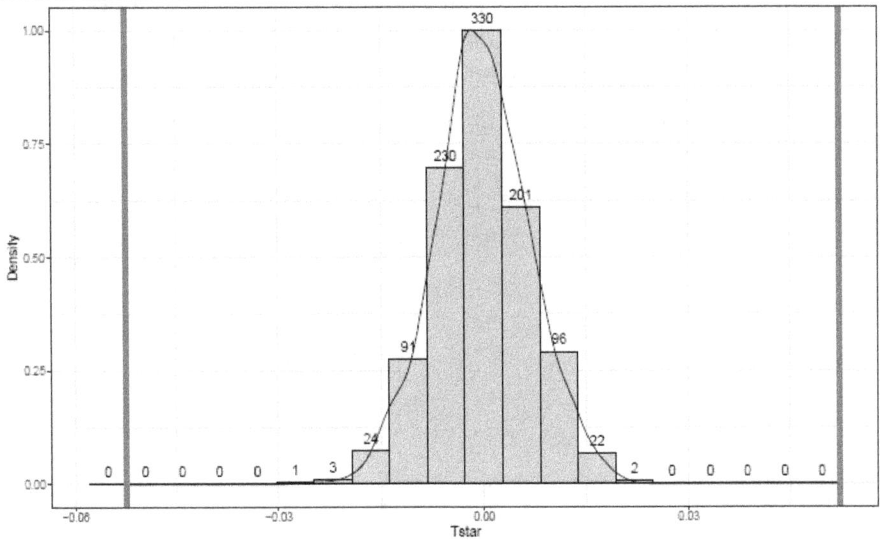

Figure 7.9: Permutation distribution of the slope coefficient in the Bozeman temperature linear regression model with bold vertical lines at ±b1 = 0.56 based on the observed estimated slope.

One other interesting aspect of exploring the permuted data sets as in Figure 7.8 is that the outlier in the late 1930s "disappears" in the permuted data sets because there were many other observations that were that warm, just none that happened around that time of the century in the real data set. This reinforces the evidence for changes over time that seem to be present in these data – old unusual years don't look unusual in more recent years (which is a pretty concerning result).

The permutation approach can be useful in situations where the normality assumption is compromised, but there are no influential points. In these situations, we might find more trustworthy p-values using permutations but only if we are working with an initial estimated regression equation that we generally trust. I personally like the permutation approach as a way of explaining what a p-value is actually measuring – the chance of seeing something like what we saw, or more extreme, if the null is true. And the previous scatterplots show what the "by chance" versions of this relationship might look like.

In a similar situation where we want to focus on confidence intervals for slope coefficients but are not completely comfortable with the normality assumption, it is also possible to generate bootstrap confidence intervals by sampling with replacement from the data set. This idea was introduced in Sections 2.8 and 2.9. This provides a 95% bootstrap confidence interval from 0.0433 to 0.061, which almost exactly matches the parametric *t*-based confidence interval. The bootstrap distributions are very symmetric (Figure 7.10). The interpretation is the same and this result reinforces the other assessments that the parametric approach is not unreasonable here except possibly for the independence assumption. These randomization approaches provide no robustness against violations of the independence assumption.

```
Tobs <- lm(meanmax ~ Year, data = bozemantemps)$coef[2]
Tobs
```

```
##       Year
## 0.05244213
```

```
B <- 1000
Tstar <- matrix(NA, nrow = B)
for (b in (1:B)){
  Tstar[b] <- lm(meanmax ~ Year, data = resample(bozemantemps))$coef[2]
}
quantiles <- qdata(Tstar, c(0.025, 0.975))
quantiles
```

```
##       2.5%      97.5%
## 0.04326952 0.06131044
```

```
tibble(Tstar) %>% ggplot(aes(x = Tstar)) +
  geom_histogram(aes(y = ..ncount..), bins = 15, col = 1, fill = "skyblue", center = 0) +
  geom_density(aes(y = ..scaled..)) +
  theme_bw() +
  labs(y = "Density") +
  geom_vline(xintercept = quantiles, col = "blue", lwd = 2, lty = 3) +
  geom_vline(xintercept = Tobs, col = "red", lwd = 2) +
  stat_bin(aes(y = ..ncount.., label = ..count..), bins = 15,
         geom = "text", vjust = -0.75)
```

7.5 TRANSFORMATIONS PART I: LINEARIZING RELATIONSHIPS

When the influential point, linearity, constant variance and/or normality assumptions are clearly violated, we cannot trust any of the inferences generated by the regression model. The violations occur on gradients from minor to really major problems. As we have seen from the examples in the previous chapters, it has been hard to find data sets that were free of all issues. Furthermore, it may seem hopeless to be able to make successful inferences in some of these situations with the previous tools. There are three potential solutions to violations of the validity conditions:

- Consider removing an offending point or two and see if this improves the results, presenting results both with and without those points to describe their impact7,
- Try to transform the response, explanatory, or both variables and see if you can force the data set to meet our SLR assumptions after transformation (the focus of this and the next section), or

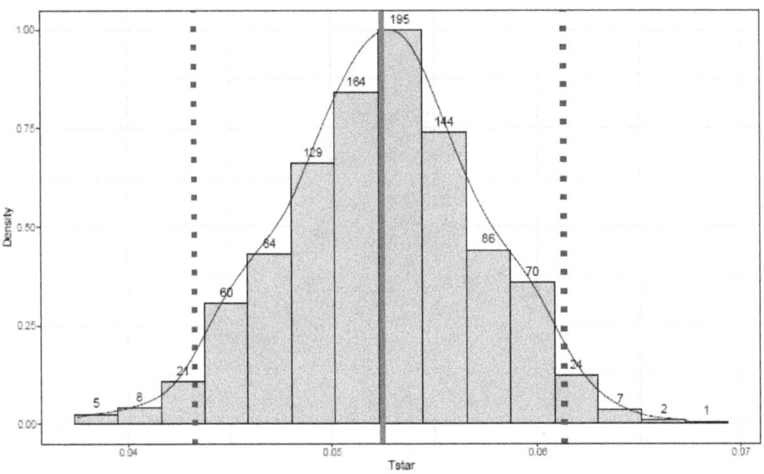

Figure 7.10: Bootstrap distribution of the slope coefficient in the Bozeman temperature linear regression model with bold dashed vertical lines delineating the 95% confidence interval and the bold solid line the observed slope of 0.052.

- Consider more advanced statistical models that can account for these issues (the focus of subsequent statistics courses, if you continue on further).

Transformations involve applying a function to one or both variables. After applying this transformation, one hopes to have alleviated whatever issues encouraged its consideration. Linear transformation functions, of the form znew = a * x + b, will never help us to fix assumptions in regression situations; linear transformations change the scaling of the variables but not their shape or the relationship between two variables. For example, in the Bozeman Temperature data example, we subtracted 1901 from the Year variable to have Year2 start at 0 and go up to 113. We could also apply a linear transformation to change Temperature from being measured in °F to °C using °C = [°F – 32] ° (5/9). The scatterplots on both the original and transformed scales are provided in Figure 7.11. All the coefficients in the regression model and the labels on the axes change, but the "picture" is still the same. Additionally, all the inferences remain the same – p-values are unchanged by linear transformations. So linear transformations can be "fun" but really are only useful if they make the coefficients easier to interpret. Here if you like temperature changes in °C for a 1 year increase, the slope coefficient is 0.029 and if you like the original change in °F for a 1 year increase,

the slope coefficient is 0.052. More useful than this is the switch into units of 100 years (so each year increase would just be 0.1 instead of 1), so that the slope is the temperature change over 100 years.

```
bozemantemps <- bozemantemps %>% mutate(meanmaxC = (meanmax - 32)*(5/9))
temp3 <- lm(meanmaxC ~ Year2, data = bozemantemps)
summary(temp1)
```

```
##
## Call:
## lm(formula = meanmax ~ Year, data = bozemantemps)
##
## Residuals:
##      Min       1Q   Median       3Q      Max
## -3.3779 -0.9300   0.1078   1.1960   5.8698
##
## Coefficients:
```

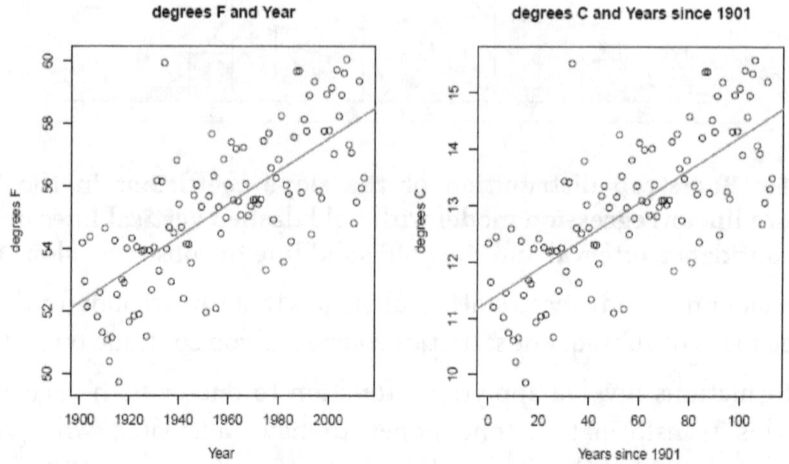

Figure 7.11: Scatterplots of Temperature (°F) versus Year (left) and Temperature (°C) vs Years since 1901 (right).

```
##              Estimate Std. Error t value Pr(>|t|)
## (Intercept) -47.35123    9.32184   -5.08 1.61e-06
## Year          0.05244    0.00476   11.02  < 2e-16
##
## Residual standard error: 1.624 on 107 degrees of freedom
## Multiple R-squared:  0.5315, Adjusted R-squared:  0.5271
## F-statistic: 121.4 on 1 and 107 DF,  p-value: < 2.2e-16
```

```
summary(temp3)
```

```
##
## Call:
## lm(formula = meanmaxC ~ Year2, data = bozemantemps)
##
## Residuals:
##     Min      1Q  Median      3Q     Max
## -1.8766 -0.5167  0.0599  0.6644  3.2610
##
## Coefficients:
##              Estimate Std. Error t value Pr(>|t|)
## (Intercept) 11.300703   0.174349   64.82   <2e-16
## Year2        0.029135   0.002644   11.02   <2e-16
##
## Residual standard error: 0.9022 on 107 degrees of freedom
## Multiple R-squared:  0.5315, Adjusted R-squared:  0.5271
## F-statistic: 121.4 on 1 and 107 DF,  p-value: < 2.2e-16
```

Table 7.1: Ladder of powers of transformations that are often used in statistical modelling.

Power	Formula	Usage
2	Y^2	seldom used
1	$y^1 = y$	no change
½	y0.5 = √y	counts and area responses
0	log(y) natural log of y	counts, normality, curves, non-constant variance
-1/2 -1 -2	y-0.5 = 1/√y y-1 = 1/y y-2 = 1/y2	uncommon for ratios seldom used

Nonlinear transformation functions are where we apply something more complicated than this shift and scaling, something like ynew = f (y), where f (·) could be a log or power of the original variable y. When we apply these sorts of transformations, interesting things can happen to our linear models and their problems. Some examples of transformations that are at least occasionally used for transforming the response variable are provided in Table 7.1, ranging from taking y to different powers from y-2 to y2. Typical transformations used in statistical modelling exist along a gradient of powers of the response variable, defined as yλ with λ being the power of the transformation of the response variable and λ

= 0 implying a log-transformation. Except for λ = 1, the transformations are all nonlinear functions of y.

There are even more transformations possible, for example y0.33 is sometimes useful for variables involved in measuring the volume of something. And we can also consider applying any of these transformations to the explanatory variable, and consider using them on both the response and explanatory variables at the same time. The most common application of these ideas is to transform the response variable using the log-transformation, at least as a starting point. In fact, the log-transformation is so commonly used (or maybe overused), that we will just focus on its use. It is so commonplace in some fields that some researchers log-transform their data prior to even plotting it. In other situations, such as when measuring acidity (pH), noise (decibels), or earthquake size (Richter scale), the measurements are already on logarithmic scales.

Actually, we have already analyzed data that benefited from a log-transformation – the log-area burned vs. summer temperature data for Montana. Figure 7.12 compares the relationship between these variables on the original hectares scale and the log-hectares scale.

```
p <- mtfires %>% ggplot(mapping = aes(x = Temperature, y = hectares)) +
  geom_point() +
  labs(title = "(a)", y = "Hectares") +
  theme_bw()

plog <- mtfires %>% ggplot(mapping = aes(x = Temperature, y = loghectares)) +
  geom_point() +
  labs(title = "(b)", y = "log-Hectares") +
  theme_bw()

grid.arrange(p, plog, ncol = 2)
```

Figure 7.12(a) displays a relationship that would be hard fit using SLR – it has a curve and the variance is increasing with increasing temperatures. With a log-transformation of Hectares, the relationship appears to be relatively linear and have constant variance (in (b)). We considered regression models for this situation previously. This shows at least one situation where a log-transformation of a response variable can linearize a relationship and reduce non-constant variance.

This transformation does not always work to “fix” curvilinear relationships, in fact in some situations it can make the relationship more nonlinear. For example, reconsider the relationship between tree diameter and tree height, which contained some curvature that we could not account for in an SLR. Figure 7.13 shows the original version of the variables and Figure 7.14 shows the same information with the response variable (height) log-transformed.

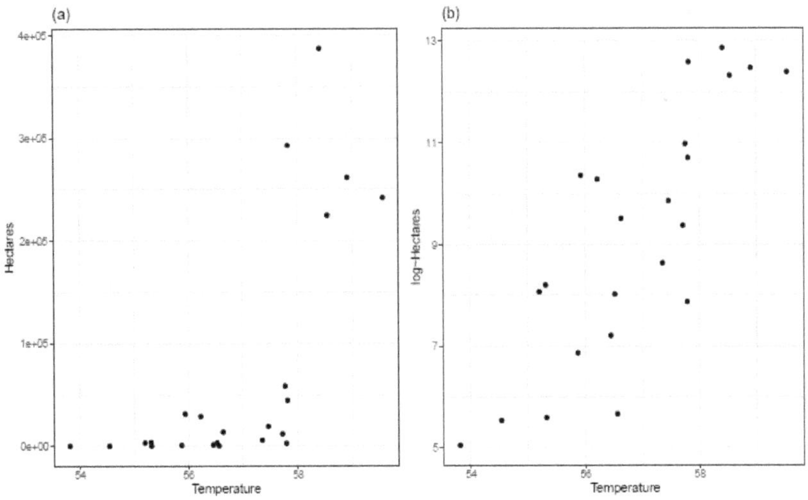

Figure 7.12: Scatterplots of Hectares (a) and log-Hectares (b) vs Temperature.

```
library(spuRs)
data(ufc)
ufc <- as_tibble(ufc)
ufc %>% slice(-168) %>% ggplot(mapping = aes(x = dbh.cm, y = height.m)) +
  geom_point() +
  geom_smooth(method = "lm") +
  geom_smooth(col = "red", lwd = 1, se = F, lty = 2) +
  theme_bw() +
  labs(title = "Tree height vs tree diameter")

ufc %>% slice(-168) %>% ggplot(mapping = aes(x = dbh.cm, y = log(height.m))) +
  geom_point() +
  geom_smooth(method = "lm") +
  geom_smooth(col = "red", lwd = 1, se = F, lty = 2) +
  theme_bw() +
  labs(title = "log-tree height vs tree diameter")
```

Figure 7.14 with the log-transformed height response seems to show a more nonlinear relationship and may even have more issues with non-constant variance. This example shows that log-transforming the response variable cannot fix all problems, even though I've seen some researchers assume it can. It is OK to try a transformation but remember to always plot the results to make sure it actually helped and did not make the situation worse.

All is not lost in this situation; we can consider two other potential uses of the log-transformation and see if they can "fix" the relationship up a bit. One option is to apply the transformation to the explanatory variable (y ~ log(x)), displayed in Figure 7.15. If the distribution of the explanatory variable is right skewed(see the boxplot on the *x*-axis), then consider log-transforming the explanatory variable. This will often reduce the leverage of those most extreme observations which can be useful. In this situation, it also seems to have been quite successful at linearizing the relationship, leaving some minor non-constant variance, but providing a big improvement from the relationship on the original scale.

The other option, especially when everything else fails, is to apply the log-transformation to both the explanatory and response variables (log(y) ~ log(x)), as displayed in Figure 7.16. For this example, the transformation seems to be better than the first two options (none and only log(y)), but demonstrates some decreasing variability with larger *x* and *y* values. It has also created a new and different curve in the relationship (see the smoothing (dashed) line start below the SLR line, then go above it, and the finish below it). In the end, we might prefer to fit an SLR model to the tree *height* vs *log(diameter)* versions of the variables (Figure 7.15).

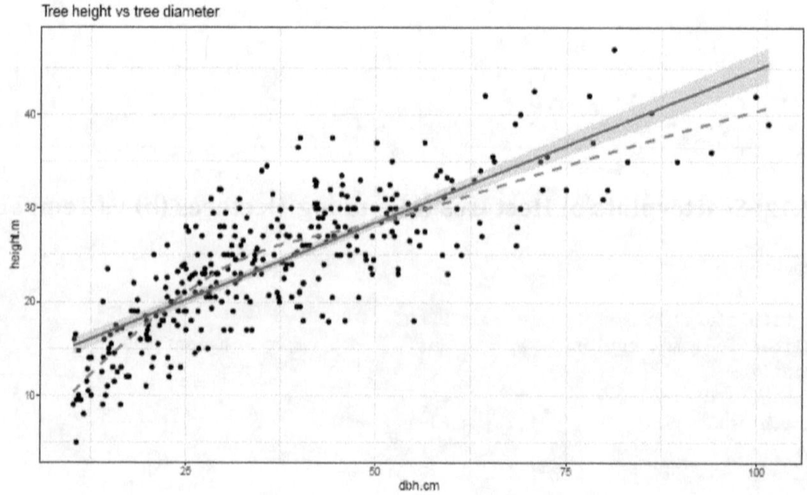

Figure 7.13: Scatterplot of tree height versus tree diameter.

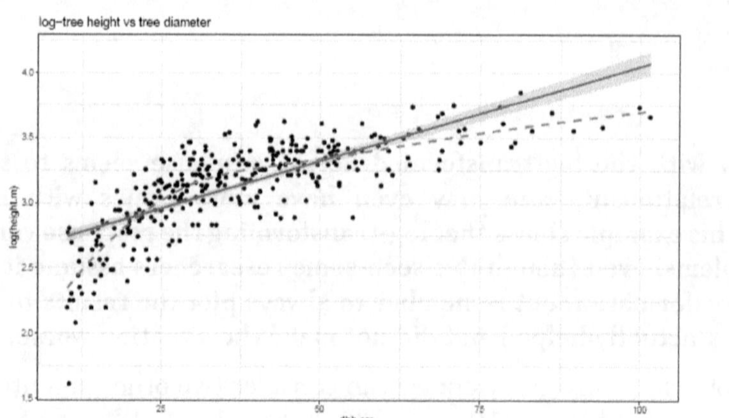

Figure 7.14: Scatterplot of log(tree height) versus tree diameter.

```
ufc %>% slice(-168) %>% ggplot(mapping = aes(x = log(dbh.cm), y = log(height.m))) +
  geom_point() +
  geom_smooth(method = "lm") +
  geom_smooth(col = "red", lwd = 1, se = F, lty = 2) +
  theme_bw() +
```

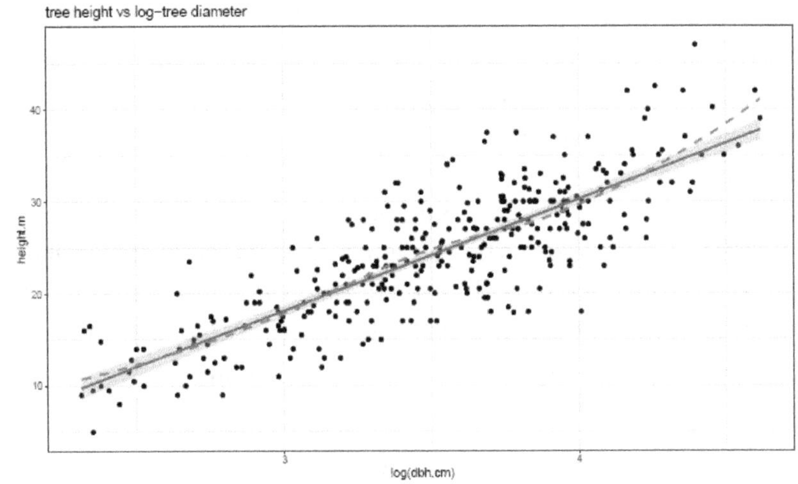

Figure 7.15: Scatterplot of tree height versus log(tree diameter).

```
  labs(title = "log-tree height vs log-tree diameter")
```

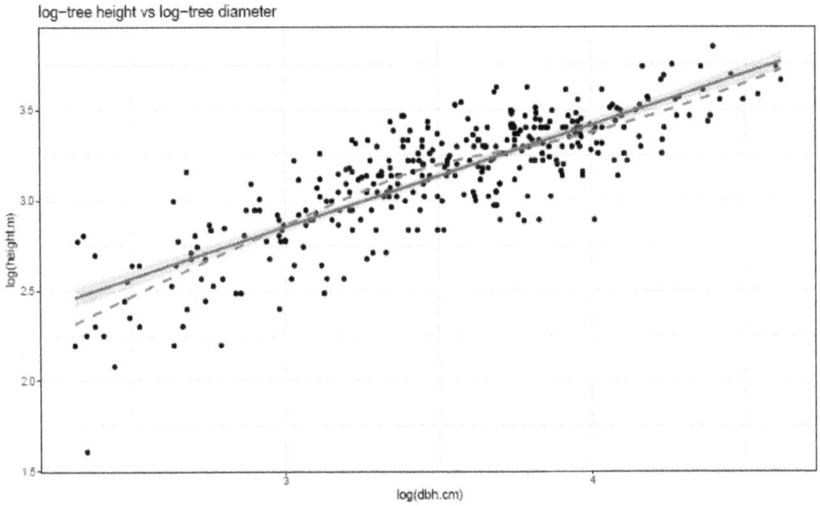

Figure 7.16: Scatterplot of log(tree height) versus log(tree diameter).

Economists also like to use log(y) ~ log(x) transformations. The log-log transformation tends to linearize certain relationships and has specific interpretations in terms of Economics theory. The log-log transformation shows up in many different disciplines as a way of obtaining a linear relationship on the log-log scale, but different fields discuss it differently. The following example shows a situation where transformations of both x and y are required and this double transformation seems to be quite successful in what looks like an initially hopeless situation for our linear modelling approach.

Data were collected in 1988 on the rates of infant mortality (infant deaths per 1,000 live births) and gross domestic product (GDP) per capita (in 1998 US dollars) from n = 207 countries. These data are available from the car Data package (Fox et al. [2022b], Fox [2003]) in a data set called UN. The four panels in Figure 7.17 show the original relationship and the impacts of log-transforming one or both variables. The only scatterplot that could potentially be modelled using SLR is the lower right panel (d) that shows the relationship between *log(infant mortality)* and *log(GDP)*. In the next section, we will fit models to some of these relationships and use our diagnostic plots to help us assess "success" of the transformations.

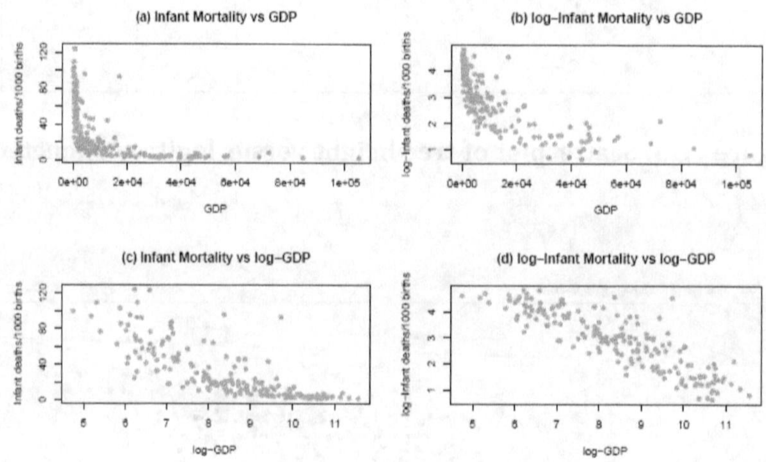

Figure 7.17: Scatterplots of Infant Mortality vs GDP under four different combinations of log-transformations.

Almost all nonlinear transformations assume that the variables are strictly greater than 0. For example, consider what happens when we apply the log function to 0 or a negative value in R:

```
log(-1)
```

```
## [1] NaN
```

```
log(0)
```

```
## [1] -Inf
```

So be careful to think about the domain of the transformation function before using transformations. For example, when using the log-transformation make sure that the data values are non-zero and positive or you will get some surprises when you go to fit your regression model to a data set with NaNs (not a number)and/ or $-\infty$'s in it. When using fractional powers (square-roots or similar), just having non-negative values are required and so 0 is acceptable.

Sometimes the log-transformations will not be entirely successful. If the relationship is monotonic (strictly positive or strictly negative but not both), then possibly another stop on the ladder of transformations in Table 7.1 might work. If the relationship is not monotonic, then it may be better to consider a more complex regression model that can accommodate the shape in the relationship or to bin the predictor, response, or both into categories so you can use ANOVA or Chi-square methods and avoid at least the linearity assumption.

Finally, remember that log in statistics and especially in R means the natural log (ln or log base e as you might see it elsewhere). In these situations, applying the log10 function (which provides log base 10) to the variables would lead to very similar results, but readers may assume you used ln if you don't state that you used log10. The main thing to remember to do is to be clear when communicating the version you are using. As an example, I was working with researchers on a study [Dieser et al., 2010] related to impacts of environmental stresses on bacterial survival. The response variable was log-transformed counts and involved smoothed regression lines fit on this scale. I was using natural logs to fit the models and then shared the fitted values from the models and my collaborators back-transformed the results assuming that I had used log10. We quickly resolved our differences once we discovered them but this serves as a reminder at how important communication is in group projects – we both said we were working with log-transformations and didn't know that we defaulted to different bases.

Generally, in statistics, it's safe to assume that everything is log base *e* unless otherwise specified

Transformations part II: Impacts on SLR interpretations: log(y), log(x), & both log(y) & log(x). The previous attempts to linearize relationships imply a desire to be able to fit SLR models. The log- transformations, when successful, provide the potential to validly apply our SLR model. There are then two options for interpretations: you can either interpret the model on the transformed scale or you can translate the SLR model on the transformed scale back to the original scale of the variables. It ends up that log-transformations have special interpretations on the original scales depending on whether the log was applied to the response variable, the explanatory variable, or both.

Scenario 1: log(y) vs x model:

First consider the $\log(y) \sim x$ situations where the estimated model is of the form $\log(y) = b0 + b1x$. When only the response is *log*-transformed, some people call this a ***semi-log model***. But many researchers will use this model without any special considerations, as long as it provides a situation where the SLR assumptions are reasonably well-satisfied. To understand the properties and eventually the interpretation of transformed-variables models, we need to try to "reverse" our transformation. If we exponentiate[8] both sidesof log(*y*) = *b*0 + *b*1*x*, we get:

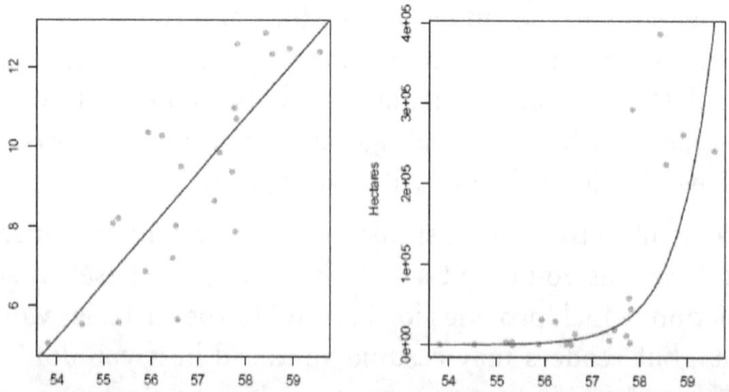

Figure 7.18: Plot of the estimated SLR (a) and implied model for the median on the original Hectares scalefor the area burned vs temperature data.

- exp(log(*y*)) = exp(*b*0 + *b*1*x*), *which is*
- y = exp(*b*0 + *b*1*x*), *which can be re-written as*
- y = exp(*b*0) exp(*b*1*x*). *This is based on the rules for* exp() *where* exp(*a* + *b*) = exp(*a*) exp(*b*).
- Now consider what happens if we increase *x* by 1 unit, going from *x* to *x* + 1, providing a new predicted
- *y* that we can call *y**: *y** = exp(*b*0) exp[*b*1(*x* + 1)]:
- *y** = **exp(*b*0) exp(*b*1*x*)** exp(*b*1). *Now note that the underlined, bold component was the y-value for x.*
- *y** = **y** exp(*b*1). *Found by replacing* exp(**b0**) exp(**b1x**) *with* **y**, *the value for x.*

So the difference in fitted values between *x* and *x* + 1 is to multiply the result for *x* (that was predicting by exp(*b*1) to get to the predicted result for *x* + 1 (called *y**). We can then use this result to form our **log(*y*)** ~ ***x* slope interpretation**: for a 1 unit increase in *x*, we observe a multiplicative change of **exp(b1)** in the response. When we compute a mean on logged variables that are symmetrically distributed (this should occur if our transformation was successful) and then exponentiate the results, the proper interpretation is that the changes are happening in the **median** of the original responses. This is the only time in the course that we will switch our inferences to medians instead of means, and we don't do this because we want to, we do it because it is result of modelling on the log(y) scale, if successful.

So there are a couple of ways to interpret these results in general:

- Log-scale interpretation of log(y) only model: for a 1 unit increase in x, we estimate a b1 unit change in the mean of log(y) or

- Original scale interpretation of log(y) only model: for a 1 unit increase in x, we estimate a exp(b1) times change in the median of y.

When we are working with regression equations, slopes can either be positive or negative and our interpretations change based on this result to either result in growth (b1 > 0) or decay (b1 < 0) in the responses as the explanatory variable is increased. As an example, consider b1 = 0.4 and exp(b1) = exp(0.4) = 1.492.

There are a couple of ways to interpret this on the original scale of the response variable y:

For b1 > 0:

- For a 1 unit increase in *x*, the median of *y* is estimated to change by 1.492 times.
- We can convert this into a **percentage increase** by subtracting 1 from exp(0.4), 1.492 - 1.0 = 0.492

and multiplying the result by 100, 0.492 * 100 = 49.2%. This is interpreted as: For a 1 unit increase in x, the median of y is estimated to increase by 49.2%.

```
exp(0.4)
```

```
## [1] 1.491825
```

For **b1 < 0**, the change on the *log*-scale is negative and that implies on the original scale that the curve decays to 0. For example, consider *b*1 = -0.3 and exp(-0.3) = 0.741. Again, there are two versions of the interpretation possible:

- For a 1 unit increase in *x*, the median of *y* is estimated to change by 0.741 times.
- For negative slope coefficients, the percentage decrease is calculated as (1 - exp(*b*1)) * 100%. For exp (-0.3) = 0.741, this is (1 - 0.741) * 100 = 25.9%. This is interpreted as: For a 1 unit increase in x, the median of *y* is estimated to decrease by 25.9%.

We suspect that you will typically prefer the "times" interpretation over the "percentage" change one for both directions but it is important to be able think about the results in terms of % change of the medians to make the scale of change more understandable. Some examples will help us see how these ideas can be used in applications.

For the area burned data set, the estimated regression model is log(hectares) = -69.8 + 1.39 · Temp. On the original scale, this implies that the model is hectares = exp(-69.8) exp(1.39 Temp). Figure 7.18 provides the log(y) scale version of the model and the model transformed to the original scale of measurement. On the log-hectares scale, the interpretation of the slope is: For a 1°F increase in summer temperature, we estimate a 1.39 log-hectares/1°F change, on average, in the log-area burned. On the original scale: A 1°F increase in temperature is related to an estimated multiplicative change in the median number of hectares burned of exp(1.39) = 4.01 times higher areas. That seems like a big rate of growth but the curve does grow rapidly as shown in panel (b), especially for values over 58°F where the area burned is starting to be really large. You can think of the multiplicative change here in the following way: the median number of hectares burned is 4 times higher at 58°F than at 57°F and the median area burned is 4 times larger at 59°F than at 58°F . . . This can also be interpreted on a % change scale: A 1°F increase in temperature is related to an estimated (4.01 - 1) * 100 = 301% increase in the median number of hectares burned.

7.6 SCENARIO 2: Y VS LOG(X) MODEL:

When only the explanatory variable is log-transformed, it has a different sort of impact on the regression model interpretation. Effectively we move the percentage change onto the x-scale and modify the first part of our slope interpretation when we consider the results on the original scale for x. Once again, we will consider the mathematics underlying the changes in the model and then work on applying it to real situations. When the explanatory variable is logged, the estimated regression model is y = b0 + b1 log(x). This models the relationship between y and x in terms of multiplicative changes in x having an effect on the average y.

To develop an interpretation on the x-scale (not log(x)), consider the impact of doubling x. This change will take us from the point (x, y = b0 + b1 log(x)) to the point (2x, y* = b0 + b1 log(2x)). Now the impact of doubling x can be simplified using the rules for logs to be:

- y* = b0 + b1 log(2x),
- y* = b0 + b1 log(x) + b1 log(2). Based on the rules for logs: log(2x) = log(x) + log(2).
- y* = y + b1 log (2)
- So if we double x, we change the mean of y by b1 log(2).

As before, there are couple of ways to interpret these sorts of results,

- log-scale interpretation of log(x) only model: for a 1 log-unit increase in x, we estimate a b1 unit change in the mean of y or
- original scale interpretation of log(x) only model: for a doubling of x, we estimate a b1 log(2)
- change in the mean of *y*. Note that both interpretations are for the mean of the *y*'s since we haven't changed the *y* ~ part of the model.

While it is not a perfect model (no model is), let's consider the model for infant mortality ~ log(GDP) in order to practice the interpretation using this type of model. This model was estimated to be infant mortality =155.77- 14.86 · log(GDP). The first (simplest) interpretation of the slope coefficient is: For a 1 log-dollar increase in GDP per capita, we estimate infant mortality to change, on average, by -14.86 deaths/1000 live births. The second interpretation is on the original GDP scale: For a doubling of GDP, we estimate infant mortality to change, on average, by -14.86 log(2) = -10.3 deaths/1000 live births. Or, the mean infant mortality is reduced by 10.3 deaths per 1000 live births for each doubling of GDP. Both versions of the model are displayed in Figure 7.19 – one on the scale the SLR model was fit (panel a) and the other on the original x-scale (panel b) that matches these last interpretations.

```
ID1 <- lm(infantMortality ~ log(ppgdp), data = UN)
summary(ID1)
```

```
##
## Call:
## lm(formula = infantMortality ~ log(ppgdp), data = UN)
##
## Residuals:
```

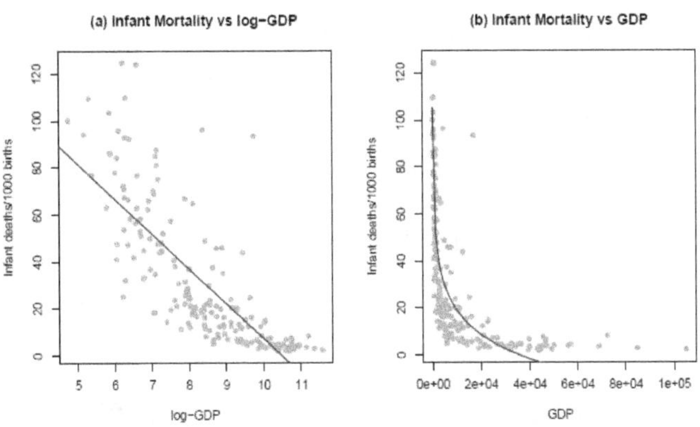

Figure 7.19: Plot of the observations and estimated SLR model (mortality ~ log(GDP)) (top) and implied model (bottom) for the infant mortality data.

```
##      Min      1Q  Median      3Q     Max
## -38.239 -11.609  -2.829   8.122  82.183
##
## Coefficients:
##             Estimate Std. Error t value Pr(>|t|)
## (Intercept) 155.7698     7.2431   21.51   <2e-16
## log(ppgdp)  -14.8617     0.8468  -17.55   <2e-16
##
## Residual standard error: 18.14 on 191 degrees of freedom
## Multiple R-squared:  0.6172, Adjusted R-squared:  0.6152
## F-statistic:   308 on 1 and 191 DF,  p-value: < 2.2e-16
```

```
-14.86*log(2)
```

```
## [1] -10.30017
```

It appears that our model does not fit too well and that there might be some non-constant variance so we should check the diagnostic plots (available in Figure 7.20) before we trust any of those previous interpretations.

```
par(mfrow = c(2,2))
plot(ID1)
```

There appear to be issues with outliers and a long right tail violating the normality assumption as it suggests a clear right skewed residual distribution. There is curvature and non-constant variance in the results as well. There are no influential points, but we are far from happy with this model and will be revisiting this example with the responses also transformed. Remember that the log-transformation of the response can potentially fix non-constant variance, normality, and curvature issues.

Scenario 3: log(y) ~ log(x) model

A final model combines log-transformations of both x and y, combining the interpretations used in the previous two situations. This model is called the log-log model and in some fields is also called the power law model. The power-law model is usually written as $y = \beta 0 x \beta 1 + \varepsilon$, where y is thought to be proportional to x raised to an estimated power of $\beta 1$ (linear if $\beta 1 = 1$ and quadratic if $\beta 1 = 2$). It is one of the modelsthat has been used in Geomorphology to model the shape of glaciated valley elevation profiles (that classic U-shape that comes with glacier-eroded mountain valleys)9. If you ignore the error term, it is possible to estimate the power-law model using our SLR approach. Consider the log-transformation of both sides of this equation starting with the power-law version:

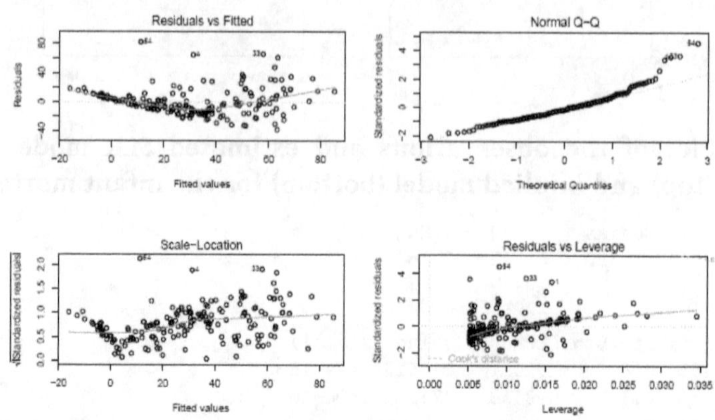

Figure 7.20: Diagnostics plots of the infant mortality model with log(GDP).

- $\log(y) = \log(\beta 0 x^{\beta 1})$,
- $\log(y) = \log(\beta 0) + \log(x^{\beta 1})$. *Based on the rules for logs:* $\log(ab) = \log(a) + \log(b)$.
- $\log(y) = \log(\beta 0) + \beta 1 \log(x)$. *Based on the rules for logs:* $\log(x^b) = b \log(x)$.

So other than $\log(\beta 0)$ in the model, this looks just like our regular SLR model with x and y both log-transformed. The slope coefficient for $\log(x)$ is the power coefficient in the original power law model and determines whether the relationship between the original x and y in $y = \beta 0x^{\beta 1}$ is linear ($y = \beta 0x^1$) or quadratic ($y = \beta 0x^2$) or even quartic ($y = \beta 0x^4$) in some really heavily glacier carved U-shaped valleys. There are some issues with "ignoring the errors" in using SLR to estimate these models [Greenwood and Humphrey, 2002] but it is still a pretty powerful result to be able to estimate the coefficients in ($y = \beta 0x^{\beta 1}$) using SLR.+

We don't typically use the previous ideas to interpret the typical log-log regression model, instead we combine our two previous interpretation techniques to generate our interpretation.

We need to work out the mathematics of doubling x and the changes in y starting with the log(y) ~ log(x)model that we would get out of fitting the SLR with both variables log-transformed:

- $\log(y) = b0 + b1 \log(x)$,
- $y = \exp(b0 + b1 \log(x))$. *Exponentiate both sides.*
- $y = \exp(b0) \exp(b1 \log(x)) = \exp(b0)x^{b1}$. *Rules for exponents and logs, simplifying.*

Now we can consider the impacts of doubling x on y, going from (x, y = exp(b0) xb1) to (2x, y∗) with

- $y* = \exp(b_0)(2x)b_1$,
- $y* = \exp(b_0)2b_1\, xb_1 = 2b_1 \exp(b_0)xb_1 = 2b_1\, y$

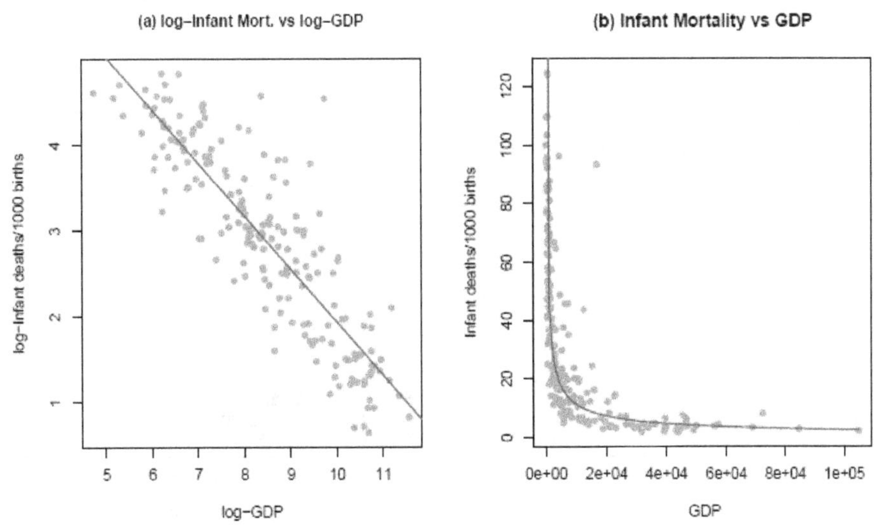

Figure 7.21: Plot of the observations and estimated SLR model log(mortality) ~ log(GDP) (left) and implied model (right) for the infant mortality data.

So doubling x leads to a multiplicative change in the median of y of 2^{b1}.

Let's apply this idea to the GDP and infant mortality data where a log(x) ~ log(y) transformation actually made the resulting relationship look like it might be close to being reasonably modelled with an SLR. The regression line in Figure 7.21 actually looks pretty good on both the estimated log-log scale (panel a) and on the original scale (panel b) as it captures the severe nonlinearity in the relationship between the two variables.

```
ID2 <- lm(log(infantMortality) ~ log(ppgdp), data = UN)
summary(ID2)
```

The estimated regression model is log(infant mortality) = 8.104 - 0.617 · log(GDP). The slope coefficient can be interpreted two ways.

- **On the log-log scale:** For a 1 log-dollar increase in GDP, we estimate, on average, a change of -0.617log(deaths/1000 live births) in infant mortality.
- **On the original scale:** For a doubling of GDP, we expect a 2b1 = 2-0.617 = 0.652 multiplicative change in the estimated median infant mortality. That is a 34.8% decrease in the estimated median infant mortality for each doubling of GDP.

The diagnostics of the log-log SLR model (Figure 7.22) show minimal evidence of violations of assumptions although the tails of the residuals are a little heavy (more spread out than a normal distribution) and there might still be a little pattern remaining in the residuals vs fitted values. There are no influential points to be concerned about in this situation.

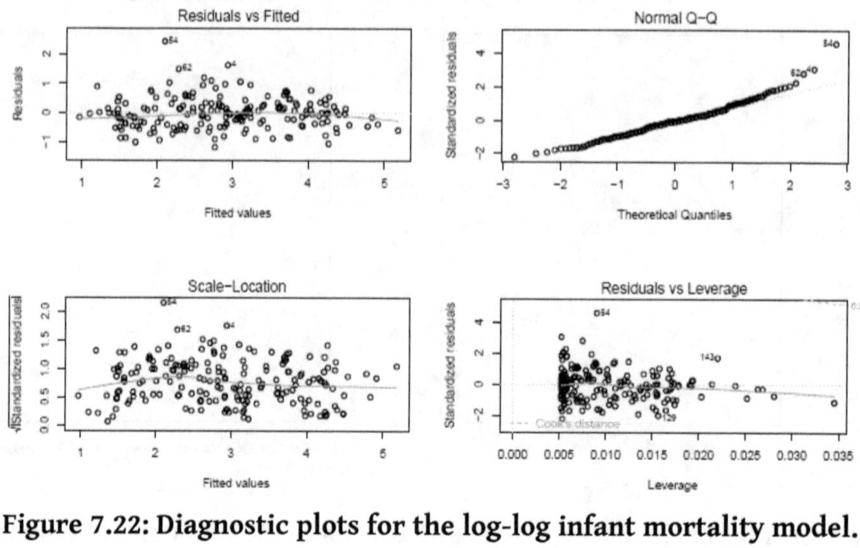

Figure 7.22: Diagnostic plots for the log-log infant mortality model.

While we will not revisit this at all except in the case-studies in Chapter 9, log-transformations can be applied to the response variable in ONE and TWO-WAY ANOVA models when we are concerned about non-constant variance and non-normality issues10. The remaining methods in this chapter return to SLR and assuming that the model is at least reasonable to consider in each situation, possibly after transformation(s). In fact, the methods in Section 7.7 are some of the most sensitive results to violations of the assumptions that we will explore.

7.7 CONFIDENCE INTERVAL FOR THE MEAN AND PREDICTION INTERVALS FOR A NEW OBSERVATION

Figure 7.7 provided a term-plot of the estimated regression line and a shaded area surrounding the estimated regression equation. Those shaded areas are based on connecting the dots on 95% confidence intervals constructed for the true mean y value across all the x-values. To formalize this idea, consider a specific value of x, and call it xv (pronounced x-new11). Then the true mean response for this subpopulation (a subpopulation is all observations we could obtain at x = xv) is given by E(Y) = μv = β0 + β1xv . To estimate the mean response at xv , we plug xv into the estimated regression equation:

- To form the confidence interval, we appeal to our standard formula of estimate $\mp t^*$ $S_{\text{Eestimate}}$. The standard error for the estimated mean at any x-value, denoted SEμ^v , can be calculated as

$$\text{SE}_{\hat{\mu}_\nu} = \sqrt{\text{SE}_{b_1}^2 (x_\nu - \bar{x})^2 + \frac{\hat{\sigma}^2}{n}}$$

where σ^2 is the squared residual standard error. This formula combines the variability in the slope estimate, SEb1 , scaled based on the distance of xv from x¯ and the variability around the regression line, σ^2. Fortunately, R's predict function can be used to provide these results for us and avoid doing this calculation by hand most of the time. The confidence interval for μv , the population mean response at xv , is

$$\hat{\mu}_\nu \mp t^*_{n-2} \text{SE}_{\hat{\mu}_\nu}.$$

In application, these intervals get wider the further we go from the mean of the *x*'s. These have interpretations that are exactly like those for the y-intercept:

For an x-value of xv , we are__% confident that the true mean of y is between LL and UL [units of y].

It is also useful to remember that this interpretation applies individually to every *x* displayed in term-plots.

A second type of interval in this situation takes on a more challenging task – to place an interval on where we think a new observation will fall, called a ***prediction interval*** (PI). This PI will need to be much wider than the CI for the mean since we need to account for both the uncertainty in the mean and the randomness in sampling a new observation from the normal distribution centered at the true mean for *x*v . The interval is centered at the estimated regression line (where else could we center it?) with the estimate denoted as *y*^v to help us see that this interval is for a **new** *y* at this *x*-value. The SE*y*^v incorporates the core of the previous SE calculation and adds in the variability of a new observation in σ^**2**:

$$\mathrm{SE}_{\hat{y}_\nu} = \sqrt{\mathrm{SE}^2_{b_1}(x_\nu - \bar{x})^2 + \frac{\hat{\sigma}^2}{n} + \hat{\sigma}^2} = \sqrt{\mathrm{SE}^2_{\hat{\mu}_\nu} + \hat{\sigma}^2}$$

The__% PI is calculated as

$$\hat{y}_\nu \mp t^*_{n-2}\mathrm{SE}_{\hat{y}_\nu}$$

and interpreted as:

We are__% sure that a new observation at ***x*v** will be between **LL** and **UL [*units of* y]**.

The formula also helps us to see that since $\mathrm{SE}_{\hat{y}_\nu} > \mathrm{SE}_{\hat{\mu}_\nu}$, **the PI will always be wider than the CI.**

As in confidence intervals, we assume that a 95% PI "succeeds" – now when it succeeds it contains the new observation – in 95% of applications of the methods and fails the other 5% of the time. Remember that for any interval estimate, the true value is either in the interval or it is not and our confidence level essentially sets our failure rate! Because PIs push into the tails of the assumed distribution of the responses, these methods are very sensitive to violations of assumptions. We should not use these if there are any concerns about violations of assumptions as they will not work as advertised (at the nominal (specified) level).

There are two ways to explore CIs for the mean and PIs for a new observation. The first is to focus on a specific x-value of interest. The second is to plot the results for all x's. To do both of these, but especially to make plots, we want to learn to use the predict function. It can either produce the estimate for a particular xv and the SEμ^v or we can get it to directly calculate the CI and PI. The first way to use it is predict(MODELNAME, se.fit = T) which will provide fitted values and SEμ^v for all observed x's. We can then use the SEμ^v to calculate SEy^v and form our own PIs. If you want CIs, run predict(MODELNAME, interval = "confidence"); if you want PIs, run predict(MODELNAME, interval = "prediction"). If you want to do prediction at an x-value that was not in the original observations, add the option new data = tibble (XVARIABLENAME_FROM_ORIGINAL_MODEL = Xnu) to the predict function call. Some examples of using the predict function follow. For example, it might be interesting to use the regression model to find a 95% CI and PI for the Beers vs BAC study for a student who would consume 8 beers. Four different applications of the predict function follow. Note that lwr and upr in the output depend on what we requested. The first use of predict just returns the estimated mean for 8 beers:

```
m1 <- lm(BAC ~ Beers, data = BB)
predict(m1, newdata = tibble(Beers = 8))
```

```
##         1
## 0.1310095
```

By turning on the se.fit = T option, we also get the SE for the confidence interval and degrees of freedom. Note that elements returned are labelled as $fit, $se.fit, etc. and provide some of the information to calculate CIs or PIs "by hand".

```
predict(m1, newdata = tibble(Beers = 8), se.fit = T)
```

```
## $fit
##         1
## 0.1310095
##
## $se.fit
## [1] 0.009204354
##
## $df
## [1] 14
##
## $residual.scale
## [1] 0.02044095
```

Instead of using the components of the intervals to make them, we can also directly request the CI or PI using the interval = ... option, as in the following two lines of code.

```
predict(m1, newdata = tibble(Beers = 8), interval = "confidence")
```

```
##         fit       lwr       upr
## 1 0.1310095 0.1112681 0.1507509
```

```
predict(m1, newdata = tibble(Beers = 8), interval = "prediction")
```

```
##         fit        lwr       upr
## 1 0.1310095 0.08292834 0.1790906
```

Based on these results, we are 95% confident that the true mean *BAC* for 8 beers consumed is between0.111 and 0.15 grams of alcohol per dL of blood. For a new student drinking 8 beers, we are 95% sure that the observed BAC will be between

0.083 and 0.179 g/dL. You can see from these results that the PI is much wider than the CI – it has to capture a new individual's results 95% of the time which is much harder than trying to capture the true mean at 8 beers consumed. For completeness, we should do these same calculations "by hand". The predict(..., se.fit = T) output provides almost all the pieces we need to calculate the CI and PI. The $fit is the estimate = μ^v = 0.131, the $se.fit is the SE for the estimate of the mean = SEμ^v = 0.0092 , $df is n - 2 = 16 - 2 = 14, and $residual.scale is σ^ = 0.02044. So we just need the t^* multiplier for 95% confidence and 14 *df*:

```
qt(0.975, df = 14) #t* multiplier for 95% CI or 95% PI
```

[1] 2.144787

The 95% CI for the true mean at **xv** = **8** is then:

```
0.131 + c(-1,1)*2.1448*0.0092
```

[1] 0.1112678 0.1507322

which matches the previous output quite well.

The 95% PI requires the calculation of $\sqrt{\text{SE}^2_{\hat{\mu}_v} + \hat{\sigma}^2} = \sqrt{(0.0092)^2 + (0.02044)^2} = 0.0224$.

```
sqrt(0.0092^2 + 0.02044^2)
```

[1] 0.02241503

The 95% PI at **xv** = **8** is

```
0.131 + c(-1,1)*2.1448*0.0224
```

[1] 0.08295648 0.17904352

These calculations are "fun" and informative but displaying these results for all *x*-values is a bit more informative about the performance of the two types of intervals and for results we might expect in this application. The calculations we just performed provide endpoints of both intervals at Beers = 8. To maket his plot, we need to create a sequence of *Beers* values to get other results for, say from 0 to 10 beers, using the seq function. The seq function requires three arguments, that the endpoints (from and to) are defined and the length.out, which defines the resolution of the grid of equally spaced points to create. Here, length.out = 30 provides 30 points evenly spaced between 0 and 10 and is more than enough to make the confidence and prediction intervals from 0 to 10 *Beers*.

```
# Creates vector of predictor values from 0 to 10
beerf <- seq(from = 0, to = 10, length.out = 30)
head(beerf, 6)
```

```
## [1] 0.0000000 0.3448276 0.6896552 1.0344828 1.3793103 1.7241379
```

```
tail(beerf, 6)
```

```
## [1]  8.275862  8.620690  8.965517  9.310345  9.655172 10.000000
```

Now we can call the predict function at the values stored in beerf to get the CIs across that range of Beersvalues:

```
BBCI <- as_tibble(predict(m1, newdata = tibble(Beers = beerf), interval = "confidence"))
head(BBCI)
```

```
## # A tibble: 6 x 3
##          fit      lwr    upr
##        <dbl>    <dbl>  <dbl>
## 1 -0.0127    -0.0398  0.0144
## 2 -0.00651   -0.0320  0.0190
## 3 -0.000312 -0.0242  0.0236
## 4  0.00588   -0.0165  0.0282
## 5  0.0121    -0.00873 0.0329
## 6  0.0183    -0.00105 0.0376
```

And the PIs:

```
BBPI <- as_tibble(predict(m1, newdata = tibble(Beers = beerf), interval = "prediction"))
head(BBPI)
```

```
## # A tibble: 6 x 3
##          fit     lwr    upr
##        <dbl>   <dbl>  <dbl>
## 1 -0.0127    -0.0642 0.0388
## 2 -0.00651   -0.0572 0.0442
## 3 -0.000312 -0.0502 0.0496
## 4  0.00588   -0.0433 0.0551
## 5  0.0121    -0.0365 0.0606
## 6  0.0183    -0.0296 0.0662
```

To visualize these results as shown in Figure 7.23, we need to work to combine some of the previous results into a common tibble, called modelresB, using the bind_cols function that allows multiple columns to be put together. Because some of the names are the same in the BBCI and BBPI objects and were awkwardly given unique names, there is an additional step to rename the columns using the rename function. The rename function changes the name to what is provided before the = for the column identified after the = (think of it like mutate except that it does not create a new variable). The layers in the plot start with adding a line for the fitted values (solid, using geom_line) based on the information in modelresB. We also introduce the geom_ribbon explicitly for the first time[12] to plot our confidence and prediction intervals. It allows plotting of a region (and its edges) defined by ymin and ymax across the values provided to x. I also wanted to include the original observations, but they are stored in a different tibble (BB), so the geom_point needs to be explicitly told to use a different data set for its contribution to the plot with data = BB along with its own local aesthetic with x and y selections from the original variables.

```
# Patch the beerf vector, fits (just one version), and intervals from BBCI and
# BBPI together with bind_cols:
modelresB <- bind_cols(beerf = tibble(beerf), BBCI, BBPI %>% select(-fit))

# Rename CI and PI limits to have more explicit column names:
modelresB <- modelresB %>% rename(lwr_CI = lwr...3, upr_CI = upr...4,
                                  lwr_PI = lwr...5, upr_PI = upr...6)
```

More importantly, note that the CI in Figure 7.23 clearly shows widening as we move further away from the mean of the x's to the edges of the observed x-values. This reflects a decrease in knowledge of the true mean as we move away from the mean of the x's. The PI also is widening slightly but not as clearly in this situation. The difference in widths in the two types of intervals becomes extremely clear when they are displayed together, with the PI much wider than the CI for any x-value.

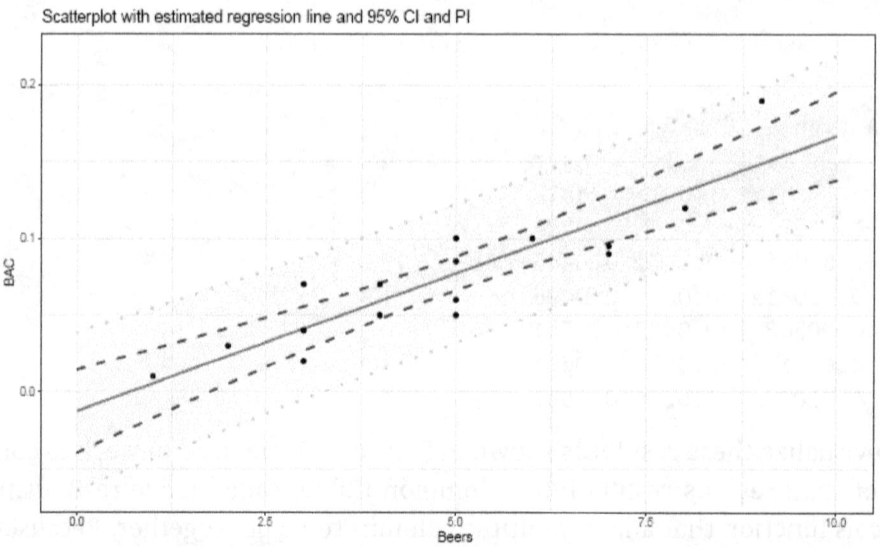

Figure 7.23: Estimated SLR for BAC data with fitted values (solid line), 95% confidence (darker, dashed lines), and 95% prediction (lighter, dotted lines) intervals.

```
modelresB %>% ggplot() +
  geom_line(aes(x = beerf, y = fit), lwd = 1) +
  geom_ribbon(aes(x = beerf, ymin = lwr_CI, ymax = upr_CI), alpha = .4,
              fill = "beige", color = "darkred", lty = 2, lwd = 1) +
  geom_ribbon(aes(x = beerf, ymin = lwr_PI, ymax = upr_PI), alpha = .1,
              fill = "gray80", color = "grey", lty = 3, lwd = 1.5) +
  geom_point(data = BB, mapping = aes(x = Beers, y = BAC)) +
  labs(y = "BAC", x = "Beers",
       title = "Scatterplot with estimated regression line and 95% CI and PI") +
  theme_bw()
```

Similarly, the 95% CI and PIs for the Bozeman yearly average maximum temperatures in Figure 7.24 provide interesting information on the uncertainty in the estimated mean temperature over time. It is also interesting to explore how many of the observations fall within the 95% prediction intervals. The PIs are for new observations, but you can see how the PIs that were constructed to contain almost all the observations in the original data set but not all of them. In fact, only 2 of the 109 observations (1.8%) fall outside the 95% PIs. Since the PI needs to be concerned with unobserved new observations it makes sense that it might contain more than 95% of the observations used to make it.

```
temp1 <- lm(meanmax ~ Year, data = bozemantemps)
Yearf <- seq(from = 1901, to = 2014, length.out = 75)

TCI <- as_tibble(predict(temp1, newdata = tibble(Year = Yearf), interval = "confidence"))
```

```
TPI <- as_tibble(predict(temp1, newdata = tibble(Year = Yearf), interval = "prediction"))

# Patch the Yearf vector, fits (just one version), and intervals from TCI and
# TPI together with bind_cols:
modelresT <- bind_cols(Yearf = tibble(Yearf), TCI, TPI %>% select(-fit))

# Rename CI and PI limits to have more explicit column names:
modelresT <- modelresT %>% rename(lwr_CI = lwr...3, upr_CI = upr...4,
                                  lwr_PI = lwr...5, upr_PI = upr...6)

modelresT %>% ggplot() +
  geom_line(aes(x = Yearf, y = fit), lwd = 1) +
  geom_ribbon(aes(x = Yearf, ymin = lwr_CI, ymax = upr_CI), alpha = .4,
              fill = "beige", color = "darkred", lty = 2, lwd = 1) +
  geom_ribbon(aes(x = Yearf, ymin = lwr_PI, ymax = upr_PI), alpha = .1,
              fill = "gray80", color = "grey", lty = 3, lwd = 1.5) +
  geom_point(data = bozemantemps, mapping = aes(x = Year, y = meanmax)) +
  labs(y = "degrees F", x = "Year",
       title = "Scatterplot with estimated regression line and 95% CI and PI") +
  theme_bw()
```

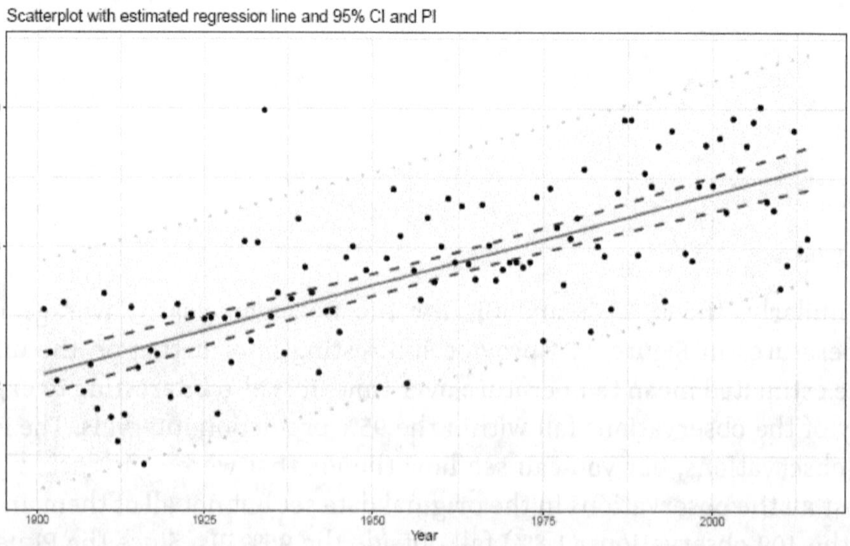

Figure 7.24: Estimated SLR for Bozeman temperature data with 95% confidence (dashed lines) and 95% prediction (lighter, dotted lines) intervals.

We can also use these same methods to do a prediction for the year after the data set ended, 2015, and in 2050:

```
predict(temp1, newdata = tibble(Year = 2015), interval = "confidence")
```

```
##        fit     lwr      upr
## 1 58.31967 57.7019 58.93744
```

```
predict(temp1, newdata = tibble(Year = 2015), interval = "prediction")
```

```
##        fit      lwr      upr
## 1 58.31967 55.04146 61.59787
```

```
predict(temp1, newdata = tibble(Year = 2050), interval = "confidence")
```

```
##        fit      lwr      upr
## 1 60.15514 59.23631 61.07397
```

```
predict(temp1, newdata = tibble(Year = 2050), interval = "prediction")
```

```
##        fit      lwr      upr
## 1 60.15514 56.80712 63.50316
```

These results tell us that we are 95% confident that the true mean yearly average maximum temperature in 2015 is (I guess "was") between 55.04°F and 61.6°F . And we are 95% sure that the observed yearly average maximum temperature in 2015 will be (I guess "would have been") between 59.2°F and 61.1°F . Obviously,2015 has occurred, but since the data were not published when the data set was downloaded in July 2016, we can probably best treat 2015 as a potential "future" observation. The results for 2050 are clearly for the future mean and a new observation13 in 2050. Note that up to 2014, no values of this response had beenobserved above 60°F and the predicted mean in 2050 is over 60°F if the trend persists. It is easy to criticize the use of this model for 2050 because of its extreme amount of extrapolation.

7.8 CONCISE SUMMARY

In this chapter, we raised our *SLR modelling* to a new level, considering inference techniques for relationships between two quantitative variables. The next chapter will build on these same techniques but add in additional explanatory variables for what is called multiple linear regression *(MLR) modelling*. For example, in the *Beers* vs *BAC* study, it would have been useful to control for the weight of the subjects since people of different sizes metabolize alcohol at different rates and body size might explain some of the variability in BAC. We still would want to study the effects of beer consumption but also would be able to control for the differences in subject's weights. Or if they had studied both male and female students, we might need to change the slope or intercept based on gender, allowing the relationship between Beers and BAC to change between these groups. That will also be handled using MLR techniques but result in two simple linear regression equations – one for each group.

In this chapter you learned how to interpret SLR models. The next chapter will feel like it is completely new initially but it actually contains very little new material, just more complicated models that use the same concepts. There will be a couple of new issues to consider for MLR and we'll need to learn how to work with categorical variables in a regression setting – but we actually fit linear models with categorical variables in Chapters 2, 3, and 4 so that isn't actually completely new either.

SLR is a simple (thus its name) tool for analyzing the relationship between two quantitative variables. It contains assumptions about the estimated regression line being reasonable and about the distribution of the responses around that line to do inferences for the population regression line. Our diagnostic plots help us to carefully assess those assumptions. If we cannot trust the assumptions, then the estimated line and any inferences for the population are un-trustworthy. Transformations can fix things so that we can use SLR to fit regression models. Transformations can complicate the interpretations on the original, untransformed scale but have minimal impact on the interpretations on the transformed scale. It is important to be careful with the units of the variables, especially when dealing with transformations, as this can lead to big changes in the results depending on which scale (original or transformed) the results are being interpreted on.

7.9 SUMMARY OF IMPORTANT R CODE

The main components of the R code used in this chapter follow with the components to modify in lighter and/or ALL CAPS text where y is a response variable, x is an explanatory variable, and the data are in DATASETNAME.

- DATASETNAME %>% ggplot(mapping = aes(x = x, y = y)) + geom_point() + geom_smooth(method = "lm")
  - Provides a scatter plot with a regression line.
  - Add + **geom_smooth()** to add a smoothing line to help detect nonlinear relationships.
- MODELNAME <- lm(y ~ x, data = DATASETNAME)
  - Estimates a regression model using least squares.
- summary(MODELNAME)
  - Provides parameter estimates and R-squared (used heavily in Chapter 8 as well).
- par(mfrow = c(2, 2)); plot(MODELNAME)
  - Provides four regression diagnostic plots in one plot.
- confint(MODELNAME, level = 0.95)
  - Provides 95% confidence intervals for the regression model coefficients.
  - Change level if you want other confidence levels.
- plot(allEffects(MODELNAME))
  - Requires the effects package.
  - Provides a term-plot of the estimated regression line with 95% confidence interval for the mean.
- DATASETNAME <- DATASETNAME %>% mutate(log.y = log(y))
  - Creates a transformed variable called log.y – change this to be more specific to your "y" or "x".
- predict(MODELNAME, se.fit = T)
  - Provides fitted values for all observed x's with SEs for the mean.
- predict(MODELNAME, newdata = tibble(x = XNEW), interval = "confidence")
  - Provides fitted value for a specific *x* (XNEW) with CI for the mean. Replace x with name of explanatory variable.
- predict(MODELNAME, newdata = tibble(x = XNEW), interval = "prediction")
  - Provides fitted value for a specific *x* (XNEW) with PI for a new observation. Replace x with name of explanatory variable.

- qt(0.975, df = n - 2)
    - Gets the t^* multiplier for making a 95% confidence or prediction interval with n - 2 replaced by the sample size – 2.

7.10 PRACTICE PROBLEMS

7.1 Treadmill data analysis We will continue with the treadmill data set introduced in Chapter 1 and the SLR fit in the practice problems in Chapter 6. The following code will get you back to where we stopped at the end of Chapter 6:

```
treadmill <- read_csv("http://www.math.montana.edu/courses/s217/documents/treadmill.csv")
treadmill %>% ggplot(mapping = aes(x = RunTime, y = TreadMillOx)) +
  geom_point(aes(color = Age)) +
  geom_smooth(method = "lm") +
  geom_smooth(se = F, lty = 2, col = "red") +
  theme_bw()
tm <- lm(TreadMillOx ~ RunTime, data = treadmill)
summary(tm1)
```

7.1.1. Use the output to test for a linear relationship between treadmill oxygen and run time, writing out all 6+ steps of the hypothesis test. Make sure to address scope of inference and interpret the p-value.

7.1.2. Form and interpret a 95% confidence interval for the slope coefficient "by hand" using the provided multiplier:

```
qt(0.975, df = 29)
```

```
## [1] 2.04523
```

7.1.3. Use the confint function to find a similar confidence interval, checking your previous calculation.

7.1.4. Use the predict function to find fitted values, 95% confidence, and 95% prediction intervals for run times of 11 and 16 minutes.

7.1.5. Interpret the CI and PI for the 11 minute run time.

7.1.6. Compare the width of either set of CIs and PIs – why are they different? For the two different predictions, why are the intervals wider for 16 minutes than for 11 minutes?

7.1.7. The Residuals vs Fitted plot considered in Chapter 6 should have suggested slight non-constant variance and maybe a little missed nonlinearity. Perform a log-transformation of the treadmill oxygen response variable and re-fit the SLR model. Remake the diagnostic plots and discuss whether the transformation changed any of them.

7.1.8. Summarize the log(y) ~ x model and interpret the slope coefficient on the transformed and original scales, regardless of the answer to the previous question.

CHAPTER-8

MULTIPLE LINEAR REGRESSION

INTRODUCTION: AN OVERVIEW

Multiple linear regression is an advanced statistical technique that extends the principles of simple linear regression to model the complex relationships between a dependent variable and multiple independent variables simultaneously. Unlike its simpler counterpart, which considers only one predictor, multiple linear regression allows for the exploration of how several predictors collectively influence the response variable. This method is widely employed in various disciplines, including economics, social sciences, and natural sciences, where understanding the intricate interplay among multiple factors is crucial for making accurate predictions and informed decisions. By capturing the multidimensional nature of real-world phenomena, multiple linear regression enables researchers and analysts to uncover nuanced patterns, quantify the individual contributions of each predictor, and assess the overall predictive power of the model.

8.1 GOING FROM SLR TO MLR

In many situations, especially in observational studies, it is unlikely that the system is simple enough to be characterized by a single predictor variable. In experiments, if we randomly assign levels of a predictor variable we can assume that the impacts of other variables cancel out as a direct result of the random assignment. But it is possible even in these experimental situations that we can "improve" our model for the response variable by adding additional predictor variables that explain additional variation in the responses, reducing the amount of unexplained variation. This can allow more precise inferences to be generated from the model. As mentioned previously, it might be useful to know the sex or weight of the subjects in the Beers vs BAC study to account for more of the variation in the responses - this idea motivates our final topic: multiple linear regression (MLR) models. In observational studies, we can think of a suite of characteristics of observations that might be related to a response variable. For example, consider a study of yearly salaries and variables that might explain the amount people get paid. We might be most interested in seeing how incomes change based on age, but it would be hard to ignore potential differences based on sex and education level. Trying to explain incomes would likely require more than one predictor variable and we wouldn't be able to explain all the variability in the responses just based on gender and education level, but a model using those variables might still provide some useful information about each component and about age impacts on income after we adjust (control) for sex and education. The extension to MLR allows us to incorporate multiple predictors into a regression model. Geometrically, this is a way of relating many different dimensions (number of x's) to what happened in a single response variable (one dimension).

We start with the same model as in SLR except now we allow *K* different *x*'s:

$$y_i = \beta_0 + \beta_1 x_{1i} + \beta_2 x_{2i} + \ldots + \beta_K x_{Ki} + \varepsilon_i$$

Note that if K = 1, we are back to SLR. In the MLR model, there are K predictors and we still have a y-intercept. The MLR model carries the same assumptions as an SLR model with a couple of slight tweaks specific to MLR (see Section 8.2 for the details on the changes to the validity conditions).

We are able to use the least squares criterion for estimating the regression coefficients in MLR, but the mathematics are beyond the scope of this course. The lm function takes care of finding the least squarescoefficients using a very sophisticated algorithm1. The estimated regression equation it returns is:

$$\widehat{y}_i = b_0 + b_1 x_{1i} + b_2 x_{2i} + \ldots + b_K x_{Ki}$$

where each *bk* estimates its corresponding parameter *βk* .

An example of snow depths at some high elevation locations in Montana on a day in April provides a nice motivation for these methods. A random sample of *n* = 25 Montana locations (from the population of *N* = 85 at the time) were obtained from the Natural Resources Conversation Service's website (http://www.wcc.nrcs.usda.gov/snotel/Montana/montana.html) a few years ago. Information on the snow depth (Snow.Depth) in inches, daily Minimum and Maximum Temperatures (Min.Temp and Max.Temp) in °F and elevation of the site (Elevation) in feet. A snow science researcher (or spring back-country skier) might be interested in understanding *Snow depth* as a function of *Minimum Temperature, Maximum Temperature*, and *Elevation*. One might assume that colder and higher places will have more snow, but using just one of the predictor variables might leave out some important predictive information. The following code loads the data set and makes the scatterplot matrix (Figure 8.1) to allow some preliminary assessment of the pair wise relationships.

```
snotel_s <- read_csv("http://www.math.montana.edu/courses/s217/documents/snotel_s.csv")
```

```
library(GGally)
# Reorder columns slightly and only plot quantitative variables using "columns = ..."
snotel_s %>% ggpairs(columns = c(4:6,3)) +
  theme_bw()
```

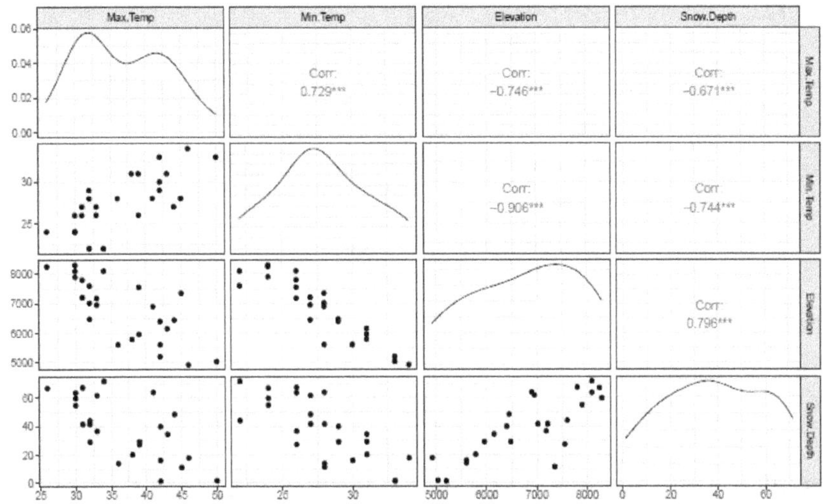

Figure 8.1: Scatterplot matrix of data from a sample of SNOTEL sites in April on four variables.

It appears that there are many strong linear relationships between the variables, with *Elevation* and *Snow Depth* having the largest magnitude, $\boldsymbol{r}$ = 0.80. Higher temperatures seem to be associated with less snow –not a big surprise so far! There might be an outlier at an elevation of 7400 feet and a snow depth below 10 inches that we should explore further.

A new issue arises in attempting to build MLR models called multicollinearity. Again, it is a not surprise that temperature and elevation are correlated but that creates a problem if we try to put them both into a model to explain snow depth. Is it the elevation, temperature, or the combination of both that matters for getting and retaining more snow? Correlation between predictor variables is called multicollinearity and makes estimation and interpretation of MLR models more complicated than in SLR. Sectiondeals with this issue directly and discusses methods for detecting its presence. For now, remember that in MLR this issue sometimes makes it difficult to disentangle the impacts of different predictor variables on the response when the predictors share information – when they are correlated.

To get familiar with this example, we can start with fitting some potential SLR models and plotting the estimated models. Figure 8.2 contains the result for the

SLR using *Elevation* and results for two temperature based models are in Figures 8.3 and 8.4. *Snow Depth* is selected as the obvious response variable both due to skier interest and potential scientific causation (snow can't change elevation but elevation could be the driver of snow deposition and retention).

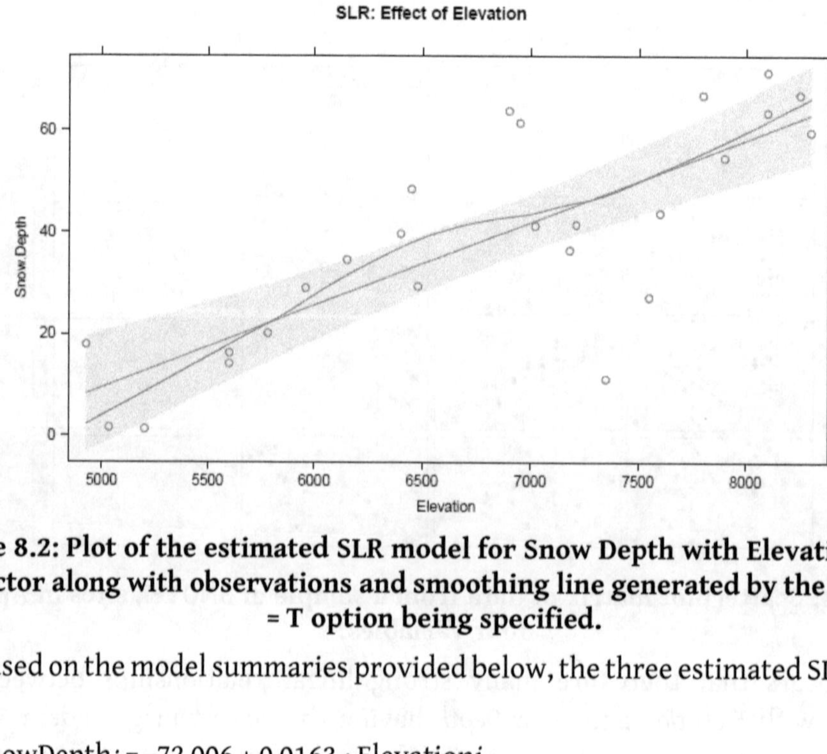

Figure 8.2: Plot of the estimated SLR model for Snow Depth with Elevation as the predictor along with observations and smoothing line generated by the residuals = T option being specified.

Based on the model summaries provided below, the three estimated SLR models are:

$\widehat{\text{SnowDepth}}_i$ = -72.006 + 0.0163 · Elevation$_i$,

$\widehat{\text{SnowDepth}}_i$ = 174.096 - 4.884 · MinTemp$_i$, and

$\widehat{\text{SnowDepth}}_i$ = 122.672 - 2.284 · MaxTemp$_i$.

The term-plots of the estimated models reinforce our expected results, showing a positive change in Snow Depth for higher Elevations and negative impacts for increasing temperatures on Snow Depth. These plots are made across the observed range2 of the predictor variable and help us to get a sense of the total impacts of predictors. For example, for elevation in Figure 8.2, the smallest observed value was 4925 feet and the largest was 8300 feet. The regression line goes from estimating a mean snow depth of 8 inches to 63 inches. That gives you some practical idea of the size of the estimated Snow Depth change for the changes in Elevation observed in the data. Putting this together, we can say that there was around a 55 inch change in predicted snow depths for a close to 3400 foot increase in elevation. This helps make the slope coefficient of 0.0163 in the model more easily understood.

Remember that in SLR, the range of *x* matters just as much as the units of *x* in determining the practical importance and size of the slope coefficient. A value of 0.0163 looks small but is actually at the heart of a pretty interesting model for predicting snow depth. A one foot change of elevation is "tiny" here relative to changes in the response so the slope coefficient can be small and still amount to big changes in the predicted response across the range of values of x. If the *Elevation* had been recorded in thousands of feet, then the slope would have been estimated to be $0.0163 * 1000 = 16.3$ inches change in mean *Snow Depth* for a 1000foot increase in elevation.

The plots of the two estimated temperature models in Figures 8.3 and 8.4 suggest a similar change in the responses over the range of observed temperatures. Those predictors range from $22^{\circ}F$ to $34^{\circ}F$ (minimum temperature) and from $26^{\circ}F$ to $50^{\circ}F$ (maximum temperature). This tells us a $1^{\circ}F$ increase in either temperature is a greater proportion of the observed range of each predictor than a 1 unit (foot) increase in elevation, so the two temperature variables will generate larger apparent magnitudes of slope coefficients. But having large slope coefficients is no guarantee of a good model – in fact, the elevation model has the highest R^2 value of these three models even though its slope coefficient looks tiny compared to the other models.

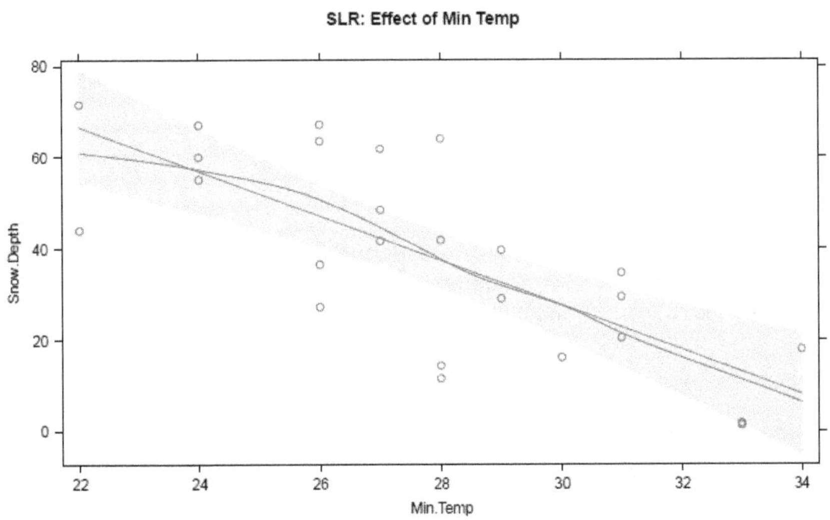

Figure 8.3: Plot of the estimated SLR model using Min Temp as predictor.

```
m1 <- lm(Snow.Depth ~ Elevation, data = snotel_s)
m2 <- lm(Snow.Depth ~ Min.Temp, data = snotel_s)
m3 <- lm(Snow.Depth ~ Max.Temp, data = snotel_s)
library(effects)
plot(allEffects(m1, residuals = T), main = "SLR: Effect of Elevation")
plot(allEffects(m2, residuals = T), main = "SLR: Effect of Min Temp")
plot(allEffects(m3, residuals = T), main = "SLR: Effect of Max Temp")
```

```
summary(m1)
```

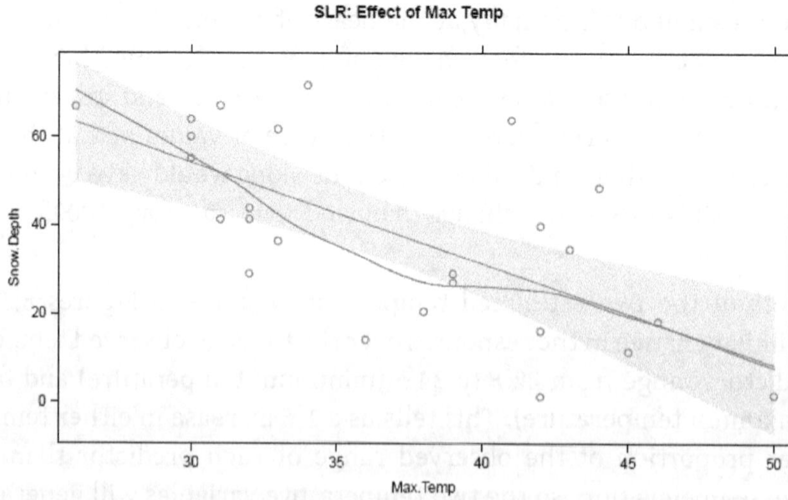

Figure 8.4: Plot of the estimated SLR model using Max Temp as predictor.

```
## Call:
## lm(formula = Snow.Depth ~ Elevation, data = snotel_s)
##
## Residuals:
##      Min      1Q  Median      3Q     Max
## -36.416  -5.135  -1.767   7.645  23.508
##
## Coefficients:
##                Estimate Std. Error t value Pr(>|t|)
## (Intercept) -72.005873  17.712927  -4.065 0.000478
## Elevation     0.016275   0.002579   6.311 1.93e-06
##
## Residual standard error: 13.27 on 23 degrees of freedom
## Multiple R-squared:  0.634,  Adjusted R-squared:  0.618
## F-statistic: 39.83 on 1 and 23 DF,  p-value: 1.933e-06
```

```
summary(m2)
```

```
##
## Call:
## lm(formula = Snow.Depth ~ Min.Temp, data = snotel_s)
##
## Residuals:
##     Min      1Q  Median      3Q     Max
## -26.156 -11.238   2.810   9.846  26.444
##
## Coefficients:
##             Estimate Std. Error t value Pr(>|t|)
## (Intercept) 174.0963    25.5628   6.811 6.04e-07
## Min.Temp     -4.8836     0.9148  -5.339 2.02e-05
##
## Residual standard error: 14.65 on 23 degrees of freedom
## Multiple R-squared:  0.5534, Adjusted R-squared:  0.534
## F-statistic:  28.5 on 1 and 23 DF,  p-value: 2.022e-05
```

```
summary(m3)
```

```
##
## Call:
## lm(formula = Snow.Depth ~ Max.Temp, data = snotel_s)
##
## Residuals:
##      Min       1Q  Median      3Q     Max
## -26.447 -10.367  -4.394  10.042  34.774
##
## Coefficients:
##             Estimate Std. Error t value Pr(>|t|)
## (Intercept) 122.6723    19.6380   6.247 2.25e-06
## Max.Temp     -2.2840     0.5257  -4.345 0.000238
##
## Residual standard error: 16.25 on 23 degrees of freedom
## Multiple R-squared:  0.4508, Adjusted R-squared:  0.4269
## F-statistic: 18.88 on 1 and 23 DF,  p-value: 0.0002385
```

Since all three variables look like they are potentially useful in predicting snow depth, we want to consider if an MLR model might explain more of the variability in *Snow Depth*. To fit an MLR model, we use the same general format as in previous topics but with adding "+" between any additional predictors[3] we want to add to the model, y ~ x1 + x2 + ... + xk:

```
m4 <- lm(Snow.Depth ~ Elevation + Min.Temp + Max.Temp, data = snotel_s)
summary(m4)
```

```
##
## Call:
## lm(formula = Snow.Depth ~ Elevation + Min.Temp + Max.Temp, data = snotel_s)
##
## Residuals:
##     Min      1Q  Median      3Q     Max
## -29.508  -7.679  -3.139   9.627  26.394
##
## Coefficients:
##               Estimate Std. Error t value Pr(>|t|)
## (Intercept) -10.506529  99.616286  -0.105   0.9170
## Elevation     0.012332   0.006536   1.887   0.0731
## Min.Temp     -0.504970   2.042614  -0.247   0.8071
## Max.Temp     -0.561892   0.673219  -0.835   0.4133
##
## Residual standard error: 13.6 on 21 degrees of freedom
## Multiple R-squared:  0.6485, Adjusted R-squared:  0.5983
## F-statistic: 12.91 on 3 and 21 DF,  p-value: 5.328e-05
```

```
plot(allEffects(m4, residuals = T), main = "MLR model with Elev, Min, & Max Temps")
```

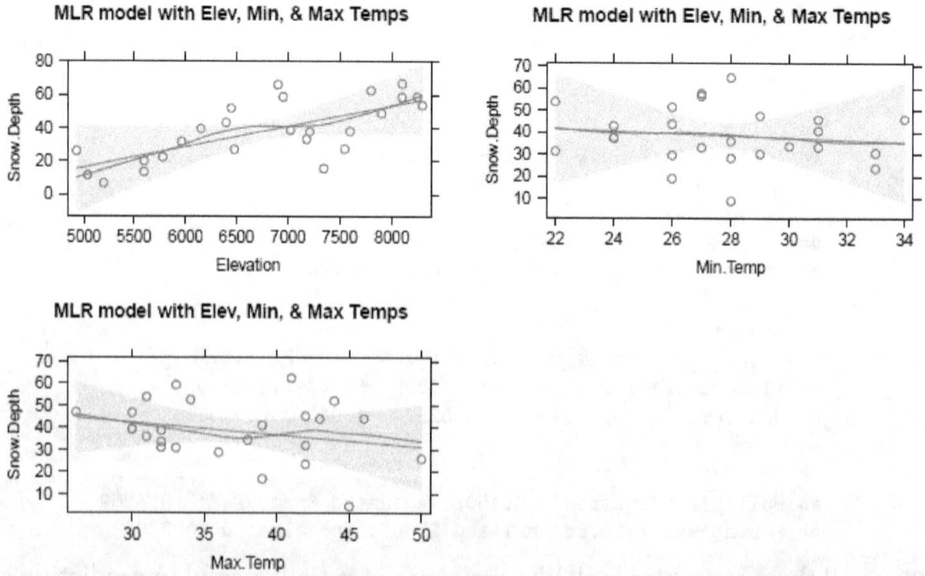

Figure 8.5: Term-plots for the MLR for Snow Depth based on Elevation, Min Temp and Max Temp. Compare this plot that comes from one MLR model to Figures 8.2, 8.3, and 8.4 for comparable SLR models. Note the points in these panels are the partial residuals that are generated after controlling for the other two of the three variables as explained below.

Based on the output, the estimated MLR model is

$$\widehat{\text{SnowDepth}}_i = -10.51 + 0.0123 \cdot \text{Elevation}_i - 0.505 \cdot \text{MinTemp}_i - 0.562 \cdot \text{MaxTemp}_i$$

The direction of the estimated slope coefficients were similar but they all changed in magnitude as compared to the respective SLRs, as seen in the estimated term-plots from the MLR model in Figure 8.5.

There are two ways to think about the changes from individual SLR slope coefficients to the similar MLR results here.

- Each term in the MLR is the result for estimating each slope after controlling for the other two variables (and we will always use this sort of interpretation any time we interpret MLR effects). For the *Elevation* slope, we would say that the slope coefficient is "corrected for" or "adjusted for" the variability that is explained by the temperature variables in the model.
- Because of multicollinearity in the predictors, the variables might share information that is useful for explaining the variability in the response variable, so the slope coefficients of each predictor get perturbed because the model cannot separate their effects on the response. This issue disappears when the predictors are uncorrelated or even just minimally correlated.

There are some ramifications of multicollinearity in MLR:

- Adding variables to a model might lead to almost no improvement in the overall variability explained by the model.
- Adding variables to a model can cause slope coefficients to change signs as well as magnitudes.
- Adding variables to a model can lead to inflated standard errors for some or all of the coefficients (this is less obvious but is related to the shared information in predictors making it less clear what slope coefficient to use for each variable, so more uncertainty in their estimation).
- In extreme cases of multicollinearity, it may even be impossible to obtain some or any coefficient estimates.

These seem like pretty serious issues and they are but there are many, many situations where we proceed with MLR even in the presence of potentially correlated predictors. It is likely that you have heard or read about inferences from models that are dealing with this issue – for example, medical studies often report the increased risk of death from some behaviour or trait after controlling for gender, age, health status, etc. In many research articles, it is becoming common practice to report the slope for a variable that is of most interest with it in the model alone (SLR) and in models after adjusting for the other variables that are expected to matter. The "adjusted for other variables" results are built with MLR or related multiple-predictor models like MLR.

8.2 VALIDITY CONDITIONS IN MLR

But before we get too excited about any results, we should always assess our validity conditions. For MLR, they are similar to those for SLR:

- Quantitative variables condition:
  - The response and all predictors need to be quantitative variables. This condition is relaxed to allow a categorical predictor in two ways in Sections 8.9 and 8.11.
- Independence of observations:
  - This assumption is about the responses – we must assume that they were collected in a fashion so that they can be assumed to be independent. This implies that we also have independent random errors.
  - This is not an assumption about the predictor variables!
- Linearity of relationship (NEW VERSION FOR MLR!):
  - Linearity is assumed between the response variable and **each** explanatory variable (y and each x).

- We can check this three ways:
- Make plots of the response versus each explanatory variable:
- Only visual evidence of a curving relationship is a problem here. Transformations of individual explanatory variables or the response are possible. It is possible to not find a problem in this plot that becomes more obvious when we account for variability that is explained by other variables in the partial residuals.

⦿ Examine the Residuals vs Fitted plot:

- When using MLR, curves in the residuals vs. fitted values suggest a missed curving relationship with at least one predictor variable, but it will not be specific as to which one is non-linear. Revisit the scatterplots to identify the source of the issue.

⦿ Examine partial residuals and smoothing line in term-plots.

- Turning on the residuals = T option in the effects plot allows direct assessment of residuals vs each predictor after accounting for others. Look for clear patterns in the partial residuals[4] that the smoothing line is also following for potential issues with the linearity assumption.

⦿ Multicollinearity effects checked for:

- Issues here do not mean we cannot proceed with a given model, but it can impact our ability to trust and interpret the estimated terms. Extreme issues might require removing some highly correlated variables prior to really focusing on a model.
- Check a scatterplot or correlation matrix to assess the potential for shared information in different predictor variables.
- Use the diagnostic measure called a variance inflation factor (VIF) discussed in Section 8.5 (we need to develop some ideas first to understand this measure).

⦿ Equal (constant) variance:

- Same as before since it pertains to the residuals.

⦿ Normality of residuals:

- Same as before since it pertains to the residuals.

⦿ No influential points:

- Leverage is now determined by how unusual a point is for multiple explanatory variables.

- The ***leverage*** values in the Residuals vs Leverage plot are scaled to add up to the *degrees of freedom (df) used for the model* which is the number of explanatory variables (K) plus 1, so $K + 1$.
- The scale of leverages depends on the complexity of the model through the *df* and the sample size.
- The interpretation is still that the larger the leverage value, the more leverage the point has.
- The mean leverage is always *(model used df)/n = (K+1)/n* – so focus on the values with above average leverage.
- For example, with $K = 3$ and $n = 20$, the average leverage is $4/20 = 1/5$.
- High leverage points whose response does not follow the pattern defined by the other observations (now based on patterns for multiple x's with the response) will be influential.
- Use the Residual's vs Leverage plot to identify problematic points. Explore further with Cook's D continuing to provide a measure of the influence of each observation.
- The rules and interpretations for Cook's D are the same as in SLR (over 0.5 is possibly influential and over 1 is definitely influential).

While not a condition for use of the methods, a note about random assignment and random sampling is useful here in considering the scope of inference of any results. To make inferences about a population, we need to have a representative sample. If we have randomly assigned levels of treatment variables(s), then we can make causal inferences to subjects like those that we could have observed. And if we both have a representative sample and randomization, we can make causal inferences for the population. It is possible to randomly assign levels of variable(s) to subjects and still collect additional information from other explanatory (sometimes called ***control***) variables. The causal interpretations would only be associated with the explanatory variables that were randomly assigned even though the model might contain other variables. Their interpretation still involves noting all the variables included in the model, as demonstrated below. It is even possible to include interactions between randomly assigned variables and other variables – like drug dosage and sex of the subjects. In these cases, causal inference could apply to the treatment levels but noting that the impacts differ based on the non-randomly assigned variable.

For the Snow Depth data, the conditions can be assessed as:

- Quantitative variables condition:
    - These are all clearly quantitative variables.

- Independence of observations:
    - The observations are based on a random sample of sites from the population and the sites are spread around the mountains in Montana. Many people would find it to be reasonable to assume that the sites are independent of one another but others would be worried that sites closer together in space might be more similar than they are to far-away observations (this is called spatial correlation). I have been in a heated discussion with statistics colleagues about whether spatial dependency should be considered or if it is valid to ignore it in this sort of situation. It is certainly possible to be concerned about independence of observations here but it takes more advanced statistical methods to actually assess whether there is spatial dependency in these data. Even if you were going to pursue models that incorporate spatial correlations, the first task would be to fit this sort of model and then explore the results. When data are collected across space, you should note that there might be some sort of spatial dependency that could violate the independence assumption.

To assess the remaining assumptions, we can use our diagnostic plots. The same code as before will provide diagnostic plots. There is some extra code (par(...)) added to allow us to add labels to the plots (sub.caption = "" and title(main="...", outer=TRUE)) to know which model is being displayed since we have so many to discuss here. We can also employ a new approach, which is to simulate new observations from the model and make plots to compare simulated data sets to what was observed. The simulate function from Chapter 2 can be used to generate new observations from the model based on the estimated coefficients and where we know that the assumptions are true. If the simulated data and the observed data are very different, then the model is likely dangerous to use for inferences because of this mismatch. This method can be used to assess the linearity, constant variance, normality of residuals, and influential points aspects of the model. It is not something used in every situation, but is especially helpful if you are struggling to decide if what you are seeing in the diagnostics is just random variability or is really a clear issue. The regular steps in assessing each assumption are discussed first.

```
par(mfrow = c(2,2), oma = c(0,0,2,0))
plot(m4, pch = 16, sub.caption = "")
title(main="Diagnostics for m4", outer=TRUE)
```

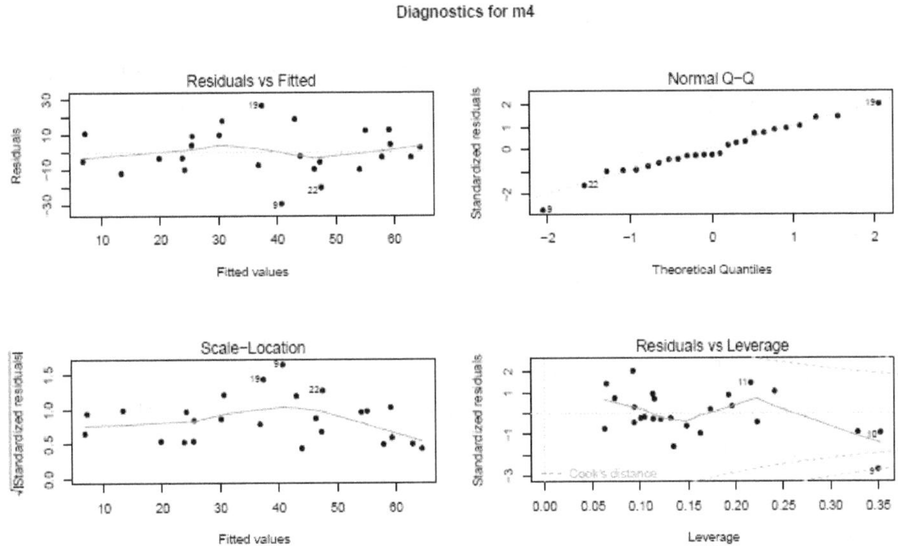

Figure 8.6: Diagnostic plots for model m4: Snow.Depth ~ Elevation + Min.Temp + Max.Temp.

- Linearity of relationship (NEW VERSION FOR MLR!):
  - Make plots of the response versus each explanatory variable:
    - In Figure 8.1, the plots of each variable versus snow depth do not clearly show any nonlinearity except for a little dip around 7000 feet in the plot vs *Elevation*.
- Examine the Residuals vs Fitted plot in Figure 8.6:
  - Generally, there is no clear curvature in the Residuals vs Fitted panel and that would be an acceptable answer. However, there is some pattern in the smoothing line that could suggest a more complicated relationship between at least one predictor and the response. This also resembles the pattern in the *Elevation* vs. *Snow depth* panel in Figure 8.1 so that might be the source of this "problem". This suggests that there is the potential to do a little bit better but that it is not unreasonable to proceed on with the MLR, ignoring this little wiggle in the diagnostic plot.
- Examine partial residuals as seen in Figure 8.5:
  - In the term-plot for elevation from this model, there is a slight pattern in the partial residuals between 6,500 and 7,500 feet. This was also apparent in the original plot and suggests a slight nonlinearity in the pattern of responses versus this explanatory variable.

- Multicollinearity effects checked for:
  - The predictors certainly share information in this application (correlations between -0.67 and-0.91) and multicollinearity looks to be a major concern in being able to understand/separate the impacts of temperatures and elevations on snow depths.
  - See Section 8.5 for more on this issue in these data.
- Equal (constant) variance:
  - While there is a little bit more variability in the middle of the fitted values, this is more an artifact of having a smaller data set with a couple of moderate outliers that fell in the same range of fitted values and maybe a little bit of missed curvature. So there is not too much of an issue with this condition.
- Normality of residuals:
  - The residuals match the normal distribution fairly closely the QQ-plot, showing only a little deviation for observation 9 from a normal distribution and that deviation is extremely minor. There is certainly no indication of a violation of the normality assumption here.
- No influential points:
  - With $K = 3$ predictors and $n = 25$ observations, the average leverage is $4/25 = 0.16$. This gives us a scale to interpret the leverage values on the *x*-axis of the lower right panel of our diagnostic plots.
  - There are three higher leverage points (leverages over 0.3) with only one being influential (point 9) with Cook's D close to 1.
    - Note that point 10 had the same leverage but was not influential with Cook's D less than 0.5.
  - We can explore both of these points to see how two observations can have the same leverage and different amounts of influence.

The two flagged points, observations 9 and 10 in the data set, are for the sites "Northeast Entrance" (to Yellowstone) and "Combination". We can use the MLR equation to do some prediction for each observation and calculate residuals to see how far the model's predictions are from the actual observed values for these sites. For the Northeast Entrance, the *Max.Temp* was 45, the *Min.Temp* was 28, and the *Elevation* was 7350 as you can see in this printout of just the two rows of the data set available by slicing rows 9 and 10 fromsnotel_s:

```
snotel_s %>% slice(9,10)
```

```
## # A tibble: 2 x 6
##      ID Station               Snow.Depth Max.Temp Min.Temp Elevation
##   <dbl> <chr>                      <dbl>    <dbl>    <dbl>     <dbl>
## 1    18 Northeast Entrance          11.2       45       28      7350
## 2    53 Combination                 14         36       28      5600
```

The estimated *Snow Depth* for the *Northeast Entrance* site (observation 9) is found using the estimated model with

SnowDepth9 = -10.51 + 0.0123 · Elevation9 - 0.505 · MinTemp9 - 0.562 · MaxTemp9

= -10.51 + 0.0123 ∗ **7350** - 0.505 ∗ **28** - 0.562 ∗ **45**

= 40.465 inches, but the observed snow depth was actually y9 = 11.2 inches. The observed residual is then

$$e_9 = y_9 - \widehat{y}_9 = 11.2 - 40.465 = -29.265 \text{ inches.}$$

So the model "misses" the snow depth by over 29 inches with the model suggesting over 40 inches of snow but only 11 inches actually being present5.

-10.51	+ 0.0123*7350 - 0.505*28 - 0.562*45
## [1]	40.465
11.2 -	40.465
## [1]	-29.265

This point is being rated as influential (Cook's D ≈ 1) with a leverage of nearly 0.35 and a standardized residual (y-axis of Residuals vs. Leverage plot) of nearly -3. This suggests that even with this observation impacting/distorting the slope coefficients (that is what influence means), the model is still doing really poorly at fitting this observation. We'll drop it and re-fit the model in a second to see how the slopes change. First, let's compare that result to what happened for data point 10 ("Combination") which was just as high leverage but not identified as influential.

The estimated snow depth for the *Combination* site is

SnowDepth10 = -10.51 + 0.0123 · Elevation10 - 0.505 · MinTemp10 - 0.562 · MaxTemp10

= -10.51 + 0.0123 ∗ 5600 - 0.505 ∗ 28 - 0.562 ∗ 36

= 23.998 inches.

The observed snow depth here was *y*10 = 14.0 inches so the observed residual is then

$$e_{10} = y_{10} - \widehat{y}_{10} = 14.0 - 23.998 = -9.998 \text{ inches.}$$

This results in a standardized residual of around -1. This is still a "miss" but not as glaring as the previous result and also is not having a major impact on the model's estimated slope coefficients based on the small Cook's D value.

```
-10.51 + 0.0123*5600 - 0.505*28 - 0.562*36
```

```
## [1] 23.998
```

```
14 - 23.998
```

```
## [1] -9.998
```

Note that any predictions using this model presume that it is trustworthy, but the large Cook's D on one observation suggests we should consider the model after removing that observation. We can re-run the model without the 9th observation using the data set snotel_s %>% slice(-9).

```
m5 <- lm(Snow.Depth ~ Elevation + Min.Temp + Max.Temp, data = snotel_s %>% slice(-9))
summary(m5)
```

```
##
## Call:
## lm(formula = Snow.Depth ~ Elevation + Min.Temp + Max.Temp, data = snotel_s %>%
##     slice(-9))
##
## Residuals:
##      Min       1Q   Median       3Q      Max
## -29.2918  -4.9757  -0.9146   5.4292  20.4260
##
## Coefficients:
##               Estimate Std. Error t value Pr(>|t|)
## (Intercept) -1.424e+02  9.210e+01  -1.546  0.13773
## Elevation    2.141e-02  6.101e-03   3.509  0.00221
## Min.Temp     6.722e-01  1.733e+00   0.388  0.70217
## Max.Temp     5.078e-01  6.486e-01   0.783  0.44283
##
## Residual standard error: 11.29 on 20 degrees of freedom
## Multiple R-squared:  0.7522, Adjusted R-squared:  0.715
## F-statistic: 20.24 on 3 and 20 DF,  p-value: 2.843e-06
```

```
plot(allEffects(m5, residuals = T), main = "MLR model with NE Ent. Removed")
```

The estimated MLR model with n = 24 after removing the influential "NE Entrance" observation is $\text{SnowDepth}_i = -142.4 + 0.0214 \cdot \text{Elevation}i + 0.672 \cdot \text{MinTemp}_i + 0.508 \cdot \text{MaxTemp}_i$

Something unusual has happened here: there is a positive slope for both temperature terms in Figure 8.7 that both contradicts reasonable expectations (warmer temperatures are related to higher snow levels?) and our original SLR results. So what happened? First, removing the influential point has drastically changed the slope coefficients (remember that was the definition of an influential point). Second, when there are predictors that share information, the results can be somewhat unexpected for some or all the predictors when they are all in the model together. Note that the Elevation term looks like what we might expect and seems to have a big impact on the predicted Snow Depths. So when the temperature variables are included in the model they might be functioning to explain some differences in sites that the Elevation term could not explain. This is where our "adjusting for" terminology comes into play. The unusual-looking slopes for the temperature effects can be explained by interpreting them as the estimated change in the response for changes in temperature after we control for the impacts of elevation. Suppose that Elevation explains most of the variation in *Snow Depth* except for a few sites where the elevation cannot explain all the variability and the site characteristics happen to show higher temperatures and more snow (or lower temperatures and less snow). This could be because warmer areas might have been hit by a recent snow storm while colder areas might have been missed (this is just one day and subject to spatial and temporal fluctuations in precipitation patterns). Or maybe there is another factor related to having marginally warmer temperatures that are accompanied by more snow (maybe the lower snow sites for each elevation were so steep that they couldn't hold much snow but were also relatively colder?). Thinking about it this way, the temperature model components could provide useful corrections to what *Elevation* is providing in an overall model and explain more variability than any of the variables could alone. It is also possible that the temperature variables are not needed in a model with *Elevation* in it, are just "explaining noise", and should be removed from the model. Each of the next sections take on various aspects of these issues and eventually lead to a general set of modelling and model selection recommendations to help you work in situations as complicated as this. Exploring the results for this model assumes we trust it and, once again, we need to check diagnostics before getting too focused on any particular results from it.

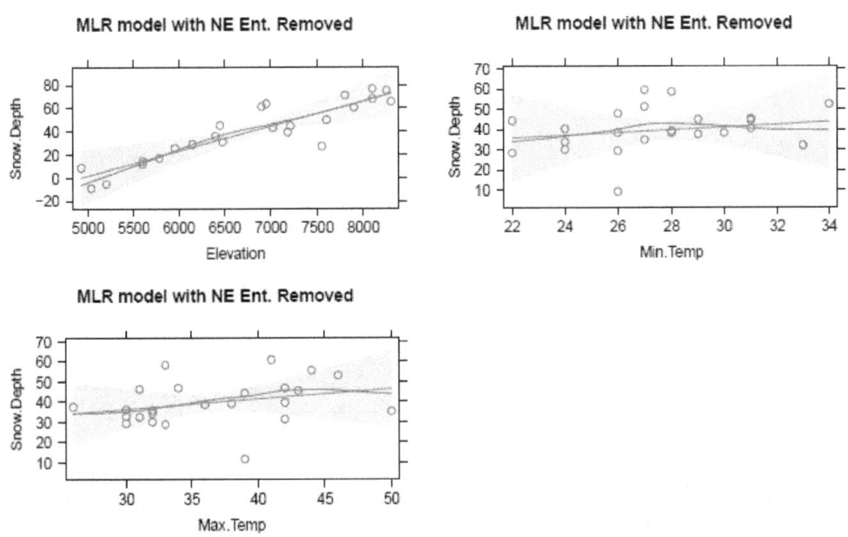

Figure 8.7: Term-plots for the MLR for Snow Depth based on Elevation, Min Temp, and Max Temp with Northeast entrance observation removed from data set (n = 24).

The Residuals vs. Leverage diagnostic plot in Figure 8.8 for the model fit to the data set without NE Entrance (now n = 24) reveals a new point that is somewhat influential (point 22 in the data set has Cook's D ≈ 0.5). It is for a location called "Bloody "6 which has a leverage of nearly 0.2 and a standardized residual of nearly -3. This point did not show up as influential in the original version of the data set with the same model but it is now. It also shows up as a potential outlier. As we did before, we can explore it a bit by comparing the model predicted snow depth to the observed snow depth. The predicted snow depth for this site (see output below for variable values) is

$$SnowDepth_{22} = -142.4 + 0.0214 * \mathbf{7550} + 0.672 * \mathbf{26} + 0.508 * \mathbf{39} = 56.45 \text{ inches.}$$

The observed snow depth was 27.2 inches, so the estimated residual is -39.25 inches. Again, this point is potentially influential and an outlier. Additionally, our model contains results that are not what we would have expected *a priori*, so it is not unreasonable to consider removing this observation to be able to work towards a model that is fully trustworthy.

```
par(mfrow = c(2,2), oma = c(0,0,2,0))
plot(m5, pch = 16, sub.caption = "")
title(main="Diagnostics for m5", outer=TRUE)
```

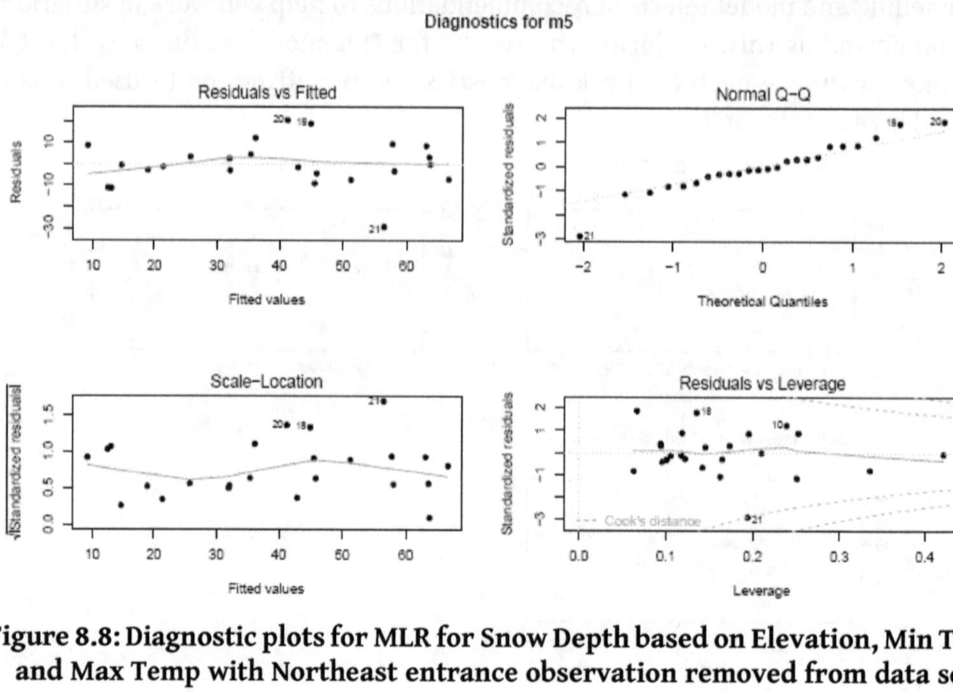

Figure 8.8: Diagnostic plots for MLR for Snow Depth based on Elevation, Min Temp and Max Temp with Northeast entrance observation removed from data set.

```
##
## Call:
## lm(formula = Snow.Depth ~ Elevation + Min.Temp + Max.Temp, data = snotel_s %>%
##     slice(-c(9, 22)))
##
## Residuals:
##     Min      1Q  Median      3Q     Max
## -14.878  -4.486   0.024   3.996  20.728
##
## Coefficients:
##               Estimate Std. Error t value Pr(>|t|)
## (Intercept) -2.133e+02  7.458e+01  -2.859   0.0100
## Elevation    2.686e-02  4.997e-03   5.374 3.47e-05
## Min.Temp     9.843e-01  1.359e+00   0.724   0.4776
## Max.Temp     1.243e+00  5.452e-01   2.280   0.0343
##
## Residual standard error: 8.832 on 19 degrees of freedom
## Multiple R-squared:  0.8535, Adjusted R-squared:  0.8304
## F-statistic:  36.9 on 3 and 19 DF,  p-value: 4.003e-08
```

This worry-some observation is located in the 22nd row of the original data set:

```
snotel_s %>% slice(22)
```

```
## # A tibble: 1 x 6
##      ID Station              Snow.Depth Max.Temp Min.Temp Elevation
##   <dbl> <fct>                     <dbl>    <dbl>    <dbl>     <dbl>
```

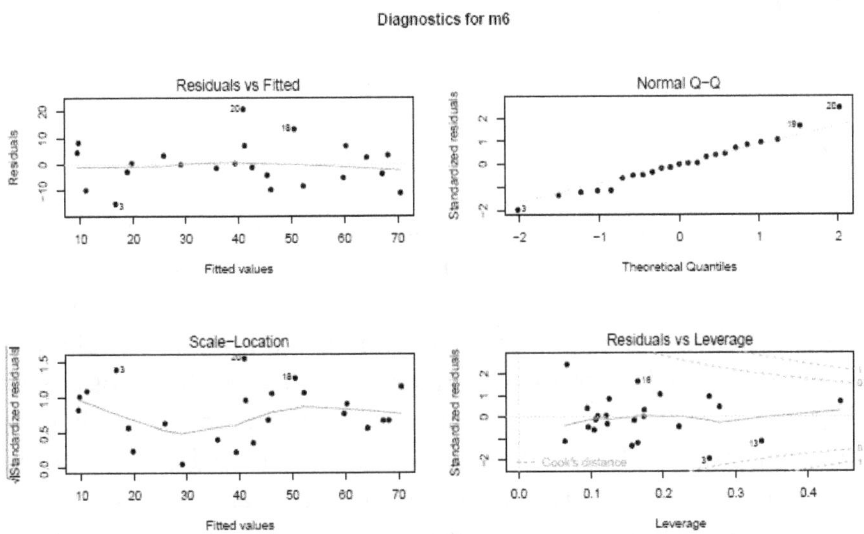

Figure 8.9: Diagnostic plots for MLR for Snow Depth based on Elevation, Min Temp and Max Temp with two observations removed (n = 23).

```
## 1    36 Bloody [Redact.]         27.2       39       26      7550
```

With the removal of both the "Northeast Entrance" and "Bloody " sites, there are n = 23 observations remaining. This model (m6) seems to contain residual diagnostics (Figure 8.9) that are finally generally reasonable.

```
m6 <- lm(Snow.Depth ~ Elevation + Min.Temp + Max.Temp, data = snotel_s %>% slice(-c(9,22)))
summary(m6)
par(mfrow = c(2,2), oma = c(0,0,2,0))
plot(m6, pch = 16, sub.caption = "")
title(main="Diagnostics for m6", outer=TRUE)
```

It is hard to suggest that there any curvature issues and the slight variation in the Scale-Location plot is mostly due to few observations with fitted values around 30 happening to be well approximated by the model. The normality assumption is generally reasonable and no points seem to be overly influential on this model (finally!).

The term-plots (Figure 8.10) show that the temperature slopes are both positive although in this model Max.Temp seems to be more "important" than Min.Temp. We have ruled out individual influential points as the source of un-expected directions in slope coefficients and the more likely issue is multicollinearity – in a model that includes Elevation, the temperature effects may be positive, again acting with the Elevation term to generate the best possible predictions of the observed responses. Throughout this discussion, we have mainly focused on the slope coefficients and diagnostics. We have other tools in MLR to more quantitatively assess and compare different regression models that are considered in the next sections.

```
plot(allEffects(m6, residuals = T), main = "MLR model with n = 23")
```

As a final assessment of this model, we can consider simulating a set of n=23 responses from this model and then comparing that data set to the one we just analyzed. This does not change the predictor variables, but creates two new versions of the response called Simulated Snow and SimulatedSnow2 in the following code chunk which are plotted in Figure 8.11. In exploring two realizations of simulated responses from the model, the results look fairly similar to the original data set. This model appeared to have reasonable assumptions and the match between simulated responses and the original ones reinforces those previous assessments. When the match is not so close, it can reinforce or create concern about the way that the assumptions have been assessed using other tools.

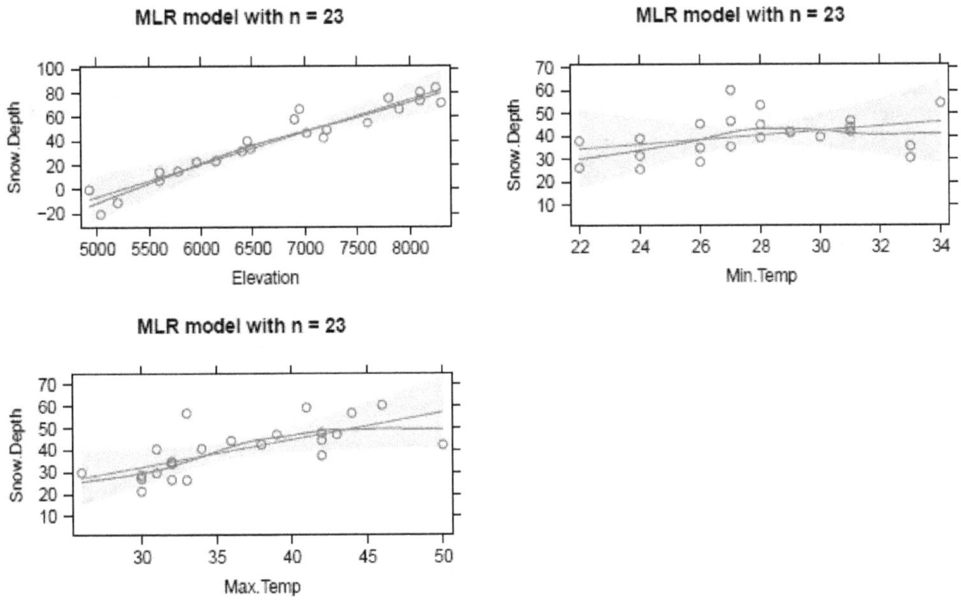

Figure 8.10: Term-plots for the MLR for Snow Depth based on Elevation, Min Temp and Max Temp with two observations removed.

```
set.seed(307)
snotel_final <- snotel_s %>% slice(-c(9,22))
snotel_final <- snotel_final %>%
  #Creates first and second set of simulated responses
  mutate(SimulatedSnow = simulate(m6)[[1]],
         SimulatedSnow2 = simulate(m6)[[1]]
         )
```

```
r1 <- snotel_final %>% ggplot(aes(x = Elevation, y = Snow.Depth)) +
  geom_point() +
  theme_bw() +
  labs(title = "Real Responses")
r2 <- snotel_final %>% ggplot(aes(x = Max.Temp, y = Snow.Depth)) +
  geom_point() +
  theme_bw() +
  labs(title = "Real Responses")
r3 <- snotel_final %>% ggplot(aes(x = Min.Temp, y = Snow.Depth)) +
  geom_point() +
  theme_bw() +
  labs(title = "Real Responses")

s1 <- snotel_final %>% ggplot(aes(x = Elevation, y = SimulatedSnow)) +
  geom_point(col = "forestgreen") +
  theme_bw() +
  labs(title = "First Simulated Responses")
```

```
s2 <- snotel_final %>% ggplot(aes(x = Max.Temp, y = SimulatedSnow)) +
  geom_point(col = "forestgreen") +
  theme_bw() +
  labs(title = "First Simulated Responses")
s3 <- snotel_final %>% ggplot(aes(x = Min.Temp, y = SimulatedSnow)) +
  geom_point(col = "forestgreen") +
  theme_bw() +
  labs(title = "First Simulated Responses")

s12 <- snotel_final %>% ggplot(aes(x = Elevation, y = SimulatedSnow2)) +
  geom_point(col = "skyblue") +
  theme_bw() +
  labs(title = "Second Simulated Responses")
s22 <- snotel_final %>% ggplot(aes(x = Max.Temp, y = SimulatedSnow2)) +
  geom_point(col = "skyblue") +
  theme_bw() +
  labs(title = "Second Simulated Responses")
s32 <- snotel_final %>% ggplot(aes(x = Min.Temp, y = SimulatedSnow2)) +
  geom_point(col = "skyblue") +
  theme_bw() +
  labs(title = "Second Simulated Responses")

grid.arrange(r1, r2, r3, s1, s2, s3, s12, s22, s32, ncol = 3)
```

8.3 INTERPRETATION OF MLR TERMS

Since these results (finally) do not contain any highly influential points, we can formally discuss interpretations of the slope coefficients and how the term-plots (Figure 8.10) aid our interpretations. Term-plots in MLR are constructed by holding all the other quantitative variables[7] at their mean and generating predictions and %95 CIs for the mean response across the levels of observed values for each predictor variable. This idea also help us to work towards interpretations of each term in an MLR model. For example, for *Elevation*, the term-plot starts at an elevation around 5000 feet and ends at an elevation around 8000 feet. To generate that line and CIs for the mean snow depth at different elevations, the MLR model of

SnowDepthi = -213.3 + 0.0269 · Elevationi + 0.984 · MinTempi + 1.243 · MaxTempi is used, but we need to have "something" to put in for the two temperature variables to predict Snow Depth for different Elevations. The typical convention is to hold the "other" variables at their means to generate these plots. This tactic also provides a way of interpreting each slope coefficient. Specifically, we can interpret the Elevation slope as: For a 1 foot increase in Elevation, we estimate the mean Snow Depth to increase by 0.0269 inches, holding the minimum and maximum temperatures constant. More generally, the slope interpretation in an MLR is: For a 1 [units of xk] increase in xk, we estimate the mean of y to change by bk [units of y], after controlling for [list of other explanatory variables in model].

To make this more concrete, we can recreate some points in the Elevation term-plot. To do this, we first need the mean of the "other" predictors, Min.Temp and Max. Temp.

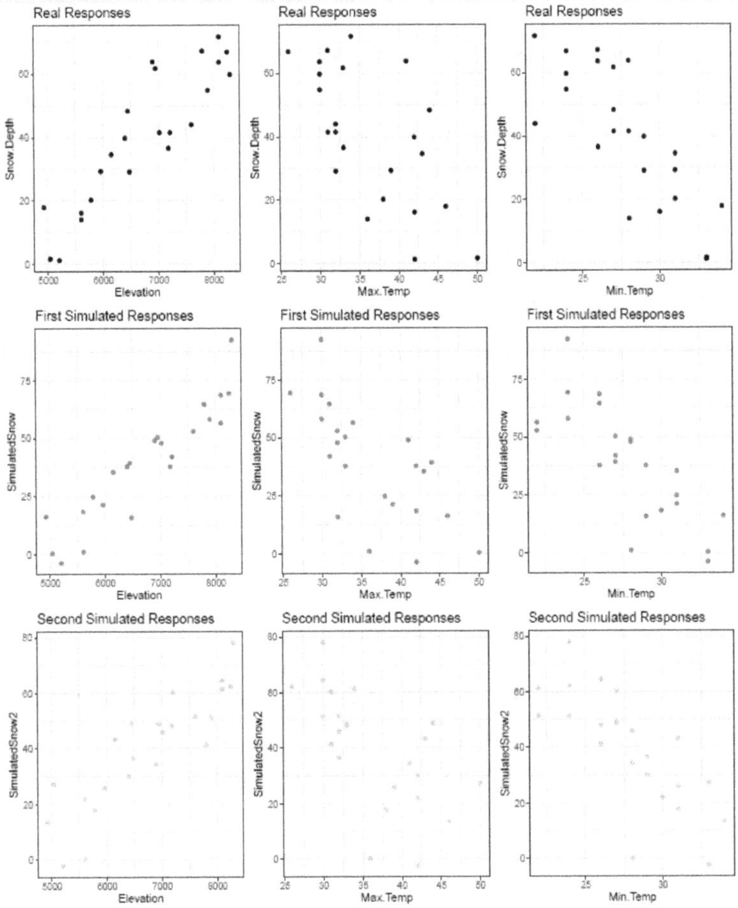

Figure 8.11: Plot of the original responses versus the three predictors (n = 23 data set) in the top row and two sets of simulated responses versus the predictors in the bottom two rows.

```
mean(snotel_final$Min.Temp)
```

```
## [1] 27.82609
```

```
mean(snotel_final$Max.Temp)
```

```
## [1] 36.3913
```

We can put these values into the MLR equation and simplify it by combining like terms, to an equation that is in terms of just Elevation given that we are holding Min.Temp and Max.Temp at their means:

$$\text{SnowDepthi} = -213.3 + 0.0269 \cdot \text{Elevationi} + 0.984 * 27.826 + 1.243 * 36.391$$

$$= -213.3 + 0.0269 \cdot \text{Elevationi} + 27.38 + 45.23$$

$$= -140.69 + 0.0269 \cdot \text{Elevationi}.$$

So at the means on the two temperature variables, the model looks like an SLR with an estimated y-intercept of -140.69 (mean *Snow Depth* for *Elevation* of 0 if temperatures are at their means) and an estimated slope of 0.0269. Then we can plot the predicted changes in y across all the values of the predictor variable (*Elevation*) while holding the other variables constant. To generate the needed values to define a line, we can plug various *Elevation* values into the simplified equation:

For an elevation of 5000 at the average temperatures, we predict a mean snow depth of $-140.69 + 0.0269 * 5000 = -6.19$ inches.

For an elevation of 6000 at the average temperatures, we predict a mean snow depth of $-140.69 + 0.0269 * 6000 = 20.71$ inches.

For an elevation of 8000 at the average temperatures, we predict a mean snow depth of $-140.69 + 0.0269 * 8000 = 74.51$ inches.

We can plot this information (Figure 8.12) using the geom_point function to show the points we calculated and the geom_line function to add a line that connects the dots. In the geom point, the size option issued to make the points a little easier to see.

Note that we only needed 2 points to define the line but need a denser grid of elevations if we want to add the 95% CIs for the true mean snow depth across the different elevations since they vary as a function of the distance from the mean of the explanatory variables.

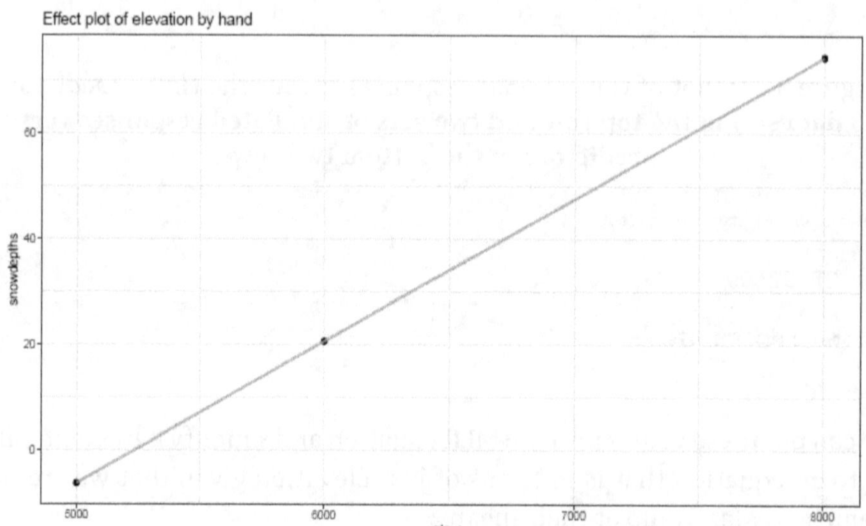

Figure 8.12: Term-plot for Elevation "by-hand", holding temperature variables constant at their means.

```
# Making own effect plot:
modelres2 <- tibble(elevs = c(5000, 6000, 8000), snowdepths = c(-6.19, 20.71, 74.51))
modelres2 %>% ggplot(mapping = aes(x = elevs, y = snowdepths)) +
  geom_point(size = 2) +
  geom_line(lwd = 1, alpha = .75, col = "tomato") +
  theme_bw() +
  labs(title = "Effect plot of elevation by hand")
```

The partial residuals in MLR models[8] highlight the relationship between each predictor and the response after the impacts of the other variables are incorporated. To do this, we start with the raw residuals, $e_i = y_i - \hat{y}_i$, which is the left-over part of the responses after accounting for all the predictors. If we add the component of interest to explore (say $b_k x_{kj}$) to the residuals, e_i, we get $e_i + b_k x_{kj} = y_i - \hat{y}_i + b_k x_{kj} = y_i - (b_0 + b_1x_{1i} + b_2x_{2i} + \ldots + b_k x_{ki} + \ldots + b_K x_{Ki}) + b_k x_{kj} = y_i - (b_0 + b_1x_{1i} + b_2x_{2i} + \ldots + b_{k-1}x_{k-1,i} + b_{k+1}x_{k+1,i} + \ldots + b_K x_{Ki})$. This new residual is a partial residual (also known as "component-plus-residuals" to indicate that we put the residuals together with the component of interest to create them). It contains all of the regular residual as well as what would be explained by $b_k x_{kj}$ given the other variables in the model. Some choose to plot these partial residuals or to center them at 0 and, either way, plot them versus the component, here x_{kj}. In effects plots, partial residuals are vertically scaled to match the height that the term-plot has created by holding the other predictors at their means so they can match the y-axis of the lines of the estimated terms based on the model. However they are vertically located, partial residuals help to highlight missed patterns left in the residuals that might be related to a particular predictor.

To get the associated 95% CIs for an individual term, we could return to using the predict function for the MLR, again holding the temperatures at their mean values. The predict function is sensitive and needs the same variable names as used in the original model fitting to work. First we create a "new" data set using the seq function to generate the desired grid of elevations and the rep function[9] to repeat the means of the temperatures for each of elevation values we need to make the plot. The code creates a specific version of the predictor variables that is stored in newdata1 that is provided to the predict function so that it will provide fitted values and CIs across different elevations with temperatures held constant.

```
elevs <- seq(from = 5000, to = 8000, length.out = 30)
newdata1 <- tibble(Elevation = elevs, Min.Temp = rep(27.826,30),
                   Max.Temp = rep(36.3913,30))
newdata1
```

```
## # A tibble: 30 x 3
##    Elevation Min.Temp Max.Temp
##        <dbl>    <dbl>    <dbl>
##  1     5000      27.8     36.4
##  2     5103.     27.8     36.4
##  3     5207.     27.8     36.4
##  4     5310.     27.8     36.4
##  5     5414.     27.8     36.4
##  6     5517.     27.8     36.4
##  7     5621.     27.8     36.4
##  8     5724.     27.8     36.4
##  9     5828.     27.8     36.4
## 10     5931.     27.8     36.4
## # ... with 20 more rows
## # i Use `print(n = ...)` to see more rows
```

The first 10 predicted snow depths along with 95% confidence intervals for the mean, holding temperatures at their means, are:

```
predict(m6, newdata = newdata1, interval = "confidence") %>% head(10)
```

```
##           fit        lwr      upr
## 1  -6.3680312 -24.913607 12.17754
## 2  -3.5898846 -21.078518 13.89875
## 3  -0.8117379 -17.246692 15.62322
## 4   1.9664088 -13.418801 17.35162
## 5   4.7445555  -9.595708 19.08482
## 6   7.5227022  -5.778543 20.82395
## 7  10.3008489  -1.968814 22.57051
## 8  13.0789956   1.831433 24.32656
## 9  15.8571423   5.619359 26.09493
## 10 18.6352890   9.390924 27.87965
```

So we could do this with any model **for each predictor** variable to create term-plots, or we can just use the all Effects function to do this for us. This exercise is useful to complete once to understand what is being displayed in term-plots but using the all Effects function makes getting these plots much easier. There are two other model components of possible interest in this model. The slope of 0.984 for *Min.Temp* suggests that for a $1^{\circ}F$ increase in *Minimum Temperature,* we estimate a 0.984 inch change in the mean*Snow Depth,* after controlling for *Elevation* and *Max. Temp* at the sites. Similarly, the slope of 1.243 for the *Max. Temp* suggests that for a $1^{\circ}F$ increase in *Maximum Temperature,* we estimate a 1.243 inch change in the mean *Snow Depth,* holding *Elevation* and *Min. Temp* constant. Note that there are a variety of ways to note that each term in an MLR is only a particular value given the other variables in the model. We can use words such as "holding the other variables constant" or "after adjusting for the other variables" or "in a model with. . . " or "for observations with similar values of the other variables but a difference of 1 unit in the predictor..". The main point is to find words that reflect that this single slope coefficient might be different if we had a different overall model and the only way to interpret it is conditional on the other model components.

Term-plots have a few general uses to enhance our regular slope interpretations. They can help us assess how much change in the mean of y the model predicts over the range of each observed x. This can help you to get a sense of the "practical" importance of each term. Additionally, the term-plots show 95% confidence intervals for the mean response across the range of each variable, holding the other variables at their means. These intervals can be useful for assessing the precision in the estimated mean at different values of each predictor. However, note that you should not use these plots for deciding whether the term should be retained in the model – we have other tools for making that assessment. And one last note about term-plots – they do not mean that the relationships are really linear between the predictor and response variable being displayed. The model forces the relationship to be linear even if that is not the real functional form. Term-plots are not diagnostics for the model unless you add the partial residuals, the lines are just summaries of the model you assumed was correct! Any time we do linear regression, the inferences are contingent upon the model we chose. We know our model is not perfect, but we hope that it helps us learn something about our research question(s) and, to trust its results, we hope it matches the data fairly well.

To both illustrate the calculation of partial residuals and demonstrate their potential utility, a small simulated example is considered. These are simulated data to help to highlight these patterns but are not too different than results that can be seen in some real applications. This situation has a response of simulated cholesterol levels with (also simulated) predictors of age, exercise level, and healthiness level with a sample size of n = 100. First, consider the plot of the response versus each of the predictors in Figure 8.13. It appears that age might be positively related to the response, but exercise and healthiness levels do not appear to be related to the response. But it is important to remember that the response is made up of potential contributions that can be explained by each predictor and unexplained variation, and so plotting the response versus each predictor may not allow us to see the real relationship with each predictor.

```
a1 <- d1 %>% ggplot(mapping = aes(x = Age, y = CholLevel)) +
  geom_point() +
  theme_bw()
e1 <- d1 %>% ggplot(mapping = aes(x = ExAmount, y = CholLevel)) +
  geom_point() +
  theme_bw()
h1 <- d1 %>% ggplot(mapping = aes(x = HealthLevel, y = CholLevel)) +
  geom_point() +
  theme_bw()
grid.arrange(a1, e1, h1, ncol = 3)
```

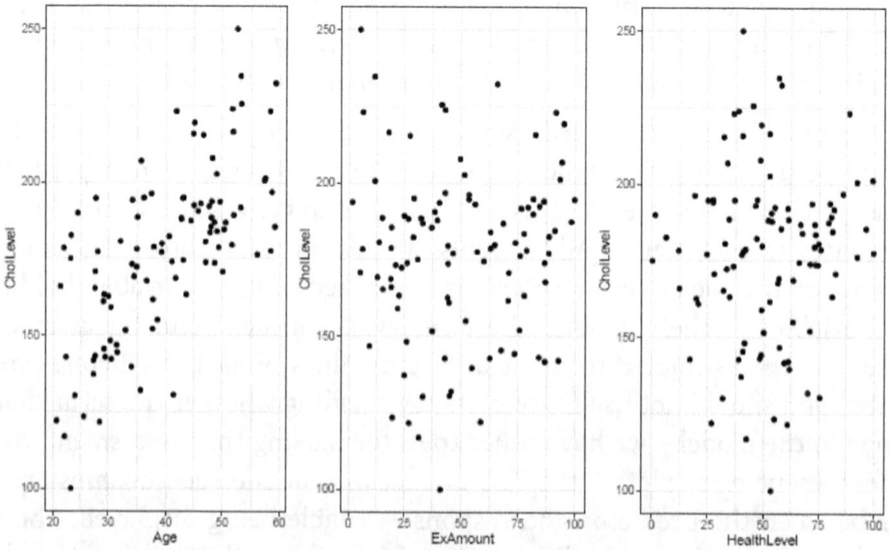

Figure 8.13: Scatterplot of Cholesterol level versus three predictors (simulated data).

```
##                Estimate Std. Error   t value     Pr(>|t|)
## (Intercept) 94.54572326 4.63863859 20.382214 1.204735e-36
## Age          3.50787191 0.14967450 23.436670 1.679060e-41
## ExAmount     0.07447965 0.04029175  1.848508 6.760692e-02
## HealthLevel -1.16373873 0.07212890 -16.134153 4.339546e-29
```

In the summary it appears that each predictor might be related to the response given the other predictors in the model with p-values of <0.0001, 0.068, and < 0.0001 for Age, Exercise, and Healthiness, respectively.

In Figure 8.14, we can see more of the story here by exploring the partial residuals versus each of the predictors. There are actually quite clear relationships for each partial residual versus its predictor. For Age and HealthLevel, the relationship after adjusting for other predictors is clearly positive and linear. For ExAmount there is a clear relationship but it is actually curving, so would violate the linearity assumption.

It is interesting that none of these were easy to see or even at all present in plots of the response versus individual predictors. This demonstrates the power of MLR methods to adjust/control for other variables to help us potentially more clearly see relationships between individual predictors and the response, or at least their part of the response.

```
plot(allEffects(sim1, residuals = T), grid = T)
```

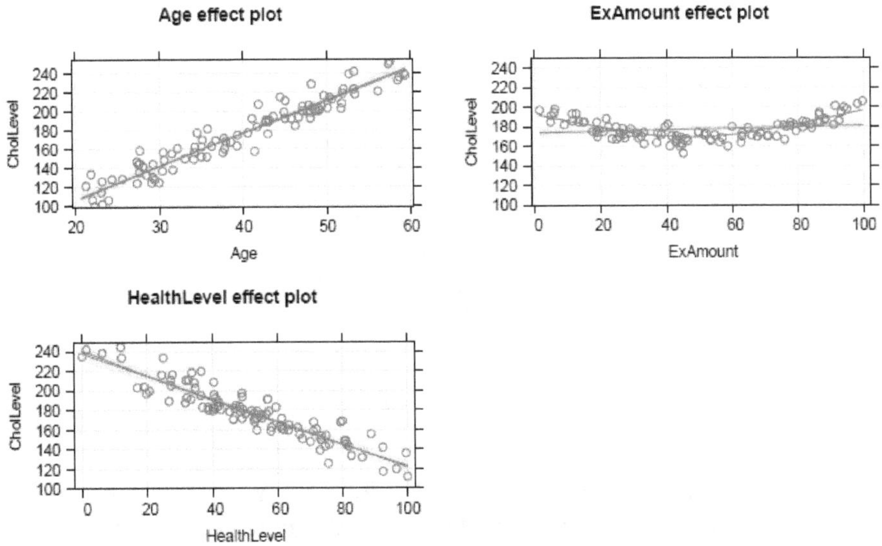

Figure 8.14: Term-plots with partial residuals for Cholesterol level versus three predictors (simulated data).

For those that are interested in these partial residuals, we can re-construct some of the work that the effects package does to provide them. As noted above, we need to take our regular residuals and addback in the impacts of a predictor of interest to calculate the partial residuals. The regular residuals can be extracted using the residuals function on the estimated model and the contribution of, say, the ExAmount predictor is found by taking the values in that variable times its estimated slope coefficient, b2 = 0.07447965.Plotting these partial residuals versus ExAmount as in Figure 8.15 provides a plot that is similar to the second term-plot except for differences in the y-axis. The y-axis in term-plots contains an additional adjustment but the two plots provide the same utility in diagnosing a clear missed curve in the partial residuals that is related to the ExAmount. Methods to incorporate polynomial functions of the predictor are simple extensions of the lm work we have been doing but are beyond the scope of this material – but you should always be checking the partial residuals to assess the linearity assumption with each quantitative predictor and if you see a pattern like this, seek out additional statistical resources such as the Statistical Sleuth (Ramsey and Schafer [2012]) or a statistician for help.

```
d1 <- d1 %>% mutate(partres = residuals(sim1) + ExAmount * 0.07447965)
d1 %>% ggplot(mapping = aes(x = ExAmount, y = partres)) +
  geom_point() +
  geom_smooth(method = "lm", se = F) +
  geom_smooth(se = F, col = "darkred", lty = 2, lwd = 1) +
  theme_bw() +
  labs(y = "Partial Residual")
```

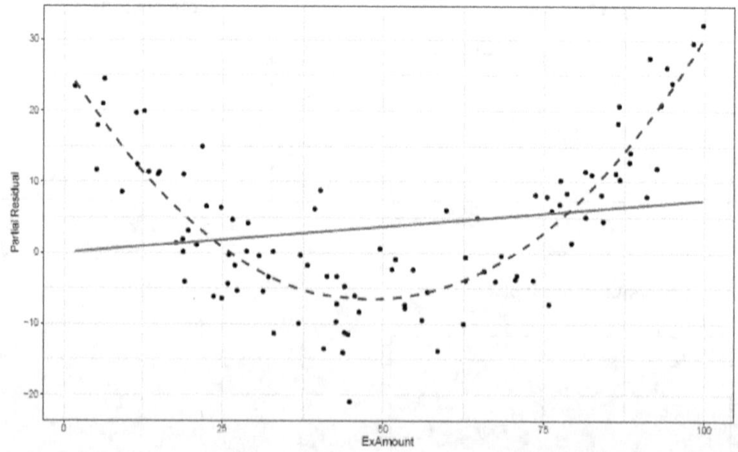

Figure 8.15: Plot of partial residual for Ex- Amount with the solid line for the MLR fit for this model component and the dashed line for the smoothing line that highlights the curvilinear relationship that the model failed to account for.

8.4 COMPARING MULTIPLE REGRESSION MODELS

With more than one variable, we now have many potential models that we could consider. We could include only one of the predictors, all of them, or combinations of sets of the variables. For example, maybe the model that includes *Elevation* does not "need" both *Min.Temp* and *Max.Temp*? Or maybe the model isn't improved over an SLR with just *Elevation* as a predictor. Or maybe none of the predictors are "useful"? In this section, we discuss some general model comparison issues and a metric that can be used to pick among a suite of different models (often called a set of ***candidate models*** to reflect that they are all potentially interesting and we need to compare them and possibly pick one).

It is certainly possible the researchers may have an *a priori* reason to only consider a single model. For example, in a designed experiment where combinations of, say, three different predictors are randomly assigned, the initial model with all three predictors may be sufficient to address all the research questions of interest. One advantage in these situations is that the variable combinations can be created to prevent multicollinearity among the predictors and avoid that complication in interpretations. However, this is more the exception than the rule. Usually, there are competing predictors or questions about whether some predictors matter more than others. This type of research always introduces the potential for multicollinearity to complicate the interpretation of each predictor in the presence of others. Because of this, multiple models are often considered, where "unimportant" variables are dropped from the model. The assessment of "importance" using p-values will be discussed in Section 8.6, but for now we will consider other reasons to pick one model over another.

There are some general reasons to choose a particular model:

- Diagnostics are better with one model compared to others.
- One model predicts/explains the responses better than the others (***R***^{2}).
- *A priori* reasons to "use" a particular model, for example in a designed experiment or it includes variable(s) whose estimated slopes directly address the research question(s), even if the variables are not "important" in the model.
- Model selection "criteria" suggest one model is better than the others10.

It is OK to consider multiple reasons to select a model but it is dangerous to "shop" for a model across many possible models – a practice which is sometimes called data-dredging and leads to a high chance of spurious results from a single model that is usually reported based on this type of exploration. Just like in other discussions of multiple testing issues previously, if you explore many versions of a model, maybe only keeping the best ones, this is very different from picking one model (and tests) a priori and just exploring that result.

As in SLR, we can use the R2 (the coefficient of determination) to measure the percentage of the variation in the response variable that the model explains. In MLR, it is important to remember that R2 is now an overall measure for the model and not specific to a single variable. It is comparable to other models including those fit with only a single predictor (SLR). So to meet criterion (2), we could simply find the model with the largest R2 value, finding the model that explains the most variation in the responses. Unfortunately for this idea, when you add more "stuff" to a regression model (even "unimportant" predictors), the R2 will always go up. This can be seen by considering

$$R^2 = \frac{\text{SS}_{\text{regression}}}{\text{SS}_{\text{total}}} \quad \text{where } \text{SS}_{\text{regression}} = \text{SS}_{\text{total}} - \text{SS}_{\text{error}} \text{ and } \text{SS}_{\text{error}} = \Sigma(y - \widehat{y})^2$$

Because adding extra variables to a linear model will only make the fitted values better, not worse, the SSerror will always go down if more predictors are added to the model. If SSerror goes down and SStotal is fixed, then adding extra variables will always increase SSregression and, thus, increase R2. This means that R2 is only useful for selecting models when you are picking between two models of the same size (same number of predictors). So we mainly use it as a summary of model quality once we pick a model, not a method of picking among a set of candidate models. Remember that R2 continues to have the property of being between 0 and 1 (or 0% and 100%) and that value refers to the proportion (percentage) of variation in the response explained by the model, whether we are using it for SLR or MLR.

However, there is an adjustment to the R2 measure that makes it useful for selecting among models. The adjusted measure is called the adjusted R2. The

R2measure adds a penalty for adding more variables to the model, providing the potential for this measure to decrease if the extra variables do not really benefit the model. The measure is calculated as

$$R^2_{\text{adjusted}} = 1 - \frac{\text{SS}_{\text{error}}/df_{\text{error}}}{\text{SS}_{\text{total}}/(N-1)} = 1 - \frac{\text{MS}_{\text{error}}}{\text{MS}_{\text{total}}},$$

which incorporates the degrees of freedom for the model via the error degrees of freedom which go down as the model complexity increases. This adjustment means that just adding extra useless variables (variables that do not explain very much extra variation) do not increase this measure. That makes this measure useful for model selection since it can help us to stop adding unimportant variables and find a "good" model among adjusted a set of candidates. Like the regular R2, larger values are better. The downside to R^2 is that it is no longer a percentage of variation in the response that is explained by the model; it can be less than 0 and so has no interpretable scale. It is just "larger is better". It provides one method for building a model (different from using p-values to drop unimportant variables as discussed below), by fitting a set of adjusted candidate models containing different variables and then picking the model with the largest R2 .You will want to interpret this new measure on a percentage scale, but do not do that. It is a just a measure to help you pick a model and that is all it is!

One other caveat in model comparison is worth mentioning: make sure you are comparing models for the same responses. That may sound trivial and usually it is. But when there are missing values in the data set, especially on some explanatory variables and not others, it is important to be careful that the *y*'s do not change between models you are comparing. This relates to our *Snow Depth* modelling because responses adjusted were being removed due to their influential nature. We can't compare $\boldsymbol{R^2}$ or $\boldsymbol{R^2}$for $n = 25$ to a model when $n = 23$ – it isn't a fair comparison on either measure since they based on the total variability which is= changing as the responses used change.

In the MLR (or SLR) model summaries, both the $\boldsymbol{R^2}$ and $\boldsymbol{R^2}$are available. Make sure you are able to pick out the correct one. For the reduced data set ($n = 23$) *Snow Depth* models, the pertinent part of the model summary for the model with all three predictors is in the last three lines:

```
m6 <- lm(Snow.Depth ~ Elevation + Min.Temp + Max.Temp,
         data = snotel_s %>% slice(-c(9,22)))
summary(m6)
```

There is a value for Multiple R-Squared of0.8535, this is the R2 value and suggests that the model Adjusted with Elevation, Min and Max temperatures explains 85.4% of the variation in Snow Depth. The R2adjustedis0.8304 and is available further to the right labelled as Adjusted R-Squared. We repeated this for a suite of different models for this same n = 23 data set and found the following results in Table 8.1. The top R2adjusted model is the model with Elevation and Max. Temp, which beats out the model with all three variables onR2adjusted. Note that the top R2 model is the model with three predictors, but the most complicated model will always have that characteristic.

```
## Call:
## lm(formula = Snow.Depth ~ Elevation + Min.Temp + Max.Temp, data = snotel_s %>%
##     slice(-c(9, 22)))
##
## Residuals:
##     Min      1Q  Median      3Q     Max
## -14.878  -4.486   0.024   3.996  20.728
##
## Coefficients:
##               Estimate Std. Error t value Pr(>|t|)
## (Intercept) -2.133e+02  7.458e+01  -2.859   0.0100
## Elevation    2.686e-02  4.997e-03   5.374 3.47e-05
## Min.Temp     9.843e-01  1.359e+00   0.724   0.4776
## Max.Temp     1.243e+00  5.452e-01   2.280   0.0343
##
## Residual standard error: 8.832 on 19 degrees of freedom
## Multiple R-squared:  0.8535, Adjusted R-squared:  0.8304
```

Table 8.1: Model comparisons for Snow Depth data, sorted by model complexity.

Model	K	R2	R^2 adjusted	R^2 adjusted Rank
SD ~ Elevation	1	0.8087	0.7996	3
SD ~ Min.Temp	1	0.6283	0.6106	5
SD ~ Max.Temp	1	0.4131	0.3852	7
SD ~ Elevation + Min.Temp	2	0.8134	0.7948	4
SD ~ Elevation + Max.Temp	2	0.8495	0.8344	1
SD ~ Min.Temp + Max.Temp	2	0.6308	0.5939	6
SD ~ Elevation + Min.Temp + Max.Temp	3	0.8535	0.8304	2

The top adjusted R2 model contained Elevation and Max.Temp and has an R2 of 0.8495, so we can say that the model with Elevation and Maximum Temperature explains 84.95% percent of the variation in adjusted Snow Depth and also that this model was selected based on the R_2. One of the important features of $R_{2adjusted}$ is available in this example – adding variables often does not always increase its value even though R2 does increase with any addition. In Section 8.13 we consider a competitor for this model selection criterion that may “work” a bit better and be extendable into more complicated modelling situations; that measure is called the AIC .

8.5 GENERAL RECOMMENDATIONS FOR MLR INTERPRETATIONS AND VIFS

There are some important issues to remember11 when interpreting regression models that can result in common mistakes.

- **Don't claim to "hold everything constant" for a single individual:**

Mathematically this is a correct interpretation of the MLR model but it is rarely the case that we could have this occur in real applications. Is it possible to increase the *Elevation* while holding the *Max.Temp* constant? We discussed making term-plots doing exactly this – holding the other variables constant at their means. If we interpret each slope coefficient in an MLR conditionally then we can craft interpretations such as: For locations that have a *Max.Temp* of, say, 45°F and *Min. Temp* of, say, 30°F, a 1 foot increase in *Elevation* tends to be associated with a 0.0268 inch increase in *Snow Depth* on average. This does not try to imply that we can actually make that sort of change but that given those other variables, the change for that variable is a certain magnitude.

- **Don't interpret the regression results causally (or casually?)...**

Unless you are analyzing the results of a designed experiment (where the levels of the explanatory variable(s) were randomly assigned) you cannot state that a change in that *x* **causes** a change in *y*, especially for a given individual. The multicollinearity in predictors makes it especially difficult to put too much emphasis on a single slope coefficient because it may be corrupted/modified by the other variables being in the model. In observational studies, there are also all the potential lurking variables that we did not measure or even confounding variables that we did measure but can't disentangle from the variable used in a particular model. While we do have a complicated mathematical model relating various *x*'s to the response, do not lose that fundamental focus on causal vs non-causal inferences based on the design of the study.

- Be cautious about doing prediction in MLR – you might be doing extrapolation!

It is harder to know if you are doing extrapolation in MLR since you could be in a region of the *x*'s that no observations were obtained. Suppose we want to predict the *Snow Depth* for an *Elevation* of 6000 and *Max.Temp* of 30. Is this extrapolation based on Figure 8.16? In other words, can you find any observations "nearby" in the plot of the two variables together? What about an *Elevation* of 6000 and a *Max.Temp* of 40? The first prediction is in a different proximity to observations than the second one... In situations with more than two explanatory variables it becomes even more challenging to know whether you are doing extrapolation and the problem grows as the number of dimensions to search increases... In fact, in complicated MLR models we typically do not know whether there are observations "nearby" if we are doing predictions for unobserved combinations of our predictors. Note that Figure 8.16 also reinforces our potential collinearity problem between *Elevation* and *Max.Temp* with higher elevations being strongly associated with lower temperatures.

⊙ Don't think that the sign of a coefficient is special...

Adding other variables into the MLR models can cause a switch in the coefficients or change their magnitude or make them go from "important" to "unimportant" without changing the slope too much. This is related to the conditionality of the relationships being estimated in MLR and the potential for sharing of information in the predictors when it is present.

⊙ Multicollinearity in MLR models:

When explanatory variables are not independent (related) to one another, then including/excluding one variable will have an impact on the other variable. Consider the correlations among the predictors in the SNOTEL data set or visually displayed in Figure 8.17:

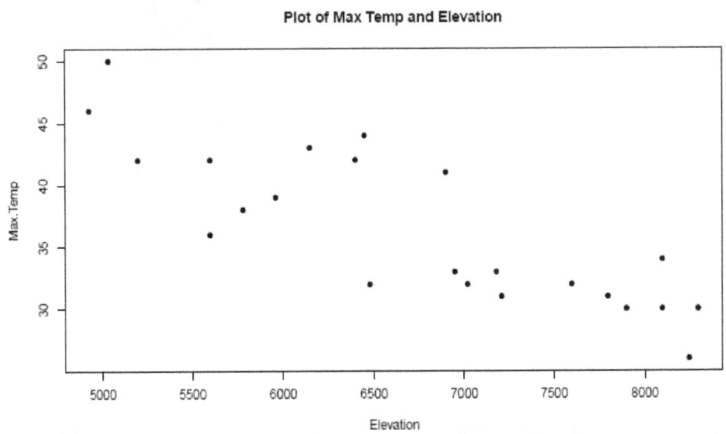

Figure 8.16: Scatterplot of observed Elevations and Maximum Temperatures for SNOTEL data.

```
library(corrplot)
par(mfrow = c(1,1), oma = c(0,0,1,0))
corrplot.mixed(cor(snotel_s %>% slice(-c(9,22)) %>% select(3:6)),
               upper.col = c(1, "orange"), lower.col = c(1, "orange"))
round(cor(snotel_s %>% slice(-c(9,22)) %>% select(3:6)), 2)
```

The predictors all share at least moderately strong linear relationships. For example, the r = -0.91 between Min.Temp and Elevation suggests that they contain very similar information and that extends to other pairs of variables as well. When variables share information, their addition to models may not improve the performance of the model and actually can make the estimated coefficients unstable, creating uncertainty in the correct coefficients because of the shared information. It seems that Elevation is related to Snow Depth but maybe it is because it has lower Minimum Temperatures? So you might wonder how we can find the "correct" slopes when they are sharing information in the response variable. The short answer is that we can't. But we do use Least Squares to find coefficient estimates as we did before – except that we have to remember that these estimates are conditional on other variables in the model for our interpretation since they impact one another within the model. It ends up that the uncertainty of pinning those variables down in the presence of shared information leads to larger SEs for all the slopes. And that we can actually measure how much each of the SEs are inflated because of multicollinearity with other variables in the model using what are called Variance Inflation Factors (or VIFs).

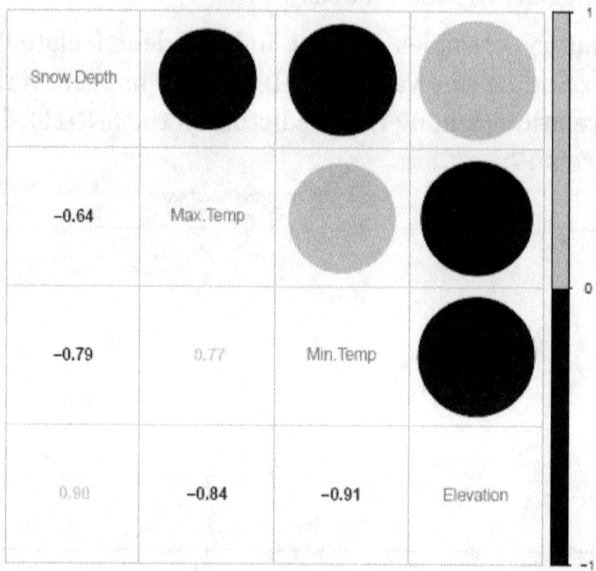

Figure 8.17: Plot of correlation matrix in the snow depth data set with influential points removed

##		Snow.Depth	Max.Temp	Min.Temp	Elevation
##	Snow. Depth	1.00	-0.64	-0.79	0.90
##	Max.Temp	-0.64	1.00	0.77	-0.84
##	Min.Temp	-0.79	0.77	1.00	-0.91
##	Elevation	0.90	-0.84	-0.91	1.00

VIFs provide a way to assess the multicollinearity in the MLR model that is caused by including specific variables. The amount of information that is shared between a single explanatory variable and the others can be found by regressing that variable on the others and calculating R2 for that model. The code for this regression is something like: lm(X1 ~ X2 + X3 + ... + XK), which regresses X1 on X2 through XK. The1 - R2 from this regression is the amount of independent information in X1 that is not explained by (orkrelated to) the other variables in the model. The VIF for each variable is defined using this quantity as (VIFk = 1/ (1 - R2) for variable k. If there is no shared information (R2 = 0), then the VIF will be 1. But if the information is completely shared with other variables (R2 = 1), then the VIF goes to infinity (1/0).Basically, large VIFs are bad, with the rule of thumb that values over 5 or 10 are considered "large" values indicating high (over 5) or extreme (over 10) multicollinearity in the model for that particular variable, both indicating that slope coefficients are dangerous to interpret in that model. We use this scale to determine if multicollinearity is a definite problem for a variable of interest. But any value of the VIF over 1 indicates some amount of multicollinearity is present. Additionally, the is also very interesting as it is the number of times larger than the SE for the slope for variable k is due to collinearity with other variables in the model. The square-root scale is the most useful scale to understand VIFs and allows you to make your own assessment of whether you think the multicollinearity is "important" based on how inflated the SEs are in a particular situation. An example will show how to easily get these results and where the results come from.

In general, the easy way to obtain VIFs is using the vif function from the car package (Fox et al. [2022b], Fox [2003]). It has the advantage of also providing a reasonable result when we include categorical variables in models (Sections 8.9 and 8.11). We apply the vif function directly to a model of interest and it generates values for each explanatory variable.

```
library(car)
vif(m6)
```

```
## Elevation  Min.Temp  Max.Temp
##  8.164201  5.995301  3.350914
```

Not surprisingly, there is an indication of problems with multicollinearity in two of the three variables in the model with the largest issues identified for *Elevation* and *Min.Temp*. Both of their VIFs exceed 5 indicating high levels of multicollinearity impacting those terms in the model. On the square-root scale, the VIFs show more interpretation utility.

```
sqrt(vif(m6))
```

```
## Elevation  Min.Temp  Max.Temp
##  2.857307  2.448530  1.830550
```

The result for *Elevation* of 2.86 suggests that the SE for *Elevation* is 2.86 times larger than it should be because of multicollinearity with other variables in the model. Similarly, the *Min.Temp* SE is 2.45 times larger and the *Max.Temp* SE is 1.83 times larger. Even the result for *Max.Temp* suggests an issue with multicollinearity even though it is below the cut-off for noting high or extreme issues with shared information. All of this generally suggests issues with multicollinearity in the model and that we need to be cautious in interpreting any slope coefficients from this model because they are all being impacted by shared information in the predictor variables to some degree or another.

In order to see how the VIF is calculated for *Elevation*, we need to regress *Elevation* on *Min.Temp* and *Max.Temp*. Note that this model is only fit to find the percentage of variation in elevation explained by the temperature variables. It ends up being 0.8775 – so a high percentage of *Elevation* can be explained by the linear model using min and max temperatures.

```
# VIF calc:
elev1 <- lm(Elevation ~ Min.Temp + Max.Temp, data = snotel_s %>% slice(-c(9,22)))
summary(elev1)
```

```
##
## Call:
## lm(formula = Elevation ~ Min.Temp + Max.Temp, data = snotel_s %>%
##     slice(-c(9, 22)))
##
## Residuals:
##      Min       1Q   Median       3Q      Max
## -1120.05  -142.99    14.45   186.73   624.61
##
## Coefficients:
##             Estimate Std. Error t value Pr(>|t|)
## (Intercept) 14593.21     699.77  20.854 4.85e-15
## Min.Temp     -208.82      38.94  -5.363 3.00e-05
## Max.Temp      -56.28      20.90  -2.693    0.014
##
## Residual standard error: 395.2 on 20 degrees of freedom
## Multiple R-squared:  0.8775, Adjusted R-squared:  0.8653
## F-statistic: 71.64 on 2 and 20 DF,  p-value: 7.601e-10
```

Using this result, we can calculate

$$\text{VIF}_{\text{elevation}} = \frac{1}{1 - R^2_{\text{elevation}}} = \frac{1}{1 - 0.8775} = \frac{1}{0.1225} = 8.16$$

```
1 - 0.8775
```

```
## [1] 0.1225
```

```
1/0.1225
```

```
## [1] 8.163265
```

Note that when we observe small VIFs (close to 1), that provides us with confidence that multicollinearity is not causing problems under the surface of a particular MLR model and that we can trust that the coefficients will not change dramatically based on whether the other terms in the model are removed. Also note that we can't use the VIFs to do anything about multicollinearity in the models – it is just a diagnostic to understand the magnitude of the problem.

8.6 MLR INFERENCE: PARAMETER INFERENCES USING THE T-DISTRIBUTION

I have been deliberately vague about what an important variable is up to this point, and chose to focus on some bigger modelling issues. Now we turn our attention to one of the most common tasks in any basic statistical model - assessing whether a particular observed result is more unusual than we would expect by chance if it really wasn't related to the response. The previous discussions of estimation in MLR models informs our interpretations of the tests. The t-tests for slope coefficients are based on our standard recipe - take the estimate, divide it by its standard error and then, assuming the statistic follows a t-distribution under the null hypothesis, find a p-value. This tests whether each true slope coefficient, βk , is 0 or not, in a model that contains the other variables. Again, sometimes we say "after adjusting for" the other x's or "conditional on" the other x's in the model or "after allowing for". . . as in the slope coefficient interpretations above. The main point is that you should not interpret anything related to slope coefficients in MLR without referencing the other variables that are in the model! The tests for the slope coefficients assess H0 : βk = 0, which in words is a test that there is no linear relationship between explanatory variable k and the response variable, y, in the population, given the other variables in model. The typical alternative hypothesis is H0 : βk /= 0. In words, the alternative hypothesis is that there is some linear relationship between explanatory variable k and the response variable, y, in the population, given the other variables in the model. It is also possible to test for positive or negative slopes in the alternative, but this is rarely the first concern, especially when MLR slopes can occasionally come out in unexpected directions.

The test statistic for these hypotheses is $t = \frac{b_k}{SE_k}$ and, if our assumptions hold, follows a t-distribution with n - K - 1 df where K is the number of predictor variables in the model. We perform the test for each slope coefficient, but the test is conditional on the other variables in the model - the order the variables are fit in does not change t-test results. For the Snow Depth example with Elevation and Maximum Temperature as predictors, the pertinent output is in the four columns of the Coefficient table that is the first part of the model summary we've been working with. You can find the estimated slope (Estimate column), the SE of the slopes (Std. Error column), the t-statistics (t value column), and the p-values (Pr(>|t|) column). The degrees of freedom for the t-distributions show up below the coefficients and the df=20 here. This is because n = 23 and K = 2, so df = 23 - 2 - 1 = 20.

```
m5 <- lm(Snow.Depth ~ Elevation + Max.Temp, data = snotel_s %>% slice(-c(9,22)))
summary(m5)
```

```
##
## Call:
## lm(formula = Snow.Depth ~ Elevation + Max.Temp, data = snotel_s %>%
##     slice(-c(9, 22)))
##
## Residuals:
##     Min      1Q  Median      3Q     Max
## -14.652  -4.645   0.518   3.744  20.550
##
## Coefficients:
##               Estimate Std. Error t value Pr(>|t|)
## (Intercept) -1.675e+02  3.924e+01  -4.269 0.000375
## Elevation    2.407e-02  3.162e-03   7.613 2.48e-07
## Max.Temp     1.253e+00  5.385e-01   2.327 0.030556
##
## Residual standard error: 8.726 on 20 degrees of freedom
## Multiple R-squared:  0.8495, Adjusted R-squared:  0.8344
## F-statistic: 56.43 on 2 and 20 DF,  p-value: 5.979e-09
```

The hypotheses for the *Maximum Temperature* term (*Max.Temp*) are:

- **H0 :** β Max. Temp = 0 given that Elevation is in the model vs
- **HA:** β Max. Temp /= 0 given that Elevation is in the model.

The test statistic is t = 2.327 with df = 20 (so under the null hypothesis the test statistic follows a t20-distribution).

The output provides a p-value of 0.0306 for this test. We can also find this using pt:

```
2*pt(2.327, df = 20, lower.tail = F)
```

```
## [1] 0.03058319
```

The chance of observing a slope for *Max.Temp* as extreme or more extreme than assuming there really is no linear relationship between *Max.Temp* and *Snow Depth* (in a model with *Elevation*), is about 3% so this presents moderate evidence against the null hypothesis, in favour of retaining this term in the model.

Conclusion

There is moderate evidence against the null hypothesis of no linear relationship between *Max.Temp* and *Snow Depth* (t_{20} = 2.33, p-value = 0.03), once we account for *Elevation*, so we can conclude that there likely is a linear relationship between them given *Elevation* in the population of SNOTEL sites in Montana on this day and we should retain this term in the model. Because we cannot randomly assign the temperatures to sites, we cannot conclude that temperature causes changes in the snow depth – in fact it might even be possible for a location to have different temperatures because of different snow depths. The inferences do pertain to the population of SNOTEL sites on this day because of the random sample from the population of sites.

Similarly, we can test for *Elevation* after controlling for the *Maximum Temperature*:

***H*0 : βElevation = 0 vs *HA* : βElevation /= 0**, given that *Max. Temp* is in the model: *t* = 7.613 (*df* = 20) with a p-value of 0.00000025 or just < 0.00001.

So there is strong evidence against the null hypothesis of no linear relationship between *Elevation* and *Snow Depth*, once we adjust for *Max.Temp* in the population of SNOTEL sites in Montana on this day, so we would conclude that they are linearly related and that we should retain the *Elevation* predictor in the model with *Max. Temp*. There is one last test that is of dubious interest in almost every situation – to test that the y-intercept(β0) in an MLR is 0. This tests if the true mean response is 0 when all the predictor variables are set to 0.I see researchers reporting this p-value frequently and it is possibly the most useless piece of information in the regression model summary. Sometimes less educated statistics users even think this result is proof of something interesting or are disappointed when the p-value is not small. Unless you want to do some prediction and are interested in whether the mean response when all the predictors are set to 0 is different from 0, this test should not be reported or, if reported, is certainly not very interesting12. But we should at least go through the motions on this test once so you don't make the same mistakes:

***H*0 : β0 = 0 vs *HA* : β0 /= 0** in a model with *Elevation* and *Maximum Temperature*. t = -4.269, with an assumption that the test statistic follows a t20-distribution under the null hypothesis, and the p-value = 0.000375.

There is strong evidence against the null hypothesis that the true mean *Snow Depth* is 0 when the *Maximum Temperature* is 0 and the *Elevation* is 0 in the population of SNOTEL sites, so we could conclude that the true mean Snow Depth is different from 0 at these values of the predictors. To reinforce the general uselessness of this test, think about the combination of *x*'s – is that even physically possible in Montana (or the continental US) in April?

Remember when testing slope coefficients in MLR, that if we find weak evidence against the null hypothesis, it does not mean that there is no relationship or even no linear relationship between the variables, but that there is insufficient evidence against the null hypothesis of no linear relationship once we account for the other variables in the model. If you do not find a small p-value for a variable, you should either be cautious when interpreting the coefficient, or not interpret it. Some model building strategies would lead to dropping the term from the model but sometimes we will have models to interpret that contain terms with larger p-values. Sometimes they are still of interest but the weight on the interpretation isn't as heavy as if the term had a small p-value – you should remember that you can't prove that coefficient is different from 0 in that model. It also may mean that you don't know too much about its specific value. Confidence intervals will help us pin down where we think the true slope coefficient might be located, given the other variables in the model, and so are usually pretty interesting to report, regardless of how you approached model building and possible refinement.

Confidence intervals provide the dual uses of inferences for the location of the true slope and whether the true slope seems to be different from 0. The confidence intervals here have our regular format of estimate ∓ margin of error. Like the previous tests, we work with *t*-distributions with *n* - *K* - 1 degrees of freedom. Specifically the 95% confidence interval for slope coefficient *k* is

$$b_k \mp t^*_{n-K-1}\mathrm{SE}_{b_k}$$

The2a 1 [unit of xk] increase in xk, we are 95% confident that the true mean of y changes by between LL and UL [units of Y] in the population, after adjusting for the other x's [list them!].

We can either calculate these intervals as we have many times before or rely on the confint function to do this:

```
confint(m5)
```

```
##                      2.5 %      97.5 %
## (Intercept) -249.37903311 -85.67576239
## Elevation      0.01747878   0.03067123
## Max.Temp       0.13001718   2.37644112
```

So for a 1°*F* increase in *Maximum Temperature*, we are 95% confident that the true mean *Snow Depth* will change by between 0.13 and 2.38 inches in the population, after adjusting for the *Elevation* of the sites. Similarly, for a 1 foot increase in *Elevation*, we are 95% confident that the true mean *Snow Depth* will change by between 0.0175 and 0.0307 inches in the population, after adjusting for the *Maximum Temperature* of the sites.

8.7 OVERALL F-TEST IN MULTIPLE LINEAR REGRESSION

In the MLR summary, there is an F -test and p-value reported at the bottom of the output. For the model with Elevation and Maximum Temperature, the last row of the model summary is:

```
## F-statistic: 56.43 on 2 and 20 DF, p-value: 5.979e-09
```

This test is called the overall F-test in MLR and is very similar to the F -test in a reference-coded One-Way ANOVA model. It tests the null hypothesis that involves setting every coefficient except the y-intercept to 0 (so all the slope coefficients equal 0). We saw this reduced model in the One-Way material when we considered setting all the deviations from the baseline group to 0 under the null hypothesis. We can frame this as a comparison between a full and reduced model as follows:

Full Model: $y_i = \beta_0 + \beta_1 x_{1i} + \beta_2 x_{2i} + \cdots + \beta_K x_{Ki} + \varepsilon_i$

Reduced Model: $y_i = \beta_0 + 0x_{1i} + 0x_{2i} + \cdots + 0x_{Ki} + \varepsilon_i$

The reduced model estimates the same values for all *y*'s, $\widehat{y}_i = \bar{y} = b_0$ and corresponds to the null hypothesis of:

- ***H0* : No explanatory variables should be included in the model:** $\beta_1 = \beta_2 = \cdots = \beta_K = 0$. The full model corresponds to the alternative:
- ***HA* : At least one explanatory variable should be included in the model: Not all** β_k's = 0 for

  $(k = 1, \ldots, K)$.

Note that β_0 is not set to 0 in the reduced model (under the null hypothesis) – it becomes the true mean of *y* for all values of the *x*'s since all the predictors are multiplied by coefficients of 0.

The test statistic to assess these hypotheses is $F = MS_{model}/MSE$, which is assumed to follow an *F* - distribution with *K* numerator *df* and $n - K - 1$ denominator *df* under the null hypothesis. The output provides us with $F(2, 20) = 56.43$ and a p-value of $5.979 * 10^{-9}$ (p-value < 0.00001) and strong evidence against the null hypothesis. Thus, there is strong evidence against the null hypothesis that the true slopes for the two predictors are 0 and so we would conclude that at least one of the two slope coefficients (*Max.Temp*'s or *Elevation*'s) is different from 0 in the population of SNOTEL sites in Montana on this date. While this test is a little bit interesting and a good indicator of something interesting existing in the model, the moment you see this result, you want to know more about each predictor variable. If neither predictor variable is important, we will discover that in the *t*-tests for each coefficient and so our general recommendation is to start there.

The overall F-test, then, is really about testing whether there is something good in the model somewhere. And that certainly is important but it is also not too informative. There is one situation where this test is really interesting, when there is only one predictor variable in the model (SLR). In that situation, this test provides exactly the same p-value as the *t*-test. *F* -tests will be important when we are mixing categorical and quantitative predictor variables in our MLR models (Section 8.12), but the overall *F* -test is of **very** limited utility.

8.8 CASE STUDY: FIRST YEAR COLLEGE GPA AND SATS

Many universities require students to have certain test scores in order to be admitted into their institutions. They obviously must think that those scores are useful predictors of student success to use them in this way. Quality assessments of recruiting classes are also based on their test scores. The Educational Testing Service (the company behind such fun exams as the SAT and GRE) collected a data set to validate their SATon n = 1000 students from an unnamed Midwestern university; the data set is available in the open intro package [Çetinkaya Rundel et al., 2022] in the satgpa data set. It is unclear from the documentation whether random sample was collected, in fact it looks like it certainly wasn't a random sample of all incoming students at a large university (more later). What potential issues would arise if a company was providing a data set to show the performance of their test and it was not based on a random sample?

We will proceed assuming they used good methods in developing their test (there are sophisticated statistical models underlying the development of the SAT and GRE) and also in obtaining a data set for testing out the performance of their tests that is at least representative of the students (or some types of students) at this university. They[13] provided information on the *SAT Verbal* (satv) and *Math* (satm)percentiles (these are not the scores but the ranking percentile that each score translated to in a particular year), *High School GPA* (hsgpa), *First Year* of college *GPA* (fygpa), *Gender* (gender of the students coded 1 and 2 with possibly 1 for males and 2 for females – the documentation was also unclear this). Should gender[even] be displayed in a plot with correlations since it is a categorical variable?14 Our interests here are in whether the two SAT percentiles are (together?) related to the first year college GPA, describing the size of their impacts and assessing the predictive potential of SAT-based measures for first year in college GPA. There are certainly other possible research questions that can be addressed with these data but this will keep us focused.

```
library(openintro)
data(satgpa)
satgpa <- as_tibble(satgpa)
satgpa <- satgpa %>% rename(gender = sex , #Renaming variables
                            satv = sat_v, satm = sat_m, satsum = sat_sum,
                            hsgpa = hs_gpa,
                            fygpa = fy_gpa)

satgpa %>%
  select(-4) %>%
  ggpairs() +
  theme_bw()
```

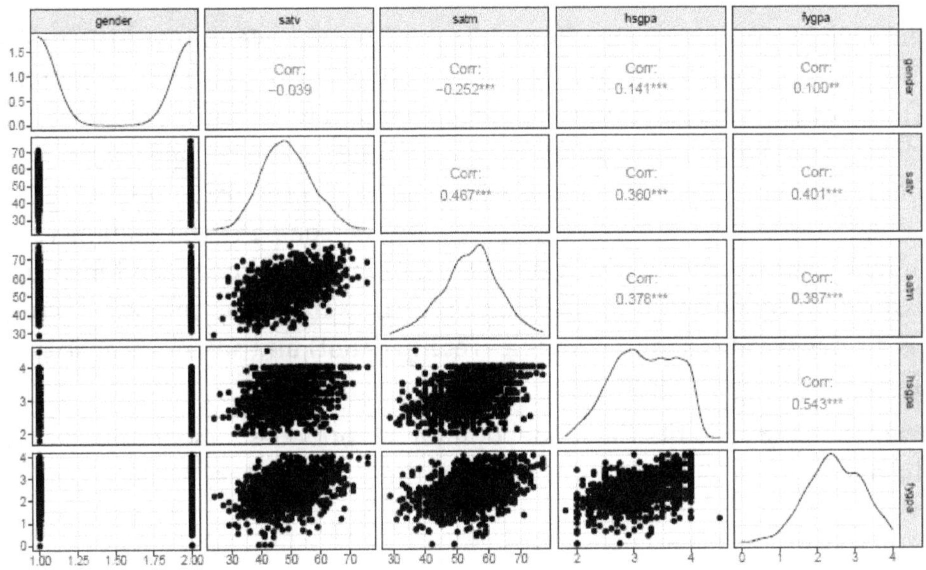

Figure 8.18: Scatterplot matrix of SAT and GPA data set.

There are positive relationships in Figure 8.18 among all the pre-college measures and the *college GPA* but none are above the moderate strength level. The *hsgpa* has a highest correlation with first year of college results but its correlation is not that strong. Maybe together in a model the SAT percentiles can also be useful? Also note this plot shows an odd *hsgpa* of 4.5 that probably should be removed[15] if that variable is going to be used (*hsgpa* was not used in the following models so the observation remains in the data).

In MLR, the modelling process is a bit more complex and often involves more than one model, so we will often avoid the 6+ steps in testing initially and try to generate a model we can use in that more specific process. In this case, the first model of interest using the two SAT percentiles, fygpai = β0 + βsatvsatvi + βsatmsatmi + εi, looks like it might be worth interrogating further so we can jump straight into considering the 6+ steps involved in hypothesis testing for the two slope coefficients to address our RQ about assessing the predictive ability and relationship of the SAT scores on first year college GPA. We will use *t*-based inferences, assuming that we can trust the assumptions and the initial plots get us some idea of the potential relationship.

Note that this is not a randomized experiment but we can assume that it is representative of the students at that single university. We would not want to extend these inferences to other universities (who might be more or less selective) or to students who did not get into this university and, especially, not to students that failed to complete the first year. The second and third constraints point to a severe limitation in this research– only students who were accepted, went to, and finished one year at this university could be studied. Lower SAT percentile students might not have been allowed in or may not have finished the first year and higher SAT students might have been attracted to other more prestigious institutions. So the scope of inference is just limited to students that were invited and chose to attend this institution and successfully completed one year of courses. It is hard to know if the SAT "works" when the inferences are so restricted in who they might apply to. . . But you could see why the company that administers the SAT might want to analyze these data. Educational researchers and institutional admissions officers also often focus on predicting first year retention rates, but that is a categorical response variable (retained/not) and so not compatible with the linear models considered here.

The following code fits the model of interest, provides a model summary, and the diagnostic plots, allowing us to consider the tests of interest:

```
gpa1 <- lm(fygpa ~ satv + satm, data = satgpa)
summary(gpa1)
```

```
##
## Call:
## lm(formula = fygpa ~ satv + satm, data = satgpa)
##
## Residuals:
##      Min       1Q   Median       3Q      Max
## -2.19647 -0.44777  0.02895  0.45717  1.60940
##
## Coefficients:
##             Estimate Std. Error t value Pr(>|t|)
## (Intercept) 0.007372   0.152292   0.048    0.961
## satv        0.025390   0.002859   8.879  < 2e-16
## satm        0.022395   0.002786   8.037 2.58e-15
##
## Residual standard error: 0.6582 on 997 degrees of freedom
## Multiple R-squared:  0.2122, Adjusted R-squared:  0.2106
## F-statistic: 134.2 on 2 and 997 DF,  p-value: < 2.2e-16
```

```
par(mfrow = c(2,2), oma = c(0,0,2,0))
plot(gpa1, sub.caption = "")
title(main="Diagnostics for GPA model with satv and satm", outer=TRUE)
```

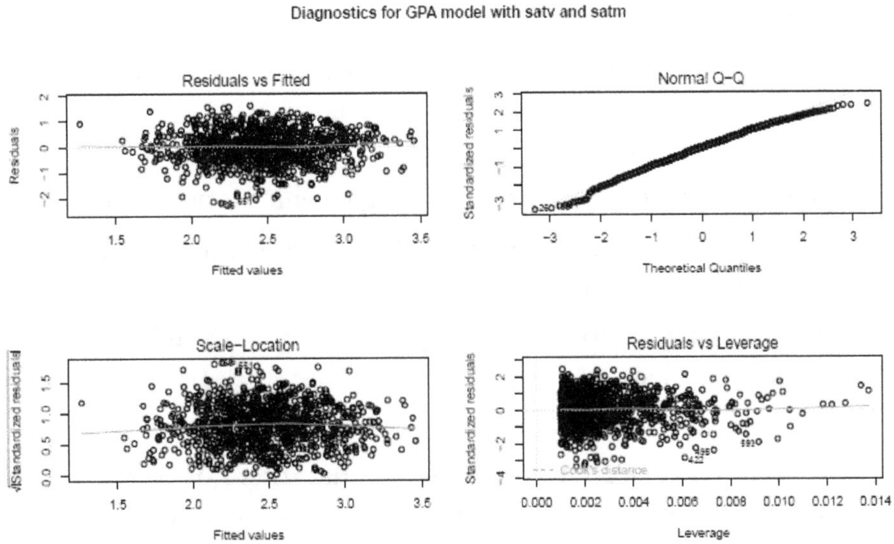

Figure 8.19: Diagnostic plots for the fygpa ~ satm + satm model.

1. Hypotheses of interest:

- H_0 : βsatv = 0 given *satm* in the model vs H_A : βsatv /= 0 given *satm* in the model.
- H_0 : βsatm = 0 given *satv* in the model vs H_A : βsatm /= 0 given *satv* in the model.

2. Plot the data and assess validity conditions:

- Quantitative variables condition:
    - The variables used here in this model are quantitative. Note that *Gender* was plotted in the previous scatterplot matrix and is not quantitative – we will explore its use later.
- Independence of observations:
    - With a sample from a single university from (we are assuming) a single year of students, there is no particular reason to assume a violation of the independence assumption. If there was information about students from different years being included or maybe even from different colleges in the university in a single year, we might worry about systematic differences in the GPAs and violations of the independence assumption. We can't account for either and there is possibly not a big difference in the GPAs across colleges to be concerned about, especially with a sample of students from a large university.

- Linearity of relationships:
  - The initial scatterplots (Figure 8.18) do not show any clear nonlinearities with each predictor used in this model.
  - The Residuals vs Fitted and Scale-Location plots (Figure 8.19) do not show much more than a football shape, which is our desired result.
  - The partial residuals are displayed in Figure 8.20 and do not suggest any clear missed curvature.
    - Together, there is no suggestion of a violation of the linearity assumption.
- Multicollinearity checked for:
  - The original scatterplots suggest that there is some collinearity between the two SAT percentiles with a correlation of 0.47. That is actually a bit lower than one might expect and suggests that each score must be measuring some independent information about different characteristics of the students.
  - VIFs also do not suggest a major issue with multicollinearity in the model with the VIFs for both variables the same at 1.27816. This suggests that both SEs are about 13% larger than they otherwise would have been due to shared information between the two predictor variables.

```
vif(gpa1)
```

```
##     satv     satm
## 1.278278 1.278278
```

```
sqrt(vif(gpa1))
```

```
##    satv    satm
## 1.13061 1.13061
```

- Equal (constant) variance:
  - There is no clear change in variability as a function of fitted values so no indication of a violation of the constant variance of residuals assumption.
- Normality of residuals:
  - There is a minor deviation in the upper tail of the residual distribution from normality. It is not pushing towards having larger values than a normal distribution would generate so should not cause us any real problems with inferences from this model. Note that this upper limit is

likely due to using GPA as a response variable and it has an upper limit. This is an example of a potentially censored variable. For a continuous variable it is possible that the range of a measurement scale doesn't distinguish among subjects who differ once they pass a certain point. For example, a 4.0 high school student is likely going to have a high first year college GPA, on average, but there is no room for variability in college GPA up, just down once you are at the top of the GPA scale. For students more in the middle of the range, they can vary up or down. So in some places you can get symmetric distributions around the mean and in others you cannot. There are specific statistical models for these types of responses that are beyond our scope. In this situation, failing to account for the censoring may push some slopes toward 0 a little because we can't have responses over 4.0 in college GPA to work with.

- No influential points:
    - There are no influential points. In large data sets, the influence of any point is decreased and even high leverage and outlying points can struggle to have any impacts at all on the results.

So we are fairly comfortable with all the assumptions being at least not clearly violated and so the inferences from our model should be relatively trustworthy.

3. Calculate the test statistics and p-values:

- For *satv*: $t = \ = 8.88$ with the t having $df = 997$ and p-value < 0.0001.
- For *satm*: $t = \ = 8.04$ with the t having $df = 997$ and p-value < 0.0001.

4. Conclusions:

- **For *satv*:** There is strong evidence against the null hypothesis of no linear relationship between *satv* and *fygpa* ($t_{997} = 8.88$, p-value < 0.0001) and conclude that, in fact, there is a linear relationship between *satv* percentile and the first year of college *GPA*, after controlling for the *satm* percentile, in the population of students that completed their first year at this university.
- **For *satm*:** There is strong evidence against the null hypothesis of no linear relationship between *satm* and *fygpa* ($t_{997} = 8.04$, p-value < 0.0001)and conclude that, in fact, there is a linear relationship between *satm* percentile and the first year of college *GPA*, after controlling for the *satv* percentile, in the population of students that completed their first year at this university.

5. Size:

- The model seems to be valid and have predictors with small p-values, but note how much of the variation is not explained by the model. It only explains 21.22% of the variation in the responses. So we found evidence that these variables are useful in predicting the responses, but are they useful enough to use for decisions on admitting students? By quantifying the size of the estimated slope coefficients, we can add to the information about how potentially useful this model might be. The estimated MLR model is

$$\text{fygpa}_i = 0.00737 + 0.0254 \cdot \text{satv}_i + 0.0224 \cdot \text{satm}_i$$

- So for a 1 percent increase in the *satv* percentile, we estimate, on average, a 0.0254 point change in *GPA*, after controlling for *satm* percentile. Similarly, for a 1 percent increase in the *satm* percentile, we estimate, on average, a 0.0224 point change in *GPA*, after controlling for *satv* percentile. While this is a correct interpretation of the slope coefficients, it is often easier to assess "practical" importance of the results by considering how much change this implies over the range of observed predictor values.

The term-plots (Figure 8.20) provide a visualization of the "size" of the differences in the response variable explained by each predictor. The *satv* term-plot shows that for the range of percentiles from around the 30th percentile to the 70th percentile, the mean first year *GPA* is predicted to go from approximately 1.9 to 3.0. That is a pretty wide range of differences in GPAs across the range of observed percentiles. This looks like a pretty interesting and important change in the mean first year GPA across that range of different SAT percentiles. Similarly, the *satm* term-plot shows that the *satm* percentiles were observed to range between around the 30th percentile and 70th percentile and predict mean GPAs between 1.95 and 2.8. It seems that the SAT Verbal percentiles produce slightly more impacts in the model, holding the other variable constant, but that both are important variables. The %95 confidence intervals for the means in both plots suggest that the results are fairly precisely estimated – there is little variability around the predicted means in each plot. This is mostly a function of the sample size as opposed to the model itself explaining most of the variation in the responses.

```
plot(allEffects(gpa1, residuals = T))
```

The confidence intervals also help us pin down the uncertainty in each estimated slope coefficient. As always, the "easy" way to get 95% confidence intervals is using the confint function:

```
confint(gpa1)
```

- So, for a 1 percent increase in the *satv* percentile, we are 95% confident that the true mean *fygpa* changes between 0.0198 and 0.031 points, in the population of students who completed this year at this institution, after controlling for *satm*. The *satm* result is similar with an interval from 0.0169 and 0.0279. Both of these intervals might benefit from rescaling the interpretation to, say, a 10 percentile satv effect plot

```
##                    2.5 %     97.5 %
## (Intercept) -0.29147825 0.30622148
## satv         0.01977864 0.03100106
## satm         0.01692690 0.02786220
```

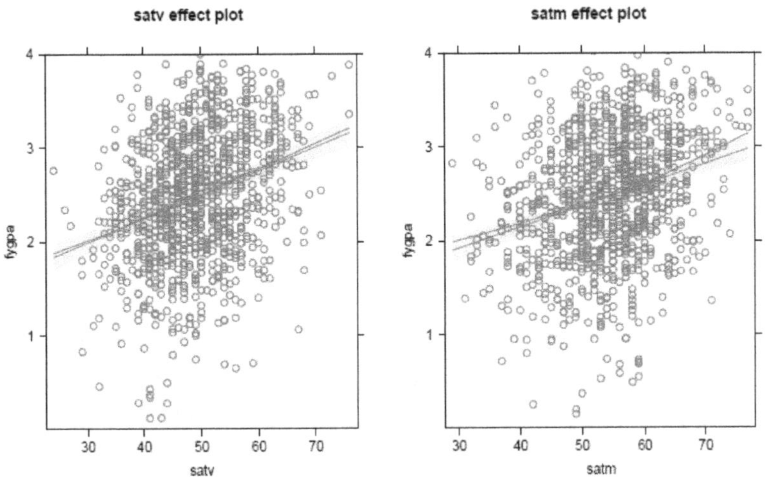

Figure 8.20: Term-plots for the fygpa ~ satv + satm model with partial residuals.

increase in the predictor variable, with the change in the *fygpa* for that level of increase of *satv* providing an interval from 0.198 to 0.31 points and for *satm* providing an interval from 0.169 to 0.279. So a boost of 10% in either exam percentile likely results in a noticeable but not huge average *fygpa* increase.

6. Scope of Inference:

- The term-plots also inform the types of students attending this university and successfully completing the first year of school. This seems like a good, but maybe not great, institution with few students scoring over the 75th percentile on either SAT Verbal or Math (at least that ended up in this data set). This result makes questions about their sampling mechanism re-occur as to who this data set might actually be representative of...
- Note that neither inference is causal because there was no random assignment of SAT percentiles to the subjects. The inferences are also limited to students who stayed in school long enough to get a GPA from their first year of college at this university.

One final use of these methods is to do prediction and generate prediction intervals, which could be quite informative for a student considering going to this university who has a particular set of SAT scores. For example, suppose that the student is interested in the average *fygpa* to expect with *satv* at the 30th percentile and *satm* at the 60th percentile. The predicted mean value is

$$\begin{aligned}\widehat{\mu}_{\text{fygpa}_i} &= 0.00737 + 0.0254 \cdot \text{satv}_i + 0.0224 \cdot \text{satm}_i \\ &= 0.00737 + 0.0254 * 30 + 0.0224 * 60 = 2.113.\end{aligned}$$

This result and the 95% confidence interval for the mean student *fygpa* at these scores can be found using the predict function as:

```
predict(gpa1, newdata = tibble(satv = 30, satm = 60))
```

```
##       1
## 2.11274
```

```
predict(gpa1, newdata = tibble(satv = 30,satm = 60), interval = "confidence")
```

```
##       fit      lwr      upr
## 1 2.11274 1.982612 2.242868
```

For students at the 30th percentile of satv and 60th percentile of satm, we are 95% confident that the true mean first year GPA is between 1.98 and 2.24 points. For an individual student, we would want the 95% prediction interval:

```
predict(gpa1, newdata = tibble(satv = 30, satm = 60), interval = "prediction")
```

```
##       fit       lwr      upr
## 1 2.11274 0.8145859 3.410894
```

For a student with *satv* = 30 and *satm* = 60, we are 95% sure that their first year GPA will be between 0.81 and 3.4 points. You can see that while we are very certain about the mean in this situation, there is a lot of uncertainty in the predictions for individual students. The PI is so wide as to almost not be useful.

To support this difficulty in getting a precise prediction for a new student, review the original scatter plots and partial residuals: there is quite a bit of vertical variability in first year *GPAs* for each level of any of the predictors. The residual SE, $\widehat{\sigma}$, is also informative in this regard – remember that it is the standard deviation of the residuals around the regression line. It is 0.6582, so the SD of new observations around the line is 0.66 GPA points and that is pretty large on a GPA scale. Remember that if the residuals meet our assumptions and follow a normal distribution around the line, observations within 2 or 3 SDs of the mean would be expected which is a large range of GPA values. Figure 8.21 remakes both term-plots, holding the other predictor at its mean, and adds the 95% prediction intervals to show the difference in variability between estimating the mean and pinning down the value of a new observation. The R code is very messy and rarely needed, but hopefully this helps reinforce the differences in these two types of intervals – to make them in MLR, you have to fix all but one of the predictor variables and we usually do that by fixing the other variables at their means.

```
# Remake effects plots with added 95% PIs
dv1 <- tibble(satv = seq(from = 24, to = 76, length.out = 50), satm = rep(54.4, 50))

mv1 <- as_tibble(predict(gpa1, newdata = dv1, interval = "confidence"))
pv1 <- as_tibble(predict(gpa1, newdata = dv1, interval = "prediction"))

mres_GPA_v <- bind_cols(dv1, mv1, pv1 %>% select(-fit))

# Rename CI and PI limits to have more explicit column names:
mres_GPA_v <- mres_GPA_v %>% rename(lwr_CI = lwr...4, upr_CI = upr...5,
                                    lwr_PI = lwr...6, upr_PI = upr...7)

v1 <- mres_GPA_v %>% ggplot() +
  geom_line(aes(x = satv, y = fit), lwd = 1) +
  geom_ribbon(aes(x = satv, ymin = lwr_CI, ymax = upr_CI), alpha = .4,
              fill = "beige", color = "darkred", lty = 2, lwd = 1) +
  geom_ribbon(aes(x = satv, ymin = lwr_PI, ymax = upr_PI), alpha = .1,
              fill = "gray80", color = "grey", lty = 3, lwd = 1.5) +
  labs(y = "GPA", x = "satv Percentile",
       title = "satv Effect plot with 95% CI and PI") +
  theme_bw()
```

```
dm1 <- tibble(satv = rep(48.93, 50), satm = seq(from = 29, to = 77, length.out = 50))

mm1 <- as_tibble(predict(gpa1, newdata = dm1, interval = "confidence"))
pm1 <- as_tibble(predict(gpa1, newdata = dm1, interval = "prediction"))

mres_GPA_m <- bind_cols(dm1, mm1, pm1 %>% select(-fit))

#Rename CI and PI limits to have more explicit column names:
mres_GPA_m <- mres_GPA_m %>% rename(lwr_CI = lwr...4, upr_CI = upr...5,
                                    lwr_PI = lwr...6, upr_PI = upr...7)

m1 <- mres_GPA_m %>% ggplot() +
  geom_line(aes(x = satm, y = fit), lwd = 1) +
  geom_ribbon(aes(x = satm, ymin = lwr_CI, ymax = upr_CI), alpha = .4,
              fill = "beige", color = "darkred", lty = 2, lwd = 1) +
  geom_ribbon(aes(x = satm, ymin = lwr_PI, ymax = upr_PI), alpha = .1,
              fill = "gray80", color = "grey", lty = 3, lwd = 1.5) +
  labs(y = "GPA", x = "satm Percentile",
       title = "satm Effect plot with 95% CI and PI") +
  theme_bw()

grid.arrange(v1, m1, ncol = 2)
```

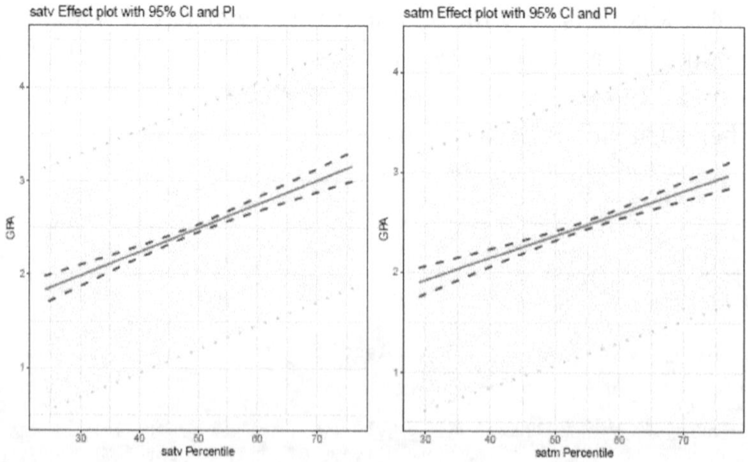

Figure 8.21: Term-plots for the fygpa ~ satv + satm model with 95% confidence intervals (dark, dashed lines) and 95% PIs (light grey, dotted lines).

8.9 DIFFERENT INTERCEPTS FOR DIFFERENT GROUPS: MLR WITH INDICATOR VARIABLES

One of the implicit assumptions up to this point was that the models were being applied to a single homogeneous population. In many cases, we take a sample from a population but that overall group is likely a combination of individuals from different sub-populations. For example, the SAT study was interested in all students at the university but that contains the obvious sub-populations based on the gender of the students. It is dangerous to fit MLR models across subpopulations but we can also use MLR models to address more sophisticated research questions by comparing groups. We will be able to compare the intercepts (mean levels) and the slopes to see if they differ between the groups. For example, does the relationship between the *satv* and *fygpa* differ for male and female students? We can add the grouping information to the scatter plot of *fygpa* vs *satv* (Figure 8.22) and consider whether there is visual evidence of a difference in the slope and/or intercept between the two groups, with men coded[17] as 1 and women coded as 2. Code below changes this variable to GENDER with more explicit labels, even though they might not be correct and the students were likely forced to choose one or the other.

It appears that the slope for females might be larger (steeper) in this relationship than it is for males. So increases in SAT Verbal percentiles for females might have more of an impact on the average first year GPA. We'll handle this sort of situation in Section 8.11, where we will formally consider how to change the slopes for different groups. In this section, we develop new methods needed to begin to handle these situations and explore creating models that assume the same slope coefficient for all groups but allow for different y-intercepts. This material ends up resembling what we did for the Two-Way ANOVA additive model.

The results for *satv* contrast with Figure 8.23 for the relationship between first year college *GPA* and *satm* percentile by gender of the students. The lines for the two groups appear to be mostly parallel and just seem to have different y-intercepts. In this section, we will learn how we can use our MLR techniques to fit a model to the entire data set that allows for different y-intercepts. The real power of this idea is that we can then also test whether the different groups have different y-intercepts – whether the shift between the groups is "real". In this example, it appears to suggest that females generally have slightly higher GPAs than males, on average, but that an increase in *satm* has about the same impact on GPA for both groups. If this difference in y-intercepts is not "real", then there appears to be no difference between the sexes in their relationship between *satm* and GPA and we can safely continue using a model that does not differentiate the two groups. We could also just subset the data set and do two analyses, but that approach will not allow us to assess whether things are "really" different between the two groups.

```
# Make 1,2 coded gender into factor GENDER
satgpa <- satgpa %>% mutate(GENDER = factor(gender))
# Make category names clear but note that level names might be wrong
levels(satgpa$GENDER) <- c("MALE", "FEMALE")

satgpa %>% ggplot(mapping = aes(x = satv, y = fygpa, color = GENDER, shape = GENDER)) +
  geom_smooth(method = "lm") +
  geom_point(alpha = 0.7) +
  theme_bw() +
  scale_color_viridis_d(end = 0.8, option = "plasma") +
  labs(title = "Scatterplot of GPA vs satv by gender")

satgpa %>% ggplot(mapping = aes(x = satm, y = fygpa, color = GENDER, shape = GENDER)) +
  geom_smooth(method = "lm") +
  geom_point(alpha = 0.7) +
  theme_bw() +
  scale_color_viridis_d(end = 0.8, option = "inferno") +
  labs(title = "Scatterplot of GPA vs satv by gender")
```

To fit one model to a data set that contains multiple groups, we need a way of entering categorical variable information in an MLR model. Regression models require quantitative predictor variables for the x's so we can't directly enter the text coded information on the gender of the students into the regression model since it contains "words" and how can multiply a word times a slope coefficient. To be able to put in "numbers" as predictors, we create what are called indicator variables18 that are made up of 0s and 1s, with the 0 reflecting one category and 1 the other, changing depending on the category of the individual in that row of the data set. The lm function does this whenever a factor variable is used as an explanatory variable. It setsup the indicator variables using a baseline category (which gets coded as a 0) and the deviation category for the other level of the variable (which gets coded as a 1). We can see how this works by exploring what happens when we put GENDER into our lm with satm, after first making sure it is categorical using the factor function and making the factor levels explicit instead of 1s and 2s.

```
SATGENDER1 <- lm(fygpa ~ satm + GENDER, data = satgpa) #Fit lm with satm and GENDER
summary(SATGENDER1)
```

```
##
## Call:
## lm(formula = fygpa ~ satm + GENDER, data = satgpa)
##
## Residuals:
##      Min       1Q   Median       3Q      Max
## -2.42124 -0.42363  0.01868  0.46540  1.66397
##
## Coefficients:
##              Estimate Std. Error t value Pr(>|t|)
## (Intercept)   0.21589    0.14858   1.453    0.147
## satm          0.03861    0.00258  14.969  < 2e-16
## GENDERFEMALE  0.31322    0.04360   7.184 1.32e-12
##
## Residual standard error: 0.6667 on 997 degrees of freedom
## Multiple R-squared:  0.1917, Adjusted R-squared:  0.1901
## F-statistic: 118.2 on 2 and 997 DF,  p-value: < 2.2e-16
```

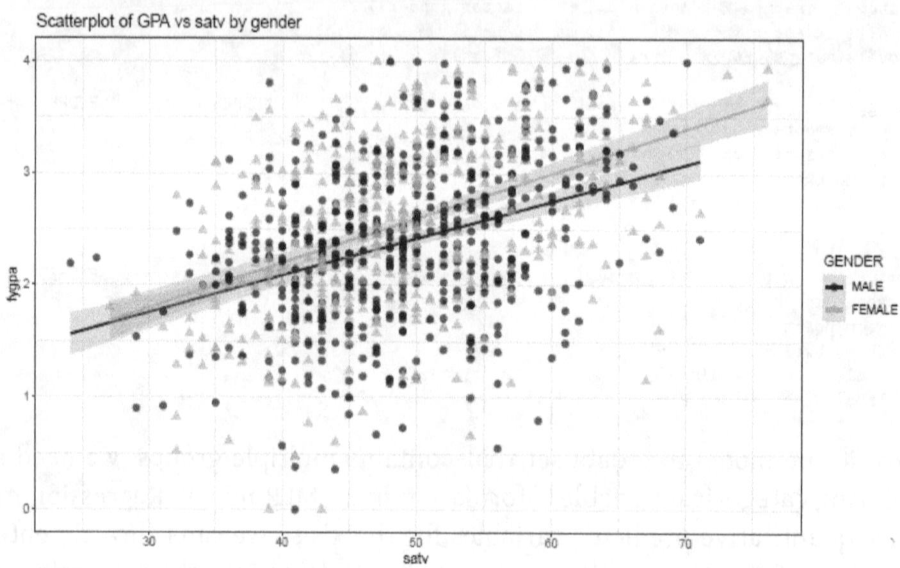

Figure 8.22: Plot of fygpa vs satv by gender of students.

The GENDER row contains information that the linear model chose *MALE* as the baseline category and *FEMALE* as the deviation category since *MALE* does not show up in the output. To see what lm is doing for us when we give it a two-level categorical variable, we can create our own "numerical" predictor that is 0 for males and 1 for females that we called GENDERINDICATOR, displayed for the first 10 observations:

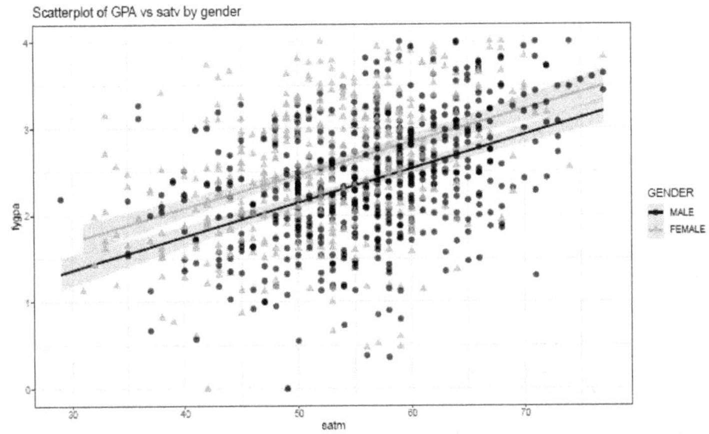

Figure 8.23: Plot of fygpa vs satm by gender of students.

```
# Convert logical to 1 for female, 0 for male using ifelse function
satgpa <- satgpa %>% mutate(GENDERINDICATOR = ifelse(GENDER == "FEMALE", 1, 0))
# Explore first 10 observations on the two versions of GENDER using the head() function
satgpa %>% select(GENDER, GENDERINDICATOR) %>% head(10)
```

```
## # A tibble: 10 x 2
##    GENDER GENDERINDICATOR
##    <fct>            <dbl>
##  1 MALE                 0
##  2 FEMALE               1
##  3 FEMALE               1
##  4 MALE                 0
##  5 MALE                 0
##  6 FEMALE               1
##  7 MALE                 0
##  8 MALE                 0
##  9 FEMALE               1
## 10 MALE                 0
```

We can define the indicator variable more generally by calling it IFemale,i to denote that it is an indicator (I) that takes on a value of 1 for observations in the category Female and 0 otherwise (Male) – changing based on the observation (i). Indicator variables, once created, are quantitative variables that take on values of 0 or 1 and we can put them directly into linear models with other x's (quantitative or categorical). If we replace the categorical GENDER variable with our quantitative GENDERINDICATOR and re-fit the model, we get:

```
SATGENDER2 <- lm(fygpa ~ satm + GENDERINDICATOR, data = satgpa)
summary(SATGENDER2)
```

```
##
## Call:
```

```
## lm(formula = fygpa ~ satm + GENDERINDICATOR, data = satgpa)
##
## Residuals:
##      Min       1Q   Median       3Q      Max
## -2.42124 -0.42363  0.01868  0.46540  1.66397
##
## Coefficients:
##                 Estimate Std. Error t value Pr(>|t|)
## (Intercept)      0.21589    0.14858   1.453    0.147
## satm             0.03861    0.00258  14.969  < 2e-16
## GENDERINDICATOR  0.31322    0.04360   7.184 1.32e-12
##
## Residual standard error: 0.6667 on 997 degrees of freedom
## Multiple R-squared:  0.1917, Adjusted R-squared:  0.1901
## F-statistic: 118.2 on 2 and 997 DF,  p-value: < 2.2e-16
```

This matches all the previous lm output except that we didn't get any information on the categories used since lm didn't know that *GENDERINDICATOR* was anything different from other quantitative predictors.

Now we want to think about what this model means. We can write the estimated model as $\text{fygpa}_i = 0.216 + 0.0386 \cdot \text{satm}i + 0.313 \cdot I\text{Female},i$

When we have a *male* observation, the indicator takes on a value of 0 so the 0.313 drops out of the model, leaving an SLR just in terms of *satm*. For a *female* student, the indicator is 1 and we add 0.313 to the previous *y*-intercept. The following works this out step-by-step, simplifying the MLR into two SLRs:

- Simplified model for *Males* (plug in a 0 for *I*Female,*i*):
  - $\text{fygpa}_i = 0.216 + 0.0386 \cdot \text{satm}i + 0.313 \cdot 0 = 0.216 + 0.0386 \cdot \text{satm}i$
- Simplified model for *Females* (plug in a 1 for *I*Female,*i*):
  - $\text{fygpa}_i = 0.216 + 0.0386 \cdot \text{satm}i + 0.313 \cdot 1$
  - $= 0.216 + 0.0386 \cdot \text{satm}i + 0.313$ (combine "like" terms to simplify the equation)
  - $= 0.529 + 0.0386 \cdot \text{satm}i$

In this situation, we then end up with two SLR models that relate *satm* to *GPA*, one model for *males* ($\text{fygpa}_i = 0.216 + 0.0386 \cdot \text{satm}i$) and one for *females* ($\text{fygpa}_i = 0.529 + 0.0386 \cdot \text{satm}i$). The only difference between these two models is in the *y*-intercept, with the *female* model's *y*-intercept shifted up from the *male* *y*-intercept by 0.313. And that is what adding indicator variables into models does in general[19] – it shifts the intercept up or down from the baseline group (here selected as *males*) to get a new intercept for the deviation group (here *females*).

To make this visually clearer, Figure 8.24 contains the regression lines that were estimated for each group. For any *satm*, the difference in the groups is the 0.313 coefficient from the GENDERFEMALE or GENDERINDICATOR row of the model summaries. For example, at *satm* = 50, the difference in terms of predicted average first year GPAs between males and females is displayed as a difference between 2.15 and 2.46. This model assumes that the slope on *satm* is the same for both groups except that they are allowed to have different *y*-intercepts, which is reasonable here because we saw approximately parallel relationships for the two groups in Figure 8.23.

Remember that lm selects baseline categories typically based on the alphabetical order of the levels of the categorical variable when it is created unless you actively use a function like relevel to change the baseline category. Here, the GENDER variable started with a coding of 1 and 2 and retained that order even with the recoding of levels that we created to give it more explicit names. Because we allow lm to create indicator variables for us, the main thing you need to do is explore the model summary and look for the hint at the baseline level that is not displayed after the name of the categorical variable.

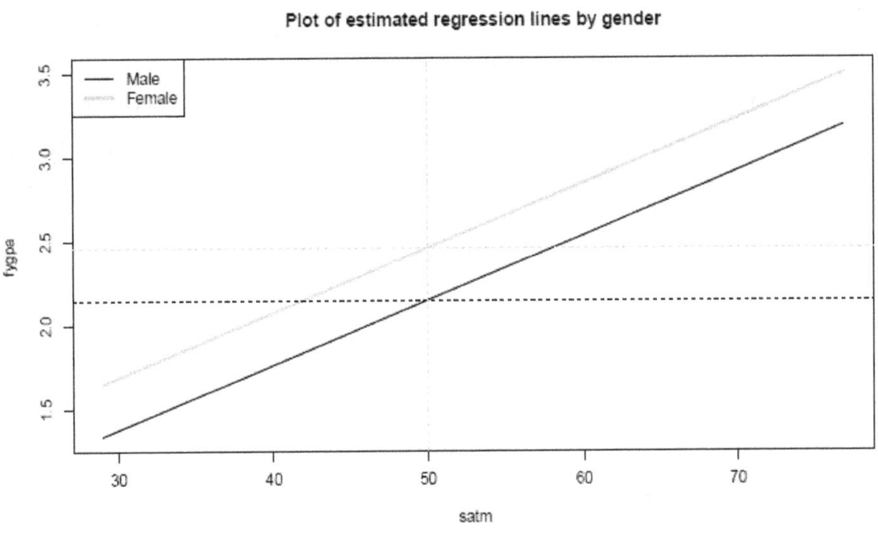

Figure 8.24: Plot of estimated model for *fygpa* vs *satm* by *GENDER* of students (female line is thicker dark line). Dashed lines aid in seeing the consistent vertical difference of 0.313 in the two estimated lines based on the model containing a different intercept for each group.

We can also work out the impacts of adding an indicator variable to the model in general in the theoretical model with a single quantitative predictor xi and indicator Ii. The model starts as in the equation below.

$$y_i = \beta_0 + \beta_1 x_i + \beta_2 I_i + \varepsilon_i$$

Again, there are two versions:

- For any observation *i* in the **baseline** category, $I_i = 0$ and the model is $y_i = \beta_0 + \beta_1 x_i + \varepsilon_i$.

- For any observation i in the **non-baseline (deviation)** category, $I_i = 1$ and the model simplifies to $y_i = (\beta_0 + \beta_2) + \beta_1 x_i + \varepsilon_i$.
    - This model has a y-intercept of $\beta_0 + \beta_2$.

The interpretation and inferences for β_1 resemble the work with any MLR model, noting that these results are "controlled for", "adjusted for", or "allowing for differences based on" the categorical variable in the model. The interpretation of β_2 is as a shift up or down in the y-intercept for the model that includes x_i. When we make term-plots in a model with a quantitative and additive categorical variable, the two reported model components match with the previous discussion – the same estimated term from the quantitative variable for all observations and a shift to reflect the different y-intercepts in the two groups. In Figure 8.25, the females are estimated to be that same 0.313 points higher on first year GPA. The males have a mean GPA slightly above 2.3 which is the predicted GPA for the average satm percentile for a male (remember that we have to hold the other variable at its mean to make each term-plot). When making the satm term-plot, the intercept is generated based on a weighted average of the intercept for the baseline category (male) of $b_0 = 0.216$ and the intercept for the deviation category (female) of $b_0 + b_2 = 0.529$ with weights of $516/1000 = 0.516$ for the estimated male intercept and $484/1000 = 0.484$ for estimated female intercept, $0.516 \cdot 0.216 + 0.484 \cdot 0.529 = 0.368$.

```
tally(GENDER ~ 1, data = satgpa)
```

```
##             1
## GENDER      1
##    MALE   516
##    FEMALE 484
```

```
plot(allEffects(SATGENDER1))
```

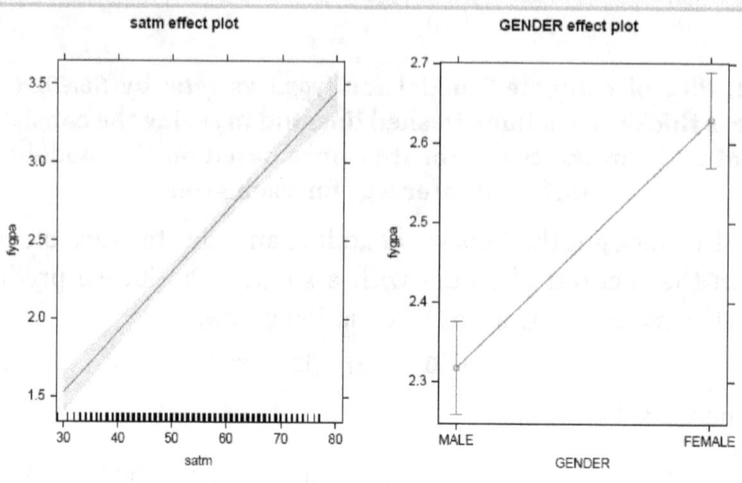

Figure 8.25: Term-plots for the estimated model for fygpa ~ satm + GENDER.

The model summary and confidence intervals provide some potential interesting inferences in these models. Again, these are just applications of MLR methods we have already seen except that the definition of one of the variables is "different" using the indicator coding idea. For the same model, the GENDER coefficient can be used to generate inferences for differences in the mean the groups, controlling for their *satm* scores.

```
##                Estimate Std. Error t value Pr(>|t|)
## GENDERFEMALE    0.31322    0.04360   7.184 1.32e-12
```

Testing the null hypothesis that $H_0 : \beta_2 = 0$ vs $H_A : \beta_2 /= 0$ using our regular *t*-test provides the opportunity to test for a difference in intercepts between the groups. In this situation, the test statistic is $t = 7.184$ and, based on a *t*997-distribution if the null is true, the p-value is < 0.0001. We have very strong evidence against the null hypothesis that there is no difference in the true *y*-intercept in a *satm* model for first year college GPA between *males* and *females*, so we would conclude that there is a difference in their true mean GPA levels controlled for *satm*. The confidence interval is also informative:

```
confint(SATGENDER1)
```

```
##                     2.5 %     97.5 %
## (Intercept)   -0.07566665 0.50744709
## satm           0.03355273 0.04367726
## GENDERFEMALE   0.22766284 0.39877160
```

We are 95% confident that the true mean GPA for females is between 0.228 and 0.399 points higher than for males, after adjusting for the *satm* in the population of students. If we had subset the data set by gender and fit two SLRs, we could have obtained the same simplified regression models for each group but we never could have performed inferences for the differences between the two groups without putting all the observations together in one model and then assessing those differences with targeted coefficients. We also would not be able to get an estimate of their common slope for *satm*, after adjusting for differences in the intercept for each group.

8.10 ADDITIVE MLR WITH MORE THAN TWO GROUPS: HEADACHE EXAMPLE

The same techniques can be extended to more than two groups. A study was conducted to explore sound tolerances using n = 98 subjects with the data available in the Headache data set from the heplots package [Fox and Friendly, 2021]. Each subject was initially exposed to a tone, stopping when the tone became definitely intolerable (DU) and that decibel level was recorded (variable called du1). Then the subjects were randomly assigned to one of four treatments: T1 (Listened again to the tone at their initial DU level, for the same amount of time they were able to tolerate it before); T2 (Same as T1, with one additional minute of exposure); T3 (Same as T2, but the subjects were explicitly instructed to use the relaxation techniques); and Control (these subjects experienced no further exposure to the noise tone until the final sensitivity measures were taken). Then the DU was measured again (variable called du2). One would expect that there would be a relationship between the upper tolerance levels of the subjects before and after treatment. But maybe the treatments impact that relationship? We can use our indicator approach to see if the treatments provide a shift to higher tolerances after accounting for the relationship between the two measurements20. The scatterplot21 of the results in Figure 8.26 shows some variation in the slopes and the intercepts for the groups although the variation in intercepts seems more prominent than differences in slopes. Note thatthe fct_relevel function was applied to the treatment variable with an option of "Control" to make the Control category the baseline category as the person who created the data set had set T1 as the baseline in the treatment variable.

```
library(heplots)
data(Headache)
Headache <- as_tibble(Headache)
Headache
```

```
## # A tibble: 98 x 6
##    type    treatment    u1   du1    u2   du2
##    <fct>   <fct>     <dbl> <dbl> <dbl> <dbl>
##  1 Migrane T3         2.34  5.3   5.8   8.52
##  2 Migrane T1         2.73  6.85  4.68  6.68
##  3 Tension T1         0.37  0.53  0.55  0.84
##  4 Migrane T3         7.5   9.12  5.7   7.88
##  5 Migrane T3         4.63  7.21  5.63  6.75
##  6 Migrane T3         3.6   7.3   4.83  7.32
##  7 Migrane T2         2.45  3.75  2.5   3.18
##  8 Migrane T1         2.31  3.25  2     3.3
##  9 Migrane T1         1.38  2.33  2.23  3.98
## 10 Tension T3         0.85  1.42  1.37  1.89
## # ... with 88 more rows
## # i Use `print(n = ...)` to see more rows
```

```
Headache <- Headache %>% mutate(treatment = factor(treatment),
                                treatment = fct_relevel(treatment, "Control")
```

```
                                )
# Make treatment a factor and Control the baseline category
Headache %>% ggplot(mapping = aes(x = du1, y = du2, color = treatment,
                                  shape = treatment)) +
  geom_smooth(method = "lm", se = F) +
  geom_point(size = 2.5) +
  theme_bw() +
  scale_color_viridis_d(end = 0.85, option = "inferno") +
  labs(title = "Scatterplot of Maximum DB tolerance before &
       after treatment (by treatment)")
```

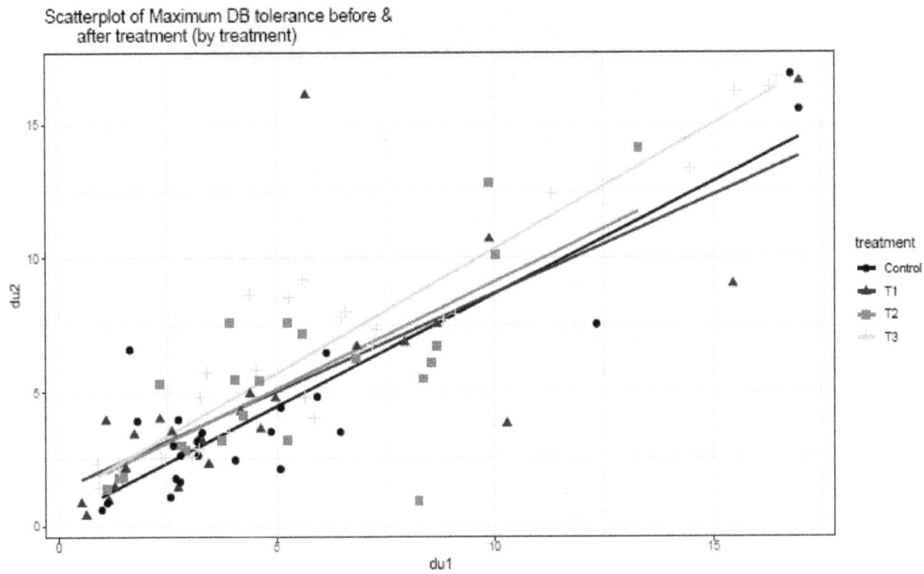

Figure 8.26: Scatterplot of post-treatment decibel tolerance (du2) vs pre-treatment tolerance (du1) by treatment level (4 groups).

This data set contains a categorical variable with 4 levels. To go beyond two groups, we have to add more than one indicator variable, defining three indicators to turn on (1) or off (0) for three of the levels of the variable with the same reference level used for all the indicators. For this example, the *Control* group is chosen as the baseline group so it hides in the background while we define indicators for the other three levels. The indicators for *T1*, *T2*, and *T3* treatment levels are:

- Indicator for *T1*: $I_{T1,i} = \begin{cases} 1 & \text{if Treatment} = T1 \\ 0 & \text{else} \end{cases}$

- Indicator for *T2*: $I_{T2,i} = \begin{cases} 1 & \text{if Treatment} = T2 \\ 0 & \text{else} \end{cases}$

- Indicator for *T3*: $I_{T3,i} = \begin{cases} 1 & \text{if Treatment} = T3 \\ 0 & \text{else} \end{cases}$

We can see the values of these indicators for a few observations and their original variable (treatment) in the following output. For *Control* all the indicators stay at 0.

Treatment	I_T1	I_T2	I_T3
T3	0	0	1
T1	1	0	0
T1	1	0	0
T3	0	0	1
T3	0	0	1
T3	0	0	1
T2	0	1	0
T1	1	0	0
T1	1	0	0
T3	0	0	1
T3	0	0	1
T2	0	1	0
T3	0	0	1
T1	1	0	0
T3	0	0	1
Control	0	0	0
T3	0	0	1

When we fit the additive model of the form y ~ x + group, the lm function takes the J categories and creates $J - 1$ indicator variables. The baseline level is always handled in the intercept. The true model will be of the form

yi = β0 + β1xi + β2ILevel2,i + β3ILevel3,i + · · · + βJ ILevelJ,i + εi

where the *ICatNamej,i*'s are the different indicator variables. Note that each indicator variable gets a coefficient associated with it and is "turned on" whenever the ith observation is in that category. At most only one of the *ICatNamej,i*'s is a 1 for any observation, so the *y*-intercept will either be β0 for the baseline group orβ0 + β*j* for *j* = 2, . . . , *J* . It is important to remember that this is an "additive" model since the effects just add and there is no interaction between the grouping variable and the quantitative predictor. To be able to trust this model, we need to check that we do not need different slope coefficients for the groups as discussed in the next section.

For these types of models, it is always good to start with a plot of the data set with regression lines for each group – assessing whether the lines look relatively parallel or not. In Figure 8.26, there are some differences in slopes – we investigate that further in the next section. For now, we can proceed with fitting the additive model with different intercepts for the four levels of treatment and the quantitative explanatory variable, du1.

```
head1 <- lm(du2 ~ du1 + treatment, data = Headache)
summary(head1)
```

```
##
## Call:
## lm(formula = du2 ~ du1 + treatment, data = Headache)
##
## Residuals:
##     Min      1Q  Median      3Q     Max
## -6.9085 -0.9551 -0.3118  1.1141 10.5364
##
## Coefficients:
##             Estimate Std. Error t value Pr(>|t|)
## (Intercept)  0.25165    0.51624   0.487   0.6271
## du1          0.83705    0.05176  16.172   <2e-16
## treatmentT1  0.55752    0.61830   0.902   0.3695
## treatmentT2  0.63444    0.63884   0.993   0.3232
## treatmentT3  1.36671    0.60608   2.255   0.0265
##
## Residual standard error: 2.14 on 93 degrees of freedom
## Multiple R-squared:  0.7511, Adjusted R-squared:  0.7404
## F-statistic: 70.16 on 4 and 93 DF,  p-value: < 2.2e-16
```

The complete estimated regression model is $\widehat{\text{du2}}_i = 0.252 + 0.837 \cdot \text{du1}i + 0.558IT1,i + 0.634IT2,i + 1.367IT3,i$ For each group, the model simplifies to an SLR as follows:

- For *Control* (baseline): $\widehat{\text{du2}}_i = 0.252 + 0.837 \cdot \text{du1}i + 0.558IT1,i + 0.634IT2,i + 1.367IT3,i$

  $= 0.252 + 0.837 \cdot \text{du1}i + 0.558 * 0 + 0.634 * 0 + 1.367 * 0$

  $= 0.252 + 0.837 \cdot \text{du1}i.$

- For *T1* : $\widehat{\text{du2}}_i = 0.252 + 0.837 \cdot \text{du1}i + 0.558IT1,i + 0.634IT2,i + 1.367IT3,i$

  $= 0.252 + 0.837 \cdot \text{du1}i + 0.558 * 1 + 0.634 * 0 + 1.367 * 0$

  $= 0.252 + 0.837 \cdot \text{du1}i + 0.558$

  $= 0.81 + 0.837 \cdot \text{du1}i.$

- Similarly for *T2* : $\widehat{\text{du2}}_i = 0.886 + 0.837 \cdot \text{du1}i$

  Finally for *T3* : $\widehat{\text{du2}}_i = 1.62 + 0.837 \cdot \text{du1}i$

To reinforce what this additive model is doing, Figure 8.27 displays the estimated regression lines for all four groups, showing the shifts in the *y*-intercepts among the groups.

The right panel of the term-plot (Figure 8.28) shows how the T3 group seems to have shifted up the most relative to the others and the Control group seems to have a mean that is a bit lower than the others, in the model that otherwise assumes that the same linear relationship holds between du1 and du2 for all the groups. After controlling for the Treatment group, for a 1 decibel increase in initial tolerances, we estimate, on average, to obtain a 0.84 decibel change in the second tolerance measurement. The R2 shows that this is a decent model for the responses, with this model explaining 75.1% percent of the variation in the second decibel tolerance measure. We should check the diagnostic plots and VIFs to check for any issues – all the diagnostics and assumptions are as before except that there is no assumption of linearity between the grouping variable and the responses. Additionally, sometimes we need to add group information to diagnostics to see if any patterns in residuals look different in different groups, like linearity or non-constant variance, when we are fitting models that might contain multiple groups.

```
plot(allEffects(head1, residuals = T), grid = T)
```

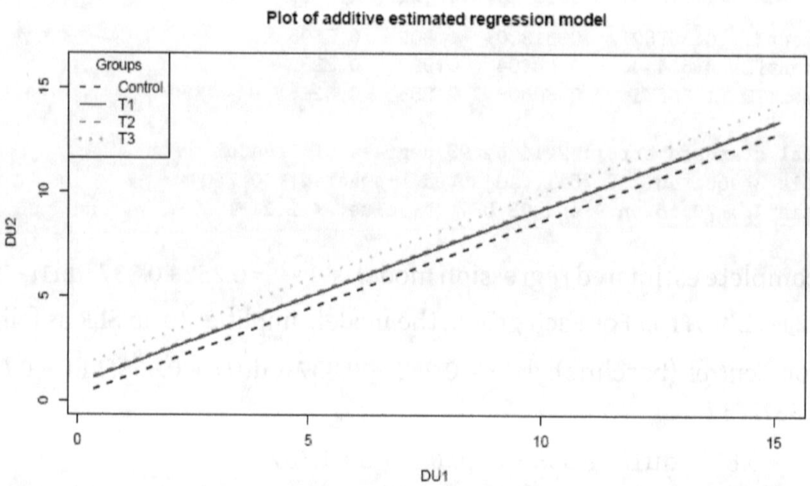

Figure 8.27: Plot of estimated noise tolerance additive model.

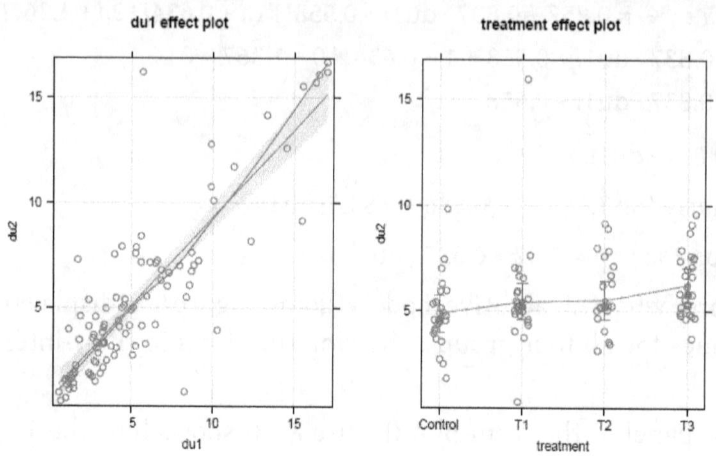

Figure 8.28: Term-plots of the additive decibel tolerance model with partial residuals.

The diagnostic plots in Figure 8.29 provides some indications of a few observations in the tails that deviate from a normal distribution to having slightly heavier tails but only one outlier is of real concern and causes some concern about the normality assumption. There is a small indication of increasing variability as a function of the fitted values as both the Residuals vs. Fitted and Scale-Location plots show some fanning out for higher values but this is a minor issue. There are no influential points here since all the Cook's D values are less than 0.5.

```
par(mfrow = c(2,2), oma = c(0,0,3,0))
plot(head1, pch = 16, sub.caption = "")
title(main="Plot of diagnostics for additive model with du1 and
     treatment for du2", outer=TRUE)
```

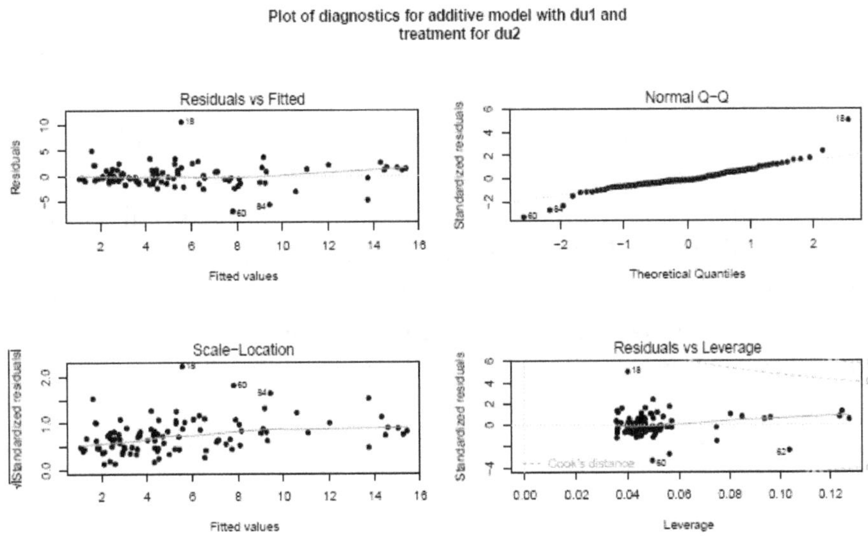

Figure 8.29: Diagnostic plots for the additive decibel tolerance model.

Additionally, sometimes we need to add group information to diagnostics to see if any patterns in residuals look different in different groups, like linearity or non-constant variance, when we are fitting models that might contain multiple groups. We can use the same scatterplot tools to make our own plot of the residuals (extracted using the residuals function) versus the fitted values (extracted using the fitted function) by groups as in Figure 8.30. This provides an opportunity to introduce faceting, where we can split our plots into panels by a grouping variable, here by the treatment applied to each subject. This can be helpful with multiple groups to be able to see each one more clearly as we avoid over plotting. The addition of + facet_grid(cols = vars(treatment)) facets the plot based on the treatment variable and puts the facets in different columns because of the cols = part of the code (rows = specifies the number of rows for the facets), labelling each panel at the top with the level being displayed of the faceting variable (vars() is needed to help ggplot find the variable). In this example, there are no additional patterns identified by making this plot although we do see some minor deviations in the fitted lines for each group, but it is a good additional check in these multi-group situations.

```
Headache <- Headache %>% mutate(resids = residuals(head1),
                                fits = fitted(head1)
                                )
Headache %>% ggplot(mapping = aes(x = fits, y = resids,
                                  color = treatment, shape = treatment)) +
  geom_smooth(method = "lm", se = F) +
  geom_point(size = 2.5) +
  theme_bw() +
  scale_color_viridis_d(end = 0.85, option = "inferno") +
  labs(title = "Scatterplot of Residuals vs Fitted by Treatment Group") +
  facet_grid(cols = vars(treatment))
```

The VIFs are different for models with categorical variables than for models with only quantitative predictors in MLR, even though we are still concerned with shared information across the predictors of all kinds. For categorical predictors, the J levels are combined to create a single measure for the predictor all together called the ***generalized VIF (GVIF)***. For GVIFs, interpretations are based on the GVIF measure to the power $1/(2 * (J - 1))$. For quantitative predictors when GVIFS are present, $J = 2$, and the power simplifies to 1/2, which is our regular square-root scale for inflation of standard errors due to multicollinearity

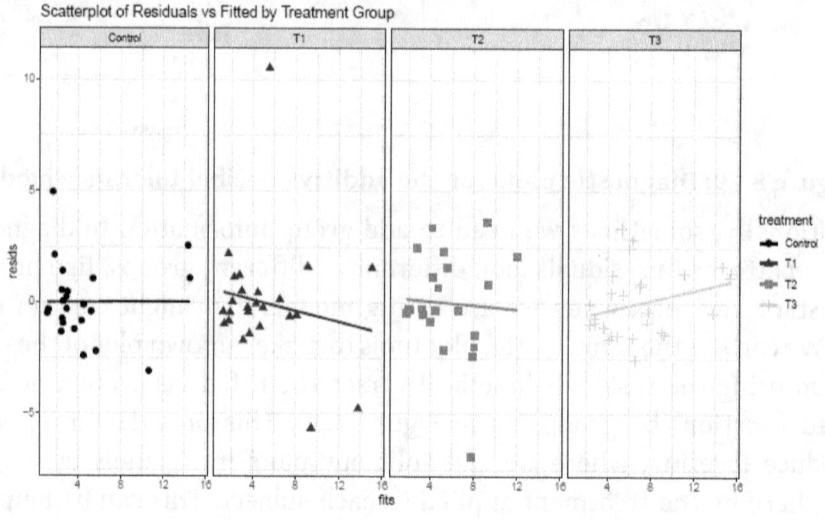

Figure 8.30: Faceted scatterplot of residuals versus fitted values by treatment group from the additive decibel tolerance model.

(so the GVIF is the VIF for quantitative predictors). For a J -level categorical predictor, the power is also 1/2 for J = 2 levels and increases for more levels. There are no rules of thumb for GVIFs for J > 2. In the following output, there are four levels, so J = 4. When raised to the requisite power, the GVIF interpretation for multi-category categorical predictors is the multiplicative increase in the SEs for the coefficients on all the indicator variables due to multicollinearity with other predictors. In this model, the SE for the quantitative predictor du1 is 1.009 times larger due to multicollinearity with other predictors and the SEs for the indicator variables for the four-level categorical treatment predictor are 1.003 times larger due to multicollinearity, both compared to what they would have been with no shared information in the predictors in the model. Neither are large, so multicollinearity is not a problem in this model.

```
vif(head1)
```

```
##                 GVIF Df GVIF^(1/(2*Df))
## du1          1.01786  1        1.008891
## treatment 1.01786  3        1.002955
```

While there are inferences available in the model output, the tests for the indicator variables are not too informative (at least to start) since they only compare each group to the baseline. In Section 8.12, we see how to use ANOVA F -tests to help us ask general questions about including a categorical predictor in the model. But we can compare adjusted R2 values with and without Treatment to see if including the categorical variable was "worth it":

```
head1R <- lm(du2 ~ du1, data = Headache)
```

```
summary(head1R)
```

```
##
## Call:
## lm(formula = du2 ~ du1, data = Headache)
##
## Residuals:
##     Min      1Q  Median      3Q     Max
## -6.9887 -0.8820 -0.2765  1.1529 10.4165
##
## Coefficients:
##             Estimate Std. Error t value Pr(>|t|)
## (Intercept)  0.84744    0.36045   2.351   0.0208
## du1          0.85142    0.05189  16.408   <2e-16
##
## Residual standard error: 2.165 on 96 degrees of freedom
## Multiple R-squared:  0.7371, Adjusted R-squared:  0.7344
## F-statistic: 269.2 on 1 and 96 DF,  p-value: < 2.2e-16
```

The adjusted R^2 in the model with both *Treatment* and *du1* is 0.7404 and the adjusted R^2 for this reduced model with just *du1* is 0.7344, suggesting the *Treatment* is useful. The next section provides a technique to be able to work with different slopes on the quantitative predictor for each group. Comparing those results to the results for the additive model allows assessment of the assumption in this section that all the groups had the same slope coefficient for the quantitative variable.

8.11 DIFFERENT SLOPES AND DIFFERENT INTERCEPTS

Sometimes researchers are specifically interested in whether the slopes vary across groups or the regression lines in the scatter plot for the different groups may not look parallel or it may just be hard to tell visually if there really is a difference in the slopes. Unless you are **very sure** that there is not an interaction between the grouping variable and the quantitative predictor, you should[22] start by fitting a model containing an interaction and then see if you can drop it. It may be the case that you end up with the simpler additive model from the previous sections, but you don't want to assume the same slope across groups unless you are absolutely sure that is the case. This should remind you a bit of the discussions of the additive and interaction models in the Two-way ANOVA material. The models, concerns, and techniques are very similar, but with the quantitative variable replacing one of the two categorical variables. As always, the scatterplot is a good first step to understanding whether we need the extra complexity that these models require.

A new example provides motivation for the consideration of different slopes and intercepts. A study was performed to address whether the relationship between nonverbal IQs and reading accuracy differs between dyslexic and non-dyslexic students. Two groups of students were identified, one group of *dyslexic* students was identified first (19 students) and then a group of gender and age similar student matches were identified(25 students) for a total sample size of $n = 44$, provided in the dyslexic3 data set from the smdata package[Merkle and Smithson, 2018]. This type of study design is an attempt to "balance" the data from the two groups on some important characteristics to make the comparisons of the groups as fair as possible. The researchers attempted to balance the characteristics of the subjects in the two groups so that if they found different results for the two groups, they could attribute it to the main difference they used to create the groups – dyslexia or not. This design, ***case-control*** or **case-comparison** where each subject with a trait is matched to one or more subjects in the "control" group would hopefully reduce confounding from other factors and then allow stronger conclusions in situations where it is impossible to randomly assign treatments to subjects. We still would avoid using "causal" language but this design is about as good as you can get when you are unable to randomly assign levels to subjects.

Using these data, we can explore the relationship between nonverbal IQ scores and reading accuracy, with reading accuracy measured as a proportion correct. The fact that there is an upper limit to the response variable attained by many students will cause complications below, but we can still learn something from our attempts to analyze these data using an MLR model. The scatterplot in Figure 8.31 seems to indicate some clear differences in the *IQ* vs *reading score* relationship between the *dys* = 0 (non-dyslexic) and *dys*= 1 (dyslexic) students (code below makes these levels more explicit in the data set). Note that the IQ isstandardized to have mean 0 and standard deviation of 1 which means that a 1 unit change in IQ score is a 1 SD change and that the *y*-intercept (for *x* = 0) is right in the center of the plot and actually interesting[23].

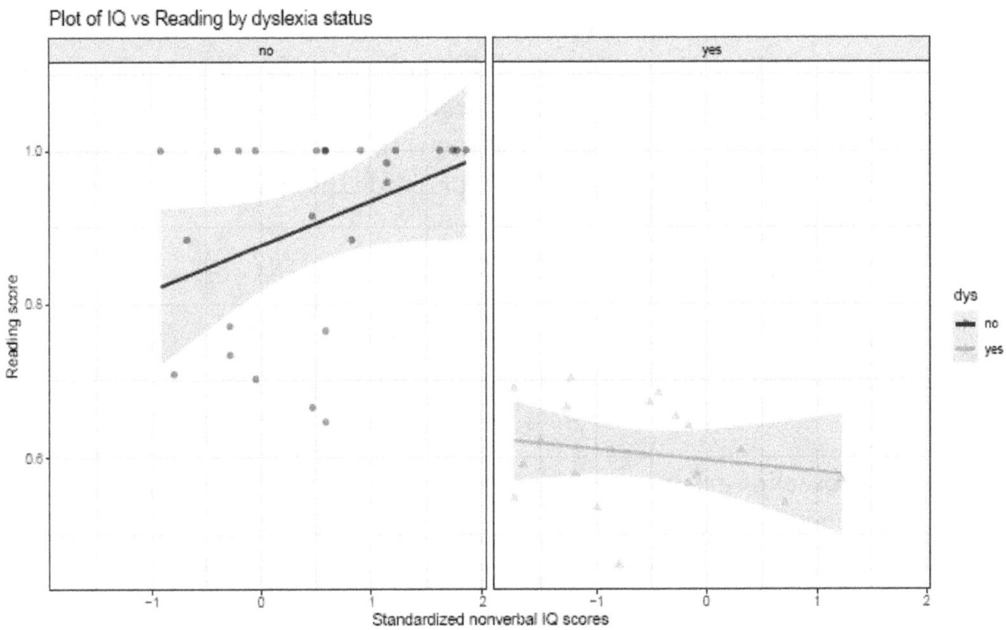

Figure 8.31: Scatterplot for reading score versus nonverbal IQ by dyslexia group.

```
library(smdata)
data("dyslexic3")
dyslexic3 <- dyslexic3 %>% mutate(dys = factor(dys))
levels(dyslexic3$dys) <- c("no", "yes")

dyslexic3 %>% ggplot(mapping = aes(x = ziq, y = score, color = dys, shape = dys)) +
  geom_smooth(method = "lm") +
  geom_point(size = 2, alpha = 0.5) +
  theme_bw() +
  scale_color_viridis_d(end = 0.7, option = "plasma") +
  labs(title = "Plot of IQ vs Reading by dyslexia status",
       x = "Standardized nonverbal IQ scores",
       y = "Reading score") +
  facet_grid(cols = vars(dys))
```

To allow for both different y-intercepts and slope coefficients on the quantitative predictor, we need to include a "modification" of the slope coefficient. This is performed using an interaction between the two predictor variables where we allow the impacts of one variable (slopes) to change based on the levels of another variable (grouping variable). The formula notation is y ~ x * group, remembering that this also includes the main effects (the additive variable components) as well as the interaction coefficients; this is similar to what we discussed in the Two-Way ANOVA interaction model. We can start with the general model for a two-level categorical variable with an interaction, which is

$$yi = \beta_0 + \beta_{1xi} + \beta_2 I_{CatName,i} + \beta_3 \boldsymbol{I}_{CatName,i} \boldsymbol{x}_i + \varepsilon_i,$$

where the new component involves the product of both the indicator and the quantitative predictor variable. The β_3 coefficient will be found in a row of output with **both** variable names in it (with the indicator level name) with a colon between them (something like x:grouplevel). As always, the best way to understand any model involving indicators is to plug in 0s or 1s for the indicator variable(s) and simplify the equations.

- For any observation in the baseline group *ICatName,i* = 0, so

$$yi = \beta_0 + \beta_{1xi} + \beta_2 I_{CatName,i} + \beta_3 \boldsymbol{I}_{CatName,i} \boldsymbol{x}_i + \varepsilon_i$$

simplifies quickly to:

$$yi = \beta_0 + \beta_{1xi} + \varepsilon_i$$

- So the baseline group's model involves the initial intercept and quantitative slope coefficient.
- For any observation in the second category *ICatName,i* = 1, so

$$yi = \beta_0 + \beta_{1xi} + \beta_2 I_{CatName,i} + \beta_3 \boldsymbol{I}_{CatName,i} \boldsymbol{x}_i + \varepsilon_i$$

is

$$yi = \beta_0 + \beta_{1xi} + \beta_2 *_1 + \beta_3 *_{\mathbf{1}} * \boldsymbol{x}_i + \varepsilon_i$$

which "simplifies" to

$$yi = (\beta_0 + \beta_2) + (\beta_1 + \beta_3)x_i + \varepsilon_i,$$

by combining like terms.

- For the second category, the model contains a modified *y*-intercept, now β0 + β2, **and** a modified slope coefficient, now β1 + β**3.**

We can make this more concrete by applying this to the dyslexia data with dys as a categorical variable for dyslexia status of subjects (levels of *no* and *yes*) and ziq the standardized IQ. The model is estimated as:

```
dys_model <- lm(score ~ ziq * dys, data = dyslexic3)
summary(dys_model)
```

```
## lm(formula = score ~ ziq * dys, data = dyslexic3)
##
## Residuals:
##      Min       1Q   Median       3Q      Max
## -0.26362 -0.04152  0.01682  0.06790  0.17740
##
## Coefficients:
##             Estimate Std. Error t value Pr(>|t|)
## (Intercept)  0.87586    0.02391  36.628  < 2e-16
## ziq          0.05827    0.02535   2.299   0.0268
## dysyes      -0.27951    0.03827  -7.304 7.11e-09
## ziq:dysyes  -0.07285    0.03821  -1.907   0.0638
##
## Residual standard error: 0.1017 on 40 degrees of freedom
```

The estimated model can be written as

$\text{Score}i = 0.876 + 0.058 \cdot \text{ZIQ}_i - 0.280I\text{yes},i - 0.073I\text{yes},i \cdot \text{ZIQ}_i$ and simplified for the two groups as:

- For the baseline (non-dyslexic, $I\text{yes},i = 0$) students:

  $\text{Score}i = 0.876 + 0.058 \cdot \text{ZIQ}_i$

- For the deviation (dyslexic, $I\text{yes},i = 1$) students:

  $\text{Score}i = 0.876 + 0.058 \cdot \text{ZIQ}_i - 0.280 * 1 - 0.073 * 1 \cdot \text{ZIQ}_i$

  $= (0.876 - 0.280) + (0.058 - 0.073) \cdot \text{ZIQ}_i,$

  which simplifies finally to:

  $\text{Score}i = 0.596 - 0.015 \cdot \text{ZIQ}_i$

- So the slope switched from 0.058 in the non-dyslexic students to -0.015 in the dyslexic students. The interpretations of these coefficients are outlined below:
  - For the non-dyslexic students: For a 1 SD increase in verbal IQ score, we estimate, on average, the reading score to go up by 0.058 "points".
  - For the dyslexic students: For a 1 SD increase in verbal IQ score, we estimate, on average, the reading score to change by -0.015 "points".

So, an expected pattern of results emerges for the non-dyslexic students. Those with higher IQs tend to have higher reading accuracy; this does not mean higher IQ's cause more accurate reading because random assignment of IQ is not possible. However, for the dyslexic students, the relationship is not what one would might expect. It is slightly negative, showing that higher verbal IQ's are related to lower reading accuracy. What we conclude from this is that we should not expect higher IQ's to show higher performance on a test like this.

Checking the assumptions is always recommended before getting focused on the inferences in the model. When interactions are present, you should not use VIFs as they are naturally inflated because the same variable is re-used in multiple parts of the model to create the interaction components. Checking the multicollinearity in the related additive model can be performed to understand shared information in the variables used in interactions. When fitting models with multiple groups, it is possible to see "groups" in the fitted values (x-axis in Residuals vs Fitted and Scale-Location plots) and that is not a problem – it is a feature of these models. You should look for issues in the residuals for each group but the residuals should overall still be normally distributed and have the same variability everywhere. It is a bit hard to see issues in Figure 8.32 because of the group differences, but note the line of residuals for the higher fitted values. This is an artifact of the upper threshold in the reading accuracy test used. As in the first year of college GPA, these observations were censored – their true score was outside the range of values we could observe – and so we did not really get a measure of how good these students were since a lot of their abilities were higher than the test could detect and they all binned up at the same value of getting all the questions correct. The relationship in this group might be even stronger if we could really observe differences in the highest level readers. We should treat the results for the non-dyslexic group with caution even though they are clearly scoring on average higher and have a different slope than the results for the dyslexic students. The QQ-plot suggests a slightly long left tail but this deviation is not too far from what might happen if we simulated from a normal distribution, so is not clear evidence of a violation of the normality assumption. The influence diagnostics do not suggest any influential points because no points have Cook's D over 0.5.

```
par(mfrow = c(2,2), oma = c(0,0,2,0))
plot(dys_model, pch = 16, sub.caption = "")
title(main="Plot of diagnostics for Dyslexia Interaction model", outer=TRUE)
```

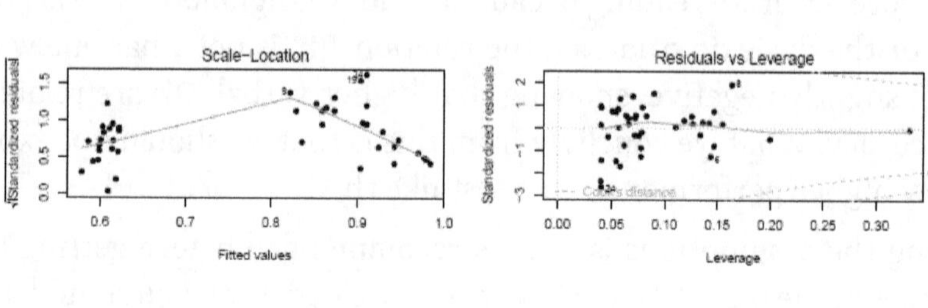

Figure 8.32: Diagnostic plots for interaction model for reading scores.

For these models, we have relaxed an earlier assumption that data were collected from only one group. In fact, we are doing specific research that is focused on questions about the differences between groups. However, these models still make assumptions that, within a specific group, the relationships are linear between the predictor and response variables. They also assume that the variability in the residuals is the same for all observations. Sometimes it can be difficult to check the assumptions by looking at the overall diagnostic plots and it may be easier to go back to the original scatter-plot or plot the residuals vs fitted values by group to fully assess the results. Figure 8.33 shows a scatter-plot of the residuals vs the quantitative explanatory variable by the groups. The variability in the residuals is a bit larger in the non-dyslexic group, possibly suggesting that variability in the reading test is higher for higher scoring individuals even though we couldn't observe all of that variability because there were so many perfect scores in this group.

```
dyslexic3 <- dyslexic3 %>% mutate(resids = residuals(dys_model),
                                  fits = fitted(dys_model)
                                  )
dyslexic3 %>% ggplot(mapping = aes(x = fits, y = resids, color = dys, shape = dys)) +
  geom_smooth(method = "lm", se = F) +
  geom_point(size = 2.5) +
  theme_bw() +
  scale_color_viridis_d(end = 0.7, option = "plasma") +
  labs(title = "Scatterplot of Residuals vs Fitted by Group")
```

If we feel comfortable enough with the assumptions to trust the inferences here (this might be dangerous), then we can consider what some of the model inferences provide us in this situation. For example, the test for $H0 : \beta 3 = 0$ vs $HA : \beta 3 /= 0$ provides an interesting comparison. Under the null hypothesis, the two groups would have the same slope so it provides an opportunity to directly consider whether the relationship (via the slope) is different between the groups in their respective populations. We find $t = -1.907$ which, if the assumptions are true, follows a t(40)-distribution under the null hypothesis. This test statistic has a corresponding p-value of 0.0638. So it provides some evidence against the null hypothesis of no difference in the slopes between the two groups but it isn't strong evidence against it. There are serious issues (like getting the wrong idea about directions of relationships) if we ignore a potentially important interaction and some statisticians would recommend retaining interactions even if the evidence is only moderate for its inclusion in the model. For the original research question of whether the relationships differ for the two groups, we only have marginal evidence to support that result. Possibly with a larger sample size or a reading test that only a few students could get 100% on, the researchers might have detected a more pronounced difference in the slopes for the two groups.

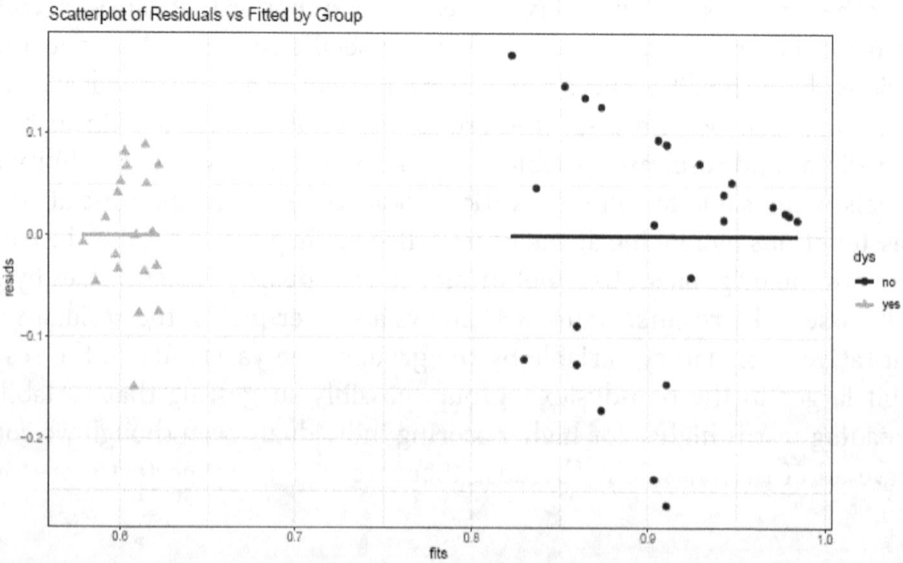

Figure 8.33: Plot of Residuals vs Fitted from interaction dyslexia data model with groups indicated.

In the presence of a categorical by quantitative interaction, term-plots can be generated that plot the results for each group on the same display or on separate facets for each level of the categorical variable. The first version is useful for comparing the different lines and the second version is useful to add the partial residuals and get a final exploration of model assumptions and ranges of values where predictor variables were observed in each group. The term-plots basically provide a plot of the "simplified" SLR models for each group. In Figure 8.34 we can see noticeable differences in the slopes and intercepts. Note that testing for differences in intercepts between groups is not very interesting when there are different slopes because if you change the slope, you have to change the intercept. The plot shows that there are clear differences in the means even though we don't have a test to directly assess that in this complicated of a model24. Figure 8.35 splits the plots up and adds partial residuals to the plots. The impact on the estimated model for the perfect scores in the non-dyslexic subjects is very prominent as well as the difference in the relationships between the two variables in the two groups.

```
plot(allEffects(dys_model), ci.style = "bands", multiline = T, lty = c(1,2), grid = T)
```

```
plot(allEffects(dys_model, residuals = T), lty = c(1,2), grid = T)
```

It certainly appears in the plots that IQ has a different impact on the mean score in the two groups (even though the p-value only provided marginal evidence in support of the interaction). To reinforce the potential dangers of forcing the same slope for both groups, consider the additive model for these data. Again, this just shifts one group off the other one, but both have the same slope. The following model summary and term-plots (Figure 8.36) suggest the potentially dangerous conclusion that can come from assuming a common slope when that might not be the case.

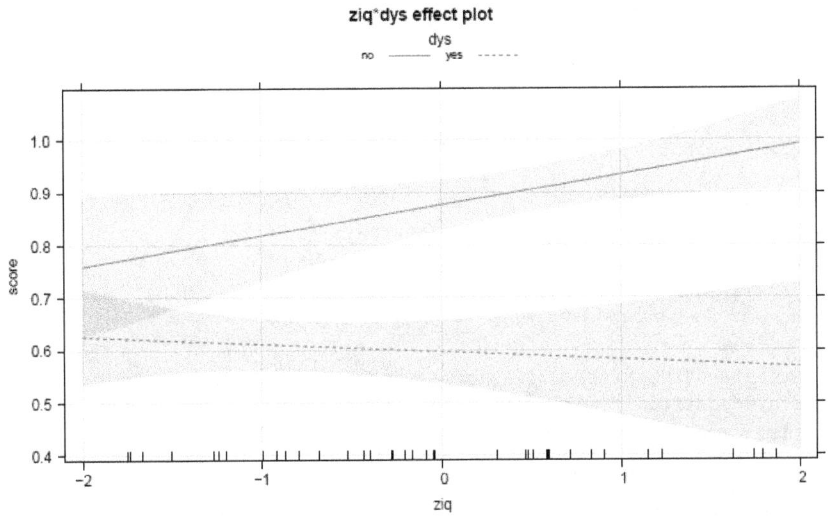

Figure 8.34: Term-plots for interaction model for reading scores using the multiline = T option to overlay the results for the two groups on one plot.

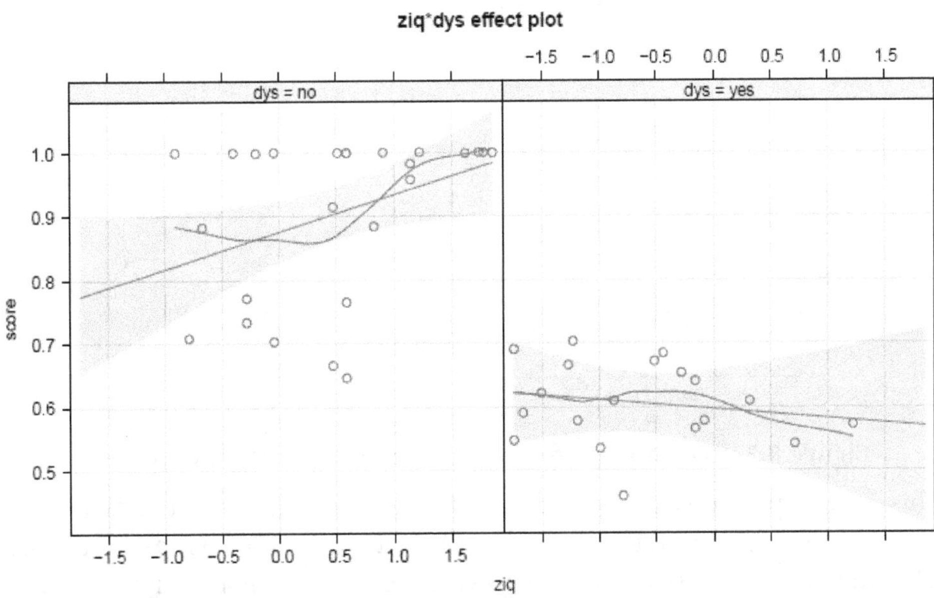

Figure 8.35: Term-plots for interaction model for reading scores with partial residuals and the results for the two groups in different panels of the plot.

```
dys_modelR <- lm(score ~ ziq + dys, data = dyslexic3)

summary(dys_modelR)
```

```
##
## Call:
## lm(formula = score ~ ziq + dys, data = dyslexic3)
##
## Residuals:
##      Min       1Q   Median       3Q      Max
## -0.26062 -0.05565  0.02932  0.07577  0.13217
##
## Coefficients:
##             Estimate Std. Error t value Pr(>|t|)
## (Intercept)  0.89178    0.02312  38.580  < 2e-16
## ziq          0.02620    0.01957   1.339    0.188
## dysyes      -0.26879    0.03905  -6.883 2.41e-08
##
## Residual standard error: 0.1049 on 41 degrees of freedom
## Multiple R-squared:  0.6858, Adjusted R-squared:  0.6705
## F-statistic: 44.75 on 2 and 41 DF,  p-value: 4.917e-11
```

```
plot(allEffects(dys_modelR, residuals = T))
```

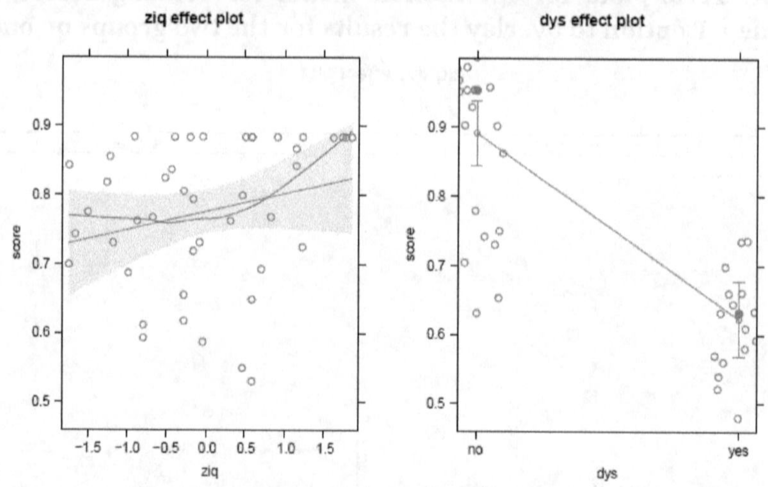

Figure 8.36: Term-plots for additive model for reading scores.

This model provides little evidence against the null hypothesis that IQ is not linearly related to reading score for all students ($t41$ = 1.34, p-value = 0.188), adjusted for dyslexia status, but strong evidence against the null hypothesis of no difference in the true y-intercepts ($t41$ = -6.88, p-value < 0.00001) after adjusting for the verbal IQ score.

Since the IQ term has a large p-value, we could drop it from the model – leaving a model that only includes the grouping variable:

```
dys_modelR2 <- lm(score ~ dys, data = dyslexic3)
summary(dys_modelR2)
```

```
##
## Call:
## lm(formula = score ~ dys, data = dyslexic3)
##
## Residuals:
##      Min       1Q   Median      3Q     Max
## -0.25818 -0.04510  0.02514  0.09520  0.09694
##
## Coefficients:
##             Estimate Std. Error t value Pr(>|t|)
## (Intercept)  0.90480    0.02117  42.737   <2e-16
## dysyes      -0.29892    0.03222  -9.278    1e-11
##
## Residual standard error: 0.1059 on 42 degrees of freedom
## Multiple R-squared:  0.6721, Adjusted R-squared:  0.6643
## F-statistic: 86.08 on 1 and 42 DF,  p-value: 1e-11
```

```
plot(allEffects(dys_modelR2, residuals = T), grid = T)
```

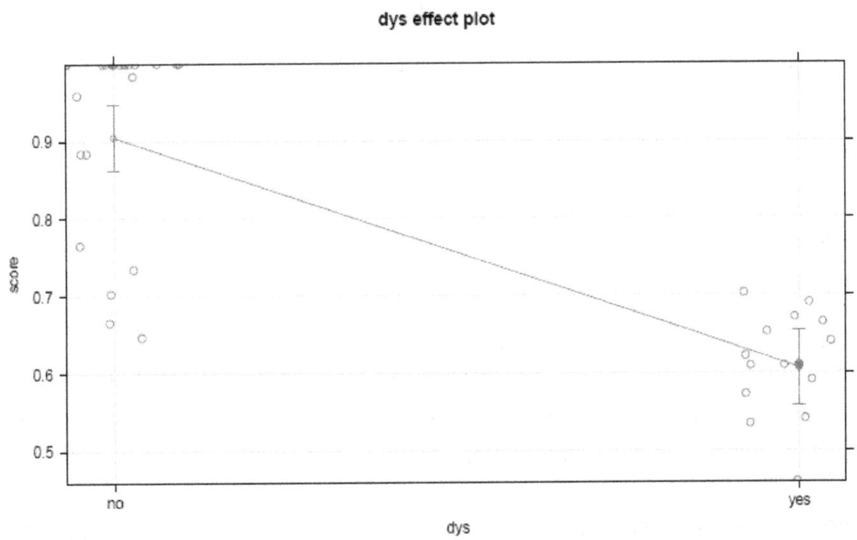

Figure 8.37: Term-plot for dyslexia status only model for reading scores.

These results, including the term-plot in Figure 8.37, suggest a difference in the mean reading scores between the two groups and maybe that is all these data really say... This is the logical outcome if we decide that the interaction is not important in this data set. In general, if the interaction is dropped, the interaction model can be reduced to considering an additive model with the categorical and quantitative predictor variables. Either or both of those variables could also be considered for removal, usually starting with the variable with the larger p-value, leaving a string of ever-simpler models possible if large p-values are continually encountered25.

It is useful to note that the last model has returned us to the first model we encountered in Chapter 2 where we were just comparing the means for two groups. However, the researchers probably were not seeking to make the discovery that dyslexic students have a tougher time than non-dyslexic students on a reading test but sometimes that is all that the data support. The key part of this sequence of decisions was how much evidence you think a p-value of 0.06 contains...

For more than two categories in a categorical variable, the model contains more indicators to keep track of but uses the same ideas. We have to deal with modifying the intercept and slope coefficients for **every** deviation group so the task is onerous but relatively repetitive. The general model is:

$$\begin{aligned} y_i = \beta_0 \quad & +\beta_1 x_i + \beta_2 I_{\text{Level } 2,i} + \beta_3 I_{\text{Level } 3,i} + \cdots + \beta_J I_{\text{Level } J,i} \\ & +\beta_{J+1} I_{\text{Level } 2,i}\, x_i + \beta_{J+2} I_{\text{Level } 3,i}\, x_i + \cdots + \beta_{2J-1} I_{\text{Level } J,i}\, x_i + \varepsilon_i. \end{aligned}$$

Specific to the audible tolerance/headache data that had four groups. The model with an interaction present is

$$\begin{aligned} \text{du2}_i = \beta_0 \quad & +\beta_1 \cdot \text{du1}_i + \beta_2 I_{T1,i} + \beta_3 I_{T2,i} + \beta_4 I_{\text{T3},i} \\ & +\beta_5 I_{T1,i} \cdot \text{du1}_i + \beta_6 I_{T2,i} \cdot \text{du1}_i + \beta_7 I_{\text{T3},i} \cdot \text{du1}_i + \varepsilon_i. \end{aligned}$$

Based on the following output, the estimated general regression model is

$$\begin{aligned} \widehat{\text{du2}}_i = 0.241 \quad & +0.839 \cdot \text{du1}_i + 1.091 I_{T1,i} + 0.855 I_{T2,i} + 0.775 I_{T3,i} \\ & -0.106 I_{T1,i} \cdot \text{du1}_i - 0.040 I_{T2,i} \cdot \text{du1}_i + 0.093 I_{T3,i} \cdot \text{du1}_i. \end{aligned}$$

Then we could work out the specific equation for **each group** with replacing their indicator variable in two places with 1s and the rest of the indicators with 0. For example, for the *T1* group:

$$\begin{aligned} \widehat{\text{du2}}_i & = 0.241 \quad +0.839 \cdot \text{du1}_i + 1.091 \cdot 1 + 0.855 \cdot 0 + 0.775 \cdot 0 \\ & \qquad\qquad -0.106 \cdot 1 \cdot \text{du1}_i - 0.040 \cdot 0 \cdot \text{du1}_i + 0.093 \cdot 0 \cdot \text{du1}_i \\ \widehat{\text{du2}}_i & = 0.241 \quad +0.839 \cdot \text{du1}_i + 1.091 - 0.106 \cdot \text{du1}_i \\ \widehat{\text{du2}}_i & = 1.332 \quad +0.733 \cdot \text{du1}_i. \end{aligned}$$

Or we can let the term-plots (Figures 8.38 and 8.39) show us all four different simplified models. Here we can see that all the slopes "look" to be pretty similar. When the interaction model is fit and the results "look" like the additive model, there is a good chance that we will be able to avoid all this complication and just use the additive model without missing anything interesting. There are two different options for displaying interaction models. Version 1 (Figure 8.38) has a different panel for each level of the categorical variable and Version 2 (Figure 8.39) puts all the lines on the same plot. In this case, neither version shows much of a difference and Version 2 overlaps so much that you can't see all the groups. In these situations, it can be useful to make the term-plots twice, once with multiline = T and once multiline = F, and then select the version that captures the results best.

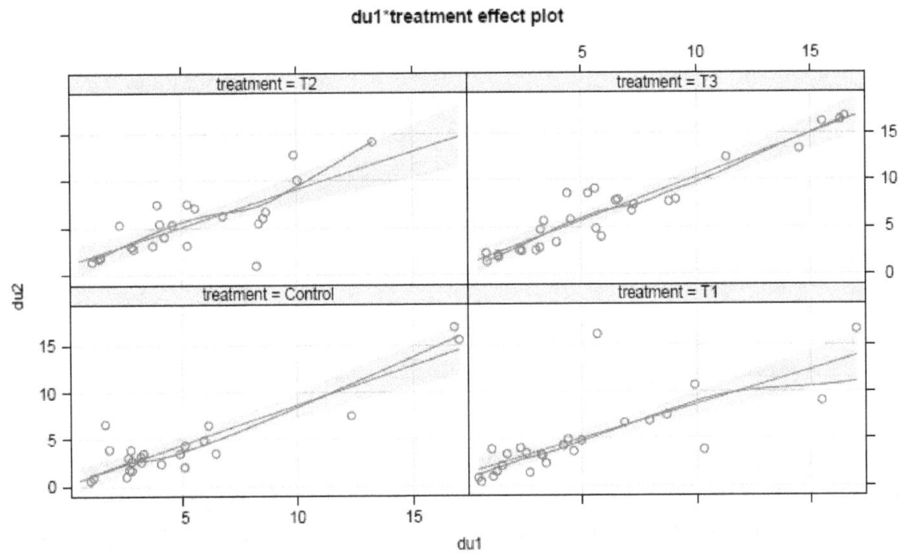

Figure 8.38: Term-plot for decibel tolerance interaction model with partial residuals (version 1).

```
head2 <- lm(du2 ~ du1 * treatment, data = Headache)
summary(head2)
```

```
##
## Call:
## lm(formula = du2 ~ du1 * treatment, data = Headache)
##
## Residuals:
##     Min      1Q  Median      3Q     Max
## -6.8072 -1.0969 -0.3285  0.8192 10.6039
##
## Coefficients:
##                   Estimate Std. Error t value Pr(>|t|)
## (Intercept)        0.24073    0.68331   0.352    0.725
## du1                0.83923    0.10289   8.157 1.93e-12
## treatmentT1        1.09084    0.95020   1.148    0.254
## treatmentT2        0.85524    1.14770   0.745    0.458
## treatmentT3        0.77471    0.97370   0.796    0.428
## du1:treatmentT1 -0.10604    0.14326  -0.740    0.461
## du1:treatmentT2 -0.03981    0.17658  -0.225    0.822
## du1:treatmentT3  0.09300    0.13590   0.684    0.496
##
## Residual standard error: 2.148 on 90 degrees of freedom
## Multiple R-squared:  0.7573, Adjusted R-squared:  0.7384
## F-statistic: 40.12 on 7 and 90 DF,  p-value: < 2.2e-16
```

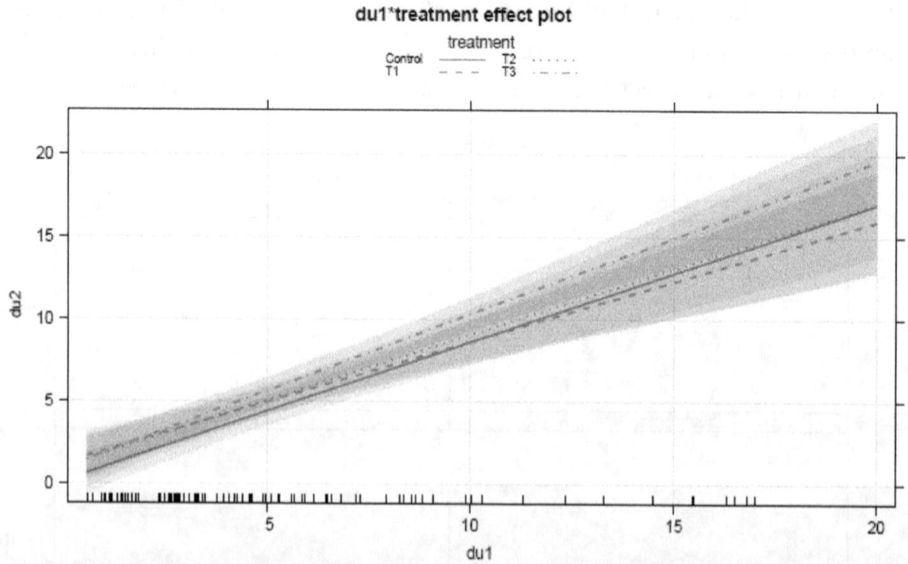

Figure 8.39: Term-plot for decibel tolerance interaction model (version 2). This plot is not printed in colour because it is impossible to distinguish the four groups whether in colour or black and white, although the lty= c(1:4) that provides four different line types does help a bit.

```
plot(allEffects(head2, residuals = T), grid = T) #version 1
plot(allEffects(head2), multiline = T, ci.style = "bands", grid = T,
     lty = c(1:4), lwd = 2) #version 2
```

In situations with more than 2 levels, the *t*-tests for the interaction or changing *y*-intercepts are not informative for deciding if you really need different slopes or intercepts for all the groups. They only tell you if a specific group is potentially different from the baseline group and the choice of the baseline is arbitrary. To assess whether we really need to have varying slopes or intercepts with more than two groups we need to develop *F*-tests for the interaction part of the model.

8.12 F-TESTS FOR MLR MODELS WITH QUANTITATIVE AND CATEGORICAL VARIABLES AND INTERACTIONS

For models with multi-category (J > 2) categorical variables we need a method for deciding if all the extra complexity present in the additive or interaction models is necessary. We can appeal to model selection methods such as the adjusted R2 that focus on balancing model fit and complexity but interests often move to trying to decide if the differences are more extreme than we would expect by chance if there were no group differences in intercepts or slopes. Because of the multi-degree of freedom aspects of the use of indicator variables (J - 1 variables for a J level categorical variable), we have to develop tests that combine and assess information across multiple "variables" – even though these indicators all pertain to a single original categorical variable. ANOVA F -tests did exactly this sort of thing in the One and Two-Way ANOVA models and can do that for us here. There are two models that we perform tests in – the additive and the interaction models. We start with a discussion of the tests in an interaction setting since that provides us the first test to consider in most situations to assess evidence of whether the extra complexity of varying slopes is really needed. If we don't "need" the varying slopes or if the plot really does have lines for the groups that look relatively parallel, we can fit the additive model and either assess evidence of the need for different intercepts or for the quantitative predictor – either is a reasonable next step. Basically this establishes a set of nested models (each model is a reduced version of another more complicated model higher in the tree of models and we can move down the tree by setting a set of slope coefficients to 0) displayed in Figure 8.40. This is based on the assumption that we would proceed through the model, dropping terms if the p-values are large ("not significant" in the diagram) to arrive at a final model.

If the initial interaction test suggests the interaction is important, then no further refinement should be considered and that model should be explored (this was the same protocol suggested in the 2-WAY ANOVA situation, the other place where we considered interactions). If the interaction is not deemed important based on the test, then the model should be re-fit using both variables in an additive model. In that additive model, both variables can be assessed conditional on the other one. If both have small p-values, then that is the final model and should be explored further. If either the categorical or quantitative variable have large p-values, then they can be dropped from the model and the model re-fit with only one variable in it, usually starting with dropping the component with the largest p-value if both are not "small". Note that if there is only a categorical variable remaining, then we would call that linear model a One-Way ANOVA (quantitative response and J group categorical explanatory) and if the only remaining variable is quantitative, then a SLR model is being fit. If that final variable has a large p-value in either model, it can be removed and all that is left to describe the responses is a mean-only model. Otherwise the single variable model is the final model. Usually we will not have to delve deeply into this tree of models and might stop earlier in the tree if that fully addresses our research question, but it is good to consider the potential paths that an analysis could involve before it is started if model refinement is being considered.

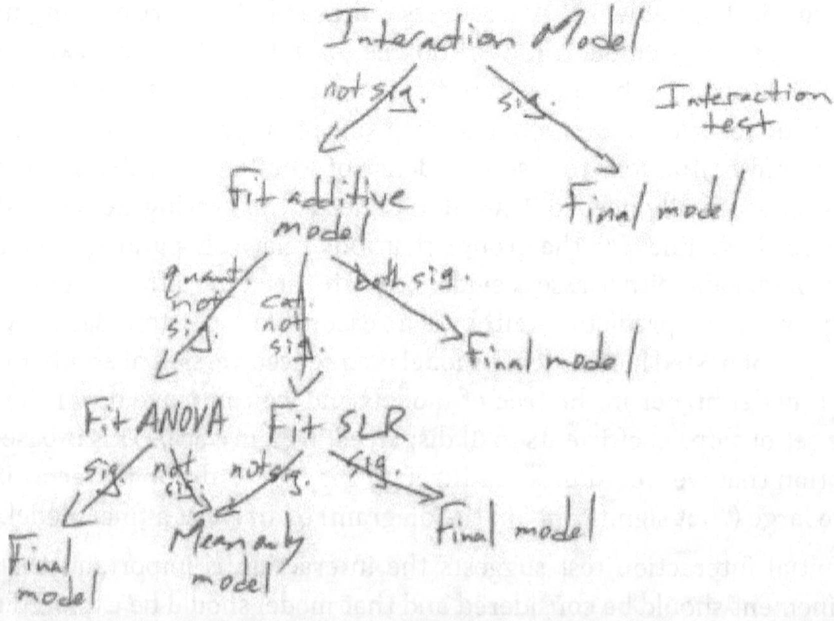

Figure 8.40: Diagram of models to consider in an interaction model. "Sig." means a small p-value for that term and "not sig." means a large one – with apologies for using the "s"-word here.

To perform the first test (after checking that assumptions are not problematic, of course), we can apply the Anova function from the car package to an interaction model[26]. It will provide three tests, one for each variable by themselves, which are not too interesting, and then the interaction test. This will result in an F-statistic that, if the assumptions are true, will follow an $F(J - 1, n - 2J)$-distribution under the null hypothesis. This tests the hypotheses:

- **H0 :** The slope for x is the same for all J groups in the population vs
- **HA :** The slope for x in at least one group differs from the others in the population.

This test is also legitimate in the case of a two-level categorical variable ($J = 2$) and then follows an $F(1, n - 4)$-distribution under the null hypothesis. With $J = 2$, the p-value from this test matches the results for the t-test (t_{n-4}) for the single slope-changing coefficient in the model summary output. The noise tolerance study, introduced in Section 8.10, provides a situation for exploring the results in detail.

With the $J = 4$ level categorical variable (*Treatment*), the model for the second noise tolerance measurement (*du2*) as a function of the interaction between *Treatment* and initial noise tolerance (*du1*) is

$$du2_i = \beta_0 + \beta_1 \cdot du1_i + \beta_2 I_{T1,i} + \beta_3 I_{T2,i} + \beta_4 I_{T3,i}$$
$$+ \beta_5 I_{T1,i} \cdot du1_i + \beta_6 I_{T2,i} \cdot du1_i + \beta_7 I_{T3,i} \cdot du1_i + \varepsilon_i.$$

We can re-write the previous hypotheses in one of two more specific ways:

- **H0**: The slope for *du1* is the same for all four *Treatment* groups in the population OR
- **H0 :** $\beta 5 = \beta 6 = \beta 7 = 0$
  - This defines a null hypothesis that all the deviation coefficients for getting different slopes for the different treatments are 0 in the population.
- HA : The slope for *du1* is NOT the same for all four *Treatment* groups in the population (at least one group has a different slope) OR
- *HA* : At least one of $\beta 5$, $\beta 6$, $\beta 7$ is different from 0 in the population.
  - The alternative states that at least one of the deviation coefficients for getting different slopes for the different *Treatments* is not 0 in the population.

In this situation, the results for the test of these hypotheses is in the row labelled du1:treatment in the Anova output. The ANOVA table below shows a test statistic of $F = 0.768$ with the numerator df of 3, coming from $J - 1$, and the denominator df of 90, coming from $n - 2J = 98 - 2 * 4 = 90$ and also provided in the Residuals row in the table, leading to an $F(3, 90)$-distribution for the test statistic under the null hypothesis. The p-value from this distribution is 0.515, showing little to no evidence against the null hypothesis, so does not suggest that the slope coefficient for du1 in explaining du2 is different for at least one of the Treatment groups in the population.

```
Anova(head2)
```

```
## Anova Table (Type II tests)
##
## Response: du2
##                 Sum Sq Df  F value Pr(>F)
## du1            1197.78  1 259.5908 <2e-16
## treatment        23.90  3   1.7265 0.1672
## du1:treatment    10.63  3   0.7679 0.5150
## Residuals       415.27 90
```

Without evidence to support using an interaction, we should consider both the quantitative and categorical variables in an additive model. The ANOVA table for the additive model contains two interesting tests. One test is for the quantitative variable discussed previously. The other is for the categorical variable, assessing whether different *y*-intercepts are needed. The additive model here is

$$du2_i = \beta_0 + \beta_1 \cdot du1_i + \beta_2 I_{T\,1,i} + \beta_3 I_{T\,2,i} + \beta_4 I_{T\,3,i} + \varepsilon_i.$$

The hypotheses assessed in the ANOVA test for treatment are:

- *H*0 : The *y*-intercept for the model with *du1* is the same for all four *Treatment* groups in the population OR
- *H*0 : $\beta 2 = \beta 3 = \beta 4 = 0$
  - This defines a null hypothesis that all the deviation coefficients for getting different *y*-intercepts for the different *Treatments* are 0 in the population.
- **HA:** The y-intercepts for the model with du1 is NOT the same for all four Treatment groups in the population (at least one group has a different y-intercept) OR
- **HA:** At least one of β2, β3, β4 is different from 0 in the population.
  - The alternative states that at least one of the deviation coefficients for getting different *y*-intercepts for the different *Treatments* is not 0 in the population.

The F -test for the categorical variable in an additive model follows F (J - 1, n - J - 1)-distribution under the null hypothesis. For this example, the test statistic for Treatment follows an F (3, 93)-distribution under the null hypothesis. The observed test statistic has a value of 1.74, generating a p-value of 0.164. So we would find weak evidence against the null hypothesis and so does not suggest some difference in y-intercepts between the treatment groups, in a model with du1, in the population. We could interpret this in the fashion we used initially in MLR by stating this result as: there is little evidence against the null hypothesis of no difference in the mean du2 for the Treatment groups after controlling for du1 so we would conclude that there is possibly no difference between the groups controlled for du1.

```
head1 <- lm(du2 ~ du1 + treatment, data = Headache)
Anova(head1)
```

```
## Anova Table (Type II tests)
##
## Response: du2
##            Sum Sq Df  F value Pr(>F)
## du1        1197.8  1 261.5491 <2e-16
## treatment    23.9  3   1.7395 0.1643
## Residuals   425.9 93
```

In the same ANOVA table, there is a test for the *du1* model component. This tests *H*0 : $\beta 1 = 0$ vs

HA : β1 /= 0 in a model with different *y*-intercepts for the different treatment groups. If we remove this term from the model, all we are left with is different

y-intercepts for the groups. A model just with different y-intercepts is typically called a One-Way ANOVA model. Here, there it appears that the quantitative variable is needed in the model after controlling for the different y-intercepts for different treatments since it has a small p-value ($F(1,93) = 261.55$ or $t(93) = 16.172$, p-value<0.0001). Note that this interpretation retains the conditional wording regardless of whether the other variable had a small p-value or it did not. If you want an unconditional interpretation for a variable, then you will need to refit the model without the other variable(s) after deciding that they are not important.

8.13 AICS FOR MODEL SELECTION

There are a variety of techniques for selecting among a set of potential models or refining an initially fit MLR model. Hypothesis testing can be used (in the case where we have nested models either by adding or deleting a single term at a time) or comparisons of adjusted R2 across different potential models (which is valid for nested or non-nested model comparisons). Diagnostics should play a role in the models considered and in selecting among models that might appear to be similar on a model comparison criterion. In this section, a new model selection method is introduced that has stronger theoretical underpinnings, a slightly more interpretable scale, and, often, better performance in picking an optimal27 model than the adjustedR2. The measure is called the AIC (Akaike's An Information Criterion28, [Akaike, 1974]). It is extremely popular, but sometimes misused, in some fields such as Ecology, and has been applied in almost every other potential application area where statistical models can be compared. Burnham and Anderson [2002] have been responsible for popularizing the use of AIC for model selection, especially in Ecology. The AIC is an estimate of the distance (or discrepancy or divergence) between a candidate model and the true model, on a log-scale, based on a measure called the Kullback-Leibler divergence. The models that are closer (have a smaller distance) to the truth are better and we can compare how close two models are to the truth, picking the one that has a smaller distance (smaller AIC) as better. The AIC includes a component that is on the log-scale, so negative values are possible and you should not be disturbed if you are comparing large magnitude negative numbers – just pick the model with the smallest AIC score.

The AIC is optimized (smallest) for a model that contains the optimal balance of simplicity of the model with quality of fit to the observations. Scientists are driven to different degrees by what is called the principle of parsimony: that simpler explanations (models) are better if everything else is equal or even close to equal. In this case, it would mean that if two models are similarly good on AIC, then select the simpler of the two models since it is more likely to be correct in general than the more complicated model. The AIC is calculated as AIC = −2log(Likelihood) + 2m, where the Likelihood provides a measure of fit of the model (we let R calculate it for us) and gets smaller for better fitting models and m = (number of estimated β's + 1). The value m is called the model degrees of freedom for AIC calculations and relates to how many total parameters are estimated. Note that it is a different measure of degrees of freedom than used in ANOVA F -tests. The main things to understand about the formula for the AIC is that as m increases, the AIC will go up and that as the fit improves, the likelihood will increase (so -2log-likelihood will get smaller)29.

There are some facets of this discussion to keep in mind when comparing models. More complicated models always fit better (we saw this for the R2 measure, as the proportion of variation explained always goes up if more "stuff" is put into the model even if the "stuff" isn't useful). The AIC resembles the adjusted R2 in that it incorporates the count of the number of parameters estimated. This allows the AIC to make sure that enough extra variability is explained in the responses to justify making the model more complicated (increasing m). The optimal model on AIC has to balance adding complexity and increasing quality of the fit. Since this measure provides an estimate of the distance or discrepancy to the "true model", the model with the smallest value "wins" – it is top-ranked on the AIC. Note that the top-ranked AIC model will often not be the best fitting model since the best fitting model is always the most complicated model considered. The top AIC model is the one that is estimated to be closest to the truth, where the truth is still unknown. . .

To help with interpreting the scale of AICs, they are often reported in a table sorted from smallest to largest values with the AIC and the "delta AIC" or, simply, ΔAIC reported. The

ΔAIC = AICmodel - AICtopModel

and so provides a value of 0 for the top-ranked AIC model and a measure of how much worse on the AIC scale the other models are. A rule of thumb is that a 2 unit difference on AICs (ΔAIC = 2) is moderate evidence of a difference in the models and more than 4 units (ΔAIC > 4) is strong evidence of a difference. This is more based on experience than a distinct reason or theoretical result but seems to provide reasonable results in most situations. Often researchers will consider any models within 2 AIC units of the top model(ΔAIC < 2) as indistinguishable on AICs and so either select the simplest model of the choices or report all the models with similar "support", allowing the reader to explore the suite of similarly supported potential models. It is important to remember that if you search across too many models, even with the AIC to support your model comparisons, you might find a spuriously top model. Individual results that are found by exploring many tests or models have higher chances to be spurious and results found in this manner are difficult to replicate when someone repeats a similar study. For these reasons, there is a set of general recommendations that have been developed for using AICs:

- Consider a suite of models (often pre-specified and based on prior research in the area of interest) and find the models with the top (in other words, smallest) AIC results.
  - The suite of candidate models need to contain at least some good models. Selecting the best of a set of BAD models only puts you at the top of $%#%-mountain, which is not necessarily a good thing.
- Report a table with the models considered, sorted from smallest to largest AICs (ΔAICs from smaller to larger) that includes a count of number of parameters estimated[30], the AICs, and ΔAICs.
  - Remember to incorporate the mean-only model in the model selection results. This allows you to compare the top model to one that does not contain any predictors.
- Interpret the top model or top models if a few are close on the AIC-scale to the top model.
- DO NOT REPORT P-VALUES OR CALL TERMS "SIGNIFICANT" when models were selected using AICs.
  - Hypothesis testing and AIC model selection are not compatible philosophies and testing in models selected by AICs invalidates the tests as they have inflated Type I error rates. The AIC results are your "evidence" – you don't need anything else. If you wanted to report p-values, use them to select your model.
  - You can describe variables as "important" or "useful" and report confidence intervals to aid in interpretation of the terms in the selected model(s) but need to avoid performing hypothesis tests with the confidence intervals.
  - Remember that the selected model is not the "true" model – it is only the best model *according to AIC* among the set of models *you provided.*
  - AICs assume that the model is specified correctly up to possibly comparing different predictor variables. Perform diagnostic checks on your initial model and the top model and do not trust AICs when assumptions are clearly violated (p-values are similarly not valid in that situation).

When working with AICs, there are two options. Fit the models of interest and then run the AIC function on each model. This can be tedious, especially when we have many possible models to consider. We can make it easy to fit all the potential candidate models that are implied by a complicated starting model by using the dredge function from the MuMIn package [Bartoń, 2022]. The name (dredge) actually speaks to what ***fitting all possible models*** really engages – what is called ***data dredging***. The term is meant to refer to considering way too many models for your data set, probably finding something good from the process, but maybe identifying something spurious since you looked at so many models. Note that if you take a hypothesis testing approach where you plan to remove any terms with large p-values in this same situation, you are really considering all possible models as well because you could have removed some or all model components. Methods that consider all possible models are probably best used in exploratory analyses where you do not know if any or all terms should be important. If you have more specific research questions, then you probably should try to focus on comparisons of models that help you directly answer those questions, either with AIC or p-value methods.

The dredge function provides an automated method of assessing all possible simpler models based on an initial (full) model. It generates a table of AIC results, ΔAICs, and also shows when various predictors are in or out of the model for all reduced models possible from an initial model. For quantitative predictors, the estimated slope is reported when that predictor is in the model. For categorical variables and interactions with them, it just puts a "+" in the table to let you know that the term is in the models. Note that you must run the options(na.action = "na. fail") code to get dredge to work31.

To explore the AICs and compare their results to the adjusted R2 that we used before for model selection, we can revisit the Snow Depth data set with related results found in Section 8.4 and Table 8.1. In that situation we were considering a "full" model that included Elevation, Min.Temp, and Max.Temp as potential predictor variables after removing two influential points. And we considered all possible reduced models from that "full"32 model. Note that the dredge output adds one more model that adjusted R2 can't consider– the mean-only model that contains no predictor variables. In the following output it is the last model in the output (worst ranked on AIC). Including the mean-only model in these results helps us "prove" that there is support for having something in the model, but only if there is better support for other models than this simplest possible model.

In reading dredge output33 as it is constructed here, the models are sorted by top to bottom AIC values (smallest AIC to largest). The column delta is for the ΔAICs and shows a 0 for the first row, which is the top-ranked AIC model. Here it is for the model with Elevation and Max.Temp but not including Min.Temp. This was also the top ranked model from adjusted R2, which is reproduced in the adjRsq column. The AICYou can simply rundredge(m6, rank = "AIC") to get just the AIC results is calculated using the previous formula based on the df and logLik columns. The df is also a useful column for comparing models as it helps you see how complex each model is. For example, the top model used up 4 model df (three β's and the residual error variance) and the most complex model that included four predictor variables used up 5 model df.

```
library(MuMIn)
options(na.action = "na.fail") #Must run this code once to use dredge
snotel2R <- snotel_s %>% slice(-c(9,22))
m6 <- lm(Snow.Depth ~ Elevation + Min.Temp + Max.Temp, data = snotel2R)
dredge(m6, rank = "AIC", extra = c("R^2", adjRsq = function(x) summary(x)$adj.r.squared))
```

##	Global model call: lm(formula = Snow.Depth ~ Elevation + Min. Temp	+ Max.Temp, data = snotel2R)
##	---	
##	Model selection table	
##	(Int) Elv Max.Tmp Min.Tmp R^2 adjRsq df logLik AIC	delta weight
##	4 -167.50 0.02408 1.2530 0.8495 0.8344 4 -80.855 169.7	0.00 0.568
##	8 -213.30 0.02686 1.2430 0.9843 0.8535 0.8304 5 -80.541 171.1	1.37 0.286
##	2 -80.41 0.01791 0.8087 0.7996 3 -83.611 173.2	3.51 0.098
##	6 -130.70 0.02098 1.0660 0.8134 0.7948 4 -83.322 174.6	4.93 0.048
##	5 179.60 -5.0090 0.6283 0.6106 3 -91.249 188.5	18.79 0.000
##	7 178.60 -0.2687 -4.6240 0.6308 0.5939 4 -91.170 190.3	20.63 0.000
##	3 119.50 -2.1800 0.4131 0.3852 3 -96.500 199.0	29.29 0.000
##	1 40.21 0.0000 0.0000 2 -102.630 209.3	39.55 0.000
##	Models ranked by AIC(x)	

You can use the table of results from dredge to find information to compare the estimated models. There are two models that are clearly favoured over the others with ΔAICs for the model with *Elevation* and *Max.Temp* of 0 and for the model with all three predictors of 1.37. The ΔAIC for the third ranked model(contains just *Elevation*) is 3.51 suggesting clear support for the top model over this because of a difference of 3.51 AIC units to the truth. The difference between the second and third ranked models also provides relatively strong support for the more complex model over the model with just *Elevation*. And the mean-only model had a ΔAIC of nearly 40 – suggesting extremely strong evidence for the top model versus using no predictors. So we have pretty clear support for models that include the *Elevation* and *Max.Temp* variables (in both top models) and some support for also including the *Min.Temp*, but the top model did not require its inclusion. It is also possible to think about the AICs as a result on a number line from "closest to the truth" to "farthest" for the suite of models considered, as shown in Figure 8.41.

We could add further explorations of the term-plots and confidence intervals for the slopes from the top or, here, possibly top two models. We would not spend any time with p-values since we already used the AIC to assess evidence related to the model components and they are invalid if we model select prior to reporting them. We can quickly compare the slopes for variables that are shared in the two models since they are both quantitative variables using the output. It is interesting that the Elevation and Max.Temp slopes change little with the inclusion of Min. Temp in moving from the top to second ranked model (0.02408 to 0.0286 and 1.253 to 1.243).

This was an observational study and so we can't consider causal inferences here as discussed previously. Generally, the use of AICs does not preclude making causal statements but if you have randomized assignment of levels of an explanatory variable, it is more philosophically consistent to use hypothesis testing methods in that setting. If you went to the effort to impose the levels of a treatment on the subjects, it also makes sense to see if the differences created are beyond what you might expect by chance if the treatment didn't matter.

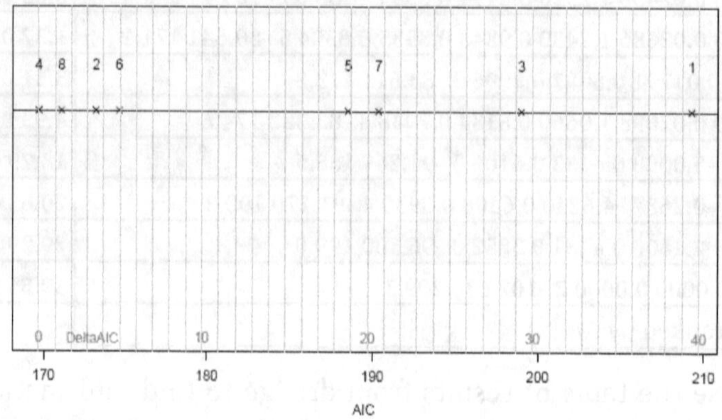

Figure 8.41: Display of AIC results on a number line with models indicated by their number in the dredge output. Note that the actual truth is unknown but further left in the plot corresponds to the models that are estimated to be closer to the truth and so there is stronger evidence for those models versus the others.

8.14 CASE STUDY: FORCED EXPIRATORY VOLUME MODEL SELECTION USING AICS

Researchers were interested in studying the effects of smoking by children on their lung development by measuring the forced expiratory volume (FEV, measured in Liters) in a representative sample of children (n = 654) between the ages of 3 and 19; this data set is available in the FEV data set in the coneproj package(Meyer and Liao [2021], Liao and Meyer [2014]). Measurements on the age (in years) and height (in inches) as well as the sex and smoking status of the children were made. We would expect both the age and height to have positive relationships with FEV (lung capacity) and that smoking might decrease the lung capacity but also that older children would be more likely to smoke. So the height and age might be confounded with smoking status and smoking might diminish lung development for older kids – resulting in a potential interaction between age and smoking. The sex of the child might also matter and should be considered or at least controlled for since the response is a size-based measure. This creates the potential for including up to four variables (age, height, sex, and smoking status) and possibly the interaction between age and smoking status. Initial explorations suggested that modelling the log-FEV would be more successful than trying to model the responses on the original scale. Figure 8.42 shows the suggestion of different slopes for the smokers than non-smokers and that there aren't very many smokers under 9 years old in the data set. So we will start with a model that contains an *age* by *smoking* interaction and include *height* and *sex* as additive terms. We are not sure if any of these model components will be needed, so the simplest candidate model will be to remove all the predictors and just have a mean-only model (FEV ~ 1). In between the mean-only and most complicated model are many different options where we can drop the interaction or drop the additive terms or drop the terms involved in the interaction if we don't need the interaction.

```
library(coneproj)
data(FEV)
FEV <- as_tibble(FEV)
FEV <- FEV %>% mutate(sex = factor(sex), #Make sex and smoke factors, log.FEV
                      smoke = factor(smoke),
                      log.FEV = log(FEV))
```

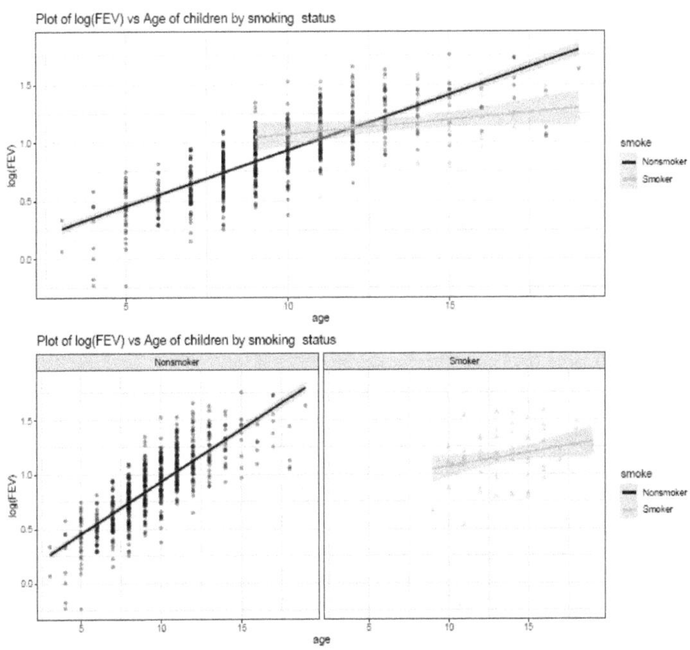

Figure 8.42: Scatterplot of log(FEV) vs Age by smoking status, with both combined (top) and faceted (bottom) versions.

```
levels(FEV$sex) <- c("Female","Male") #Make sex labels explicit
levels(FEV$smoke) <- c("Nonsmoker","Smoker") #Make smoking status labels explicit

p1 <- FEV %>% ggplot(mapping = aes(x = age, y = log.FEV, color = smoke, shape = smoke)) +
  geom_point(size = 1.5, alpha = 0.5) +
  geom_smooth(method = "lm") +
  theme_bw() +
  scale_color_viridis_d(end = 0.8) +
  labs(title = "Plot of log(FEV) vs Age of children by smoking  status",
       y = "log(FEV)")

p2 <- FEV %>% ggplot(mapping = aes(x = age, y = log.FEV, color = smoke, shape = smoke)) +
  geom_point(size = 1.5, alpha = 0.5) +
  geom_smooth(method = "lm") +
  theme_bw() +
  scale_color_viridis_d(end = 0.8) +
  labs(title = "Plot of log(FEV) vs Age of children by smoking  status",
       y = "log(FEV)") +
  facet_grid(cols = vars(smoke))

grid.arrange(p1, p2, nrow = 2)
```

To get the needed results, start with the full model – the most complicated model you want to consider. It is good to check assumptions before considering reducing the model as they rarely get better in simpler models and the AIC is only appropriate to use if the model assumptions are not clearly violated. As suggested above, our "fullish" model for the log(FEV) values is specified as log(FEV) ~ height + age * smoke + sex.

```
fm1 <- lm(log.FEV ~ height + age * smoke + sex, data = FEV)
summary(fm1)
```

```
##
## Call:
## lm(formula = log.FEV ~ height + age * smoke + sex, data = FEV)
##
## Residuals:
##      Min       1Q   Median       3Q      Max
## -0.62926 -0.08783  0.01136  0.09658  0.40751
##
## Coefficients:
##                   Estimate Std. Error t value Pr(>|t|)
## (Intercept)      -1.919494   0.080571 -23.824  < 2e-16
## height            0.042066   0.001759  23.911  < 2e-16
## age               0.025368   0.003642   6.966 8.03e-12
## smokeSmoker       0.107884   0.113646   0.949  0.34282
## sexMale           0.030871   0.011764   2.624  0.00889
## age:smokeSmoker  -0.011666   0.008465  -1.378  0.16863
##
## Residual standard error: 0.1454 on 648 degrees of freedom
## Multiple R-squared:  0.8112, Adjusted R-squared:  0.8097
## F-statistic: 556.8 on 5 and 648 DF,  p-value: < 2.2e-16
```

```
par(mfrow = c(2,2), oma = c(0,0,2,0))
plot(fm1, pch = 16, sub.caption = "")
title(main="Diagnostics for full FEV model", outer=TRUE)
```

The diagnostic plots suggest that there are a few outlying points (Figure 8.43) but they are not influential and there is no indication of violations of the constant variance assumption. There is a slight left skew with a long left tail to cause a very minor concern with the normality assumption but not enough to be concerned about our inferences from this model. If we select a different model(s),we would want to check its diagnostics and make sure that the results do not look noticeably worse than these do.

```
plot(allEffects(fm1, residuals = T), grid = T)
```

Figure 8.44 provides our first term-plot with multiple predictors and an interaction. Each term is interpreted using our "conditional" language for any of the other two panels. So we could explore the impacts of height on log-fev after controlling for sex as well as age, smoking status, and the age by smoking interaction. This mirrors how we would interpret any of the coefficients or confidence intervals from this full model (we are not doing hypothesis tests here).

The AIC function can be used to generate the AIC values for a single or set of candidate models. It will also provide the model degrees of freedom used for each model if run the function on multiple models. For example, suppose that they want to compare fm1 to a model without the interaction term in the model, called fm1R. You need to fit both models and then apply the AIC function to them with commas between the model names:

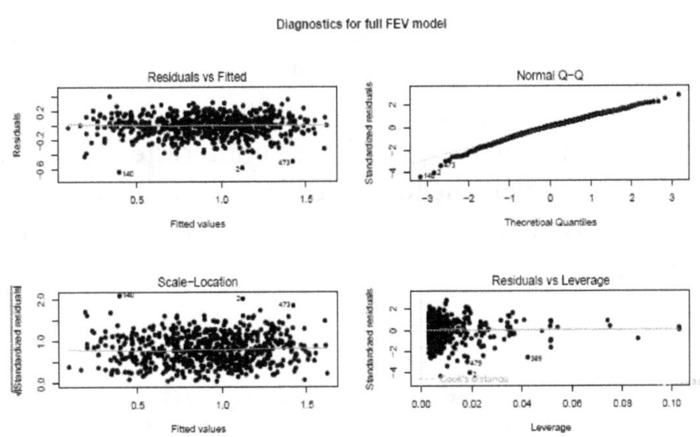

Figure 8.43: Diagnostics for the log(FEV) model that includes height, sex, and an interaction between age and smoking status (the full model).

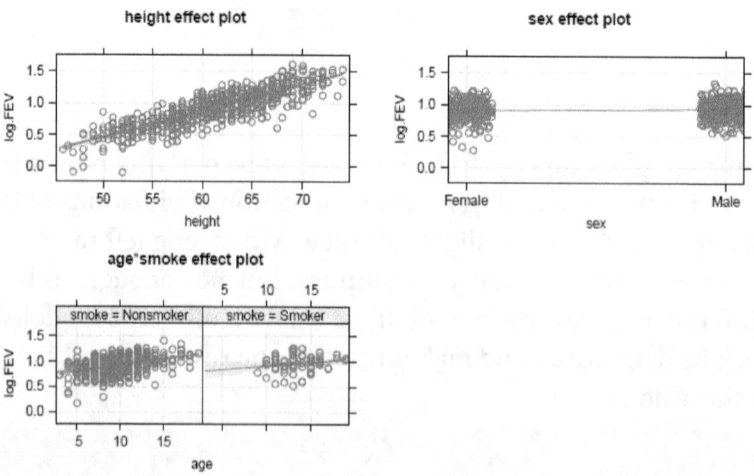

Figure 8.44: Term-plots for the log(FEV) model that includes height, sex, and an interaction between age and smoking status (the full model), with partial residuals.

```
fm1R <- lm(log.FEV ~ height + age + smoke + sex, data = FEV)
AIC(fm1, fm1R)
```

```
##      df       AIC
## fm1   7 -658.5178
## fm1R  6 -658.6037
```

These results tells us that the fm1R model (the one without the interaction) is better (more negative) on the AIC by 0.09 AIC units. Note that this model does not "fit" as well as the full model, it is just the top AIC model – the AIC results suggest that it is slightly closer to the truth than the more complicated model but with such a small difference there is similar support and little evidence of a difference between the two models. This provides only an assessment of the difference between including or excluding the interaction between *age* and *smoking* in a model with two other predictors. We are probably also interested in whether the other terms are needed in the model. The full suite of results from dredge provide model comparisons that help us to assess the presence/absence of each model component including the interaction.

```
options(na.action = "na.fail") #Must run this code once to use dredge
dredge(fm1, rank = "AIC",
       extra = c("R^2", adjRsq = function(x) summary(x)$adj.r.squared))
```

```
## Global model call: lm(formula = log.FEV ~ height + age * smoke + sex, data = FEV)
## ---
## Model selection table
##         (Int)     age     hgh sex smk age:smk     R^2  adjRsq df   logLik    AIC   delta weight
## 16 -1.944000 0.02339 0.04280   +   +          0.81060 0.80950  6  335.302 -658.6    0.00  0.414
## 32 -1.919000 0.02537 0.04207   +   +        + 0.81120 0.80970  7  336.259 -658.5    0.09  0.397
## 8  -1.940000 0.02120 0.04299   +              0.80920 0.80830  5  332.865 -655.7    2.87  0.099
## 12 -1.974000 0.02231 0.04371       +          0.80880 0.80790  5  332.163 -654.3    4.28  0.049
## 28 -1.955000 0.02388 0.04315       +        + 0.80920 0.80800  6  332.802 -653.6    5.00  0.034
## 4  -1.971000 0.01982 0.04399                  0.80710 0.80650  4  329.262 -650.5    8.08  0.007
## 7  -2.265000         0.05185   +              0.79640 0.79580  4  311.594 -615.2   43.42  0.000
## 3  -2.271000         0.05212                  0.79560 0.79530  3  310.322 -614.6   43.96  0.000
## 15 -2.267000         0.05190   +   +          0.79640 0.79550  5  311.602 -613.2   45.40  0.000
## 11 -2.277000         0.05222       +          0.79560 0.79500  4  310.378 -612.8   45.85  0.000
## 30 -0.067780 0.09493           +   +        + 0.64460 0.64240  6  129.430 -246.9  411.74  0.000
## 26 -0.026590 0.09596               +        + 0.62360 0.62190  5  110.667 -211.3  447.27  0.000
## 14 -0.015820 0.08963           +   +          0.62110 0.61930  5  108.465 -206.9  451.67  0.000
## 6   0.004991 0.08660           +              0.61750 0.61630  4  105.363 -202.7  455.88  0.000
## 10  0.022940 0.09077               +          0.60120 0.60000  4   91.790 -175.6  483.02  0.000
## 2   0.050600 0.08708                          0.59580 0.59520  3   87.342 -168.7  489.92  0.000
## 13  0.822000                   +   +          0.09535 0.09257  4 -176.092  360.2 1018.79  0.000
## 9   0.888400                       +          0.05975 0.05831  3 -188.712  383.4 1042.03  0.000
## 5   0.857400                   +              0.02878 0.02729  3 -199.310  404.6 1063.22  0.000
## 1   0.915400                                  0.00000 0.00000  2 -208.859  421.7 1080.32  0.000
## Models ranked by AIC(x)
```

There is a lot of information in the output and some of the needed information in the second set of rows, so we will try to point out some useful features to consider. The left columns describe the models being estimated. For example, the first row of results is for a model with an intercept (Int), *age* (age) , *height* (hgh), *sex*(sex), and *smoking*(smk). For *sex* and *smoking*, there are "+"s in the output row when they are included in that model but no coefficient since they are categorical variables. There is no interaction between *age* and *smoking* in the top ranked model. The top AIC model has an R^2 = 0.8106, adjusted R^2 of 0.8095, *model df*= 6 (from an intercept, four slopes, and the residual variance), log-likelihood (logLik) = 335.302, an AIC= -658.6 and ΔAIC of 0.00. The next best model adds the interaction between *age* and *smoking*, resulting in increases in the R^2, adjusted R^2, and *model df*, but increasing the AIC by 0.09 units (ΔAIC = 0.09). This suggests that these two models are essentially equivalent on the AIC because the difference is so small and this comparison was discussed previously. The simpler model is a little bit better on AIC so you could focus on it or on the slightly more complicated model – but you should probably note that the evidence is equivocal for these two models.

The comparison to other potential models shows the strength of evidence in support of all the other model components. The intercept-only model is again the last in the list with the least support on AICs with aΔAIC of 1080.32, suggesting it is not worth considering in comparison with the top model. **Comparing the mean-only model to our favourite model on AICs is a bit like the overall F -test we considered in Section** 8.7 **because it compares a model with no predictors to a complicated model.** Each model with just one predictor included is available in the table as well, with the top single predictor model based on *height* having a

ΔAIC of 43.96. So we certainly need to pursue something more complicated than SLR based with such strong evidence for the more complex models versus the single predictor models at over 40 AIC units different. Closer to the top model is the third-ranked model that includes *age, height*, and *sex*. It has a ΔAIC of 2.87 so we would say that these results present marginal support for the top two models over this model. It is the simplest model of the top three but not close enough to be considered in detail.

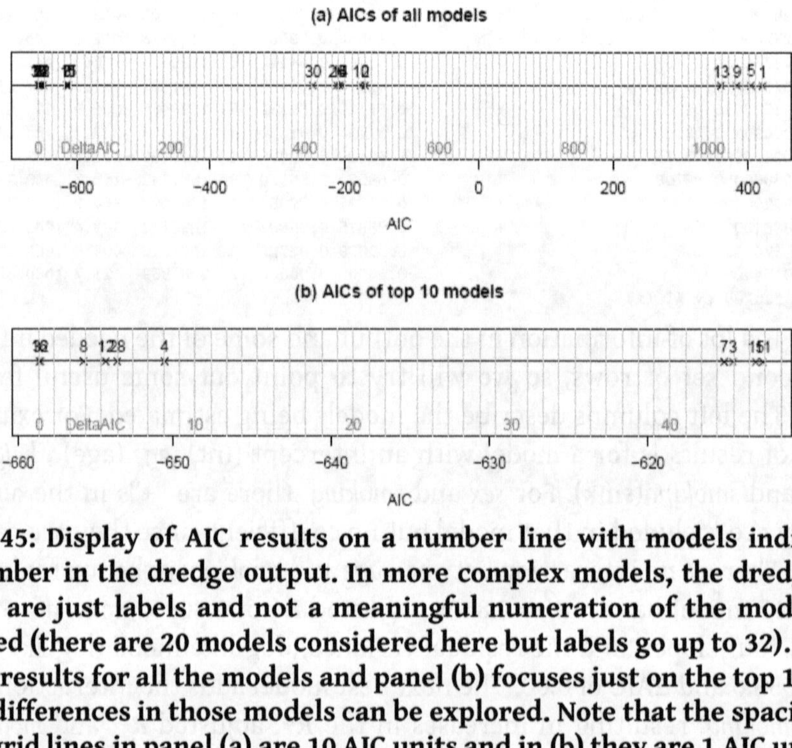

Figure 8.45: Display of AIC results on a number line with models indicated by their number in the dredge output. In more complex models, the dredge model numbers are just labels and not a meaningful numeration of the models being considered (there are 20 models considered here but labels go up to 32). Panel (a) presents results for all the models and panel (b) focuses just on the top 10 models so some differences in those models can be explored. Note that the spacing of the vertical grid lines in panel (a) are 10 AIC units and in (b) they are 1 AIC unit apart.

The dredge results also provides the opportunity to compare the model selection results from the adjusted R^2 compared to the AIC. The AIC favours the model without an interaction between *age* and *smoking* whereas the adjusted R^2 favours the most complicated model considered here that included an *age* and *smoking* interaction. The AIC provides units that are more interpretable than adjusted R^2 even though the scale for the AIC is a bit mysterious as distances from the unknown true model with possibly negative distances.

The top AIC model (and possibly the other similar models) can then be explored in more detail. You should not then focus on hypothesis testing in this model. Hypothesis testing so permeates the use of statistics that even after using AICs many researchers are pressured to report p-values for model components. Some of this could be confusion caused when people first learned these statistical methods because when we teach you statistics we show you how to use various methods, one after another, and forget to mention that you should not use **every** method we taught you in **every** analysis. Confidence intervals and term-plots are useful for describing the different model components and making inferences for the estimated sizes of differences in the population. These results should not be used for deciding if terms are "significant" when the models (and their components) have already been selected using measures like the AIC or adjusted R^2. But you can discuss the estimated model components to go with how you arrived at having them in the model. In this situation, the top model is estimated to be **log(FEV)** i = **-1.94 + 0.043 · Height**$_i$ + **0.0234 · Age**$_i$ - **0.046*I*Smoker,*i*** + **0.0293*I*Male,*i*** based on the estimated coefficients provided below. Using these results and the term-plots (Figure 8.46) we see that in this model there are positive slopes for Age and Height on log-FEV, a negative coefficient for smoking (Smoker), and a positive coefficient for sex (Males). There is some multicollinearity impacting the estimates for height and age based on having VIFs near 3 but these are not extreme issues. We could go further with interpretations such as for the age term: For a 1 year increase in age, we estimate, on average, a 0.0234 log-liter increase in FEV, after controlling for the height, smoking status, and sex of the children. We can even interpret this on the original scale since this was a log(y) response model using the same techniques as in Section 7.6. If we exponentiate the slope coefficient of the quantitative variable, exp(0.0234) = 1.0237.This provides the interpretation on the original FEV scale, for a 1 year increase in age, we estimate a 2.4% increase in the median FEV, after controlling for the height, smoking status, and sex of the children. The only difference from Section 7.6 when working with a log(y) model now is that we have to note that the model used to generate the slope coefficient had other components and so this estimate is after adjusting for them.

```
fm1R$coefficients
```

```
## (Intercept)      height         age smokeSmoker     sexMale
## -1.94399818  0.04279579  0.02338721 -0.04606754  0.02931936
```

```
vif(fm1R)
```

```
##   height      age    smoke      sex
## 2.829728 3.019010 1.209564 1.060228
```

```
confint(fm1R)
```

```
##                     2.5 %       97.5 %
## (Intercept) -2.098414941 -1.789581413
## height       0.039498923  0.046092655
## age          0.016812109  0.029962319
## smokeSmoker -0.087127344 -0.005007728
## sexMale      0.006308481  0.052330236
```

Like any statistical method, the AIC works better with larger sample sizes and when assumptions are not clearly violated. It also will detect important variables in models more easily when the effects of the predictor variables are strong. Along with the AIC results, it is good to report the coefficients for your top estimated model(s), confidence intervals for the coefficients and/or term-plots, and R^2. This provides a useful summary of the reasons for selecting the model(s), information on the importance of the terms within the model, and a measure of the variability explained by the model. The R^2 is not used to select the model, but after selection can be a nice summary of model quality. For fm1R, the $R^2 = 0.8106$ suggesting that the selected model explains 81% of the variation in log-*FEV* values.

The AICs are a preferred modelling strategy in some fields such as Ecology. As with this and many other methods discussed in this book, it is sometimes as easy to find journal articles with mistakes in using statistical methods as it is to find papers doing it correctly. After completing this material, you have the potential to have the knowledge and experience of two statistics classes and now are better trained than some researchers that frequently use these methods. This set of tools can be easily mis-applied. Try to make sure that you are thinking carefully through your problem before jumping to the statistical results. Make a graph first, think carefully about your study design and variables collected and what your models of interest might be, what assumptions might be violated based on the data collection story, and then start fitting models. Then check your assumptions and only proceed on with any inference if the conditions are not clearly violated. The AIC provides an alternative method for selecting among different potential models and they do not need to be nested (a requirement of hypothesis testing methods used to sequentially simplify models).

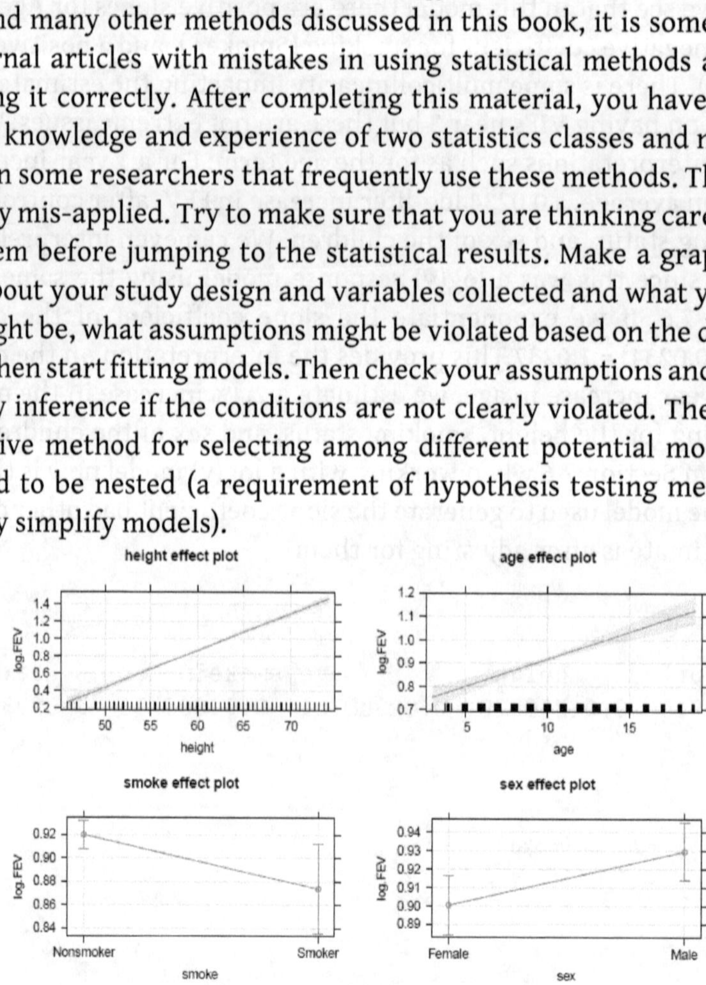

Figure 8.46: Term-plots for the top AIC model for log(FEV) that includes height, age, smoking status, and sex in the model.

The automated consideration of all possible models in the dredge function should not be considered in all situations but can be useful in a preliminary model exploration study where no clear knowledge exists about useful models to consider. Where some knowledge exists of possible models of interest *a priori*, fit those models and use the AIC function to get AICs to compare. Reporting the summary of AIC results beyond just reporting the top model(s) that were selected for focused exploration provides the evidence to support that selection – not p-values!

8.15 CONCISE SUMMARY

This chapter explored the most complicated models we're going to explore. MLR models can incorporate features of SLR and ANOVAs. The MLR's used in this chapter highlight the flexibility of the linear modelling framework to move from two-sample mean models to multi-predictor models with interactions of categorical and quantitative variables. It is useful to use the pertinent names for the simpler models, but at this point we could have called everything we are doing fitting linear models. The power of the linear model involves being able to add multiple predictor variables to the model and handle categorical predictors using indicator variables. All this power comes with some responsibility in that you need to know what you are trying to fit and how to interpret the results provided. We introduced each scenario working from simple to the most complicated version of the models, trying to motivate when you would encounter them, and the specific details of the interpretations of each type of model. In Chapter 9, case studies are used to review the different methods discussed with reminders of how to identify and interpret the particular methods used.

When you have to make modelling decisions, you should remember the main priorities in modelling. First, you need to find a model that can address research question(s) of interest. Second, find a model that is trustworthy by assessing the assumptions in the model relative to your data set. Third, report the logic and evidence that was used to identify and support the model. All too often, researchers present only a final model with little information on how they arrived at it. You should be reporting the reasons for decisions made and the evidence supporting them, whether that is using p-values or some other model selection criterion. For example, if you were considering an interaction model and the interaction was dropped and an additive model is re-fit and interpreted, the evidence related to the interaction test should still be reported. Similarly, if a larger MLR is considered and some variables are removed, the evidence (reason) for those removals should be provided. Because of multicollinearity in models, you should never remove more than one quantitative predictor at a time or else you could remove two variables that are important but were "hiding" when both were included in the model.

8.16 SUMMARY OF IMPORTANT R CODE

There is very little "new" R code in this chapter since all these methods were either used in the ANOVA or SLR chapters. The models are more complicated but are built off of methods from previous chapters. In this code, y is a response variable, x1, x2, ..., xK are quantitative explanatory variables, group factor is a factor variable and the data are in DATASETNAME.

- DATASETNAME %>% ggplot(mapping = aes(x = x, y = y)) + geom_point() +geom_smooth(method = "lm")
  - Provides a scatter plot with a regression line.
  - Add + **geom_smooth()** to add a smoothing line to help detect nonlinear relationships.
  - Add **color = groupfactor** to aesthetic to color points and lines based on a grouping variable.
  - Add + **facet_grid(cols = vars(groupfactor))** to facet by groups.
- MODELNAME <- lm(y ~ x1 + x2 +...+ xK, data = DATASETNAME)
  - Estimates an MLR model using least squares with K quantitative predictors.
- MODELNAME <- lm(y ~ x1 * groupfactor, data = DATASETNAME)
  - Estimates an interaction model between a quantitative and categorical variable, providing different slopes and intercepts for each group.
- MODELNAME <- lm(y ~ x1 + groupfactor, data = DATASETNAME)
  - Estimates an additive model with a quantitative and categorical variable, providing different intercepts for each group.
- summary(MODELNAME)
  - Provides parameter estimates, overall F-test, $\boldsymbol{R^2}$, and adjusted $\boldsymbol{R^2}$.
- par(mfrow = c(2, 2)); plot(MODELNAME)
  - Provides four regression diagnostic plots in one plot.
- confint(MODELNAME, level = 0.95)
  - Provides 95% confidence intervals for the regression model coefficients.
  - Change level if you want other confidence levels.
- plot(allEffects(MODELNAME))
  - Requires the effects package.
  - Provides a plot of the estimated regression lines with 95% confidence interval for the mean.

- vif(MODELNAME)
  - Requires the car package.
  - Provides VIFs for an MLR model. Only use in additive models - not meaningful for models with interactions present.
- predict(MODELNAME, se.fit = T)
  - Provides fitted values for all observed *x*'s with SEs for the mean.
- predict(**MODELNAME, newdata = tibble(x1 = X1_NEW, x2 = X2_NEW, ..., xK =XK_NEW, interval = "confidence")**
  - Provides fitted value for specific values of the quantitative predictors with CI for the mean.
- predict(MODELNAME, newdata = tibble(x1 = X1_NEW, x2 = X2_NEW, ..., xK = XK_NEW, interval = "prediction")
  - Provides fitted value for specific values of the quantitative predictors with PI for a new observation.
- Anova(MODELNAME)
  - Requires the car package.
  - Use to generate ANOVA tables and *F* -tests useful when categorical variables are included in either the additive or interaction models.
- **AIC(MODELNAME_1, MODELNAME_2)**
  - Use to get AIC results for two candidate models called MODELNAME_1 and MODELNAME_2.
- options(na.action = "na.fail") dredge(FULL_MODELNAME, rank = "AIC")
  - Requires the MuMIn package.
  - Provides AIC and delta AIC results for all possible simpler models given a full model calledFULL_MODELNAME.

8.17 PRACTICE PROBLEMS

8.1 Treadmill data analysis The original research goal for the treadmill data set used for practice problems in the last two chapters was to replace the costly treadmill oxygen test with a cheap to find running time measurement but there were actually quite a few variables measured when the run time was found – maybe we can replace the treadmill test result with a combined prediction built using a few variables using the MLR techniques. The following code will get us re-started in this situation.

```
treadmill <- read_csv("http://www.math.montana.edu/courses/s217/documents/treadmill.csv")
tm1 <- lm(TreadMillOx ~ RunTime, data = treadmill)
```

8.1.1. Fit the MLR that also includes the running pulse (RunPulse), the resting pulse (RestPulse), body weight (BodyWeight), and Age (Age) of the subjects. Report and interpret the R2 for this model.

8.1.2. Compare the R2 and the adjusted R2 to the results for the SLR model that just had RunTime in the model. What do these results suggest?

8.1.3. Interpret the estimated RunTime slope coefficients from the SLR model and this MLR model. Explain the differences in the estimates.

8.1.4. Find the VIFs for this model and discuss whether there is an issue with multicollinearity noted in these results.

8.1.5. Report the value for the overall F -test for the MLR model and interpret the result.

8.1.6. Drop the variable with the largest p-value in the MLR model and re-fit it. Compare the resulting R2and adjusted R2 values to the others found previously.

8.1.7. Use the dredge function as follows to consider some other potential reduced models and report the top two models according to adjusted R2 values. What model had the highest R2? Also discuss and compare the model selection results provided by the delta AICs here.

```
library(MuMIn)
options(na.action = "na.fail") #Must run this code once to use dredge
dredge(MODELNAMEFORFULLMODEL, rank = "AIC",
       extra = c("R^2", adjRsq = function(x) summary(x)$adj.r.squared))
```

8.1.8. For one of the models, interpret the *Age* slope coefficient. Remember that only male subjects between 38 and 57 participated in this study. Discuss how this might have impacted the results found as compared to a more general population that could have been sampled from.

8.1.9. The following code creates a new three-level variable grouping the ages into low, middle, and high for those observed. The scatterplot lets you explore whether the relationship between treadmill oxygen and run time might differ across the age groups.

Based on the plot, do the lines look approximately parallel or not?

8.1.10. Fit the MLR that contains a RunTime by Ageb interaction – do not include any other variables. Compare the R2 and adjusted R2 results to previous models.

8.1.11. Find and report the results for the F -test that assesses evidence relative to the need for different slope coefficients.

8.1.12. Write out the overall estimated model. What level was R using as baseline? Write out the simplified model for two of the age levels. Make an effects plot and discuss how it matches the simplified models you generated.

8.1.13. Fit the additive model with Run Time and predict the mean treadmill oxygen values for subjects with run times of 11 minutes in each of the three Ageb groups.

8.1.14. Find the F -test results for the binned age variable in the additive model. Report and interpret those results.

```
treadmill <- treadmill %>% mutate(Ageb = factor(cut(Age, breaks = c(37, 44.5, 50.5, 58))))
summary(treadmill$Ageb)
treadmill %>% ggplot(mapping = aes(x = RunTime, y = TreadMillOx,
                              color = Ageb, shape = Ageb)) +
  geom_point(size = 1.5, alpha = 0.5) +
  geom_smooth(method = "lm") +
  theme_bw() +
  scale_color_viridis_d(end = 0.8) +
  facet_grid(rows = vars(Ageb))
```

CHAPTER-9

CASE STUDIES

9.1 OVERVIEW OF MATERIAL COVERED

At the beginning of the text, we provided a schematic of methods that you would learn about that was (probably) gibberish. Hopefully, revisiting that same diagram (Figure 9.1) will bring back memories of each of the chapters. One common theme was that categorical variables create special challenges whether they are explanatory or response variables.

Type of response vanable

Type/ number explanatory

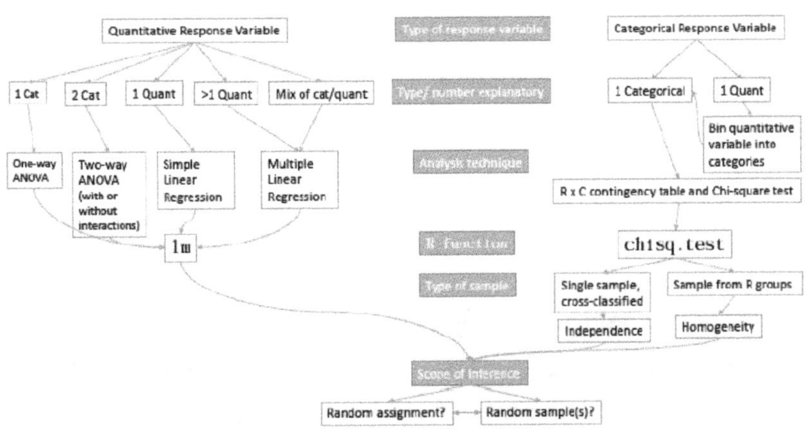

Figure 9.1: Schematic of methods covered.

Every scenario with a quantitative response variable was handled using linear models. The last material on multiple linear regression modelling tied ha.ck to the One-Way and Two-\,Vay ANOVA models as categorical variables were added to the models. As both a review and to emphasize the connections, let's connect some of the different versions of the general linear model that we considered .

If we start with the One-Way ANOVA, the referenced-coded model was written out as:

$$y_{ij} = \alpha + \tau_j + \varepsilon_{ij} .$$

We didn't want to introduce indicator variables at that early stage of the material, but we can now write out the same model using our indicator variable approach from Chapter 8 for a J -level categorical explanatory variable using J - 1 indicator variables as:

$$y_i = \beta_0 + \beta_1 I_{Level\ 2,i} + \beta_2 I_{Level\ 3,i} + \cdots + \beta_{J-1} I_{Level\ J,i} + \varepsilon_i.$$

We now know how the indicator variables are either 0 or 1 for each observation and only one takes in the value 1 (is "turned on") at a time for each response. We can then equate the general notation from Chapter 8 with our specific One-Way ANOVA (Chapter 3) notation as follows:

For the baseline category, the mean is:

$$\alpha = \beta_0$$

- The mean for the baseline category was modelled using α which is the intercept term in the output that we called β_0 in the regression models.
- For category j, the mean is:
  - From the One-Way ANOVA model:

$$\alpha + \tau_j$$

- From the regression model where the only indicator variable that is 1 is $I_{Level\ j,i}$:

$$\beta_0 + \beta_1 I_{Level\ 2,i} + \beta_2 I_{Level\ 3,i} + \cdots + \beta_J I_{Level\ J,i}$$

$$= \beta_0 + \beta_{j-1} \cdot 1$$

$$= \beta_0 + \beta_{j-1}$$

So with intercepts being equal, $\beta_{j-1} = \tau_j$.

The ANOVA reference-coding notation was used to focus on the coefficients that were "turned on" and their interpretation without getting bogged down in the full power (and notation) of general linear models.

The same equivalence is possible to equate our work in the Two-Way ANOVA interaction model,

$$y_{ijk} = \alpha + \tau_j + \gamma_k + \omega_{jk} + \varepsilon_{ijk} ,$$

with the regression notation from the MLR model with an interaction:

$$y_i = \beta_0 + \beta_1 x_i + \beta_2 I_{Level\ 2,i} + \beta_3 I_{Level\ 3,i} + \cdots + \beta_J I_{Level\ J,i} + \beta_{J+1} I_{Level\ 2,i}\, x_i$$

$$+ \beta_{J+2} I_{Level\ 3,i}\, x_i + \cdots + \beta_{2J-1} I_{Level\ J,i}\, x_i + \varepsilon_i$$

If one of the categorical variables only had two levels, then we could simply replace *xi* with the pertinent indicator variable and be able to equate the two versions of the notation. That said, we won't attempt that here. And if both variables have more than 2 levels, the number of coefficients to keep track of grows rapidly. The great increase in complexity of notation to fully writing out the indicator variables in the regression approach with interactions with two categorical variables is the other reason we explored the Two-Way ANOVA using a "simplified" notation system even though lm used the indicator approach to estimate the model. The Two-Way ANOVA notation helped us distinguish which coefficients related to main effects and the interaction, something that the regression notation doesn't make clear.

In the following four sections, you will have additional opportunities to see applications of the methods considered here to real data. The data sets are taken directly from published research articles, so you can see the potential utility of the methods we've been discussing for handling real problems. They are focused on biological applications because most come from a particular journal (*Biology Letters*) that encourages authors to share their data sets, making our re-analyses possible. Use these sections to review the methods from earlier in the book and to see some hints about possible extensions of the methods you have learned.

9.2 THE IMPACT OF SIMULATED CHRONIC NITROGEN DEPOSITION ON THE BIOMASS AND N2-FIXATION ACTIVITY OF TWO BOREAL FEATHER MOSS– CYANOBACTERIA ASSOCIATIONS

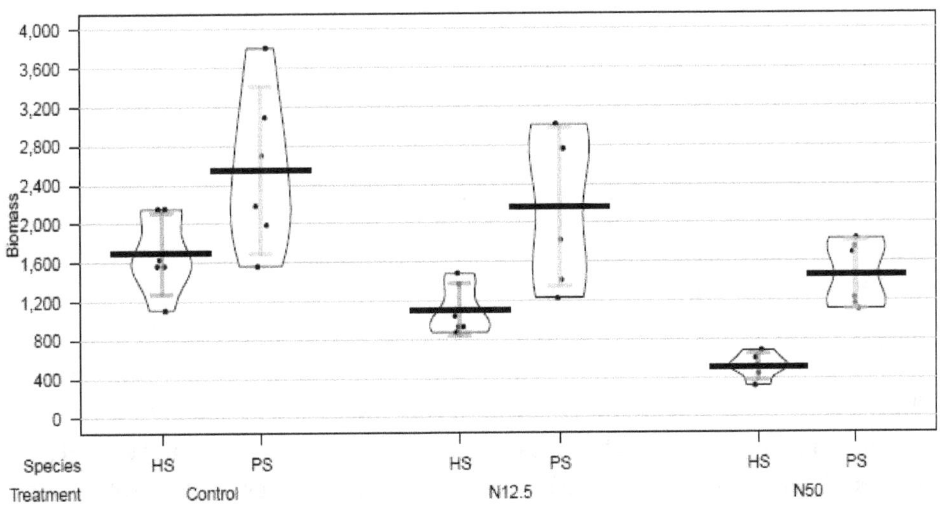

Figure 9.2: Pirate-plot of biomass responses by treatment and species.

In a 16-year experiment, Gundale et al. [2013] studied the impacts of Nitrogen (N) additions on the mass of two feather moss species (*Pleurozium schreberi* (PS) and *Hylocomium* (HS)) in the Svartberget Experimental Forest in Sweden. They used a randomized block design: here this means that within each of 6 blocks (pre-specified areas that were divided into three experimental units or plots of area 0.1 hectare), one of the three treatments were randomly applied. ***Randomized block designs*** involve randomization of levels within blocks or groups as opposed to ***completely randomized designs*** where each ***experimental unit***(the subject or plot that will be measured) could be randomly assigned to any treatment. This is done in agricultural studies to control for systematic differences across the fields by making sure each treatment level is used in each area or ***block*** of the field. In this example, it resulted in a balanced design with six replicates at each combination of *Species* and *Treatment*.

The three treatments involved different levels of N applied immediately after snow melt, *Control* (no additional N – just the naturally deposited amount), 12.5 kg N ha¯1yr¯1 (*N12.5*), and 50 kg N ha¯1yr¯1(*N50*). The researchers were interested in whether the treatments would have differential impacts on the two species of moss growth. They measured a variety of other variables, but here we focus on the estimated *biomass* per hectare (mg/ha) of the *species* (*PS* or *HS*), both measured for each plot within each block, considering differences across the *treatments* (*Control*, *N12.5*, or *N50*). The pirate-plot in Figure 9.2 provides some initial information about the responses. Initially there seem to be some differences in the combinations of groups and some differences in variability in the different groups, especially with much more variability in the *control* treatment level and more variability in the *PS* responses than for the *HS* responses.

```
csv("http://www.math.montana.edu/courses/s217/documents/gundalebachnordin_2
```

```
<- gdn %>% mutate(Species = factor(Species),
                 Treatment = factor(Treatment)
                 )
rary(yarrr)
ateplot(Massperha ~ Species + Treatment, data = gdn, inf.method = "ci",
        inf.disp = "line", theme = 2, ylab = "Biomass", point.o = 1,
        pal = "southpark")
```

The Two-WAY ANOVA model that contains a *species* by *treatment* interaction is of interest (this has a quantitative response variable of *biomass* and two categorical predictors of *species* and *treatment*)[1]. We can make an interaction plot to focus on the observed patterns of the means across the combinations of levels as provided in Figure 9.3. The interaction plot suggests a relatively additive pattern of differences between *PS* and *HS* across the three treatment levels. However, the variability seems to be quite different based on this plot as well.

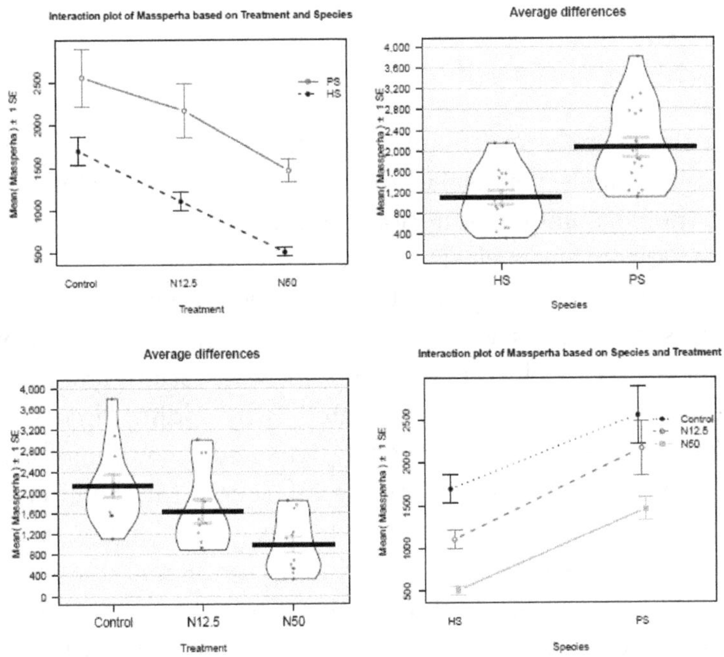

Figure 9.3: Interaction plot of biomass responses by treatment and species.

```
library(catstats) #Or directly using:
#source("http://www.math.montana.edu/courses/s217/documents/intplotfunctions_v3.R")
```

```
intplotarray(Massperha ~ Species * Treatment, data = gdn, col = viridis(4)[1:3],
             lwd = 2, cex.main = 1)
```

Based on the initial plots, we are going to be most concerned about the equal variance assumption. We can fit the interaction model and explore the diagnostic plots to verify that we have a problem.

```
m1 <- lm(Massperha ~ Species * Treatment, data = gdn)
summary(m1)
```

```
##
## Call:
## lm(formula = Massperha ~ Species * Treatment, data = gdn)
##
## Residuals:
##    Min     1Q Median     3Q    Max
## -992.6 -252.2  -64.6  308.0 1252.9
##
## Coefficients:
##                           Estimate Std. Error t value Pr(>|t|)
## (Intercept)                1694.80     211.86   8.000 6.27e-09
## SpeciesPS                   859.88     299.62   2.870  0.00745
## TreatmentN12.5             -588.26     299.62  -1.963  0.05893
## TreatmentN50              -1182.91     299.62  -3.948  0.00044
## SpeciesPS:TreatmentN12.5    199.42     423.72   0.471  0.64130
## SpeciesPS:TreatmentN50       88.29     423.72   0.208  0.83636
##
## Residual standard error: 519 on 30 degrees of freedom
## Multiple R-squared:  0.6661, Adjusted R-squared:  0.6104
## F-statistic: 11.97 on 5 and 30 DF,  p-value: 2.009e-06
```

```
par(mfrow = c(2,2), oma = c(0,0,2,0))
plot(m1, pch = 16, sub.caption = "")
title(main="Initial Massperha 2-WAY model", outer=TRUE)
```

There is a clear problem with non-constant variance showing up in a fanning shape[2] in the Residuals versus Fitted and Scale-Location plots in Figure 9.4. Interestingly, the normality assumption is not an issue as the residuals track the 1-1 line in the QQ-plot quite closely so hopefully we will not worsen this result by using a transformation to try to address the non-constant variance issue. The independence assumption is violated in two ways for this model by this study design – the blocks create clusters or groups of observations and the block should be accounted for (they did this in their models by adding *block* as a categorical variable to their models). Using blocked designs and accounting for the blocks in the model will typically give more precise inferences for the effects of interest, the treatments randomized within the blocks. Additionally, **there are two measurements on each plot** within block, one for *SP* and one for *HS* and these might be related (for example, high *HS* biomass might be associated with high or low *SP*) so putting both observations into a model **violates the independence assumption** at a second level. It takes more advanced statistical models (called linear mixed models) to see how to fully deal with this, for now it is important to recognize the issues. The more complicated models provide similar results here and include the *treatment* by *species* interaction we are going to explore, they just add to this basic model to account for these other issues.

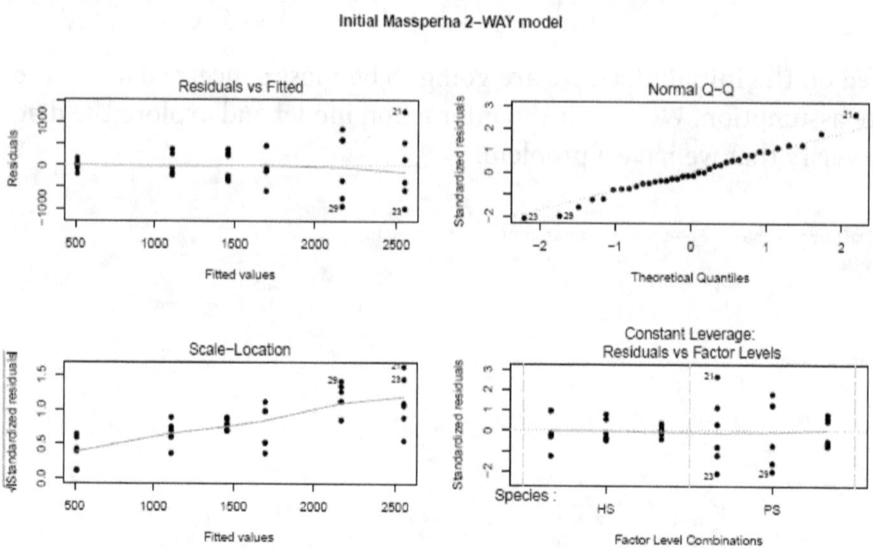

Figure 9.4: Diagnostic plots of treatment by species interaction model for Biomass.

Remember that before using a log-transformation, you always must check that the responses are strictly greater than 0:

```
summary(gdn$Massperha)
```

```
##     Min.  1st Qu.   Median     Mean  3rd Qu.     Max.
##    319.1   1015.1   1521.8   1582.3   2026.6   3807.6
```

The minimum is 319.1 so it is safe to apply the natural log-transformation to the response variable (Biomass) and repeat the previous plots:

```
gdn <- gdn %>% mutate(logMassperha = log(Massperha))
par(mfrow = c(2,1))
pirateplot(logMassperha ~ Species + Treatment, data = gdn, inf.method = "ci",
           inf.disp = "line", theme = 2, ylab = "log-Biomass", point.o = 1,
           pal = "southpark", main = "(a)")
intplot(logMassperha ~ Species * Treatment, data = gdn, col = viridis(4)[1:3],
        lwd = 2, main = "(b)")
```

The variability in the pirate-plot in Figure 9.5(a) appears to be more consistent across the groups but the lines appear to be a little less parallel in the interaction plot Figure 9.5(b) for the log-scale response. That is not problematic but suggests that we may now have an interaction present – it is hard to tell visually sometimes. Again, fitting the interaction model and exploring the diagnostics is the best way to assess the success of the transformation applied.

The log(Mass per ha) version of the response variable has little issue with changing variability present in the residuals in Figure 9.6 with much more similar variation in the residuals across the fitted values. The normality assumption is leaning toward a slight violation with too little variability in the right tail and so maybe a little bit of a left skew. This is only a minor issue and fixes the other big issue (clear non-constant variance), so this model is at least closer to giving us trustworthy inferences than the original model. The model presents moderate evidence against the null hypothesis of no *Species* by *Treatment* interaction on the log-biomass (F (2, 30) = 4.2, p-value = 0.026). This suggests that the effects on the log-biomass of the treatments differ between the two species. The mean log-biomass is lower for *HS* than *PS* with the impacts of increased nitrogen causing *HS* mean log-biomass to decrease more rapidly than for *PS*. In other words, increasing nitrogen has more of an impact on the resulting log-biomass for *HS* than for *PS*. The highest mean log-biomass rates were observed under the control conditions for both species making nitrogen appear to inhibit growth of these species.

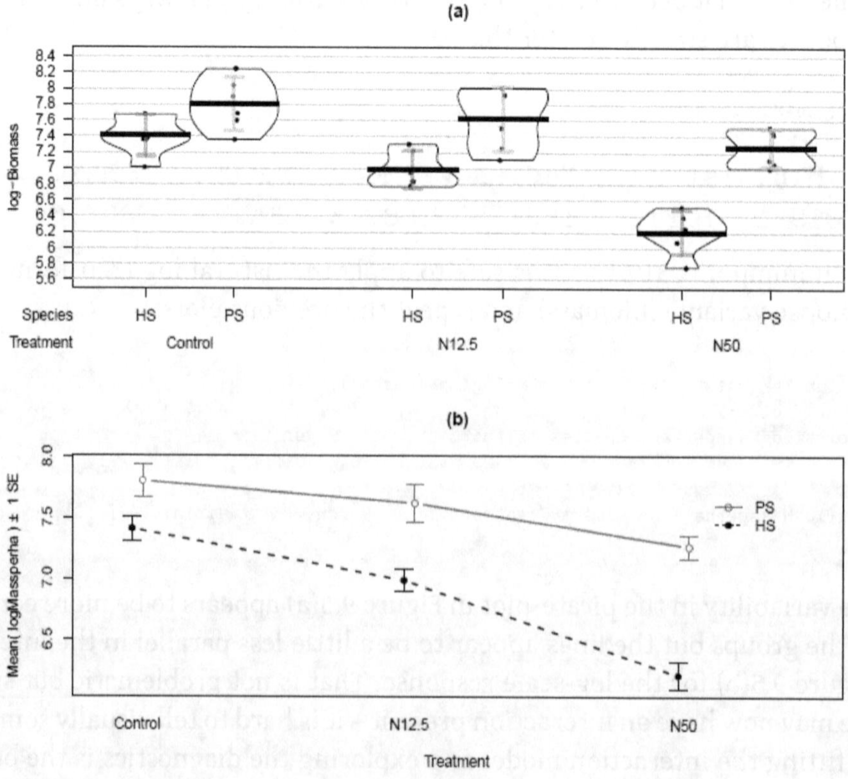

Figure 9.5: Pirate-plot and interaction plot of the log-Biomass responses by treatment and species.

```
m2 <- lm(logMassperha ~ Species * Treatment, data = gdn)
summary(m2)
```

```
##
## Call:
## lm(formula = logMassperha ~ Species * Treatment, data = gdn)
##
## Residuals:
##      Min       1Q   Median       3Q      Max
## -0.51138 -0.16821 -0.02663  0.23925  0.44190
##
## Coefficients:
##                         Estimate Std. Error t value Pr(>|t|)
## (Intercept)               7.4108     0.1160  63.902  < 2e-16
## SpeciesPS                 0.3921     0.1640   2.391  0.02329
## TreatmentN12.5           -0.4228     0.1640  -2.578  0.01510
## TreatmentN50             -1.1999     0.1640  -7.316 3.79e-08
## SpeciesPS:TreatmentN12.5  0.2413     0.2319   1.040  0.30645
## SpeciesPS:TreatmentN50    0.6616     0.2319   2.853  0.00778
##
## Residual standard error: 0.2841 on 30 degrees of freedom
## Multiple R-squared:  0.7998, Adjusted R-squared:  0.7664
## F-statistic: 23.96 on 5 and 30 DF,  p-value: 1.204e-09
```

```
library(car)
Anova(m2)
```

```
## Anova Table (Type II tests)
##
## Response: logMassperha
##                    Sum Sq Df F value    Pr(>F)
## Species            4.3233  1  53.577 3.755e-08
## Treatment          4.6725  2  28.952 9.923e-08
## Species:Treatment  0.6727  2   4.168   0.02528
## Residuals          2.4208 30
```

```
par(mfrow = c(2,2), oma = c(0,0,2,0))
plot(m2, pch = 16, sub.caption = "")
title(main="log-Massperha 2-WAY model", outer=TRUE)
```

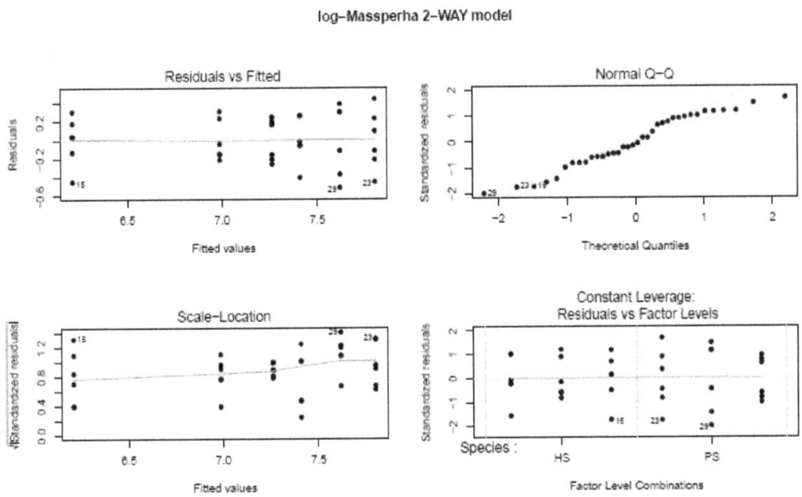

Figure 9.6: Diagnostic plots of treatment by species interaction model for log-Biomass.

The researchers actually applied a log(*y* + 1) transformation to all the variables. This was used because one of their many variables had a value of 0 and so they added 1 to avoid analyzing a -∞ response. This was not needed for most of their variables because most did not attain the value of 0. Adding a small value to observations and then log-transforming is a common but completely arbitrary practice and the choice of the added value can impact the results. Sometimes considering a square-root transformation can accomplish similar benefits as the log-transform and be applied safely to responses that include 0s. Or more complicated statistical models can be used that allow 0s in responses and still account for the violations of the linear model assumptions – see a statistician or continue exploring more advanced statistical methods for ideas in this direction.

The term-plot in Figure 9.7 provides another display of the results with some information on the results for each combination of the species and treatments. Retaining the interaction because of moderate evidence in the interaction test suggests that the treatments caused different results for the different species. And it appears that there are some clear differences among certain combinations such as the mean for *PS-Control* is clearly larger than for *HS -N50*. The researchers were probably really interested in whether the *N12.5* results differed from *Control* for *HS* and whether the *species* differed at *Control* sites. As part of performing all pair-wise comparisons, we can assess those sorts of detailed questions. This sort of follow-up could be considered in any Two-Way ANOVA model but will be most interesting in situations where there are important interactions.

```
library(effects)
plot(allEffects(m2), multiline = T, lty = c(1,2), ci.style = "bars", grid = T)
```

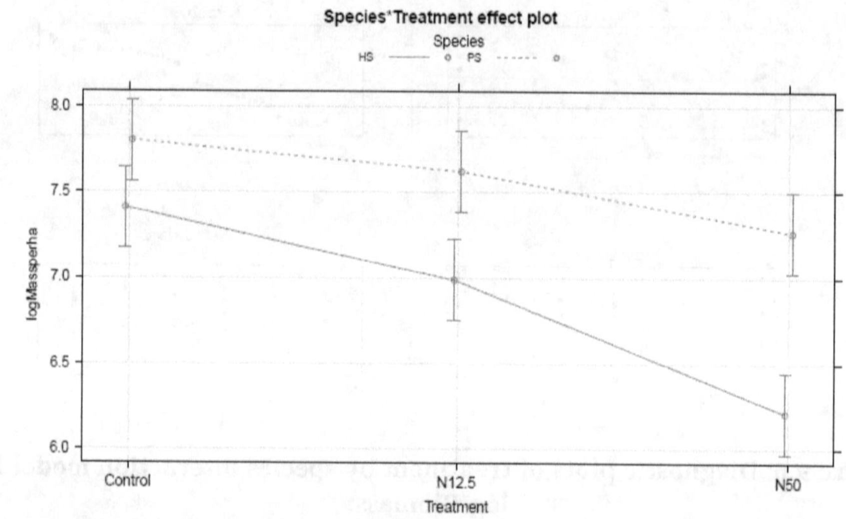

Figure 9.7: Term-plot of the interaction model for log-biomass.

Follow-up Pairwise Comparisons:

Given at least moderate evidence against the null hypothesis of no interaction, many researchers would like more details about the source of the differences. We can re-fit the model with a unique mean for each combination of the two predictor variables, fitting a One-Way ANOVA model (here with six levels) and using Tukey's HSD to provide safe inferences for differences among pairs of the true means. There aresix groups corresponding to all combinations of Species (HS, PS) and treatment levels (Control, N12.5, and N50) provided in the new variable SpTrt by the interaction function with new levels of HS.Control, PS.Control, HS.N12.5, PS.N12.5, HS.N50, and PS.N50. The One-Way ANOVA F -test ($F(5, 30) = 23.96$, p-value < 0.0001) suggests that there is strong evidence against the null hypothesis of no difference in the true mean log-biomass among the six treatment/species combinations and so we would conclude that at least one differs from the others. Note that the One-Way ANOVA table contains the test for at least one of those means being different from the others; the interaction test above was testing a more refined hypothesis – does the effect of treatment differ between the two species? As in any situation with a small p-value from the overall One-Way ANOVA test, the pair-wise comparisons should be of interest.

```
# Create new variable:
gdn <- gdn %>% mutate(SpTrt = interaction(Species, Treatment))
levels(gdn$SpTrt)
```

```
## [1] "HS.Control" "PS.Control" "HS.N12.5"   "PS.N12.5"   "HS.N50"
## [6] "PS.N50"
```

```
newm2 <- lm(logMassperha ~ SpTrt, data = gdn)
Anova(newm2)
```

```
## Anova Table (Type II tests)
##
## Response: logMassperha
##           Sum Sq Df F value    Pr(>F)
## SpTrt     9.6685  5  23.963 1.204e-09
## Residuals 2.4208 30
```

```
library(multcomp)
PWnewm2 <- glht(newm2, linfct = mcp(SpTrt = "Tukey"))
confint(PWnewm2)
```

```
##
##   Simultaneous Confidence Intervals
##
## Multiple Comparisons of Means: Tukey Contrasts
##
##
## Fit: lm(formula = logMassperha ~ SpTrt, data = gdn)
##
## Quantile = 3.0421
## 95% family-wise confidence level
##
##
## Linear Hypotheses:
##                              Estimate lwr      upr
## PS.Control - HS.Control == 0  0.39210 -0.10682  0.89102
## HS.N12.5 - HS.Control == 0   -0.42277 -0.92169  0.07615
## PS.N12.5 - HS.Control == 0    0.21064 -0.28827  0.70956
## HS.N50 - HS.Control == 0     -1.19994 -1.69886 -0.70102
## PS.N50 - HS.Control == 0     -0.14620 -0.64512  0.35272
## HS.N12.5 - PS.Control == 0   -0.81487 -1.31379 -0.31596
## PS.N12.5 - PS.Control == 0   -0.18146 -0.68037  0.31746
## HS.N50 - PS.Control == 0     -1.59204 -2.09096 -1.09312
## PS.N50 - PS.Control == 0     -0.53830 -1.03722 -0.03938
## PS.N12.5 - HS.N12.5 == 0      0.63342  0.13450  1.13234
## HS.N50 - HS.N12.5 == 0       -0.77717 -1.27608 -0.27825
## PS.N50 - HS.N12.5 == 0        0.27657 -0.22235  0.77549
## HS.N50 - PS.N12.5 == 0       -1.41058 -1.90950 -0.91166
## PS.N50 - PS.N12.5 == 0       -0.35685 -0.85576  0.14207
## PS.N50 - HS.N50 == 0          1.05374  0.55482  1.55266
```

We can also generate the Compact Letter Display (CLD) to help us group up the results.

```
cld(PWnewm2)
```

```
## HS.Control PS.Control    HS.N12.5    PS.N12.5       HS.N50      PS.N50
##       "bd"        "d"         "b"        "cd"          "a"        "bc"
```

And we can add the CLD to an interaction plot to create Figure 9.8. Researchers often use displays like this to simplify the presentation of pair-wise comparisons. Sometimes researchers add bars or stars to provide the same information about pairs that are or are not detectably different. The following code creates the plot of these results using our intplot function and the cld = T option.

```
intplot(logMassperha ~ Species * Treatment, cld = T, cldshift = 0.16, data = gdn, lwd = 2,
        main = "Interaction Plot with CLD from Tukey's HSD on One-Way ANOVA")
```

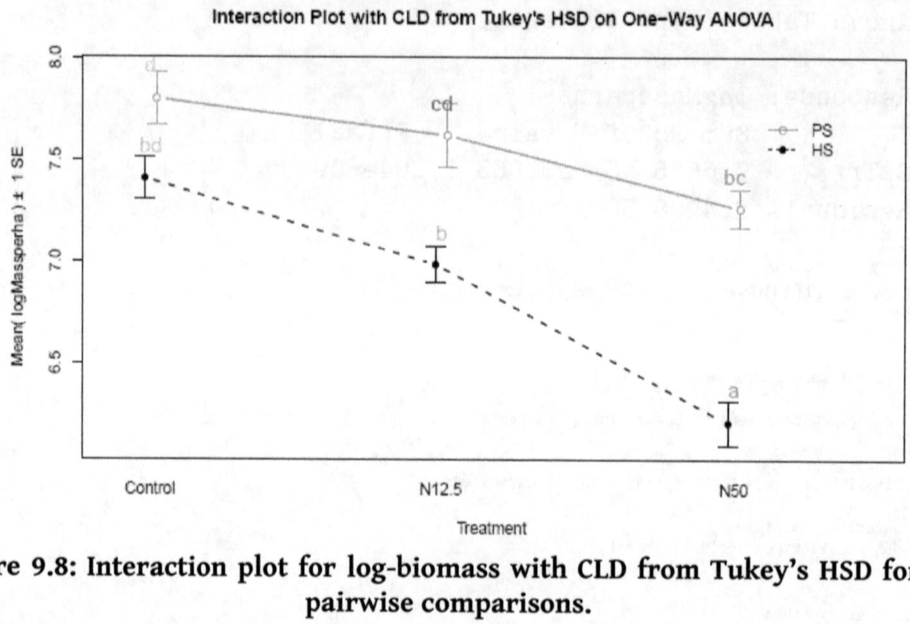

Figure 9.8: Interaction plot for log-biomass with CLD from Tukey's HSD for all pairwise comparisons.

These results suggest that HS-N50 is detectably different from all the other groups (letter "a"). The rest of the story is more complicated since many of the sets contain overlapping groups in terms of detectable differences. Some specific aspects of those results are most interesting. The mean log-biomasses were not detectably different between the species in the control group (they share a "d"). In other words, without treatment, there is little to no evidence against the null hypothesis of no difference in how much of the two species are present in the sites. For N12.5 and N50 treatments, there are detectable differences between the Species. These comparisons are probably of the most interest initially and suggest that the treatments have a different impact on the two species, remembering that in the control treatments, the results for the two species were not detectably different. Further explorations of the sizes of the differences that can be extracted from selected confidence intervals in the Tukey's HSD results printed above. Because these results are for the log-scale responses, we could exponentiate coefficients for groups that are deviations from the baseline category and interpret those as multiplicative changes in the median relative to the baseline group, but at the end of this amount of material, I thought that might stop you from reading on any further.

9.3 ANTS LEARN TO RELY ON MORE INFORMATIVE ATTRIBUTES DURING DECISION-MAKING

In Sasaki and Pratt [2013], a set of ant colonies were randomly assigned to one of two treatments to study whether the ants could be "trained" to have a preference for or against certain attributes for potential nest sites. The colonies were either randomly assigned to experience the repeated choice of two identical colony sites except for having an inferior light or entrance size attribute. Then the ants were allowed to choose between two nests, one that had a large entrance but was dark and the other that had a small entrance but was bright. 54 of the 60 colonies that were randomly assigned to one of the two treatments completed the experiment by making a choice between the two types of sites. The data set and some processing code follows.

The first question is what type of analysis is appropriate here. Once we recognize that there are two categorical variables being considered (Treatment group with two levels and After choice with two levels SmallBright or LargeDark for what the colonies selected), then this is recognized as being within our Chi- square testing framework. The random assignment of colonies (the subjects here) to treatment levels tells us that the Chi-square Homogeneity test is appropriate here and that we can make causal statements about the effects of the Treatment groups.

```
sasakipratt <- read_csv("http://www.math.montana.edu/courses/s217/documents/sasakipratt.csv")
```

```
sasakipratt <- sasakipratt %>% mutate(group = factor(group),
                                      after = factor(after),
                                      before = factor(before)
                                      )
levels(sasakipratt$group) <- c("Light", "Entrance")
levels(sasakipratt$after) <- c("SmallBright", "LargeDark")
levels(sasakipratt$before) <- c("SmallBright", "LargeDark")
plot(after ~ group, data = sasakipratt, col = cividis(2))
```

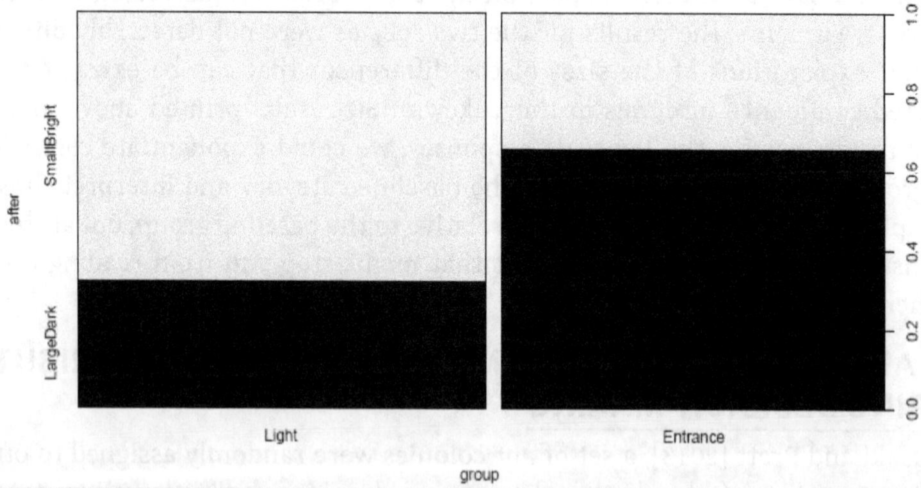


```
library(mosaic)
tally(~ group + after, data = sasakipratt)
```

```
##           after
## group      SmallBright LargeDark
##   Light             19         9
##   Entrance           9        17
```

```
table1 <- tally(~ group + after, data = sasakipratt, margins = F)
```

The null hypothesis of interest here is that there is no difference in the distribution of responses on After – the rates of their choice of den types – between the two treatment groups in the population of all ant colonies like those studied. The alternative is that there is some difference in the distributions of After between the groups in the population.

To use the Chi-square distribution to find a p-value for the X2 statistic, we need all the expected cell counts to be larger than 5, so we should check that. Note that in the following, the correct = F option issued to keep the function from slightly modifying the statistic used that occurs when overall sample sizes are small.

```
chisq.test(table1, correct = F)$expected
```

```
##           after
## group      SmallBright LargeDark
##   Light       14.51852  13.48148
##   Entrance    13.48148  12.51852
```

Our expected cell count condition is met, so we can proceed to explore the results of the parametric test:

```
chisq.test(table1, correct = F)
```

```
##
##  Pearson's Chi-squared test
##
## data:  table1
## X-squared = 5.9671, df = 1, p-value = 0.01458
```

The X^2 statistic is 5.97 which, if our assumptions hold, should approximately follow a Chi-square distribution with $(R - 1) * (C - 1) = 1$ degrees of freedom under the null hypothesis. The p-value is 0.015, suggesting that there is moderate to strong evidence against the null hypothesis and we can conclude that there is a difference in the distribution of the responses between the two treated groups in the population of all ant colonies that could have been treated. Because of the random assignment, we can say that the treatments caused differences in the colony choices. These results cannot be extended to ants beyond those being studied by these researchers because they were not randomly selected.

Further exploration of the standardized residuals can provide more insights in some situations, although here they are similar for all the cells:

```
chisq.test(table1, correct = F)$residuals
```

```
##             after
## group        SmallBright LargeDark
##   Light         1.176144 -1.220542
##   Entrance     -1.220542  1.266616
```

When all the standardized residual contributions are similar, that suggests that there are differences in all the cells from what we would expect if the null hypothesis were true. Basically, that means that what we observed is a bit larger than expected for the *Light* treatment group in the *SmallBright* choice and lower than expected in *LargeDark* – those treated ants preferred the small and bright den. And for the *Entrance* treated group, they preferred the large entrance, dark den at a higher rate than expected if the null is true and lower than expected in the small entrance, bright location.

The researchers extended this basic result a little further using a statistical model called logistic regression, which involves using something like a linear model but with a categorical response variable (well– it actually only works for a two-category response variable). They also had measured which of the two types of dens that each colony chose before treatment and used this model to control for that choice. So the actual model used in their paper contained two predictor variables – the randomized treatment received that we explored here and the prior choice of den type. The interpretation of their results related to the same treatment effect, but they were able to discuss it after adjusting for the colonies previous selection. Their conclusions were similar to those found with our simpler analysis. Logistic regression models are a special case of what are called generalized linear models and are a topic for the next level of statistics if you continue exploring.

9.4 MULTI-VARIATE MODELS ARE ESSENTIAL FOR UNDERSTANDING VERTEBRATE DIVERSIFICATION IN DEEP TIME

Benson and Mannion [2012] published a paleontology study that considered modelling the diversity of Sauropodomorphs across n = 26 "stage-level" time bins. Diversity is measured by the count of the number of different species that have been found in a particular level of fossils. Specifically, the counts in the Sauropodomorphs group were obtained for stages between Carnian and Maastrichtian, with the first three stages in the Triassic, the next ten in the Jurassic, and the last eleven in the Cretaceous. They were concerned about variation in sampling efforts and the ability of paleontologists to find fossils across different stages creating a false impression of the changes in biodiversity (counts of species) over time. They first wanted to see if the species counts were related to factors such as the count of dinosaur-bearing-formations (DBF) and the count of dinosaur-bearing-collections (DBC) that have been identified for each period. The thought is that if there are more formations or collections of fossils from certain stages, the diversity might be better counted (more found of those available to find) and those stages with less information available might be under-counted. They also measured the length of each stage (Duration) but did not consider it in their models since they want to reflect the diversity and longer stages would likely have higher diversity.

Their main goal was to develop a model that would control for the effects of sampling efforts and allow them to perform inferences for whether the diversity was different between the Triassic/Jurassic (grouped together) and considered models that included two different versions of sampling effort variables and one for the comparisons of periods (an indicator variable TJK = 0 if the observation is in Triassic or Jurassic or 1 if in Cretaceous), which are more explicitly coded below. They log-e transformed all their quantitative variables because the untransformed variables created diagnostic issues including influential points. They explored a model just based on the DBC predictor3 and they analyzed the residuals from that model to see if the biodiversity was different in the Cretaceous or before, finding a "p-value >= 0.0001" (I think they meant< 0.00014). They were comparing the MLR models you learned to some extended regression models that incorporated a correction for correlation in the responses over time, but we can proceed with fitting some of their MLR models and using an AIC comparison similar to what they used. There are some obvious flaws in their analysis and results that we will avoid5.

First, we start with a plot of the log-diversity vs the log-dinosaur bearing collections by period. We can see fairly strong positive relationships between the log amounts of collections and species found with potentially similar slopes for the two periods but what look like different intercepts. Especially for TJK level 1 (Cretaceous period) observations, we might need to worry about a curving relationship. Note that a similar plot can also be made using the formations version of the quantitative predictor variable and that the research questions involve whether DBF or DBC are better predictor variables.

```
bm <- read_csv("http://www.math.montana.edu/courses/s217/documents/bensonmanion.csv")
```

```
bm <- bm %>% mutate(logSpecies = log(Species),
                    logDBCs = log(DBCs),
                    logDBFs = log(DBFs),
                    TJK = factor(TJK)
                    )
levels(bm$TJK) <- c("Trias_Juras","Cretaceous")
bm %>% ggplot(mapping = aes(x = logDBCs, y = logSpecies, color = TJK, shape = TJK)) +
  geom_smooth(method = "lm") +
  geom_smooth(se = F, lty = 2) +
  geom_point(size = 2) +
  theme_bw() +
  scale_color_colorblind()
```

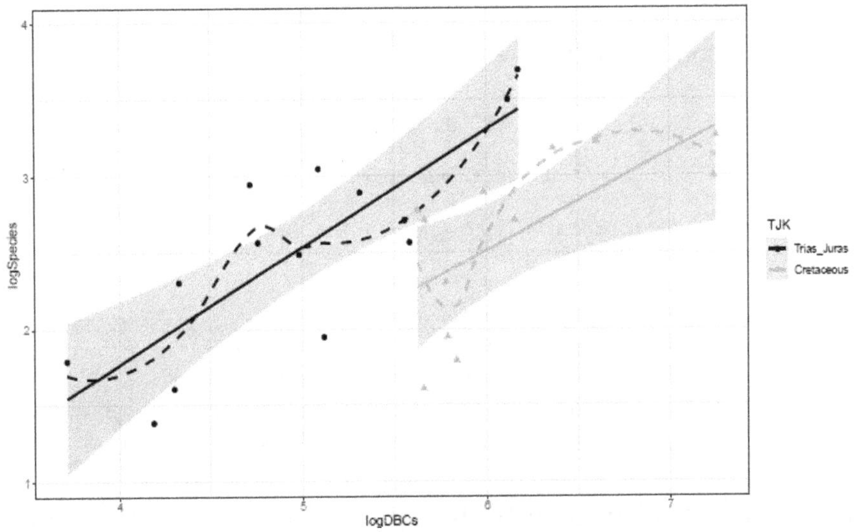

Figure 9.10: Scatterplot of log-biodiversity vs log-DBCs by TJK.

The following results will allow us to explore models similar to theirs. One "full" model they considered is:

$$\log(\text{count})_i = \beta 0 + \beta 1 \cdot \log(\text{DBC})_i + \beta_2 I_{\text{TJK},i} + \varepsilon i$$

which was compared to

$$\log(\text{count})_i = \beta 0 + \beta 1 \cdot \log(\text{DBF})_i + \beta_2 I_{\text{TJK},i} + \varepsilon i$$

as well as the simpler models that each suggests:

$\log(\text{count})_i = \beta_0 + \beta_1 \cdot \log(\text{DBC})_i + \varepsilon_i,$

$\log(\text{count})_i = \beta_0 + \beta_1 \cdot \log(\text{DBF})_i + \varepsilon_i,$

$\log(\text{count})_i = \beta_0 + \beta_2 I_{TJK,i} + \varepsilon_i,$

and $\log(\text{count})_i = \beta_0 + \varepsilon_i.$

Both versions of the models (based on *DBF* or *DBC*) start with an MLR model with a quantitative variable and two slopes. We can obtain some of the needed model selection results from the first full model using:

```
bd1 <- lm(logSpecies ~ logDBCs + TJK, data = bm)
library(MuMIn)
options(na.action = "na.fail")
dredge(bd1, rank = "AIC",
       extra = c("R^2", adjRsq = function(x) summary(x)$adj.r.squared))
```

```
## Global model call: lm(formula = logSpecies ~ logDBCs + TJK, data = bm)
## ---
## Model selection table
##    (Intrc)  lgDBC TJK      R^2   adjRsq df  logLik  AIC delta weight
## 4 -1.0890 0.7243   + 0.580900  0.54440  4 -12.652 33.3  0.00  0.987
## 2  0.1988 0.4283     0.369100  0.34280  3 -17.969 41.9  8.63  0.013
## 1  2.5690            0.000000  0.00000  2 -23.956 51.9 18.61  0.000
## 3  2.5300          + 0.004823 -0.03664  3 -23.893 53.8 20.48  0.000
## Models ranked by AIC(x)
```

And from the second model:

```
bd2 <- lm(logSpecies ~ logDBFs + TJK, data = bm)
dredge(bd2, rank = "AIC",
       extra = c("R^2", adjRsq = function(x) summary(x)$adj.r.squared))
```

```
## Global model call: lm(formula = logSpecies ~ logDBFs + TJK, data = bm)
## ---
## Model selection table
##    (Intrc)  lgDBF TJK      R^2   adjRsq df  logLik  AIC delta weight
## 4 -2.4100 1.3710   + 0.519900  0.47810  4 -14.418 36.8  0.00  0.995
## 2  0.5964 0.4882     0.209800  0.17690  3 -20.895 47.8 10.95  0.004
## 1  2.5690            0.000000  0.00000  2 -23.956 51.9 15.08  0.001
## 3  2.5300          + 0.004823 -0.03664  3 -23.893 53.8 16.95  0.000
## Models ranked by AIC(x)
```

The top AIC model is $\log(\text{count})_i = \beta_0 + \beta_1 \cdot \log(\text{DBC})_i + \beta_2 I_{TJK,i} + \varepsilon_i$ with an AIC of 33.3. The next best ranked model on AICs was $\log(\text{count})_i = \beta_0 + \beta_1 \cdot \log(\text{DBF})_i + \beta_2 I_{TJK,i} + \varepsilon_i$ with an AIC of 36.8, so 3.5 AIC units worse than the top model and so there is clear evidence to support the DBC+TJK model over the best version with DBF and all others. We put these two runs of results together in Table 9.1, re-computing all the AICs based on the top model from the first full model considered to make it easier to see this.

Table 9.1: Model comparison table.

Model	R^2	adj R^2	df	logLik	AIC	Delta AIC
log(count)i = β0 + β1 · log(DBC)i + β2ITJK,i + εi	0.5809	0.5444	4	-12.652	33.3	0
log(count)i = β0 + β1 · log(DBF)i + β2ITJK,i + εi	0.5199	0.4781	4	-14.418	36.8	3.5
log(count)i = β0 + β1 · log(DBC)i + εi	0.3691	0.3428	3	-17.969	41.9	8.6
log(count)i = β0 + β1 · log(DBF)i + εi	0.2098	0.1769	3	-20.895	47.8	14.5
log(count)i = β0 + εi	0	0	2	-23.956	51.9	18.6
log(count)i = β0 + β2ITJK,i + εi	0.0048	-0.0366	3	-23.893	53.8	20.5

Table 9.1 suggests some interesting results. By itself, TJK leads to the worst performing model on the AIC measure, ranking below a model with nothing in it (mean-only) and 20.5 AIC units worse than the top model. But the two top models distinctly benefit from the inclusion of TJK. This suggests that after controlling for the sampling effort, either through DBC or DBF, the differences in the stages captured by TJK can be more clearly observed.

So the top model in our (correct) results6 suggests that log(DBC) is important as well as different intercepts for the two periods. We can interrogate this model further but we should check the diagnostics (Figure 9.11) and consider our model assumptions first as AICs are not valid if the model assumptions are clearly violated.

```
par(mfrow = c(2,2), oma = c(0,0,2,0))
plot(bd1, pch = 16)
```

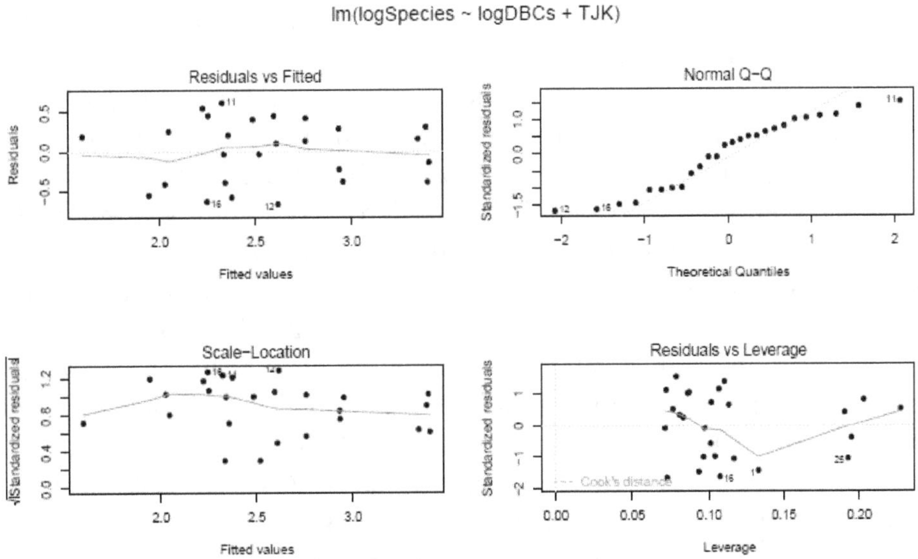

Figure 9.11: Diagnostic plots for the top AIC model.

The constant variance, linearity, and assessment of influence do not suggest any problems with those assumptions. This is reinforced in the partial residuals in Figure 9.12. The normality assumption is possibly violated but shows lighter tails than expected from a normal distribution and so should cause few problems with inferences (we would be looking for an answer of "yes, there is a violation of the normality assumption but that problem is minor because the pattern is not the problematic type of violation because both the upper and lower tails are shorter than expected from a normal distribution"). The other assumption that is violated for all our models is that the observations are independent. Between neighbouring stages in time, there would likely be some sort of relationship in the biodiversity so we should not assume that the observations are independent (this is another time series of observations). The authors acknowledged this issue but un skillfully attempted to deal with it. Because an interaction was not considered in any of the models, there also is an assumption that the results are parallel enough for the two groups. The scatterplot in Figure 9.10 suggests that using parallel lines for the two groups is probably reasonable but a full assessment really should also explore that fully to verify that there is no support for an interaction which would relate to different impacts of sampling efforts on the response across the levels of TJK.

Ignoring the violation of the independence assumption, we are otherwise OK to explore the model more and see what it tells us about biodiversity of Sauropodomorphs. The top model is estimated to belog (count)i = −1.089 + 0.724 · log (DBC)i − 0.75ITJK,i. This suggests that for the early observations (TJK= Trias_ Juras), the model is log (count)i = −1.089 + 0.724 · log (DBC)i and for the Cretaceous period (TJK = Cretaceous), the model is log (count)i = −1.089 + −0.75 + 0.724 · log (DBC)i which simplifies tolog (count)i = −1.84 + 0.724 · log (DBC)i. This suggests that the sampling efforts have the same impacts on all observations and having an increase in logDBCs is associated with increases in the mean log-biodiversity. Specifically, for a 1 log-count increase in the log-DBCs, we estimate, on average, to have a 0.724 log-count change in the mean log-biodiversity, after accounting for different intercepts for the two periods considered. We could also translate this to the original count scale but will leave it as is, because their real question of interest involves the differences between the periods. The change in the y-intercepts of -0.76 suggests that the Cretaceous has a lower average log-biodiversity by 0.75 log-count, after controlling for the log-sampling effort. This suggests that the Cretaceous had a lower corrected mean log-Sauropodomorph biodiversity(t23 = −3.41; p-value = 0.0024) than the combined results for the Triassic and Jurassic. On the original count scale, this suggests exp(−0.76) = 0.47 times (53% drop in) the median biodiversity count per stage for Cretaceous versus the prior time period, after correcting for log-sampling effort in each stage.

```r
summary(bd1)
```

```
##
## Call:
## lm(formula = logSpecies ~ logDBCs + TJK, data = bm)
##
## Residuals:
##      Min       1Q  Median       3Q      Max
## -0.6721 -0.3955  0.1149  0.2999  0.6158
##
## Coefficients:
##               Estimate Std. Error t value Pr(>|t|)
## (Intercept)    -1.0887     0.6533  -1.666   0.1092
## logDBCs         0.7243     0.1288   5.622 1.01e-05
## TJKCretaceous  -0.7598     0.2229  -3.409   0.0024
##
## Residual standard error: 0.4185 on 23 degrees of freedom
## Multiple R-squared:  0.5809, Adjusted R-squared:  0.5444
## F-statistic: 15.94 on 2 and 23 DF,  p-value: 4.54e-05
```

```r
plot(allEffects(bd1, residuals = T), grid = T)
```

Their study shows some interesting contrasts between methods. They tried to use AIC-based model selection methods across all the models but then used p-values to really make their final conclusions. This presents a philosophical inconsistency that bothers some more than others but should bother everyone. One thought is whether they needed to use AICs at all since they wanted to use p-values?

The one reason they might have preferred to use AICs is that it allows the direct comparison of

$$\log(\text{count})_i = \beta_0 + \beta_1 \log(\text{DBC})_i + \beta_2 I_{\text{TJK},i} + \varepsilon_i$$

to

$$\log(\text{count})_i = \beta_0 + \beta_1 \cdot \log(\text{DBF})_i + \beta_2 I_{\text{TJK},i} + \varepsilon_i,$$

exploring whether DBC or DBF is “better” with TJK in the model. There is no hypothesis test to compare these two models because one is not nested in the other – it is not possible to get from one model to the other by setting one or more slope coefficients to 0 so we can't hypothesis test our way from one model to the other one. The AICs suggest strong support for the model with DBC and TJK as compared to the model with DBF and TJK, so that helps us make that decision. After that step, we could rely on t-tests or ANOVA F -tests to decide whether further refinement is suggested/possible for the model with *DBC* and *TJK*. This would provide the direct inferences that they probably want and are trying to obtain from AICs along with p-values in their paper.

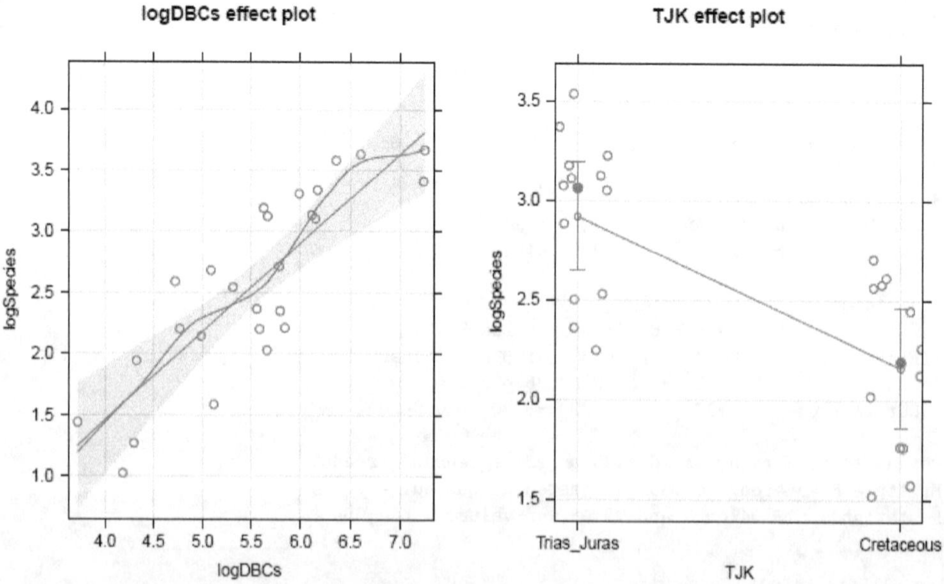

Figure 9.12: Term-plots for the top AIC model with partial residuals.

Finally, their results would actually be more valid if they had used a set of statistical methods designed for modelling responses that are counts of events or things, especially those whose measurements change as a function of sampling effort; models called Poisson rate models would be ideal for their application which are also special cases of the generalized linear models noted in the extensions for modelling categorical responses. The other aspect of the biodiversity that they measured for each stage was the duration of the stage. They never incorporated that information and it makes sense given their interests in comparing biodiversity across stages, not understanding why more or less biodiversity might occur. But other researchers might want to estimate the biodiversity after also controlling for the length of time that the stage lasted and the sampling efforts involved in detecting the biodiversity of each stage, models that are only a few steps away from those considered here. In general, this paper presents some of the pitfalls of attempting to use advanced statistical methods as well as hinting at the benefits. The statistical models are the only way to access the results of interest; inaccurate usage of statistical models can provide inaccurate conclusions. They seemed to mostly get the right answers despite a suite of errors in their work.

9.5 WHAT DO DIDGERIDOOS REALLY DO ABOUT SLEEPINESS?

In the practice problems at the end of Chapter 4, a study (Puhan et al. [2006]) related to a pre-post, two group comparison of the sleepiness ratings of subjects was introduced. They obtained n = 25 volunteers and they randomized the subjects to either get a lesson or be placed on a waiting list for lessons. They constrained the randomization based on the high/low apnoea and high/low on the Epworth scale of the subjects in their initial observations to make sure they balanced the types of subjects going into the treatment and control groups. They measured the subjects' Epworth value (daytime sleepiness, higher is more sleepy) initially and after four months, where only the treated subjects (those who took lessons) had any intervention. We are interested in whether the mean Epworth scale values changed differently over the four months in the group that got didgeridoo lessons than it did in the control group (that got no lessons). Each subject was measured twice (so the total sample size in the data set is 50) in the data set provided that is available at http://www.math.montana.edu/courses/s217/documents/epworthdata.csv.

The data set was not initially provided by the researchers, but they did provide a plot very similar to Figure 9.13. To make Figure 9.13 geom_line is used to display a line for each subject over the two time points (pre and post) of observation and indicate which group the subjects were assigned to. This allows us to see the variation at a given time across subjects and changes over time, which is critical here as this shows clearly why we had a violation of the independence assumption in these data. In the plot, you can see that there are not clear differences in the two groups at the "Pre" time but that treated group seems to have most of the lines go down to lower sleepiness ratings and that this is not happening much for the subjects in the control group. The violation of the independence assumption is diagnosable from the study design (two observations on each subject). The plot allows us to go further and see that many subjects had similar Epworth scores from pre to post (high in pre, generally high in post) once we account for systematic changes in the treated subjects that seemed to drop a bit on average.

```
epworthdata <- read_csv("http://www.math.montana.edu/courses/s217/documents/epworthdata.csv")
```

```
epworthdata <- epworthdata %>% mutate(Time = factor(Time),
                                      Group = factor(Group)
                                      )
levels(epworthdata$Time) <- c("Pre" , "Post")
levels(epworthdata$Group) <- c("Control" , "Didgeridoo")

epworthdata %>%
  ggplot(mapping = aes(x = Time, y = Epworth, group = Subject, color = Group)) +
  geom_point() +
  geom_line() +
  theme_bw() +
  scale_color_colorblind()
```

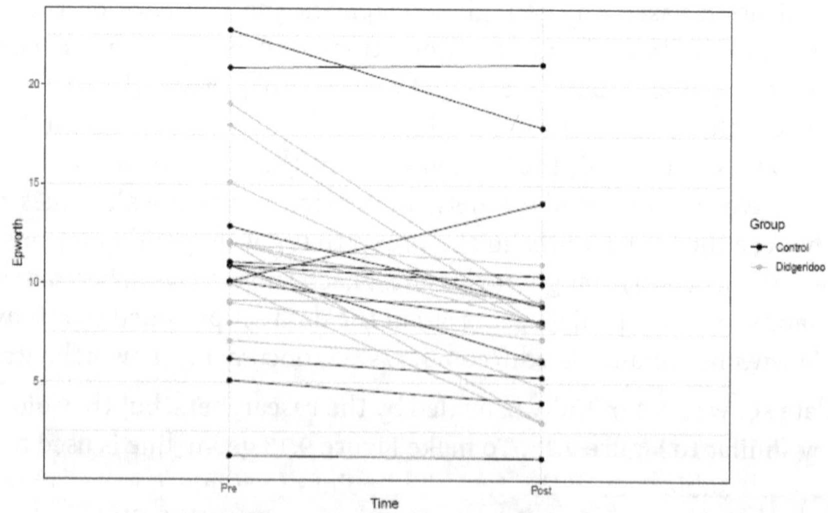

Figure 9.13: Plot of Epworth responses for each subject, initially and after four months, based on treatment groups with one line for each subject connecting observations made over time.

This plot seems to contradict the result from the following Two-Way ANOVA (that is a repeat of what you would have seen had you done the practice problem earlier in the book and the related interaction plot) –there is little to no evidence against the null hypothesis of no interaction between Time and Treatment group on Epworth scale ratings (F (1, 46) = 1.37, p-value = 0.2484 as seen in Table 9.2). But this model assumes all the observations are independent and so does not account for the repeated measures on the same subjects. It ends up that if we account for systematic differences in subjects, we can (sometimes) find the differences we are interested in more clearly. We can see that this model does not really seem to capture the full structure of the real data by comparing simulated data to the original one, as in Figure 9.14. The real data set had fairly strong relationships between the pre and post scores but this connection seems to disappear in responses simulated from the estimated Two-Way ANOVA model (that assumes all observations are independent).

Table 9.2: ANOVA table from Two-Way ANOVA interaction model.

	Sum Sq	Df	F value	Pr(>F)
Time	120.746	1	5.653	0.022
Group	8.651	1	0.405	0.528
Time: Group	29.265	1	1.370	0.248
Residuals	982.540	46		

```
library(car)
lm_int <- lm(Epworth ~ Time * Group, data = epworthdata)
Anova(lm_int)
```

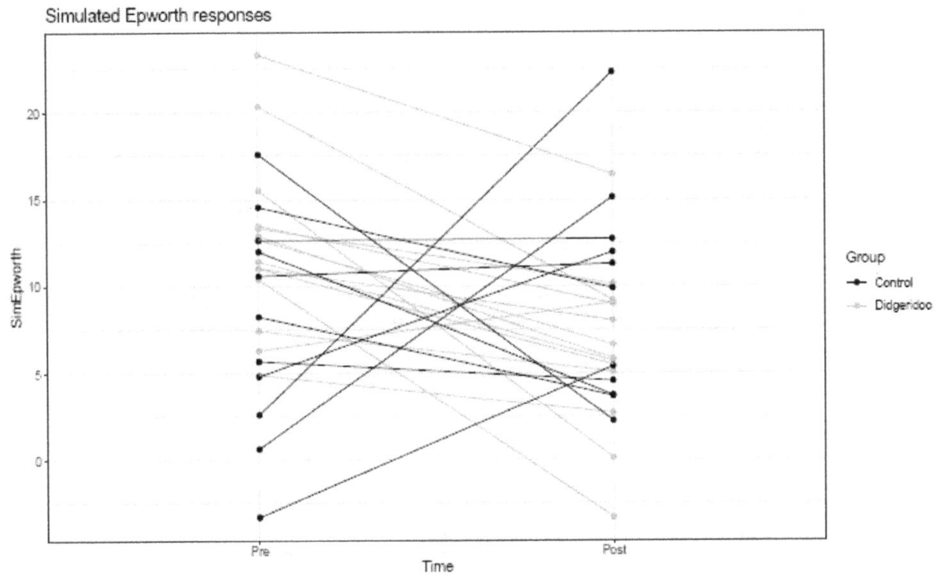

Figure 9.14: Plot of simulated data from the Two-Way ANOVA model that does not assume observations are on repeated measures on subjects to compare to the real data set. Even though the treatment levels seem to decrease on average, there is a much less clear relationship between the starting and ending values in the individuals.

If the issue is failing to account for differences in subjects, then why not add "Subject" to the model? There are two things to consider. First, we would need to make sure that "Subject" is a factor variable as the "Subject" variable is initially numerical from 1 to 25. Second, we have to deal with having a factor variable with 25 levels (so 24 indicator variables!). This is a big number and would make writing out the model and interpreting the term-plot for Subject extremely challenging. Fortunately, we are not too concerned about how much higher or lower an individual is than a baseline subject, but we do need to account for it in the model. This sort of "repeated measures" modelling is more often handled by a more complex set of extended regression models that are called linear mixed models and are designed to handle this sort of grouping variable with many levels.

But if we put the Subject factor variable into the previous model, we can use Type II ANOVA tests to test for an interaction between Time and Group (our primary research question) after controlling for subject-to-subject variation. There is a warning message about aliasing that occurs when you do this which means that it is not possible to estimate all the βs in this model (and why we more typically used mixed models to do this sort of thing). Despite this, the test for Time:Group in Table 9.3 is correct and now accounts for the repeated measures on the subject. It provides F (1, 23) = 5.43 with a p-value of 0.029, suggesting that there is moderate evidence against the null hypothesis of no interaction of time and group once we account for subject. This is a notably different result from what we observed in the Two-Way ANOVA interaction model that didn't account for repeated measures on the subjects and matches the results in the original paper closely.

Table 9.3: ANOVA table from Two-Way ANOVA interaction model.

	Sum Sq	Df	F value	Pr(>F)
Time	120.746	1	22.410	0.000
Group		0		
Subject	858.615	23	6.929	0.000
Time: Group	29.265	1	5.431	0.029
Residuals	123.924	23		

```
epworthdata <- epworthdata %>% mutate(Subject = factor(Subject))
lm_int_wsub <- lm(Epworth ~ Time * Group + Subject, data = epworthdata)
Anova(lm_int_wsub)
```

With this result, we would usually explore the term-plots from this model to get a sense of the pattern of the changes over time in the treatment and control groups. That aliasing issue means that the "effects" function also has some issues. To see the effects plots, we need to use a linear mixed model from the nlmepackage [Pinheiro et al., 2022]. This model is beyond the scope of this material, but it provides the same F -statistic for the interaction (F (1, 23) = 5.43) and the term-plots can now be produced (Figure 9.15). In library (nlme)lme_int <- lme(Epworth ~ Time * Group, random = ~1|Subject, data = epworthdata) anova(lme_int)that plot, we again see that the didgeridoo group mean for "Post" is noticeably lower than in the "Pre" and that the changes in the control group were minimal over the four months. This difference in the changes over time was present in the initial graphical exploration but we needed to account for variation in subjects to be able to detect this difference. While these results rely on more complex models than we have time to discuss here, hopefully the similarity of the results of interest should resonate with the methods we have been exploring while hinting at more possibilities if you learn more statistical methods.

##		numDF	denDF	F-value	p-value
##	(Intercept)	1	23	132.81354	<.0001
##	Time	1	23	22.41014	0.0001
##	Group	1	23	0.23175	0.6348
##	Time:Group	1	23	5.43151	0.0289

```
plot(allEffects(lme_int), multiline = T, lty = c(1,2), ci.style = "bars", grid = T)
```

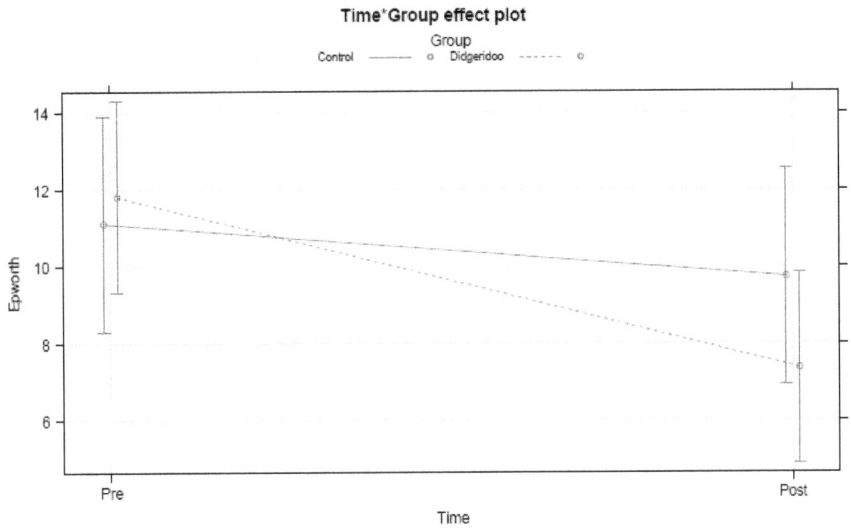

Figure 9.15: Term-plot of Time by Group interaction, results are from model that accounts for subject-to- subject variation in a mixed model.

9.6 GENERAL SUMMARY: AT A GLANCE

As we wrap up, it is important to remember that these tools are limited by the quality of the data collected. If you are ever involved in applying these statistical models, whether in a research or industrial setting, make sure that the research questions are discussed before data collection. And before data collection is started, make sure that the methods will provide results that can address the research questions. And, finally, make sure someone involved in the project knows how to perform the appropriate graphical and statistical analysis. One way to make sure you know how to analyze a data set and, often, clarify the research questions and data collection needs, is to make a simulated data set that resembles the one you want to collect and analyze it. This can highlight the sorts of questions the research can address and potentially expose issues before the study starts. With this sort of preparation, many issues can be avoided. Remember to think about reasons why assumptions of your proposed method might be violated.

You are now armed and a bit dangerous with statistical methods. If you go to use them, remember the fundamentals and find the story in the data. After deciding on any research questions of interest, graph the data and make sure that the statistical methods will give you results that make some sense based on the graphical results. In the MLR results, it is possible that graphs will not be able to completely tell you the story, but all the other methods should follow the pictures you see. Even when (or especially when) you use sophisticated statistical methods, graphical presentations are critical to helping others understand the results. We have discussed examples that involve displaying categorical and quantitative variables and even some displays that bridge both types of variables. We hope you have enjoyed this material and been able to continue to develop your interests in statistics. You will see it in many future situations both in courses in your area of study and outside of academia to try to address problems that need answers. You are also prepared to take more advanced statistics courses.

FOOTNOTES

CHAPTER 1

1. A placebo is a treatment level designed to mimic the potentially efficacious level(s) but that can have no actual effect. The placebo effect is the effect that thinking that an effective treatment was received has on subjects. There are other related issues in performing experiments like the Hawthorne or observer effect where subjects modify behaviour because they are being observed.
2. We will reserve the term "effect" for situations where we could potentially infer causal impacts on the response of the explanatory variable which occurs in situations where the levels of the explanatory variable are randomly assigned to the subjects.
3. There is a cloud version of R Studio available at https://rstudio.cloud/ that is free for limited usage and some institutions have locally hosted versions that you can use with a web-browser (check with your instructor for those options). We recommend following the steps to be able to work locally but try this option if you have issues with the installation process and need to complete an assignment or two until you get the installation sorted out.
4. I created this interactive website (https://rconnect.math.montana.edu/InstallDemo/) that contains discussions and activities related to installing and using R and RStudio.
5. The need to keep the code up-to-date as R continues to evolve is one reason that this book is locally published and that this is the 9th time it has been revised in nine years...
6. There are ways to read ".xls" and ".xlsx" files directly into R that we will explore later so you can also use that format if you Prefer.
7. If you are having trouble getting the file converted and read into R, copy and run the following code: treadmill <- read_csv("http://www.math.montana.edu/courses/s217/documents/treadmill.csv").

8. You can also use Ctrl+Enter if you like hot keys (Command+Enter on Mac OS).
9. Tibbles are R objects that can contain both categorical and quantitative variables on your n subjects with a name for each variable that is also the name of each column in a matrix. Each subject is a row of the data set. The name (supposedly) is due to the way table sounds in the accent of a particularly influential developer at RStudio who is from New Zealand.
10. The median, quartiles and whiskers sometimes occur at the same values when there are many tied observations. If you can't see all the components of the boxplot, produce the numerical summary to help you understand what happened.
11. I recently had to revisit some work from almost a decade ago (before I switched to using R Markdown) as we were working on a journal article submission that re-used some of that work and it was unclear where some results came from, so I had to do some new work that could have been avoided if I had worked in a reproducible fashion.
12. his discussion is based on materials developed for a data visualization workshop originally developed by Dr. Allison Theobold and related to the https://datacarpentry.org/ workshops.
13. Jittering typically involves adding random variability to each observation that is uniformly distributed in a range determined based on the spacing of the observations. The idea is to jitter just enough to see all the points but not too much. Because it is random noise being added, this also means that if you re-run the jitter function, the results will change if you do not set the random number seed using set.seed that is discussed more below. For more details, type help(geom_rug) in the console in RStudio. The code is unfortunately clumsy to add jittering to the rug, so a simpler option is to use geom_rug (alpha = 0.3) where the transparency is modified with the alpha option to help with identifying over plotting of lines in the rug.
14. This certainly could have waited until later, but I have now seen enough base ggplot graphs that I really like to change their overall look.

CHAPTER 2

1. You will more typically hear "data is" but that more often refers to information, sometimes even statistical summaries of data sets, than to observations made on subjects collected as part of a study, suggesting the confusion of this term in the general public. We will explore a data set in Chapter 5 related to perceptions of this issue collected by researchers at http://fivethirtyeight.com/.

2. Either try to remember "data is a plural word" or replace "data" with "things" or, as one former student suggested that helped her with this, replace "data" with "puppies" or "penguins" in your sentence and consider whether it sounds right.
3. Of particular interest to the bicycle rider might be the "close" passes and we will revisit this as a categorical response with "close" and "not close" as its two categories later.
4. Thanks to Ian Walker for allowing me to use and post these data.
5. As noted previously, we reserve the term "effect" for situations where random assignment allows us to consider causality as the reason for the differences in the response variable among levels of the explanatory variable.
6. Make sure you can work from left to right and top down to fill in the ANOVA table given just the necessary information to determine the other components or from a study description to complete the DF part of the table – there are always questions like these on exams...
7. If you've taken calculus, you will know that the curve is being constructed so that the integral from -1 to 1 is 1. If you don't know calculus, think of a rectangle with area of 1 based on its height and width. These cover the same area but the top of the region wiggles.
8. I admit that there are parts of the logic of using ggplot that are confusing to me and this is one of them – but I learned to plot in R before ggplot2 and have been growing fonder and fonder of this way of working. Now instead of searching the internet, I will just get to search my book for the code to make this version of the plot.
9. If you want to type this character in R Markdown, try `$\sim$` outside of code chunks.
10. Remember the bell-shaped curve you encountered in introductory statistics? If not, you can see some at https://en.wikipedia.org/wiki/Normal_distribution.
11. The package and function are intentionally amusingly titled but are based on ideas in the beanplot in Kampstra [2008] and provide what they call an RDI graphic – Raw data, Descriptive, and I nferential statistic in the same display.
12. The default version seems to get mis-interpreted as the box from a boxplot too easily. This display choice also matches the display style for later plots for confidence intervals in term-plots.

13. The hypothesis of no difference that is typically generated in the hopes of being rejected in favor of the alternative hypothesis, which contains the sort of difference that is of interest in the application.
14. The null model is the statistical model that is implied by the chosen null hypothesis. Here, a null hypothesis of no difference translates to having a model with the same mean for both groups.
15. Later we will shuffle other types of explanatory variables.
16. While not required, we often set our random number seed using the set.seed function so that when we re-run code with randomization in it we get the same results.
17. We'll see the shuffle function in a more common usage below; here we are creating a new variable using mutate to show the permuted results that are stored in Perm1.
18. This is a bit like getting a new convertible sports car and driving it to the grocery store – there might be better ways to get groceries, but we probably would want to drive our new car as soon as we got it.
19. This will be formalized and explained more in the next chapter when we encounter more than two groups in these same models. For now, it is recommended to start with the sample means from favstats for the two groups and then use that to sort out which direction the differencing was done in the lm output.
20. P-values are the probability of obtaining a result as extreme as or more extreme than we observed given that the null hypothesis is true.
21. In statistics, vectors are one dimensional lists of numeric elements – basically a column from a matrix of our tibble.
22. We often say "under" in statistics and we mean "given that the following is true".
23. This is another place where the code is a bit cryptic when you are starting – just copy this entire chunk of code – you only ever need to modify the lm line in this code!
24. This is a fancy way of saying "in advance", here in advance of seeing the observations.
25. Statistically, a conservative method is one that provides less chance of rejecting the null hypothesis in comparison to some other method or less than some pre-defined standard. A liberal method provides higher rates of false rejections.

26. Both approaches are reasonable. By using both tails of the distribution we can incorporate potential differences in shape in both tails of the permutation distribution.
27. P-values of 1 are the only result that provide no evidence against the null hypothesis but this still doesn't prove that the null hypothesis is true.
28. We'll leave the discussion of the CLT to your previous statistics coursework or an internet search. For this material, just remember that it has something to do with distributions of statistics looking more normal as the sample size increases.
29. The t.test function with the var.equal = T option is the more direct route to calculating this statistic (here that would be t.test(Distance ~ Condition, data = dsamp, var.equal = T)), but since we can get the result of interest by fitting a linear model, we will use that approach.
30. On exams, you might be asked to describe the area of interest, sketch a picture of the area of interest, and/or note the distribution you would use. Make sure you think about what you are trying to do here as much as learning the mechanics of how to get p-values from R.
31. In some studies, the same subject is measured in both conditions and this violates the assumptions of this procedure.
32. At this level, it is critical to learn the tools and learn where they might provide inaccurate inferences. If you explore more advanced statistical resources, you will encounter methods that can allow you to obtain valid inferences in even more scenarios.
33. Only male and female were provided as options on the survey. These data were collected as part of a project to study learning of material using online versus paper versions of this book but we focus just on the gender differences in GPA here.
34. The data are provided and briefly discussed in the Practice Problems for Chapter 3.
35. Researchers often measure multiple related response variables on the same subjects while they are conducting a study, so these would not meet the "independent studies" assumption that is used here, but we can start with the assumption of independent results across these responses as the math is easier and the results are conservative. You can consult a statistician for other related approaches that incorporate the dependency of the different responses.

36. You can correctly call octothorpes number symbols or, in the twitter verse, hashtags. For more on this symbol, see "http://blog.dictionary.com/octothorpe/". Even after reading this, I call them number symbols.
37. An unbiased estimator is a statistic that is on average equal to the population parameter.
38. Some perform bootstrap sampling in this situation by re-sampling within each of the groups. We will discuss using this technique in situations without clearly defined groups, so prefer to sample with replacement from the entire data set. It also directly corresponds to situations where the data came from one large sample and then the grouping variable of interest was measured on the n subjects.
39. The as.numeric function is also used here. It really isn't important but makes sure the output of table is sorted by observation number by first converting the orig.id variable into a numeric vector.
40. In any bootstrap sample, about 1/3 of the observations are not used at all.
41. There are actually many ways to use this information to make a confidence interval. We are using the simplest method that is called the "percentile" method.
42. We will often use this term to indicate perform a calculation using the favstats results – not that you need to go back to the data set and calculate the means and standard deviations yourself.
43. Note that this modifier is added to note less certainty than when we encounter strong evidence against the null. Also note that someone else might decide that this more like weak evidence against the null and might choose to interpret it as in the "weak" case. In cases that are near boundaries for evidence levels, it becomes difficult to find a universal answer and it is best to report that the evidence is both not strong and not weak and is somewhere in between and let the reader decide what they think it means to them. This is complicated by often needing to make decisions about next steps based on p-values where we might choose to focus on the model with a difference or without it.

CHAPTER 3

1. In Chapter 4, methods are discussed for when there are two categorical explanatory variables that is called the Two-Way ANOVA and related ANOVA tests are used in Chapter 8 for working with extensions of these models.

2. In Chapter 2, we used lm to get these estimates and focused on the estimate of the difference between the second group and the baseline – that was and still is the difference in the sample means. Now there are potentially more than two groups and we need to formalize notation to handle this more complex situation.
3. If you look closely in the code for the rest of the book, any model for a quantitative response will use this function, suggesting a common thread in the most commonly used statistical models.
4. We can and will select the order of the levels of categorical variables as it can make plots easier to interpret.
5. Suppose we were doing environmental monitoring and were studying asbestos levels in soils. We might be hoping that the mean-only model were reasonable to use if the groups being compared were in remediated areas and in areas known to have never been contaminated.
6. Make sure you can work from left to right and top down to fill in the ANOVA table given just the necessary information to determine the other components or from a study description to complete the DF part of the table – there are always questions like these on exams...
7. Any further claimed precision is an exaggeration and eventually we might see p-values that approach the precision of the computer at 2.2e-16 and anything below 0.0001 should just be reported as being below 0.0001. Also note the way that R represents small or extremely large numbers using scientific notation such as 3e-4 which is 3 · 10-4 = 0.0003.
8. This would be another type of publication bias – where researchers search across groups and only report their biggest differences and fail to report the other pairs that they compared. As discussed before, this biases the results to detecting results more than they should be and then when other researchers try to repeat the same studies and compare just, say, two groups, they likely will fail to find similar results unless they also search across many different possible comparisons and only report the most extreme. The better approach is to do the ANOVA F-test first and then Tukey's comparisons and report all these results, as discussed below.
9. You need to use this command for linear model diagnostics or you won't get the plots we want from the model. And you really just need plot(lm2) but the pch = 16 option makes it easier to see some of the points in the plots.
10. Along with multiple names, there is variation of what is plotted on the x and y axes, the scaling of the values plotted, and even the way the line is chosen to represent the 1-1 relationship, increasing the challenge of interpreting

QQ-plots. We are consistent about the x and y axis choices throughout this book and how the line is drawn but different versions of these plots do vary in what is presented, so be careful with using QQ-plots.

11. Here this means re-scaled so that they should have similar scaling to a standard normal with mean 0 and standard deviation 1. This does not change the shape of the distribution but can make outlier identification simpler – having a standardized residual more extreme than 5 or -5 would suggest a deviation from normality since we rarely see values that many standard deviations from the mean in a normal distribution. But mainly focus on the pattern in points in the QQ-plot and whether it matches the 1-1 line that is being plotted.
12. resistant procedure is one that is not severely impacted by a particular violation of an assumption. For example, the median is resistant to the impact of an outlier. But the mean is not a resistant measure as changing the value of a single point changes the mean.
13. In working with researchers on hundreds of projects, my experience has been that many conversations are often required to discover all the potential sources of issues in data sets, especially related to assessing independence of the observations. Discussing how the data were collected is sometimes the only way to understand whether violations of independence are present or not.
14. When this procedure is used with unequal group sizes it is also sometimes called Tukey-Kramer's method.
15. We often use "spurious" to describe falsely rejected null hypotheses, but they are also called false detections.
16. In more complex models, this code can be used to create pair-wise comparisons on one of many explanatory variables.
17. The plot of results usually contains all the labels of groups but if the labels are long or there many groups, sometimes the row labels are hard to see even with re-sizing the plot to make it taller in RStudio. The numerical output is useful as a guide to help you read the plot in those situations.
18. Note that this method is implemented slightly differently than explained here in some software packages so if you see this in a journal article, read the discussion carefully.
19. There is a warning message produced by the default Tukey's code here related to the algorithms used to generate approximate p-values and then the CLD, but the results seem reasonable and just a few p-values seem to vary in the second or third decimal points.

CHAPTER 4

1. We would not suggest throwing away observations to get balanced designs. Plan in advance to try to have a balanced design but analyze the responses you get.
2. Github.com is a version control system used for software development and collaborative work, which we used to allow us to make changes to it and track the modifications. This book is also written using github to allow the same connection for writing and editing it, and one location where the digital version is hosted: https://greenwoodstat.github.io/GreenwoodBookHTML/.
3. 3Copy and include this code in the first code chunk in any document where you want to use the intplot or inplotarray functions.
4. 4We will use "main effects" to refer to the two explanatory variables in the additive model even if they are not randomly assigned to contrast the terminology with having those variables involved in an interaction term in the model. It is the one place in the book where we use "effects" without worrying about the causal connotation of that word.
5. In the standard ANOVA table, $S_{SA} + S_{SB} + S_{SAB} + S_{SE} = S_{STotal}$. However, to get the tests we really desire when our designs are not balanced, a slight modification of the SS is used, using what are called Type II sums of squares and this result doesn't hold in the output you will see for additive models. This is discussed further below.
6. This does not mean that there is truly no interaction in the population but does mean that we are going to proceed assuming it is not present since we couldn't prove the null was wrong.
7. The anova results are not wrong, just not what we want in all situations.
8. Actually, the tests are only conditional on other main effects if Type II Sums of Squares are used for an interaction model, but we rarely focus on the main effect tests when the interaction is present.
9. In Multiple Linear Regression models in Chapter 8, the reasons for this wording will (hopefully) become clearer.
10. This goes beyond our considerations with character variables that have text levels but are not declared as factors in the first chapters. Those often will be modeled correctly in linear models whether they are characters or factors – but numerical variables will be modeled in a way that you did not intend for these predictors that we will discuss in Chapters 7 and 8.

11. Just so you don't think that perfect R code should occur on the first try, I have made similarly serious coding mistakes even after accumulating more than decade of experience with R. It is finding those mistakes (in time) that matters.
12. To get dosef on the x-axis in the plot, the x.var = "dosef" option was employed to force the Dose to be the variable on the x-axis.
13. We can also use select to only retain these three variables and then drop_na() to get the same result for these three variables.
14. Correctly accounting for these missing data is a complex topic and you should not always engage drop_na(), but the first step to handling missing data issues is to find out (1) if you have an issue, (2) how prevalent it is, and (3) whether it is systematic in any way – in other words (and to date myself), "knowing is half the battle" with missing data. Consult a statistician or take more advanced statistics courses to explore this challenging topic further.
15. We switched back to the anova function here as the Anova function only reports Error in Anova.lm(lm(responses ~ dropsf * brand, data = ptR)) : residual df = 0, which is fine but not as useful for understanding this issue as what anova provides.
16. This package is not on the "CRAN" repository and from time to time involves more complex installation requirements to install it from its "github" repository and some packages it depends on. In order to install this package, we usually can use the following code after installing the remotes [Csárdi et al., 2021] package in the regular way: library(remotes); remotes::install_github("mtennekes/tabplot")
17. In larger data sets, multiple subjects are displayed in each row as proportions of the rows in each category.
18. Quantitative variables are displayed with boxplot-like bounds to describe the variability in the variable for that row of responses for larger data sets.

CHAPTER 5

1. Install the tabplot package from the authors' github repository using library(remotes); remotes::install_github("mtennekes/tabplot") if you haven't already done so.
2. While randomization is typically useful in trying to "equalize" the composition of groups, a possible randomization of subjects to the groups is to put all the males into the treatment group. Sometimes we add additional constraints to randomization of subjects to treatments to guarantee that we don't get stuck with an unusual and highly unlikely assignment like that.

It is important at least to check the demographics of different treatment groups to see if anything odd occurred.

3. The vertical line, "|", in ~ y | x is available on most keyboards on the same key as "\". It is the mathematical symbol that means "conditional on" whatever follows.
4. Technically this is a "spineplot" as it generalizes the stacked bar chart based on the proportion of the total in each vertical bar.
5. Standardizing involves dividing by the standard deviation of a quantity so it has a standard deviation 1 regardless of its original variability and that is what is happening here even though it doesn't look like the standardization you are used to with continuous variables.
6. Note that in smaller data sets to get results as discussed here, use the correct = F option. If you get output that contains "...with Yate's continuity correction", a slightly modified version of this test is being used.
7. Here it allows us to relax a requirement that all the expected cell counts are larger than 5 for the parametric test (Section 5.6).
8. Doctors are faced with this exact dilemma – with little more training than you have now in statistics, they read a result like this in a paper and used to be encouraged to focus on the p-value to decide about treatment recommendations. Would you recommend the treatment here just based on the small p-value? Would having Figure 5.11 to go with the small p-value help you make a more educated decision? Recommendations for users of statistical results are starting to move past just focusing on the p-values and thinking about the practical importance and size of the differences. The potential benefits of a treatment need to be balanced with risks of complications too, but that takes us back into discussing having multiple analyses in the same study (treatment improvement, complications/not, etc.).
9. Independence tests can't be causal by their construction. Homogeneity tests could be causal or just associational, depending on how the subjects ended up in the groups.
10. A stratified random sample involves taking a simple random sample from each group or strata of the population. It is useful to make sure that each group is represented at a chosen level (for example the sample proportion of the total size). If a simple random sample of all schools had been taken, it is possible that a level could have no schools selected.

CHAPTER 6

1. There are measures of correlation between categorical variables but when statisticians say correlation they mean correlation of quantitative variables. If they are discussing correlations of other types, they will make that clear.
2. Some of the details of this study have been lost, so we will assume that the subjects were randomly assigned and that a beer means a regular sized can of beer and that the beer was of regular strength. We don't know if any of that is actually true. It would be nice to repeat this study to know more details and possibly have a larger sample size but I doubt if our institutional review board would allow students to drink as much as 9 beers0.
3. This interface with the cor function only works after you load the mosaic package.
4. The natural log (loge or ln) is used in statistics so much that the function in R log actually takes the natural log and if you want a log10 you have to use the function log10. When statisticians say log we mean natural log.
5. We will not use the "significance stars" in the plot that display with the estimated correlations. You can ignore them but we will sometimes remove them from the plot by using the more complex code of ggpairs(upper = list(continuous = GGally::wrap(ggally_cor, stars = F))).
6. The end = 0.7 is used to avoid the lightest yellow color in the gradient that is often hard to see.
7. This is related to what is called Simpson's paradox, where the overall analysis (ignoring a grouping variable) leads to a conclusion of a relationship in one direction, but when the relationship is broken down into subgroups it is in the opposite direction in each group. This emphasizes the importance of checking and accounting for differences in groups and the more complex models we are setting the stage to consider in the coming chapters.
8. Observations at the edge of the x's will be called high leverage points in Section 6.9; this point is a low leverage point because it is close to mean of the x's.
9. Even with clear scientific logic, we sometimes make choices to flip the model directions to facilitate different types of analyses. In Vsevolozhskaya et al. [2014] we looked at genomic differences based on obesity groups, even though we were really interested in exploring how gene-level differences explained differences in obesity.
10. The residuals from these methods and ANOVA are the same because they all come from linear models but are completely different from the standardized residuals used in the Chi-square material in Chapter 5.

CHAPTER 7

1. We can also write this as E(yi|xi) = μ{yi|xi} = _0 + _1xi, which is the notation you will see in books like the Statistical Sleuth [Ramsey and Schafer, 2012]. We will use notation that is consistent with how we originally introduced the methods.
2. There is an area of statistical research on how to optimally choose x-values to get the most precise estimate of a slope coefficient. In observational studies we have to deal with whatever pattern of x's we ended up with. If you can choose, generate an even spread of x's over some range of interest similar to what was used in the Beers vs BAC study to provide the best distribution of values to discover the relationship across the selected range of x-values.
3. See http://fivethirtyeight.com/features/which-city-has-the-most-unpredictable-weather/ for an interesting discussion of weather variability where Great Falls, MT had a very high rating on "unpredictability".
4. It is actually pretty amazing that there are hundreds of locations in the U.S. with nearly complete daily records for over 100 years.
5. 5All joking aside, if researchers can find evidence of climate change using conservative methods (methods that reject the null hypothesis when it is true less often than stated), then their results are even harder to ignore.
6. It took many permutations to get competitor plots this close to the real data set and they really aren't that close.
7. If the removal is of a point that is extreme in x-values, then it is appropriate to note that the results only apply to the restricted range of x-values that were actually analyzed in the scope of inference discussion. Our results only ever apply to the range of x-values we had available so this is a relatively minor change.
8. Note exp(x) is the same as e(x) but easier to read in-line and exp() is the R function name to execute this calculation.
9. You can read my dissertation if you want my take on modeling U and V-shaped valley elevation profiles that included some discussion of these models, some of which was also in Greenwood and Humphrey [2002].
10. This transformation could not be applied directly to the education growth score data in Chapter 5 because there were negative "growth" scores.
11. This silly nomenclature was inspired by De Veaux et al. [2011] Stats: Data and Models text. If you find this too cheesy, you can just call it x-vee.

12. The geom_ribbon has been used inside the geom_smooth function we have used before, but this is the first time we are drawing these intervals ourselves.
13. I have really enjoyed writing this book and enjoy updating it yearly, but hope someone else gets to do the work of checking the level of inaccuracy of this model in another 30 years.

CHAPTER 8

1. If yo u take advanced applied mathematics courses, you can learn more about the algorithms being used by lm. Everyone else only cares about the algorithms when they don't work – which is usually due to the user's inputs in these models not the algorithm itself.
2. Sometimes the effects plots ignores the edge explanatory observations with the default display. Always check the original variable summaries when considering the range of observed values. By turning on the "partial residuals" with SLR models, theplots show the original observations along with the fitted values and 95% confidence interval band. In more complex models, these displays with residuals are more complicated but can be used to assess linearity with each predictor in the model after accounting for other variables.
3. We used this same notation in the fitting the additive Two-Way ANOVA and this is also additive in terms of these variables. Interaction models are discussed later in the chapter.
4. I have not given you a formula for calculating partial residuals. We will leave that for more advanced material.
5. Imagine showing up to a ski area expecting a 40 inch base and there only being 11 inches. I'm sure ski areas are alway smore accurate than this model in their reporting of amounts of snow on the ground...
6. The site name is redacted to protect the innocence of the reader. More informationonthissite,locatedinBeaverheadCountyinMontana,isavailable at http://www.wcc.nrcs.usda.gov/nwcc/site?sitenum=355&state=mt.
7. Term-plots with additive factor variables use the weighted (based on percentage of the responses in each category) average of their predicted mean responses across their levels but we don't have any factor variables in the MLR models, yet.
8. This also applies to the additive two-way ANOVA model.

9. The seq function has syntax of seq(from = startingpoint, to = endingpoint, length.out =#ofvalues_between_start_and_end) and the rep function has syntax of rep(numbertorepeat, #oftimes).
10. Also see Section 8.13 for another method of picking among different models.
11. This section was inspired by a similar section from De Veaux et al. [2011].
12. There are some social science models where the model is fit with the mean subtracted from each predictor so all have mean 0and the precision of the y-intercept is interesting. In some cases both the response and predictor variables are "standardized" tohave means of 0 and standard deviations of 1. The interpretations of coefficients then relates to changes in standard deviations around the means. These coefficients are called "standardized betas". But even in these models where the x-values of 0 are of interest, the test for the y-intercept being 0 is rarely of interest.
13. The variables were renamed to better interface with R code and our book formatting using the rename function.
14. The answer is no – it should be converted to a factor variable prior to plotting so it can be displayed correctly by ggpairs, but was intentionally left this way so you could see what happens when numerically coded categorical variables are not carefully handled in R.
15. Either someone had a weighted GPA with bonus points, or more likely here, there was a coding error in the data set since only one observation was over 4.0 in the GPA data. Either way, we could remove it and note that our inferences for HSGPA donot extend above 4.0.
16. When there are just two predictors, the VIFs have to be the same since the proportion of information shared is the same in both directions. With more than two predictors, each variable can have a different VIF value.
17. We are actually making an educated guess about what these codes mean. Other similar data sets used 1 for males but the documentation on these data is a bit sparse. We proceed with a small potential that the conclusions regarding differences in gender are in the wrong direction.
18. Some people also call them dummy variables to reflect that they are stand-ins for dealing with the categorical information. But it seems like a harsh anthropomorphism so I prefer "indicators".
19. This is true for additive uses of indicator variables. In Section 8.11, we consider interactions between quantitative and categorical variables which has the effect of changing slopes and intercepts. The simplification ideas to produce estimated equations for each group are used there but we have to account for changing slopes by group too.

20. Models like this with a categorical variable and quantitative variable are often called ANCOVA or analysis of covariance models but really are just versions of our linear models we've been using throughout this material.
21. The scale_color_viridis_d(end = 0.85, option = "inferno") code makes the plot in a suite of four colors from theviridis package [Garnier, 2021] that attempt to be color-blind friendly.
22. The strength of this recommendation drops when you have many predictors as you can't do this for every variable, but the concern remains about an assumption of no interaction whenever you fit models without them. In more complex situations, think about variables that are most likely to interact in their impacts on the response based on the situation being studied and try to explore those.
23. Standardizing quantitative predictor variables is popular in social sciences, often where the response variable is also standardized. In those situations, they generate what are called "standardized betas" (https://en.wikipedia.org/wiki/Standardized_coefficient) that estimate the change in SDs in the response for a 1 SD increase in the explanatory variable.
24. There is a way to test for a difference in the two lines at a particular x value but it is beyond the scope of this material.
25. This is an example of what is called "step down" testing for model refinement which is a commonly used technique for arriving at a final model to describe response variables. Note that each step in the process should be reported, not just the final model that only has variables with small p-values remaining in it.
26. We could also use the anova function to do this but using Anova throughout this material provides the answers we want in the additive model and it has no impact for the only test of interest in the interaction model since the interaction is the last component in the model.
27. In most situations, it would be crazy to assume that the true model for a process has been obtained so we can never pick the "correct" model. In fact, we won't even know if we are picking a "good" model, but just the best from a set of the candidate models on a criterion. But we can study the general performance of methods using simulations where we know the true model and the AIC has some useful properties in identifying the correct model when it is in the candidate set of models. No such similar theory exists for the adjusted R2.
28. Most people now call this Akaike's (pronounced ah-kah-ee-kay) Information Criterion, but he used the AIC nomenclature to mean An Information

Criterion – he was not so vain as to name the method after himself in the original paper that proposed it. But it is now common to use "A" for his last name.

29. More details on these components of the methods will be left for more advanced material – we will focus on an introduction to using the AIC measure here.
30. Although sometimes excluded, the count of parameters should include counting the residual variance as a parameter.
31. It makes it impossible to fit models with any missing values in the data set and this prevents you from making incorrect comparisons of AICs to models with different observations.
32. We put quotes on "full" or sometimes call it the "fullish" model because we could always add more to the model, like interactions or other explanatory variables. So we rarely have a completely full model but we do have our "most complicated that we are considering" model.
33. The options in extra = ... are to get extra information displayed that you do not necessarily need. You can simply run dredge(m6, rank = "AIC") to get just the AIC results.

CHAPTER 9

1. The researchers did not do this analysis so never directly addressed this research question although they did discuss it in general ways.
2. Instructors often get asked what a problem with non-constant variance actually looks like – this is a perfect example of it!
3. This was not even close to their top AIC model so they made an odd choice.
4. I had students read this paper in a class and one decided that this was a reasonable way to report small p-values – it is WRONG. We are interested in how small a p-value might be and saying it is over a value is never useful, especially if you say it is larger than a tiny number.
5. All too often, I read journal articles that have under-utilized, under-reported, mis-applied, or mis-interpreted statistical methods and results. One of the reasons that I wanted to write this book was to help more people move from basic statistical knowledge to correct use of intermediate statistical methods and beginning to see the potential in more advanced statistical methods. It took me many years of being a statistician (after getting a PhD) just to feel armed for battle when confronted with new applications and two stat courses are not enough to get you there, but you have to start somewhere. You are only maybe two or three hundred hours

into your 10,000 hours required for mastery. This book is intended get you some solid fundamentals to build on or a few intermediate tools to use if this is your last statistics training experience.

6. They also had an error in their AIC results that is difficult to explain here but was due to an un-careful usage of the results from the more advanced models that account for autocorrelation, which seems to provide the proper ranking of models (that they ignored) but did not provide the correct differences among models.

BIBLIOGRAPHY

Hirotugu Akaike. A new look at the statistical model identification. IEEE Transactions on Automatic Control, 19:716–723, 1974.

JJ Allaire. manipulate: Interactive Plots for RStudio, 2014. URL https://CRAN.R project.org/package=ma nipulate. R package version 1.0.1.

Jeffrey B. Arnold. ggthemes: Extra Themes, Scales and Geoms for ggplot2, 2021. URL https://github.com/j rnold/ggthemes. R package version 4.2.4.

Adelchi Azzalini and Adrian W. Bowman. A look at some data on the Old Faithful geyser. Applied Statistics, 39:357–365, 1990.

Kamil Bartoń. MuMIn: Multi-Model Inference, 2022. URL https://CRAN.R-project.org/package=MuMIn. R package version 1.46.0.

Roger B.J. Benson and Philip D. Mannion. Multivariate models are essential for understanding vertebrate diversification in deep time. Biology Letters, 8:127–130, 2012. doi: 10.1098/rsbl.2011.0460. URL http://rsbl.royalsocietypublishing.org/content/8/1/127.full? sid= 79fb7eab-a445-4abc-aeb4 970794357614.

J Martin Bland and Douglas G Altman. Multiple significance tests: the bonferroni method. BMJ, 310(6973): 170, 1995. ISSN 0959-8138. doi: 10.1136/bmj.310.6973.170. URL https://www.bmj.com/content/310/6973/170.

Kenneth P. Burnham and David R. Anderson. Model selection and Multimodel Inference. Springer, NY, 2002. Original by F.L. Ramsey, D.W. Schafer; modifications by Daniel W. Schafer, Jeannie Sifneos, BerwinTurlach; vignettes contributed by Nicholas Horton, Linda Loi, Kate Aloisio, Ruobing Zhang, and with corrections by Randall Pruim. Sleuth3: Data Sets from Ramsey and Schafer's "Statistical Sleuth (3rd Ed)", 2019. URL http://r-forge.r-project.org/projects/sleuth2/. R package version 1.0-3.

Anthony P Clevenger and Nigel Waltho. Performance indices to identify attributes of highway crossing structures facilitating movement of large mammals. Biological conservation, 121(3):453–464, 2005. ISSN 0006-3207.

E. Crampton. The growth of the odontoblast of the incisor teeth as a criterion of vitamin c intake of the guinea pig. The Journal of Nutrition, 33(5):491–504, 1947. URL http://jn.nutrition.org /content/33/5/491.full.pdf.

Gábor Csárdi, Jim Hester, Hadley Wickham, Winston Chang, Martin Morgan, and Dan Tenenbaum. remotes: R Package Installation from Remote Repositories, Including GitHub, 2021. URL https://CRAN.R- project.org/package=remotes. R package version 2.4.2.

C. Mitchell Dayton. Latent Class Scaling Analysis. SAGE Publications, Thousand Oaks, CA, 1998.

Richard D. De Veaux, Paul F. Velleman, and David E. Bock. Stats: Data and Models, 3rd edition. Pearson, 2011.

Markus Dieser, Mark C. Greenwood, and Christine M. Foreman. Carotenoid pigmentation in Antarctic heterotrophic bacteria as a strategy to withstand environmental stresses. Arctic, Antarctic, and Alpine Research, 42(4):396–405, 2010. doi: 10.1657/1938-4246-42.4.396.

Alan E. Doolittle and Catherine Welch. Gender differences in performance on a college-level acheivement test. ACT Research Report, pages 89–90, 1989.

Mine Çetinkaya Rundel, David Diez, Andrew Bray, Albert Y. Kim, Ben Baumer, Chester Ismay, Nick Paterno, and Christopher Barr. openintro: Data Sets and Supplemental Functions from OpenIntro Textbooks and Labs, 2022. URL https://CRAN.R-project.org/package=openintro. R package version 2.3.0.

Julian Faraway. faraway: Functions and Datasets for Books by Julian Faraway, 2016. URL http://people.bat h.ac.uk/jjf23/. R package version 1.0.7.

John Fox. Effect displays in R for generalised linear models. Journal of Statistical Software, 8(15):1–27, 2003. URL http://www.jstatsoft.org/v08/i15/.

John Fox and Michael Friendly. heplots: Visualizing Hypothesis Tests in Multivariate Linear Models, 2021. URL http://friendly.github.io/heplots/. R package version 1.3-9.

John Fox and Sanford Weisberg. An R-Companion to Applied Regression, Second Edition. SAGE Publications, Thousand Oaks, CA, 2011. URL http://socserv.socsci.mcmaster.ca /jfox/Books/Companion.

John Fox, Sanford Weisberg, and Brad Price. car: Companion to Applied Regression, 2022a. URL https://CRAN.R-project.org/package=car. R package version 3.1-0.

John Fox, Sanford Weisberg, and Brad Price. carData: Companion to Applied Regression Data Sets, 2022b. URL https://CRAN.R-project.org/package=carData. R package version 3.0-5.

John Fox, Sanford Weisberg, Brad Price, Michael Friendly, and Jangman Hong. effects: Effect Displays for Linear, Generalized Linear, and Other Models, 2022c. URL https://CRAN.R-project.org/package=effects. R package version 4.2-2.

Christopher Gandrud. Reproducible Research with R and R Studio, Second Edition. Chapman Hall, CRC, 2015.

Simon Garnier. viridis: Colorblind-Friendly Color Maps for R, 2021. URL https://CRAN.R-project.org/pa ckage=viridis. R package version 0.6.2.

Mark Greenwood, Stacey Hancock, and Nicole Carnegie. catstats: Statistics for Montana State University Bobcats, 2022. R package version 0.1.

Mark C. Greenwood and N.F. Humphrey. Glaciated valley profiles: An application of nonlinear regression.Computing Science and Statistics, 34:452–460, 2002.

Mark C. Greenwood, Joel Harper, and Johnnie Moore. An application of statistics in climate change: Detection of nonlinear changes in a streamflow timing measure in the Columbia and Missouri Headwaters. In P.S. Bandyopadhyay and M. Forster, editors, Handbook of the Philosophy of Science, Vol. 7: Statistics, pages 1117–1142. Elsevier, 2011.

Patricia H. Gude, J. Anthony Cookson, Mark C. Greenwood, and Mark Haggerty. Homes in wildfire-prone areas: An empirical analysis of wildfire suppression costs and climate change, 2009. URL www.headwaters economics.org.

Michael J. Gundale, Lisbet H. Bach, and Annika Nordin. The impact of simulated chronic nitrogen deposition on the biomass and N2-fixation activity of two boreal feather moss–cyanobacteria associations. Biology Letters, 9(6), 2013. ISSN 1744-9561. doi: 10.1098/rsbl.2013.0797. URL http://rsbl.royalsocietypublishing. org/content/9/6/20130797.

Torsten Hothorn, Frank Bretz, and Peter Westfall. Simultaneous inference in general parametric models.Biometrical Journal, 50(3):346–363, 2008.

Torsten Hothorn, Frank Bretz, and Peter Westfall. multcomp: Simultaneous Inference in General Parametric Models, 2022. URL https://CRAN.R-project.org/package=multcomp. R package version 1.4-20.

Stuart H. Hurlbert. Pseudoreplication and the design of ecological field experiments. Ecological Monographs, 54(2):187–211, 1984. ISSN 00129615. URL http://www.jstor.org/stable/1942661.

Owen Jones, Robert Maillardet, Andrew Robinson, Olga Borovkova, and Steven Carnie. spuRs: Functions and Datasets for "Introduction to Scientific Programming and Simulation Using R", 2018. URL https://CRAN.R-project.org/package=spuRs. R package version 2.0.2.

Peter Kampstra. Beanplot: A boxplot alternative for visual comparison of distributions. Journal of Statistical Software, Code Snippets, 28(1):1–9, 2008. URL http://www.jstatsoft.org/v28/c01/.

Stephen E.G. Lea, Paul Webley, and Catherine M. Walker. Psychological factors in consumer debt: Money management, economic socialization, and credit use. Journal of Economic Psychology, 16(4):681–701, 1995.

Xiyue Liao and Mary C. Meyer. coneproj: An R package for the primal or dual cone projections with routines for constrained regression. Journal of Statistical Software, 61(12):1–22, 2014. URL http://www.jstatsoft.or g/v61/i12/.

Rensis Likert. A technique for the measurement of attitudes. Archives of Psychology, 140:1–55, 1932.

Drew Linzer and Jeffrey Lewis. polca: An R package for polytomous variable latent class analysis. Journal of Statistical Software, 42(10):1–29, 2011.

Drew Linzer and Jeffrey Lewis. poLCA: Polytomous Variable Latent Class Analysis, 2022. URL https: //github.com/dlinzer/poLCA. R package version 1.6.0.1.

Thomas Lumley. survey: Analysis of Complex Survey Samples, 2021. URL http://r-survey.r-forge.r- project.org/survey/. R package version 4.1-1.

Ed Merkle and Michael Smithson. smdata: Data to Accompany Smithson & Merkle, 2013, 2018. URL https://CRAN.R-project.org/package=smdata. R package version 1.2.

David Meyer, Achim Zeileis, and Kurt Hornik. vcd: Visualizing Categorical Data, 2022. URL https://CRAN.R- project.org/package=vcd. R package version 1.4-10.

Mary C. Meyer and Xiyue Liao. coneproj: Primal or Dual Cone Projections with Routines for Constrained Regression, 2021. URL https://CRAN.R-project.org/package=coneproj. R package version 1.16.

Johnnie N. Moore, Joel T. Harper, and Mark C. Greenwood. Significance of trends toward earlier snowmelt runoff, Columbia and Missouri Basin headwaters, Western United States. Geophysical Research Letters, 34 (16), 2007. ISSN 1944-8007. doi: 10.1029/2007GL031022. URL http://dx.doi.org/10.1029/2007GL031022. L16402.

Erich Neuwirth. RColorBrewer: ColorBrewer Palettes, 2022. URL https://CRAN.R-project.org/package= RColorBrewer. R package version 1.1-3.

Nathaniel Phillips. yarrr: A Companion to the e-Book "YaRrr!: The Pirate's Guide to R", 2017. URL www.thepiratesguidetor.com. R package version 0.1.5.

Hans-Peter Piepho. An algorithm for a letter-based representation of all-pairwise comparisons. Journal of Computational and Graphical Statistics, 13(2):456–466, 2004.

José Pinheiro, Douglas Bates, and R Core Team. nlme: Linear and Nonlinear Mixed Effects Models, 2022. URL https://svn.r-project.org/R-packages/trunk/nlme/. R package version 3.1-157.

Randall Pruim, Daniel Kaplan, and Nicholas Horton. mosaicData: Project MOSAIC Data Sets, 2021a. URL https://github.com/ProjectMOSAIC/mosaicData. R package version 0.20.2.

Randall Pruim, Daniel T. Kaplan, and Nicholas J. Horton. mosaic: Project MOSAIC Statistics and Mathematics Teaching Utilities, 2021b. URL https://CRAN.R-project.org/package=mosaic. R package version 1.8.3.

Milo A Puhan, Alex Suarez, Christian Lo Cascio, Alfred Zahn, Markus Heitz, and Otto Braendli. Didgeridoo playing as alternative treatment for obstructive sleep apnoea syndrome: randomised controlled trial. BMJ, 332(7536):266–270, 2006. ISSN 0959-8138. doi: 10.1136/bmj.38705.470590.55. URL https://www.bmj.com/content/332/7536/266.

R Core Team. R: A Language and Environment for Statistical Computing. R Foundation for Statistical Computing, Vienna, Austria, 2022. URL https://www.R-project.org/.

Fred Ramsey and Daniel Schafer. The Statistical Sleuth: A Course in Methods of Data Analysis. Cengage Learning, 2012. ISBN 9781133490678. URL https://books.google.com /books?id=eSlLjA9TwkUC.

Brian Ripley. MASS: Support Functions and Datasets for Venables and Ripley's MASS, 2022. URL http://www.stats.ox.ac.uk/pub/MASS4/. R package version 7.3-57.

Rebekah Robinson and Homer White. tigerstats: R Functions for Elementary Statistics, 2020. URL https://CRAN.R-project.org/package=tigerstats. R package version 0.3.2.

RStudio Team. RStudio: Integrated Development Environment for R. RStudio, PBC, Boston, MA, 2022. URL http://www.rstudio.com/.

Pamela A. Santibáñez, Olivia J. Maselli, Mark C. Greenwood, Mackenzie M. Grieman, Eric S. Saltzman, Joseph R. McConnell, and John C. Priscu. Prokaryotes in the wais divide ice core reflect source and transport changes between last glacial maximum and the early holocene. Global Change Biology, 24(5): 2182–2197, 2018. doi: 10.1111/gcb.14042. URL https://onlinelibrary.wiley.com/doi/abs/10.1111/gcb.14042.

Takao Sasaki and Stephen C. Pratt. Ants learn to rely on more informative attributes during decision- making. Biology Letters, 9(6), 2013. ISSN 1744-9561. doi: 10.1098/rsbl.2013.0667. URL http://rsbl.royalsocietypublishing.org/content/9/6/20130667.

Barret Schloerke, Di Cook, Joseph Larmarange, Francois Briatte, Moritz Marbach, Edwin Thoen, Amos Elberg, and Jason Crowley. GGally: Extension to ggplot2, 2021. URL https://CRAN.R-project.org/pack age=GGally. R package version 2.1.2.

Andreas Schneck. Examining publication bias—a simulation-based evaluation of statistical tests on publication bias. PeerJ, 5:e4115, November 2017. ISSN 2167-8359. doi: 10.7717/peerj.4115. URL https://doi.org/10.7 717/peerj.4115.

Michael L. Smith. Honey bee sting pain index by body location. PeerJ, 2:e338, April 2014. ISSN 2167-8359. doi: 10.7717/peerj.338. URL https://doi.org/10.7717/peerj.338.

Martijn Tennekes and Edwin de Jonge. tabplot: Tableplot, a Visualization of Large Datasets, 2019. URL https://github.com/mtennekes/tabplothttp://www.jds-online.com/file_download/379/JDS-1108.pdf. R package version 1.4.1.

Olga A. Vsevolozhskaya, Dmitri V. Zaykin, Mark C. Greenwood, Changshuai Wei, and Qing Lu. Functional analysis of variance for association studies. PLOS ONE, 9(9):13, 2014. URL http://journals.plos.org/plos one/article?id=10.1371/journal.pone.0105074.

Ian Walker, Ian Garrard, and Felicity Jowitt. The influence of a bicycle commuter's appearance on drivers' overtaking proximities: An on-road test of bicyclist stereotypes, high-visibility clothing and safety aids in the united kingdom. Accident Analysis & Prevention, 64:69 – 77, 2014. ISSN 0001-4575. doi: https://doi.or g/10.1016/j.aap.2013.11.007. URL http://www.sciencedirect.com / science/article/pii/S0001457513004636.

Ronald L. Wasserstein and Nicole A. Lazar. The asa statement on p-values: Context, process, and purpose.The American Statistician, 70(2):129–133, 2016. doi: 10.1080/00031305.2016.1154108. URL https://doi.org/10.1080/00031305.2016 .1154108.

Taiyun Wei and Viliam Simko. corrplot: Visualization of a Correlation Matrix, 2021. URL https://github.c om/taiyun/corrplot. R package version 0.92.

Sanford Weisberg. Applied Linear Regression, Fourth Edition. Wiley, Hoboken, NJ, 2014.

Sanford Weisberg. alr4: Data to Accompany Applied Linear Regression 4th Edition, 2018. URL http://www.z.umn.edu/alr4ed. R package version 1.0.6.

Peter H. Westfall and S. Stanley Young. Resampling-Based Multiple Testing: Examples and Methods for p-value Adjustment. Wiley, New York, 1993.

Hadley Wickham, Winston Chang, Lionel Henry, Thomas Lin Pedersen, Kohske Takahashi, Claus Wilke, Kara Woo, Hiroaki Yutani, and Dewey Dunnington.

ggplot2: Create Elegant Data Visualisations Using the Grammar of Graphics, 2022a. URL https://CRAN.R-project.org/package=ggplot2. R package version 3.3.6.

Hadley Wickham, Jim Hester, and Jennifer Bryan. readr: Read Rectangular Text Data, 2022b. URL https://CRAN.R-project.org/package=readr. R package version 2.1.2.

Simon Wood. mgcv: Mixed GAM Computation Vehicle with Automatic Smoothness Estimation, 2022. URL https://CRAN.R-project.org/package=mgcv. R package version 1.8-40.

INDEX

www.ingramcontent.com/pod-product-compliance
Lightning Source LLC
LaVergne TN
LVHW062243200926
848484LV00006B/1606